American
Genealogical Resources
in German Archives

American Genealogical Resources in German Archives (AGRIGA): A Handbook

Clifford Neal Smith
Anna Piszczan-Czaja Smith

Verlag Dokumentation, Publishers, München 1977

To our daughter
Helen Inez Smith-Piszczan

CIP-Kurztitelaufnahme der Deutschen Bibliothek

Smith, Clifford Neal
American genealogical resources in German
archives / Clifford Neal Smith ; Anna Piszczan-
Czaja Smith. — München : Verlag Dokumentation,
1977.
  ISBN 3-7940-5180-7
NE: Piszczan-Czaja Smith, Anna:

The cover page shows the seal of the ratification document
of the American-Bavarian treaty on the citizenship of 1868.
From the Bayerisches Hauptstaatsarchiv Abt. II Geheimes Staatsarchiv,
Bayern Urkunde No. 3450

© 1977 by Verlag Dokumentation Saur KG, München
Druck/Binden: Friedr. Pustet Graph. Großbetrieb, Regensburg
Printed in West Germany
ISBN 3-7940-5180-7

Contents

Preface

This Handbook, frequently referred to as AGRIGA in the pages of its companion volume, entitled "Encyclopedia of German-American Genealogical Research" (New York & London: R. R. Bowker Co., 1976), is devoted to the primary source materials of German-American genealogical interest to be found in the archives of Germany. Despite its length, it cannot be asserted that the Handbook is complete; many small communal archives and those of most noble estates are not included; the vast archival remains in East Germany are missing entirely. So it is that the volume can be regarded only as a preliminary listing, awaiting the day when other researchers can explore the attics of a myriad "Rathäuser" (town halls) and the dungeons and stalls of old castles; even more, researchers must await a rapprochement between East and West, so that the archives of Prussia, Saxony, and Silesia can be made accessible.

The purpose of the Handbook is to aid in the establishment of links between Americans of German origin and their European homeland. American genealogical researchers will quickly realize that the technique for tracing the lineages of German-Americans in the United States is no different than that for other Americans — one checks census reports, land records, wills, and church records back to tidewater. In turn, German genealogical researchers will trace the lineage of an emigrant to America from his European point of departure back into the earlier mists of time, usually to the late sixteenth century, from church records, guild registers, and estate lists. Where both American and German researchers encounter most difficulty is finding the direct documentary links between the old and the new worlds. This, then, becomes the focus of the Handbook.

The Carnegie Institution of Washington was the first to recognize the importance of German, Austrian, and Swiss archives for an understanding of early American history. Under its auspices, scholars prepared guides which were for many years the basis upon which the Library of Congress Manuscript Division acquired photocopies.[1] For the genealogical researcher, these guides were sad-

6

ly inadequate, because the historian-compilers clearly ignored material of extreme value to the micro-historians. As a consequence, some items, notably the Bremen emigration files burned during the second world war, have been lost forever; had they been included in the Carnegie guides, they would almost certainly have been micro-filmed for the Library of Congress in the 1920s and 1930s.

The inadequacy of the Carnegie guides was the impetus which led in 1958–1961 to a project of the Deutschen Gesellschaft für Amerikastudien (German Society for American Studies) at the University of Cologne to compile a new inventory of archival materials having to do with America. The result has been a still-unpublished eleven-part typescript of transcendent importance, entitled "Americana in deutschen Sammlungen (ADS): Ein Verzeichnis von Materialien zur Geschichte der Vereinigten Staaten von Amerika in Archiven und Bibliotheken der Bundesrepublik Deutschlands und West-Berlin" (Americana in German Collections (ADS): A List of Materials on the History of the United States of America in Archives and Libraries of the German Federal Republic and West Berlin).[2] It is sad to note that, despite its availability since 1967, only three copies of ADS are known to these writers to be in American repositories.[3]

Those entries in ADS which appear prima facie to be of interest to genealogists have been translated in their entirety for inclusion in this Handbook. The writers hasten to add, however, that some entries, when inspected, may turn out to be statistical in nature and therefore without direct usefulness, because no names appear. Since the exact nature of some files was not clear from the ADS descriptions, the rule has been to include, rather than exclude, them.

The introduction to ADS has the following to say about entries appearing therein:

"Within each individual archive the material was listed according to "Bestände" [record class]. All files which from their titles clearly referred to the United States were selected. In addition, all files having probable reference to the United States, even though not evident from the titles, were included; examples thereof were files entitled "Emigration" (usually only summarized because of their number), "Overseas Passages," "Estates of Germans Dying Abroad," etc. Sampling of such files made it clear that they almost always contained material pertaining to America.

"Despite this relatively broad selection, the list [ADS] is by no means a complete tabulation of available materials. A more thorough listing, complete in every last detail and following up every possibility, would have been beyond the personal capabilities and financial resources available to the project. Consequently, it was the main purpose of the enterprise to provide a basic listing of important materials, thereby giving impetus to, and providing the conditions for, further scholarly work. If one were to pursue certain subject matters more intensively, it would quickly become apparent that other record classes, not listed herein, would be relevant. This is particularly true of genealogical inqiries wherein church records, notarial files, and university matriculations should be considered, but which have not been included in ADS [space added]. So, likewise, with studies of ship transportation, exports and imports, and so on, for which additional archival material, not included herein, should be consulted."

ADS lists archives to which questionnaires regarding their holdings were sent, but from which no responses were received. Researchers not finding villages of interest to them listed in the geographic index of this Handbook should consult the ADS "Fehlanzeigen" (lists of unreported archives). Researchers should also be aware that the archives of eastern France and the shipping records of Le Havre, Antwerp, and the Dutch ports on the Rhine are likely also to contain rich troves of material on German emigration, particularly for the period before 1850. These repositories await another day and another volume.

1. Carnegie Institution, Washington. [Guides to manuscript materials for the history of the United States] Washington, D.C.: Carnegie Institution, 1906–1943. 23 volumes. The pertinent volumes are: M. D. Learned, "Guide to the Materials Relating to American History in German State Archives", 1912; A. B. Faust, "Guide to the Materials for American History in Swiss and Austrian Archives", 1916.
2. Photocopies of ADS can be obtained from Firma Omnia K.G. Kraus, Weiss & Co., 8 München 54, Hanauerstrasse 30a, at a cost of about $ 300, depending upon format and rate of exchange.
3. A copy of ADS will be found in the Library of Congress Manuscript Division, usually on the desk of the specialist for German manuscripts. A second copy is in the Newberry Library, Chicago. A third copy is now in the possession of the writers.

Arrangement of Entries

Entries in this Handbook are arranged in three separate indexes: a name index, a geographic index, and a subject index. The following can be said of these indexes:

Name index. It is clear that persons having personal files worthy of individual listing in ADS have been, for the most part, unusual in one respect or another. Many have been the subjects of police surveillance, political or criminal, or wards of the state, either as convicts, orphans, or indigents. Thus it is that the individuals listed in the Handbook vary considerably from the emigrant norm — persons who have passed through the emigration process without special decisions having to be made by the German bureaucracy. A second category of names pertains to estates, either in the United States or Germany, in which heirs were thought to have been located in the other country.

Geographic index. The majority of the emigrants to America are not listed individually, but their names will be found among the many emigrant lists in the towns and villages throughout Germany from whence they came. This means that the problem of finding a specific emigrant in German archival files is almost impossible, unless one can establish his place of origin in Germany with precision. There is often a special problem, even if the town of origin is known, for in the old days before "Leibeigenschaft" (serfdom) was abolished, emigration permits may have been issued not by the officials of the town where the emigrant lived but in the estate office of the nobleman to whom the emigrant was subject. It frequently happened that subjects of a number of adjacent estates lived in market towns in the neighborhood, but these individuals remained subjects to the administrative control of their "Herren" (lords). For Jews, the situation was similar: some members of a local synagogue congregation might have had the protection of one nobleman, while other members had that of another nobleman. Fortunately, this odd administrative arrangement was done away with after the Napoleonic Wars, so that one can use the year 1810, more or less, as the cut-off date. Thereafter, with some exceptions, the emigration records are to be found among the village or "Kreis" (district) records and are so listed in this geographic index.

Faced with the needle in the haystack, what then can one hope for? The best mode of entry is through the Hamburg police records, if the emigrant left Germany through that port after 1850. Shortly before 1850 the Hanseatic ports began to realize that transportation of German emigrants to America could be a lucrative business; it had previously been pre-empted by the Dutch and English shipmasters who picked up their passengers at the European Channel ports. Hamburg and Bremen, states then quite independent of the German princes, began to bid for the business of transporting the emigrants, and it soon came to pass that the main route of emigration was via Bremen and Hamburg and the minor German ports controlled by them. Both regimes kept meticulous records of the emigrants passing through, keeping the travelers in special barracks until port authorities were certain all papers were in order and that there was no contagion among them. Thus it is that these port records are of overweening importance to German-American genealogical researchers. These records are equally important, it should be added, to researchers seeking the emigrants from Slavic lands to the East, who also passed through the port of Hamburg on their way to America.

Sad to say, the Bremen records were destroyed during the "Feuersturm" (fire storm) near the end of the second world war; fortunately, the Hamburg records remain. The entry in this Handbook for these records will be found under Hamburg [Emigration Matters]. These records comprise 399 volumes, covering the period 1850–1914, listing all persons going directly through the port of Hamburg to America, and an additional 118 volumes for emigrants passing indirectly — that is, through Hamburg and minor ports controlled by Hamburg via other European (usually British) ports. The Library of Congress Manuscript Division has microfilms of volumes for the years 1850–1873. The Genealogical Society of the Church of Jesus Christ of Latter-Day Saints, Salt Lake City, appears to have microfilms of all the volumes. If one knows the year in which an emigrant left Germany, one can — with luck — find his name, exact place of origin, and profession, and similar data on all accompanying family members. Usually, there are both chronological and index volumes for each year.

The writers hypothesize that emigrants leaving through the port of Hamburg were mainly from the eastern and southern parts of Germany — those from the Rhineland and southwestern parts of the country may have used Bremen more frequently. This hypothesis has not yet been tested, however. One should also remember that emigrants from Austria and from the Slavic countries are also most likely to have embarked at Hamburg for their journeys to America.

If researchers have been able to determine the ship upon which their immigrant ancestor came to America from American ship-entry lists, both the Bremen and Hamburg ship records will provide linkage to places of origin in Germany. Here, again the ship records of Bremen are fragmentary, those of Hamburg copious.

Subject index. In the opinion of the writers, all the documents described in the subject index, with the exception of the Berlin Document Center collections, ought to be microfilmed in their entirety and made available through the Library of Congress Manuscript Division. A great many of these items are already photocopied, as noted in the entries themselves. In particular, the records of the German deserter-immigrants 1776–1783 (see the entry American Revolution) are of exceptional value to American researchers.

How to Read the Entries in This Handbook

Each entry in this Handbook has five parts: a heading (either geographic or personal name, or subject), a short description of the contents of the file, an archive indicator (the underscored number), the archival call number, and an ADS citation. Files already to be found in photocopied form in the Library of Congress Manuscript Division are so indicated. Here are typical entries:

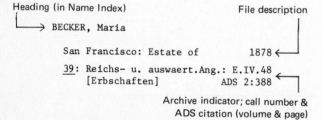

Heading (in Name Index) File description

BECKER, Maria

San Francisco: Estate of 1878

39: Reichs- u. auswaert.Ang.: E.IV.48
   [Erbschaften] ADS 2:388

Archive indicator; call number &
ADS citation (volume & page)

Note herein that San Francisco is placed in the descriptive portion of the entry, rather than after the name, indicating that Maria Becker's file probably originated in the German consulate in San Francisco, but that she may have lived anywhere within the western states assigned to the San Francisco consular district. Had she lived in San Francisco itself, the heading would have been BECKER, Maria (of San Francisco). It has not always been possible to distinguish between the two alternatives, and researchers should be alert to the possibility of error in this matter.

Heading (in Geographic Index) File description

RUMBACH (village; Kreis Dahn, Kurpfalz) RPL

Names of emigrants 18th century

201: Archivabt. Ausfautheiakten. ADS 8:63

Archive indicator; call number;
ADS citation (volume & page)

Since the political subdivisions of Germany were quite complex and shifted over time, as much geographic information has been given as was available in the ADS description — particularly necessary because of a surprising number of these subdivisions no longer exist. In this entry, Rumbach is a village, at the time in Kreis Dahn, which was then a subdivision of the Kurpfalz (Electoral Palatinate). It is now within Kreis Pirmasens (not shown above) in the modern state of Rheinland-Pfalz (RPL). Occasionally, the modern state designation indicates that the file is in an archive in this state with other historically-related files, even though, due to modern boundary rectifications, the village itself may currently be within the confines of a neighboring state. This information is of particular value to the genealogical researcher not only as a means of pinpointing the village in question, but also in determining the local political and social history pertinent to the emigrants (the Kurpfalz once belonged to the House of Wittelsbach (Bavaria), which has confessional implications for emigrants from this region).

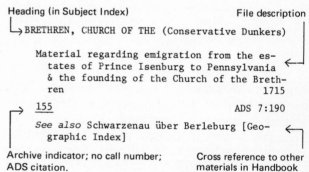

Heading (in Subject Index) File description

BRETHREN, CHURCH OF THE (Conservative Dunkers)

Material regarding emigration from the estates of Prince Isenburg to Pennsylvania & the founding of the Church of the Brethren 1715

155 ADS 7:190

See also Schwarzenau über Berleburg [Geographic Index]

Archive indicator; no call number; Cross reference to other
ADS citation. materials in Handbook

A number of the smaller archival collections, in this case that of the income office of a principality no longer existant as a political entity, have no call numbers. Such collections can be of exceptional value, however, because of the information they may contain on manumissions of serfs and early emigration permits. As noted above, only a few such collections have been described in ADS. (In the above example, were are indebted to a distinguished German genealogist, Professor Dr. W. K. Prinz von Isenburg, who no doubt takes as much interest in the descendants of the Dunkard congregations from his estates as his ancestor did in the eighteenth century.)

In the three tables which follow are listed the abbreviations used in the three indexes in this Handbook. Users will need constantly to refer to these tables.

Table I

German Political Subdivisions

The geographic entries contain a number of German terms denoting political subdivisions. These are of importance to the genealogical researcher in order to determine the administrative level of the files described therein.

Amt	Office; in its narrowest sense meaning a local office for a specific purpose, as in postoffice. Also Oberamt, Landratsamt, Obervogteiamt, Vogteiamt.
Amtsbezirk	see Bezirk
Amtsgericht	see Gericht
Arrondise-ment	A subdivision of a French departement; term used in Germany only during the Napoleonic occupation, 1807–1812.
Bezirk	Area under the administration of a lower- or middle-level bureaucracy (usually middle-level). Also Bezirksamt, Regierungsbezirk.
Canton	Lower level governmental subdivision; used in Germany only during the Napoleonic occupation, 1807–1812; in Switzerland the equivalent of a state.
Departement	An upper-level political subdivision, in France under a Prefect; used in Germany only during the Napoleonic occupation, 1807–1812.
Flecken	Historically, a large village with some municipal rights; when the village had the right hold a market, it was designated a Marktflecken.
Gemeinde	Commune; formerly a village or commune had certain rights to hold a court.
Gericht	A court of law; in the eighteenth century many estates had their own courts (Herrschaftsgericht). Also Amtsgericht, Landgericht.
Gesandtschaft	Permanent legation abroad; in eighteenth and nineteenth century Germany there were legations between principalities, as, for example, the Bavarian legation or ministry in Prussia.
Grafschaft	The area under the control of a Graf, a count, who might be of the higher or lower nobility, there being several gradations, as Reichsgraf, Landgraf, Pfalzgraf, Burggraf.
Herrschaft	A large landholding; the estate of a lord.
Herrschaftsgericht	see Gericht
Hochstift	A term often used in the Middle Ages for bishoprics, cathedral chapters, and royal abbeys.
Kirchenkreis	see Kreis
Kreis	Lower-level administrative area; a Landkreis is a rural administrative area, and a Stadtkreis is an urban administrative area. The term Kirchenkreis refers to the administration of a number of parishes.
Kurhessen	Electorate of Hesse or Electoral Hesse
Kurpfalz	Palatine Electorate, or Electoral Palatinate
Land	A state
Landgericht	see Gericht
Landgrafschaft	see Grafschaft
Landkreis	see Kreis
Landratsamt	see Amt
Oberamt	see Amt
Obervogteiamt	see Vogtei
Praefektur	Office of a prefect, the administrator of a French departement; used in Germany only during the Napoleonic occupation, 1807–1812. Also Subpraefektur, a subdivision of the office of a prefect.
Regierung	Central government; the highest level of government; a Landesregierung is the highest level of a government within a state; a Regierungsbezirk signifies a regional organization above a number of Land- or Stadtkreise.
Reichsgrafschaft	see Grafschaft
Stadtkreis	see Kreis
Subpraefektur	see Praefektur
Vogt; Vogtei	Administrative official, usually of an estate; office of such an official. Also a judge on an estate belonging to a religious order or prince, or the effective administrator of a castle.

10

Table II

Abbreviations Used for the Modern German States

B Berlin

BWL Baden-Württemberg

BYL Bayern (Bavaria)

HB Bremen

HEL Hessen (Hesse)

HH Hamburg

NL Niedersachsen (Lower Saxony)

RPL Rheinland-Pfalz (Rhineland-Palatinate)

RWL Nordrhein-Westfalen (North Rhine-Westphalia)

SHL Schleswig-Holstein

Table III

Archive Indicators

Entries in volume one contain archive indicators and call numbers whereby the location and identification of documents may be ascertained. The underscored number in each entry is the archive indicator, as follows:

Archive Indicator	Archive
1	Staatsarchiv Schleswig-Holstein, Lübeck, Sankt Annenstrasse 2
2	Staatsarchiv Schleswig, Schleswig, Schloss Gottorf
3	Kirchspielsvogtei-Archiv, Albersdorf, Schleswig-Holstein
4	Kirchspielsvogtei-Archiv, Büsum, Schleswig-Holstein
5	Stadtarchiv, Burg auf Fehmarn, Schleswig-Holstein
6	Kirchspielsarchiv, Delve/Kreis Norderdithmarschen, Schleswig-Holstein
7	Stadtarchiv, Eckernförde, Sandkrug-Schule, Kieler Landstrasse 129, Schleswig-Holstein
8	Stadtarchiv, Flensburg, Deutsches Haus-Turm, Schleswig-Holstein
9	Koogsarchiv, Hedwigenkoog, Schleswig-Holstein
10	Landschaftliches Archiv Norderdithmarschen, Heide, Kreishaus, Schleswig-Holstein
11	Archiv der Stadt und des Kirchspiels Heide, Heide, Schleswig-Holstein
12	Stadtarchiv, Heiligenhafen, Rathaus, Schleswig-Holstein

Archive Indicator	Archive
13	Kirchspielsarchiv, Hemme, Schleswig-Holstein
14	Kirchspielsarchiv, Hennstedt, Kreis Norderdithmarschen, Schleswig-Holstein
15	Stadtarchiv, Husum, Ludwig-Nissen-Haus, Schleswig-Holstein
16	Archiv der Nordergoesharde, Husum, Ludwig-Nissen-Haus, Schleswig-Holstein
17	Stadtarchiv, Kiel, Schleswig-Holstein
18	Archiv der Gesellschaft der freiwilligen Armenfreunde, Kiel, Schleswig-Holstein
19	Schleswig-Holsteinische Landesbibliothek, Kiel-Wik, Warnemünder-Strasse 16–18, Schleswig-Holstein
20	Kropper Predigerseminar, Kropp bei Schleswig, c/o Herr Pastor P. D. Rohnert, Haus Bethulia, Schleswig-Holstein
21	Dr. H. H. Pantel, Lübeck, Moltkeplatz 9, Schleswig-Holstein
22	Kirchspielsarchiv, Lunden, Schleswig-Holstein
23	Guts- und Amtsbezirksarchiv, Mehlbek, Kreis Steinburg, Schleswig-Holstein
24	Dithmarscher Landesmuseum, Landschaftsarchiv für Süderdithmarschen, Meldorf, Schleswig-Holstein
25	Kirchspielsarchiv, Norderwöhrden/Kreis Norderdithmarschen, Schleswig-Holstein
26	Stadtarchiv, Oldenburg, Schleswig-Holstein
27	Archiv der Herrschaft Hessenstein, Panker, Schleswig-Holstein
28	Landschaftsarchiv, Pellworm, Schleswig-Holstein
29	Kreisarchiv, Ratzeburg, Schleswig-Holstein
30	Stadtarchiv, Ratzeburg, Rathaus, Schleswig-Holstein
31	Domarchiv, Ratzeburg, Domhof 12, Schleswig-Holstein
32	Stadtarchiv, Rendsburg, Feuerwache, Ecke Jungfernstieg und Holstenstrasse, Schleswig-Holstein
33	Kirchspielsvogtei-Archiv, Süderhastedt, Kreis Süderdithmarschen, Schleswig-Holstein
34	Stadtarchiv, Tönning, Volksschule, Hochsteg 26, Schleswig-Holstein
35	Kirchspielsarchiv, Weddingstedt/Kreis Norderdithmarschen, Schleswig-Holstein
36	Kirchspielsarchiv, Wesselburg/Kreis Norderdithmarschen, Schleswig-Holstein
37	Stadtarchiv, Wilster/Kreis Steinburg, Altes Rathaus, Schleswig-Holstein
38	Föhrer Inselarchiv, Wyk auf Föhr, Rathaus, Schleswig-Holstein
39	Staatsarchiv, Hamburg, Rathaus

Archive Indicator	Archive

40 Staatsarchiv, Hamburg-Dienststelle Altona, Hamburg-Altona, Neues Rathaus, Platz der Republik 1

41 Staatsarchiv Hamburg-Dienststelle Harburg, Hamburg-Harburg, Rathaus, Harburger Rathausplatz

42 Commerzbibliothek, Hamburg 11, Börse

43 Staats- und Universitäts-Bibliothek, Hamburg 13, Moorweidenstrasse 40

44 Archiv der Freien Evangelisch-Lutherischen Bekenntniskirche zu St. Anschar, Hamburg 36, Anscharplatz

45 Archiv der Evangelisch-Lutherischen Auswanderermission, Hamburg 1, Reutenbergstrasse 11

46 Predigerseminar der Deutschen Baptisten, Hamburg-Horn, Rennbahnstrasse 115

47 Firma D. C. Brandt, Hamburg 1, Schauenburgerstrasse 52

48 Staatsarchiv, Bremen, Am Dobben 91

49 Bremen Senatsregistratur, Bremen, Rathaus

50 Staatsarchiv Aurich, Aurich, Von Ihering-Straase 17, Niedersachsen

51 Staatliches Archivlager, Göttingen, Merkelstrasse 3

52 Staatsarchiv, Hannover, Am Archive 1

53 Staatsarchiv, Oldenburg, Damm 43

54 Staatsarchiv, Osnabrück, Schloss-Strasse 29

55 Staatsarchiv, Stade 1, Regierungsgebäude

56 Staatsarchiv, Wolfenbüttel 1, Forstweg

57 Pfarrarchiv zu Barterode, Barterode, Niedersachsen

58 Heimatbund zwischen Dehmse und Hunte, Beckstedt/Grafschaft Hoya, Niedersachsen (In 1959 written information could be obtained from Herrn Willi Krüger, Bremen-Lesum, Louis-Segelken-Strasse 42.)

59 Gutsarchiv, Bockel, Niedersachsen

60 Stadtarchiv, Braunschweig, Steintorwall 15, Niedersachsen

61 Stadtarchiv, Bremerhaven

62 Landratsamt, Bremervörde, Niedersachsen

63 Kreismuseum, Bremervörde, Niedersachsen

64 Archiv des Fleckens Bücken, Bücken, Niedersachsen

65 Heimatsmuseum am St. Petriplatz, Buxtehude, Niedersachsen

66 Stadtarchiv, Celle, Niedersachsen

67 Stadtarchiv, Clausthal-Zellerfeld, Niedersachsen

68 Archiv des Oberbergamtes Clausthal, Clausthal-Zellerfeld, Niedersachsen

69 Pfarrarchiv, Daverden/Kreis Vechta, Niedersachsen

70 Pfarrarchiv, Dissen, Niedersachsen

71 Stadtarchiv, Dransfeld, Niedersachsen

72 Stadtarchiv, Einbeck, Niedersachsen

73 Stadtarchiv, Emden, Niedersachsen

74 Stadtarchiv, Eschershausen, Niedersachsen

75 Pfarrarchiv I, Eschershausen, Niedersachsen

76 Archiv der Inspektion Gifhorn, Gifhorn, Niedersachsen

77 Niedersächsische Staats- und Universitätsbibliothek, Göttingen, Prinzenstrasse 1, Niedersachsen

78 Pfarrarchiv, Grasdorf a.d. Leine, Niedersachsen

79 Stadtarchiv, Hannover, Koebelingerstrasse 59, Niedersachsen

80 Staatsarchiv Hamburg, Dienststelle Harburg, Hamburg-Harburg, Rathaus, Harburger Rathausplatz (see also archive indicator 41 above)

81 Archiv des Fleckens Harpstedt, Harpstedt, Niedersachsen

82 Pfarrarchiv, Hattorf, Niedersachsen

83 Stadtarchiv, Helmstedt, Niedersachsen

84 Stadtarchiv, Hildesheim, Niedersachsen

85 Pfarrarchiv, Hohnsen, Niedersachsen

86 Archiv der Familie von der Wense, Holdenstedt, Niedersachsen

87 Stadtarchiv, Hornburg, Niedersachsen

88 Pfarrarchiv, Iber, Niedersachsen

89 Pfarrarchiv, Idensen, Niedersachsen

90 Pfarrarchiv, Imbshausen, Niedersachsen

91 Pfarrarchiv, Jacobidrebber, Niedersachsen

92 Altländer Archiv in Jork, Jork, Niedersachsen

93 Stadtarchiv, Leer, Niedersachsen

94 Pfarrarchiv Wangelnstedt und Kapellenarchiv Lenne und Linnenkamp mit Emmerborn, Lenne, Niedersachsen

95 Stadtarchiv, Lüchow, Niedersachsen

96 Stadtarchiv, Lüneburg, Niedersachsen

97 Stadtarchiv, Melle, Niedersachsen

98 Stadtarchiv, Northeim, Niedersachsen

99 Guts- und Familienarchiv der Freiherrn von Oldershausen, Oldershausen, Niedersachsen

100 Kreisarchiv Land Hadeln, Otterndorf, Niedersachsen

101 Stadtarchiv, Bad Pyrmont, Niedersachsen

102 Pfarrarchiv Kirchenkreis Loccum-Stolzenau, Riessen, Niedersachsen

103 Schaumburgisches Heimatsmuseum, Rinteln, Niedersachsen

12

Archive Indicator	Archive

172 Gemeindearchiv, Ginsheim-Gustavsburg, Hessen

173 Gemeindearchiv, Goddelau, Hessen

174 Stadtarchiv, Gross-Gerau, Hessen

175 Gemeindearchiv, Kelsterbach, Hessen

176 Gemeindearchiv, Klein-Gerau, Hessen

177 Gemeindearchiv, Klein-Rohrheim, Hessen

178 Gemeindearchiv, Leeheim, Hessen

179 Gemeindearchiv, Mörfelden, Hessen

180 Gemeindearchiv, Nauheim, Hessen

181 Herrn Adolf Thiel, Raunheim, Hessen

182 Stadtarchiv, Rüsselsheim, Hessen

183 Gemeindearchiv, Trebur, Hessen

184 Gemeindearchiv, Walldorf, Hessen

185 Gemeindearchiv, Wallerstädten, Hessen

186 Gemeindearchiv, Worfelden, Hessen

187 Archiv der Stadt, Hungen, Hessen

188 Archiv der Stadt, Ilbenstadt, Hessen

189 Archiv der Stadt, Korbach, Hessen

190 Pfarrarchiv, Korbach, Hessen

191 Archiv der Stadt, Ortenberg, Hessen

192 Archiv der Stadt, Sachsenhausen, Hessen

193 Archiv der Stadt, Seligenstadt, Hessen

194 Archiv der Stadt, Ulrichstein, Hessen

195 Archiv der Stadt, Viernheim, Hessen

196 Gemeindearchiv, Wallau, Hessen

197 Stadtarchiv, Weilburg/Lahn, Schlossplatz 2, Heimatmuseum, Hessen

198 Historisches Archiv der Stadt, Wetzlar, Hessen

199 Bundesarchiv, Koblenz, Am Rhein 12, Rheinland-Pfalz

200 Staatsarchiv, Koblenz, Rheinland-Pfalz

201 Staatsarchiv, Speyer, Domplatz 6, Rheinland-Pfalz

202 Archiv der Stadt, Andernach, Rheinland-Pfalz

203 Archiv der Stadt, Annweiler, Rheinland-Pfalz

204 Archiv der Stadt, Bad Dürkheim, Rheinland-Pfalz

205 Stadtarchiv, Diez, Rheinland-Pfalz

206 Archiv der Evangelischen Kirche im Rheinland, Koblenz, Karmeliterstrasse 1—3, Rheinland-Pfalz

207 Landratsamt, Kusel, Rheinland-Pfalz

208 Archiv der Stadt, Landau/Pfalz, Rheinland-Pfalz

209 Archiv der Stadt, Landstuhl, Rheinland-Pfalz

210 Archiv der Stadt, Mainz, Rheinallee 33/10, Rheinland-Pfalz

211 Archiv der Gemeinde, Mörzheim, Rheinland-Pfalz

212 Archiv der Stadt, Montabaur, Rheinland-Pfalz

213 Fürstliches Archiv, Neuwied, Rheinland-Pfalz

214 Archiv der Gemeinde, Odernheim a.Gl., Rheinland-Pfalz

215 Archiv des Kreises, Simmern, Rheinland-Pfalz

216 Archiv der Stadt, Speyer, Rheinland-Pfalz

217 Landeskirchenarchiv, Speyer, Domplatz 5, Rheinland-Pfalz

218 Archiv der Stadt, Wachenheim, Rheinland-Pfalz

219 Archiv des Mennonitischen Geschichtsvereins, Weierhof/Pfalz, Rheinland-Pfalz

220 Heimatstelle Pfalz, Kaiserslautern, Stiftsplatz 5, Rheinland-Pfalz

221 Archiv der Stadt, Neustadt an der Weinstrasse, Rheinland-Pfalz

222 Staatsarchiv, Amberg, Archivstrasse 3, Bayern

223 Staatsarchiv, Bamberg, Hainstrasse 39, Bayern

224 Staatsarchiv, Coburg, Schloss Ehrenburg, Bayern

225 Staatsarchiv, Landshut, Schloss Trausnitz, Bayern

226 Bayerisches Hauptstaatsarchiv, Allgemeines Staatsarchiv, München, Arcisstrasse 12, Bayern

227 Geheimes Staatsarchiv, München, Leonrodstrasse 57/I

228 Kriegsarchiv, München, Leonrodstrasse 52/I

229 Staatsarchiv für Oberbayern, München, Himbselstrasse 1

230 Staatsarchiv für Niederbayern, Neuburg a.D., Schloss, Bayern

231 Staatsarchiv, Nürnberg, Archivstrasse 17, Bayern

232 Staatsarchiv, Würzburg, Residenz, Bayern

233 Fürstlich Leiningische Domänenverwaltung, Amorbach, Bayern

234 Historischer Verein für Mittelfranken, Regierung von Mittelfranken, Ansbach, Regierungsbibliothek, Bayern

235 Stadtarchiv, Augsburg, Fuggerstrasse 12, Bayern

236 Diözesan-Caritasverband, Bamberg, Geyerswörthstrasse 2, Bayern

237 Stadtarchiv, Dinkelsbühl, Bayern

238 Stadtarchiv, Erlangen, Bayern

239 Stadtarchiv, Fürth, Bayern

240 Redemptoristenkloster, Gars/Inn, Bayern

241 Stadtarchiv, Greding, Bayern

242 Stadtarchiv, Gunzenhausen, Bayern

243 Stadtarchiv, Herrieden, Bayern

244 Stadtarchiv, Kempten, Bayern

14

Name Index

van AARSSEN, --

Estate   1940-1941
<u>148</u>: R XI.   ADS 6:260

ABIGER, Karl (in St. Louis, Missouri)

Information regarding him   1878
<u>39</u>: Bestand Senat: Cl.VIII, No.X, Jg.1878 (Regis-
     ter), 20 [Nordamerika]   ADS 2:237

ACKER, Heinrich (from Edenkofen; killed [in action?]
in America)

Petition of -- Acker, teamster's widow in Eden-
     kofen [Germany] for issuance of a death certi-
     ficate for her son   1866
<u>227</u>: Rep.MA 1921: A.St.I: Nr. 1277.   ADS 9:72

ADAMS, Daniel (from America)

Naturalization [in Germany]   1874
<u>151</u>: Bestand 180: Rotenburg: Acc.1935/52: Nr. 75.
                            ADS 7:121

ADUTT, -- (a prisoner?)

Transit from America to Austria   1894
<u>39</u>: Reichs- und Auswaert. Ang.: D I C, 55.
                            ADS 2:267

AH, -- (in San Francisco)

Claims of inheritance of the Ah sisters   1865
<u>227</u>: Rep.MA 1921: A.St. I: Nr. 72.   ADS 9:65

AHLERS, J. Nic. (in Charter Oak, Iowa)

Estate   1895
<u>39</u>: Reichs- u. auswaert.Ang.: E.IV.33 (Alt.Sig.)
     [Erbschaften]   ADS 2:275

AHLERS, --

Petition of Superior Court in New York in case of
     Ahlers *versus* Broedermann   1910
<u>39</u>: Reichs- u. auswaert.Ang.: J. Ia.2.25 [Justiz-
     sachen]   ADS 2:363

AHLHEIT, Martin (farmer in Iowa; died 23 December
1884)

Location of heirs   1885-1888
<u>110</u>: Rep.III, Fach 30: Nr. 7.   ADS 4:339

AHLSPACH* David (from Froeschen, RPL; in Pennsylvania)

Father Johannes Alsbach* signs receipt for corres-
     pondence sent to Amt Pirmasens by son David
     [addressed to father?]   1771
<u>201</u>: Archivabt. Grafschaft Hanau-Lichtenberg: Akt
     Nr. 1161.
     *So spelled.   ADS 8:67

AHRENS, Heinrich Wilhelm (handworker, from Wenzen)

Incarceration in house of corrections in Bevern &
     emigration to America   1859-1862
<u>56</u>: L Neu Abt.129A: Gr.11, Nr. 87.   ADS 4:192

AICH, Herbert (American citizen)

Expulsion of brothers Herbert & Waldemar Aich 1914
<u>39</u>: Reichs- u. auswaert.Ang.: A.IV.7 [Ausweisungen]
                           ADS 2:383

AICH, Waldemar (American citizen). *See* Aich, Herbert

de AIZPURNA, Ramón (of Havana & New Orleans)

Letters to C. A. Offensandt   1833-1834
<u>48</u>: Bestand 7,76: C. A. Offensandt: Nr. 51-55.
                          ADS 3:75

ALBERTI, --, Senior (Captain, Third English-Waldeck Regiment)

Courtmartial proceedings regarding his behavior before, during, & after his Spanish imprisonment [prisoner of war?]   1779-1780 (1782)

151: Bestand 118: Nr. 984.
      DLC has.   ADS 7:101

ALBERTI, -- (an emigrant)

Denunciation made by Alberti   1840

39: Bestand Senat: Cl.VIII, No.X, Jg.1840 (Register) 2 [Auswanderer-Verhaeltnisse]   ADS 2:139

ALBRECHT, Johnn (from Neusiegelsum NL)

Emigration of convict to America   1866

50: Rep.21a: Nr. 8967.   ADS 4:8

ALBRECHT, Ludwig (died in Pomona [California?])

File on [estate?]   1899

227: Rep.MA 1921: A.St. I: Nr. 120.   ADS 9:65

ALDEN, Miss J. D. (in Rochester, New York)

Correspondence [she was related to Rendtorff family]   1922-1923

2: Abt.399 (Fam.Archiv Rendtorff): Nr. 217.
                                            ADS 1:34

ALEXANDEROWITZ, David

Request to American Consulate in Hamburg for an American death certificate for Alexanderowitz   1910

39: Reichs- u. auswaert.Ang.: S.IV.3.2 [Standesamtssachen]   ADS 2:367

ALF? -- (from Arnstorf?)

Costs for defendant -- Alf's transportation to America   1847

62: A, Fach 6, vol. 6.   ADS 4:232

ALLARDT, B. H.
Planned expulsion of Allardt & Julius Davis & prosecution for an insulting article in the *Michigan Wegweiser* about immigration into northern Michigan   1873

39: Bestand Senat: Cl.VIII, No.X, Jg.1873 (Register) 21 [Auswanderer-Deputation]   ADS 2:223

ALLARDT, H. H.

Investigation requested of J. Davis & Allardt, immigrant emissaries for Michigan   1874

39: Bestand Senat: Cl.VIII, No.X, Jg.1874 (Register) 6 [Auswanderer-Deputation]   ADS 2:226

ALLARDT, M. H.

Expulsion of Allardt & Jul. Davis, immigration emisaries for Michigan   1876

39: Bestand Senat: Cl.VIII, No.X, Jg.1876 (Register) 8 [Auswanderer-Deputation]   ADS 2:231

ALLEN, Willard S.

Hamburg police search for Allen, wanted in America for fraud   1903

39: Reichs- u. auswaert.Ang.: P.III.1.3 [Polizeiwesen]   ADS 2:333

ALLMANN, Jacob (in Rockville [state not given])

Estate of   1886

227: Rep.MA 1921: A.St.I: Nr. 97.   ADS 9:65

ALLMANY, Karolina. *See* Allmany, Philipp

ALLMANY, Philipp

Philipp & Karolina Allmany, brother & sister, property claims   1846-1849

227: Rep.MA 1898: A.St.: Nr. 1022.   ADS 9:58

ALPHEY? Martin Andreas (born in Celle)

File on emigration permit for Martin Alphey, a person subject to military service, the son of highway superintendent -- Alphey of Selsingen   1855

62: A, Fach 6, vol. 9.   ADS 4:232

ALSBACH, Johannes. *See* Ahlspach, David

ALT, --

Estate   1941
148: R XI.   ADS 6:260

ALTHOEN, Anna Maria (in New York)

Financial condition of   1878
227: Rep.MA 1921: A.St.I: Nr. 80.   ADS 9:65

ALTMANN, Charlotte (in San Francisco)

Estate of   1909

39: Reichs- u. auswaert.Ang.: E.IV.4 [Erbschaften]   ADS 2:356

ALVES, Frieda Aug. Joh., born Bleistein (in America)

Transmittal of money to Mrs. Alves, former ward of Hamburg orphanage   1895

39: Reichs- u. auswaert.Ang.: W.I.2a [alt.Sig.] [Waisenhausang.]   ADS 2:279

ALVES, Frieda A. J., born Bleistein. *See also* Bleistein, Clara

AMANN, Adolf (in USA)

Transmittal of historical material between a certain Adolf Amann and a correspondent in the Salzburger colonies of Georgia   1829

227: Rep.Bayer.& Pfaelz. Gesandtschaft Paris: Nr. 10610.   ADS 9:50

AMANN, Adolf. *See also* Amman, Adolph

AMBROS, Alois Melchior (died in San Francisco)

Estate of   1891

227: Rep. MA 1921: A.St.I: Nr. 109.   ADS 9:65

AMENDT, Georg (from Gross-Gerau, Hesse)

Application for emigration to North America
                                          1851-1852

174: Abt.XI,4.   ADS 7:226

AMMAN, Adolph (in America)

The forwarding of packages to the USA, particularly to a certain Adolph Amman   1829

227: Rep. Bayerische Gesandtschaft London: Nr. 770.
                                             ADS 4:46

AMMAN, Adolph. *See also* Amann, Adolf

AMMON, Joh[ann](in San Francisco)

Estate of   1889

227: Rep. MA 1921: A.St.I: Nr. 101.   ADS 9:65

AMSINCK, Gustav (died in New York, 8 June 1909)

Estate of   1909-1932

39: Bestand Familienarchive: Fam. Merck: IV.3. Konv.9.   ADS 2:489

AMTSBERG, Ernst

New York: Information regarding   1878

39: Bestand Senat: Cl.VIII, No.X, Jg.1878 (Register) 24 [Nordamerika]   ADS 2:237

AMWEG, Jacob. *See* Anweg, Jakob

ANDERS, -- (sailor)

San Francisco: estate of   1914

39: Reichs- u. auswaert.Ang.: E.IV.58 [Erbschaften]
                                             ADS 2:384

ANDERSEN, Julie, born Hartmann

New Orlean [Consulate?] reports deaths of Julie Andersen & Martha Louise Anna Schill   1873

39: Bestand Senat: Cl.VIII, No.X, Jg.1873 (Register) 45 [Nordamerika]   ADS 2:226

ANDES, -- (in America)

Estate of   1892

227: Rep. MA 1921: A.St.I: Nr. 112.   ADS 9:65

ANDLER, F. (in Galveston, Texas)

Inquiry by private person in Germany regarding Andler & H. A. Haeusinger   1900

39: Reichs- u. auswaert.Ang.: N.Ib.1.7 [Nachforschungen]   ADS 2:310

ANDRE, Ph[ilipp]. *See* Hammann, Christoph

ANDREAE, Heinrich

Papers regarding   1840

2: Abt.65, Nr. 4860, Fasc. 4.
DLC has [see under Kiel]   ADS 1:15

ANDREAE, Ida

Papers regarding   1840

2: Abt.65: Nr. 4860, Fasc. 4.
DLC has [see under Kiel]   ADS 1:15

ANDREAE, Johann (in Camden, New Jersey)

Correspondence   1902

120: Akte Frank Buchman (D.II.B).   ADS 6:75

ANDRESSEN, F. M. Chr. (of San Francisco)

Death of, & estate of (formerly of Eutin, SHL) [File may have been destroyed]   1864

2: Re. Eutin: 25.1.B.: Lit.A, Nr. 17 [Nachrichten]
DLC has?   ADS 1:11

ANGELRODT, -- (in St. Louis, Missouri)

Letter & money from Angelrodt for Hamburg residents who lost possessions in fire   1843

39: Bestand Senat: Cl.VIII, No.X, Jg.1843 (Register).   ADS 2:145

ANGERMANN, Betty, born Rendsburg

Washington: Apparent disinheritance of Betty Angermann by her great-uncle Raphael Hirsch   1878

39: Bestand Senat: Cl.VIII, No.X, Jg.1878 (Register) 48 [Nordamerika]   ADS 2:238

ANGERSTEIN, Johanne Maria Ernestine

Sentencing for loose living; transportation to
   America [Goettingen]   1821-1831

52: Han.Des.80: Hildesheim I: E.DD.38b: Nr. 1406.
                                           ADS 4:90

ANKENBRAND, Anton (in America?)

Petition of Johann Ankenbrand, of Oberessfeld, Ldg.
   Koenigshofen, for remission of part of his son
   Anton Ankenbrand's fine & court costs   1847

232: Regierungs-Praesidialakten: Sachen 206.
                                           ADS 9:137

ANSKOHL, Max (in Litchfield, Illinois)

Estate of   1909

39: Reichs- u. auswaert.Ang.: E.IV.25 [Erbeschaften]
                                           ADS 2:356

ANTES, Johann Heinrich (Moravian Brethren; in Pennsylvania?)

Letters to [Count] Zinzendorf   1741-1745

500: Herrnhut: Archiv: Rep.14.A: Nr. 24. Not in ADS

ANTON, Adam (from Offenbach)

Investigation of the killing of Anton & the taking
   of testimony in America   1838-1842

227: Rep.MA 1898: A.St.: Nr. 6257.   ADS 9:58

ANWEG, Jakob (in Cincinnati, Ohio)

Search for him   1899

225: Rep.168: Verz.1: Fasz.19: Nr. 1-2; 4. ADS 9:15

APPEL, Andreas III (from Astheim, HEL)

Register of citizens, 1770-1866, contains notation
   regarding his emigration to America   1863

161: Abt.XI: Abschnitt 4.   ADS 7:211

APPEL, Peter (of Zennhusen, SHL)

Petition for release from citizenship   1867

13: Alphabetischen Sammlung.   ADS 1:49

APPUHN, Johann Heinrich (farmhand; from Klein Rhueden)
   See Cleve, Johann Heinrich

ARENSON, Johanna

New York: Information regarding   1877

39: Bestand Senat: Cl.VIII, No.X, Jg.1877 (Register) 19 [Nordamerika]   ADS 2:234

ARFMANN, Christ. Heinr.

Emigration file contains letters from friends &
   relatives in New York   1854

55: Han.Des.74: Blumenthal: IX.J.1: Fach 301, Nr.
   10.   ADS 4:154

ARMAND, Adolf (in San Francisco)

Application for copy of will   1881

227: Rep.MA 1921: A.St.I: Nr. 86.   ADS 9:65

ARMGARDT, Robert (in Cincinnati)

Estate of   1911

39: Reichs- u. auswaert.Ang.: E.IV.100 [Erb-
   schaften]   ADS 2:369

ARNDT, -- (in Seattle [Washington])

Estate of   1915

39: Reichs- u. auswaert.Ang.: E.IV.35 [Erb-
   schaften]   ADS 2:388

ARNECKE, Friedr[ich] (in Oregon)

Estate of   1910

39: Reichs- u. auswaert.Ang.: E.IV.19: [Erb-
   schaften]   ADS 2:362

ARTMAIER, Johann (in America)

Trusteeship for [during absence abroad]   1891

227: Rep.MA 1921: A.St.I: Nr. 107.   ADS 9:65

ASAM, -- (in America?)

Hamburg court seeks accounting papers from a third-
   party creditor in case of Gebr[uder] Phillips,
   Jr. [of Aachen] versus Asam   1898

39: Reichs- u. auswaert.Ang.: J.Ib.74 [Justiz-
   sachen]   ADS 2:297

ASSUS, -- (from Wisper, HEL)

Emigration of the Assus & Hartmann families to
   America   1854

152: Abt.231: Langenschwalbach: Nr. 2872.
                                           ADS 7:133

AST, --

Chicago consular fees in case of Ast versus Ast
                                                1900

39: Reichs- u. auswaert.Ang.: J.Ib.1.2 [Justiz-
   sachen]   ADS 2:309

ASTOR, John Jacob (in New York)

Karl Lederle, of Steinmauern, inquires regarding
Astor's estate   1854

39: Bestand Senat: Cl.VIII, No.X, Jg.1854 (Register).   ADS 2:166

ASTOR, --. *See* Emmerich-Astor, --

ATZBERGER, Elizabeth (widow of innkeeper)

File on [estate?]   1890

227: Rep.MA 1921: A.St.I: Nr. 104.   ADS 9:65

AUDE, Adelheid, born Rendtorff (in New York)

Correspondence   1911-1920

2: Abt.399 (Fam.Archiv Rendtorff): Nr. 211.
                                               ADS 1:34

AUFINGER, Jakob (in America)

His life & place of residence   1887

227: Rep.MA 1921: A.St.I: Nr. 1285.   ADS 9:72

AUGUSTIN, Karl Wilh[elm]

Transportation to America of Augustin, "a lazy
drunken bum" [Goettingen]   1841-1842

52: Han.Des.80: Hildesheim I: E.DD.38b: Nr. 1440.
                                               ADS 4:90

AUMUELLER, Konrad (of Altenhassel)

Accused of enticing other citizens to emigrate
                                               1754

151: Bestand 86: Nr. 51981.   ADS 7:88

AUSTRIAN, Falk (died in 1887 in Chicago)

Estate of   1889

227: Rep.MA 1921: A.St.I: Nr. 102.   ADS 9:65

BACH, Jakob (of Crumstaedt, Landkreis Gross-Gerau
                                               HEL)

Emigration of Jewish person & wife   1938

167: Abmelderegister 1938-1957.   ADS 7:217

Indemnities paid to Jewish person who emigrated to
New York in 1938

167: Wiedergutmachungsakten [after 1945] ADS 7:217

BACHMANN, Agnes (wife of George; in Musselshell,
Montana, & Monrovia, California)

Letter to George Bachmann [nephew] in Bremervoerde
[from Musselshell, Montana]   1902

63   ADS 4:238

Letter to George Bachmann [nephew] in Bremervoerde
[from Monrovia, California]   1935

63   ADS 4:239

Two letters to George Bachmann [nephew] in Bremervoerde [from Monrovia, California]   1937

63   ADS 4:239

BACHMANN, August

Letter from New York to brother in Bremervoerde;
requests emigration certificate [apparently not
obtained before he emigrated]   1854

63   ADS 4:234

Letter from Union City, California, formerly in
New York, to relatives in Bremervoerde   1854

63   ADS 4:235

Letter from Humbug Creek, Siscigon? County, California, to parents in Bremervoerde   1859

63   ADS 4:235

Letter from Salmon River, Sawyer's Bar, north of
Salmon River, Klamath County, California, to
parents in Bremervoerde   1862

63   ADS 4:235

Letter from Bannock City, Boise County, Idaho, to
parents in Bremervoerde   1864

63   ADS 4:235

Letter from Sawyer's Bar, Klamath County, California, to parents & brothers & sisters in
Bremervoerde   1866

63   ADS 4:235

Letter from Sawyer's Bar, Klamath County, California, to brother in Bremervoerde   1867
Letter [from same address] to parents in Bremervoerde   1872

63   ADS 4:235

Letter from Mosinee, Wisconsin, to brother in
Bremervoerde; another brother, George, also in
Mosinee   1877

63   ADS 4:236

Letter from New York to father in Bremervoerde
                                               1880

63   ADS 4:236

Letter from Mosinee, Wisconsin, to relatives in
Bremervoerde; mentions brother George in Mosinee
                                               1881

63   ADS 4:236

Letter from Mosinee, Wisconsin, to hephew in
Bremervoerde; mentions daughter Beatrice   1902

63   ADS 4:236

Newspaper clipping from Mosinee, Wisconsin, regarding death on 14 May 1931; immigrated to New York
in 1869; returned for short time to Germany, &

in Mosinee, Wisconsin, after 1875

63   ADS 4:236-237

Letter from Mosinee, Wisconsin, by daughter Beatrice Bachmann-Knoedler to relatives in Bremervoerde   1931

63   ADS 4:237

BACHMANN, Beatrice

Letter from Mosinee, Wisconsin, to relatives in
Bremervoerde   1905

63   ADS 4:238

BACHMANN, Beatrice. *See also* Knoedler, Beatrice,
born Bachmann

BACHMANN, Earl A (in Mosinee, Wisconsin)

Letter to cousin George in Bremervoerde   1934

63   ADS 4:238

BACHMANN, Georg

Letter from New York to August Bachmann   1874

63   ADS 4:237

Letter from Mosinee, Wisconsin, to father in Bremervoerde   1880; 1881

63   ADS 4:237

Letter from Musselshell, Montana, to brother in
Bremervoerde; asks for payment of his inheritance   1895

63   ADS 4:237

Letter from Musselshell, Montana, to brother in
Bremervoerde regarding payment of inheritance
                                           1896

63   ADS 4:238

BACHMANN, Karl (died in New York)

File on [estate?]   1898

227: Rep.MA 1921: A.St.I: Nr. 224.   ADS 9:66

BACHMANN, -- (master miller; from Baselmuehle; in
Philadelphia)

Guardianship for the children of Bachmann & the
transmittal of their property to Philadelphia
                                       1847-1848

227: Rep.MA 1898: A.St.: Nr. 6265.   ADS 9:58

BADE, Peter (from Bergenheim; in America)

Property settlement   1856

227: Rep.MA 1898: A.St.: Nr. 1032.   ADS 9:58

BADEN, D. H. (pastor; in Brooklyn, New York)

Correspondence   1886

120: Akte Frank Buchman (D.II.B).   ADS 6:75

BADER, J. G.

Sentenced to lengthy penitentiary sentence in
Frankfurt/Main for political activities; expulsion to America   1838

200: Abt.403: Oberpraesidium Koblenz (Rheinprovinz): Nr. 2528.   ADS 8:21

BAER, -- (Doctor)

Estate of   1943

148: R XI.   ADS 6:260

BAERTLING Family (of Einbeck, NL)

Genealogical studies regarding

72: Harland's Nachlass: Manuscripte Nr. 9: Vol.
VIII.   ADS 4:250

BAESEL, Heinrich (from Jettenbach; in North America)

Search for emigrant as heir [of estate in Germany]
                                           1856

201: Archivabt. Bezirksaemter: Kusel: Akt Nr. 32.
                                         ADS 8:64

BAETCKE, Friedrich Heinrich Carl (peddler)

Expulsion from Hamburg as Social Democrat & emigration to America   1880-1881

39: Polizeibeoherde-Polit.Pol.: S 149/111: Bl.
29-35.   ADS 2:414

BAETKE, Johann Wilhelm Heinrich (deliverer of bread)

Expulsion from Hamburg as Social Democrat & emigration to America   1880-1881

39: Polizeibehoerde-Polit.Pol.: S 149/110: Bl.
36-43.   ADS 2:414

BAETTGER, --

New York: Estate of   1895

39: Reichs- u. auswaert.Ang.: E.IV.44 [Alt.Sig.]
[Erbschaften]   ADS 2:275

BAHN, E. (in Rochester, Indiana)

Bahn suggests method of ventilating immigrant
ships   1855

39: Bestand Senat: Cl.VIII, No.X, Jg.1855 (Register).   ADS 2:169

BALBIERER, Philipp Nickel (from Waldmohr, Ducy of Saarbruecken)

Emigration of Bablierer, Valentin Blum, & Eva Elisabetha Blum, married to Nickel Lang, all from Waldmohr, to Pennsylvania
1769-1786; 1737-1778

201: Archivabt. Herzogtum Zweibruecken: Akt Zwei-bruecken I: Nr. 1359a 2.   ADS 8:56

BALCKE, Heinrich

Request for death certificate   1864

39: Bestand Senat: Cl.VIII, No.X, Jg.1864 (Regis-ter) I [Amerikanische V.S.]   ADS 2:185

BALLINGER, G. (in Chicago)

Appeal, at the request of Ballinger, in the case of Josef Schmidt, convicted of arson & murder
1891

227: Rep.MA 1921: A.V. V: Nr. 15745.   ADS 9:90

BANDER, Hans (a Swiss citizen)

Extradition to Switzerland from the USA [via Ham-burg?]   1913

39: Reichs- u. auswaert.Ang.: A.Ia.11 [Ausliefer-ungen]   ADS 2:377

BARANOWSKI, --

Estate of   1941-1942
148: R XI.   ADS 6:260

BARGE, Jakob (died 1908 in Philadelphia). *See* Bertsch, Jakob

BARGENQUAST, Johann Jochim (Hemme, SHL)

Emigration to America   1873
13: Alphabetischen Sammlung.   ADS 1:49

BARIEM, W.

New York consular report on death of   1875

39: Bestand Senat: Cl.VIII, No.X, Jg.1875 (Regis-ter) 24 [Nordamerika]   ADS 2:230

BARRER, Stephan (born in Ansbach; Reformed Church minister in Pittsburgh, Pennsylvania). *See* Har-gienst, Jakob

BARTA, --

Estate of   1941
148: R XI.   ADS 6:260

BARTELS, F.

St. Louis, Missouri: Estate of   1917

39: Reichs- u. auswaert.Ang.: E.IV.21 [Erb-schaften]   ADS 2:392

BARTHLING, Catharine (in Good Harbor, Michigan)

Estate of   1903

39: Reichs- u. auswaert.Ang.: E.IV.72 [Erb-schaften]   ADS 2:330

BARTHOLOMAEUS, Carl (factory owner; in Philadelphia)

Claims of the estates of parents & grandparents by wife of Carl Bartholomaeus, who was born Metzler in Herborn, & by her mother, now the wife of Moritz Dobel in Tulpehoken, Pennsylvania   1791

152: Abt.172: Landratsamt Dillenburg: Heft 4. DLC has.   ADS 7:129

BARTUM, Marie (from Klein-Rhueden). *See* Cleve, Johann Heinrich

BARTUM, Wilhelmine (spinster; from Klein-Rhueden). *See* Cleve, Johann Heinrich

BAST, Christian (in Philadelphia)

Inquiry by private person in Germany regarding
1902

39: Reichs- u. auswaert.Ang.: N.Ib.1.2 [Nachforsch-ungen]   ADS 2:326

BATTEIGER, Valentin (from Westheim; wine & fruit dealer in Philadelphia)

File on   1818-1819

227: Rep. Bayerische Gesandtschaft London: Nr. 854.   ADS 9:47

BAUER, Albert (pick & shovel laborer from Clausthal-Zellerfeld)

Emigration to New York of Bauer & Carl August Her-mann Richter   1853

68: VI.2.b: Fach 162, Nr. 36.   ADS 4:246

BAUER, Georg L. (university professor; from Chicago; died in Munich)

Estate of   1898
227: Rep.MA 1921: A.St.I: Nr. 231.   ADS 9:66

BAUER, Johann (farmer; in Park Falls [state not giv-en])

Estate of   1897-1899
225: Rep.168: Verz.1: Fasz.19, Nr. 1-2.   ADS 9:15

BAUER-BAJOR, L.

Estradition to Hungary from USA via Hamburg   1912

39: Reichs- u. auswaert.Ang.: A.Ia.27 [Auslieferungen]
                                          ADS 2:372

BAUMANN, August (printer)

Expulsion from Hamburg as Social Democrat & emigration to America   1880-1881

39: Polizeibehoerde-Polit.Pol.: S 149/54: Bl. 1-35.
                                          ADS 2:412

BAUMANN, A. (rag collector from Moordorf)

Emigration to America   1863

50: Rep.21a: Nr. 9018.   ADS 4:8

BAUMANN, Rosalie, born Lueht

New York: Estate of   1896

39: Reichs- u. auswaert.Ang.: E.IV.47 [alt.Sig.]
    [Erbschaften]   ADS 2:282

BAUMGARTEN, Johann Christoph W. (born in Hamburg)

Citizenship of prisoner aboard the Savannah (ship
    of the Confederate States of America)   1861

39: Bestand Senat: Cl.VI, No.16p: Vol.4a: Fasc. 6
    [Heimatsverhaeltnisse]   ADS 2:43

BAUMGARTEN, Wilh[elm]

New York: Information regarding   1877

39: Bestand Senat: Cl.VIII, No.X, Jg.1877 (Register) 33 [Nordamerika]   ADS 2:235

BAUR, -- (non-commissioned officer in Hessian Jaeger
Corps during American Revolution)

Interrogation of, during investigation of Trenton
    debacle   1778 (1782)

151: Bestand 4: Kriegssachen: 328: Nr. 141; 142.
     DLC has.   ADS 7:37

BAUTZ, Johann Philipp (of Hochstadt, Hesse-Hanau)

Emigration [to Carolinas?]   1741

151: Bestand 80: Lit.B: Nr. 1 [Unterthanen]
                                          ADS 7:79

BAYE, Christian Ludwig (servant; from Hallensen).
See Voss, August

BAYE, Gustav F. (bookkeeper)

Search in Germany at request of Chicago police
                                          1909

39: Reichs- u. auswaert.Ang.: P.III.2.5 [Polizeiwesen]
                                          ADS 2:359

BAYR, Ver. (died in Baltimore [Maryland])

Property of   1893-1896

227: Rep.MA 1921: A.St.I: Nr. 156.   ADS 9:66

BECHER, Ernst Theodor (cigar worker)

Expulsion from Hamburg as Social Democrat & emigration to America   1881

39: Polizeibehoerde-Polit.Pol.: S 149/419: Bl. 73-
    79.   ADS 2:418

BECHTEL, Johannes (of Moravian Brethren; in Pennsylvania?). See Wiegner, Christ.; Wade, John

BECHTHOLD, Tobias (of Herrschaft Weilburg)

Recruitment of Bechthold, 15 years of age, illegitimate child, into British military service
    [probably for the 60th English Regiment of Royal
    Americans]   1780?

152: Abt.150: Herrschaft Weilburg: Nr. 2665.
                                          ADS 7:127

BECK, -- (woman pastor; in Cleveland). See Friedel,
C. [could refer to the wife of a pastor -- Beck]

BECKENSCHUETZ, Johann

Search for heirs of Beckenschuetz who was killed
    in the American war with Mexico   1854

114: Lipp.Reg.: Aelt.Regis.: Fach 222: Nr. 2.
                                          ADS 6:3

BECKER, Adam (from Raunheim, Landkreis Gross-Gerau,
HEL)

Emigration to America with wife   1880

181: Raunheimer Evangelischen Pfarrei Familienregister (Lehrer Johannes Buxbaum).   ADS 7:234

BECKER, Andreas (from Raunheim, Landkreis Gross-
Gerau, HEL)

Emigration to America with wife & 2 children 1854

181: Raunheimer Evangelischen Pfarrei Familienregister (Lehrer Johannes Buxbaum).   ADS 7:234

BECKER, Heinrich (from Moerzheim; died in New Orleans)

Death certificate for Heinrich Becker, son of Jakob
    Becker & Maria Singer [Becker] in Moerzheim [RPL]
                                          1843

211   ADS 8:79

BECKER, Karl (died Barada Township, Nebraska [county not given in ADS description])

Estate of   1914
151: Bestand 165: I: Nr. 7094.   ADS 7:111

BECKER, Margaretha, born Heilmann (from Iggersheim)

Exportation of personal property [see also Bull-
    mer, G. I.]   1851
227: Rep.MA 1898: A.St.: Nr. 1027.   ADS 9:58

BECKER, Maria

San Francisco: Estate of   1915
39: Reichs- u. auswaert.Ang.: E.IV.48 [Erbschaften]
                                                   ADS 2:388

BECKER, Michael (of Crumstaedt, Landkreis Gross-
Gerau, HEL)

Emigration to America   1884
167: Verzeichnis ueber ausgestellte Heimatscheine
     1883-1912: Abt. XI, 3.   ADS 7:217

BECKER, Peter (from Zweibruecken)

Property claim [in America?]   1844-1845
227: Rep.MA 1898: A.St.: Nr. 1025.   ADS 9:58

BECKER, Mr. & Mrs. --

San Francisco: Estate of   1911
39: Reichs- u. auswaert.Ang.: E.IV.103 [Erb-
    schaften]   ADS 2:369

BECKER, --

Petition of Joh. Christ. Becker of Massenbach for
    the payment of funds transmitted from his father,
    an emigrant in America, to the Sparkasse [savings
    bank] in Zwingenberg   1854
149: Rep.48/1: Konv.13, Fasz.29: Zwingenberg: (a).
                                                   ADS 7:14

BECKER, --

Inheritance of millions in America   1930-1933
149: Rep.49a: Nr. 565/3.   ADS 7:16

BECKMANN, E. (jailed in Trenton, New Jersey)

Inquiry from Trenton, New Jersey, as to whether
    Beckmann had been involved in counterfeiting in
    Hamburg   1890
39: Bestand Senat: Cl.VIII, No.X, Jg.1890 (Regis-
    ter) 1 [Verein.St.Am.]   ADS 2:264

BEDASSEM, J. (in America)

Mentioned in summary report on political refugees:
    participation in a plot to overturn the [German]
    government; plundering of the army in Pruem.
    Some of the participants in America were the
    following: J. Bedassem, A. Goebel, Chr. Hall,
    S. Seibel, B. Varani, J. A. Kemmerling, F. W.
    Bempel, Professor Kinkel, I. Koch, -- von Mir-
    bach, W. Groch, -- Butz, -- Post, H. Wesendonck,
    & others   n.d.
200: Abt.403: Oberpraesidium Koblenz (Rheinpro-
     vinz).   ADS 8:21

BEER, Leopold

Expulsion of Beer, an American citizen, from Ham-
    burg jurisdiction   1899
39: Reichs- u. auswaert.Ang.: A.IV.1 [Ausweis-
    ungen]   ADS 2:300

BEHN, James B.

Letter to C. A. Offensandt from Tampico; Mont-
    gomery, Alabama; New York; Baltimore; Fishkill,
    New York; Louisville [Kentucky]; St. Louis; New
    Orleans; Brooklyn   1832-1835
48: Bestand 7,76: C.A. Offensandt: Nr. 1-38.
                                                   ADS 3:75

BEHNCKE, Karsten Juergen Eduard (cigar worker)

Expulsion from Hamburg as Social Democrat & emi-
    gration to America   1881-1882
39: Polizeibehoerde-Polit.Pol.: S 149/363: Bl. 80-
    99.   ADS 2:417

BEHRENS, August

Behren's complaint against Bertha Koenig regarding
    emigration contract   1873
39: Bestand Senat: Cl.VIII, No.X, Jg.187 3 (Regis-
    ter) 17 [Auswanderer-Deputation]   ADS 2:223

BEHRENS, Heinrich (journeyman miller)

Proposed transportation of Behrens to America
    [from Uslar]   1856
52: Han.Des.80: Hildesheim I: E.DD.44a: Nr. 1720.
                                                   ADS 4:91

BEHRENS, Heinrich Friedrich August (servant; from
Muenchehof)

Emigration to America   1855-1857
56: L Neu Abt.129A: Gr.11, Nr. 65.   ADS 4:190

BEHRENS, Henry (died in New York)

Estate of   1914
151: Bestand 165: I: Nr. 7096.   ADS 7:111

BEHRENS, --

Baltimore consular fees in case of Behrens *versus* Steiner   1900

39: Reichs- u. auswaert.Ang.: J.Ib.10.1 [Justiz-
        sachen]   ADS 2:309

BEIRICH, Nik[olas?](in Atchison, Kansas)

Estate of   1908

39: Reichs- u. auswaert.Ang.: E.IV.87 [Erbschaften]
                                                            ADS 2:352

BEISEL, Konrad (1691-1768)

Information on Beisel, the leader of Ephrata [Com-
        munity] 1735-1814   after 1814

268: *Ebersbacher Geschichtsblatt*, Neue Folge, Nr.
        6 (ganzen Reihe Nr. 32), *and* Reihe Nr. 56
        (Juli 1957)   ADS 10:68

BEK, Christoph (hatter; in New York)

Letter from lawyer of Johann Jacob Springer in
        Gunzenhausen, Mittelfranken, Bavaria, asking
        that a shipment of hat material destined to Bek
        be stopped   1841

39: Bestand Senat: Cl.VIII, No.X, Jg.1841 (Regis-
        ter) 1 (Ansbach, Bavaria)   ADS 2:140

BELLOTTI, Antonio (from New Orleans)

Arrest of Bellotti in Goettingen; apparently he
        was without fixed residency   1845

52: Han.Des.80: Hildesheim I: E.DD.38b: Nr. 1456.
                                                            ADS 4:90

BELSCHBACH, Wilh[elm]

New York: Estate of   1912

39: Reichs- u. auswaert.Ang.: E.IV.55 [Erbschaften]
                                                            ADS 2:373

BEMPEL, F. W. (in America). *See* Bedassem, J.

BENDORF, Isador (from Dornheim, Landkreis Gross-
Gerau, HEL)

Emigration of family to New York   1922

168: Abmelderegister.   ADS 7:218

BENDORF, Theodor (from Dornheim, Landkreis Gross-
Gerau, HEL)

Emigration to America   1907

168: Abmelderegister.   ADS 7:218

BENEKE, Alfred (in Havana, Cuba, & New York)

Correspondence with members of family in Hamburg
                                                        1842-1848

39: Bestand Familienarchiv: Familie Beneke.
                                                        ADS 2:486-487

BENEKE, Alfred (in New York). *See also* von Melle,
Emil

BENSEL, Christian (from Amt Birkenfeld, Duchy of
Zweibruecken). *See* Wittib, Jacob Mausen

BENSSEN family (Einbeck, NL)

Genealogical studies regarding the family   n.d.

72: Harland's Nachlass: Manuscripte Nr.9: Vol.
        VIII.   ADS 4:250

BENTHIEN, Hermann

Information regarding the heirs of Benthien   1873

39: Bestand Senat: Cl.VIII, No.X, Jg.1873 (Regis-
        ter) 6 [Nordamerika]   ADS 2:224

BENVENUTI, Kuni. *See* Renvenuti, Kuni

BENZ, Johann Heinrich (from Dettingen, Bezirksamt
Heidenheim, BWL; in Salinas [state not given in
ADS description])

Report of a court decision handed down to brother
        & sister Johann Heinrich & Maria Agnes Benz
                                                        1833-1835

263: Rep.E 41-44: Minist.d.AA.II: Verzeichnis 63:
        Nr. 143.   ADS 10:42

BENZ, Maria Agnes. *See* Benz, Johann Heinrich

BERBERICH, Frank (died 16 November 1898 in San Francisco)

Estate of   1900

152: Abt.405: Wiesbaden: Rep.405/I,2: Nr. 3445.
                                                            ADS 7:142

BERCKEMEYER [also BERCKENMEYER], -- (ministerial
candidate & thereafter pastor in New York)

Call to America & collection for the congregation
        in New York   1724-1725

39: Geistl.Minis.: II,6: Protokoll lit.A, 1720-
        1745: p.26 (V); p.28r (VIII)   ADS 2:468

Call to be pastor in America   1724

39: Geistl.Minis.: III,Aln: Gebundene Akten: Nr.
        CXXXI, p.238; *see also* III,Alp: Gebundene
        Akten, Nr. CXII, p.187.   ADS 2:477

Letter from the Spiritual Ministry [Ministry of Religion] to Berckemeyer   15 December 1730

<u>39</u>: Geistl.Minis.: II,6: Protokoll lit.A, 1720-1745: p.55; p.56(4).   ADS 2:468

Letter from D. Kraeuter in London reporting death of Berckemeyer & his dispute with pastor Hartwig   14 April 1752

<u>39</u>: Geistl.Minis.: II,7: Protokoll lit.B, 1746-1758: p.263; p.264 (III).   ADS 2:477

BERCKEMEYER [also BERCKENMEYER], --. *See also* Berkemeier, --; Wolff, -- (pastor of Raritan Congregation)

BERG, Jacob (from Altenkirchen, RPL)

Confiscation of his property because of his illicit emigration to America   1774-1805

<u>201</u>: Zweibruecken III, Akt Nr. 2027.   ADS 8:59

von BERG, Jakob (of Haeuser Hof by Radheim, HEL)

Had heirs in USA   1898

<u>149</u>: Rep.49a: Nr. 565/16.   ADS 7:17

von de BERGEN, -- (Doctor; in Philadelphia)

Letter   1879

<u>39</u>: Bestand Familienarchive: Fam. Marr: A.13.   ADS 2:488

BERGER, Josef (in Brooklyn [New York])

Death of   1894

<u>227</u>: Rep.MA 1921: A.St.I: Nr. 168.   ADS 9:66

BERGGREEN, Robert (cigar worker)

Expulsion from Hamburg as Social Democrat & emigration to America   1880-1884

<u>39</u>: Polizeibehoerde-Polit.Pol.: S 149/237: Bl. 122-142.   ADS 2:415

BERGMANN, H. *See* Meline, J. F.

BERGNER, Adam Heinrich. *See* Leutbecher, Heinrich

BERINGER, K. Ph.

Expulsion to America; sentenced to lengthy penitentiary sentence in Frankfurt/Main for political activities   1838

<u>200</u>: Abt.403: Oberpraesidium Koblenz (Rheinprovinz): Nr. 2528.   ADS 8:21

BERKEFELD family (Einbeck, NL)

Genealogical studies regarding family   n.d.

<u>72</u>: Harlands Nachlass: Manuscripte Nr.9: Vol. VIII.   ADS 4:250

BERKEMEIER, -- (pastor; in New York)

Information regarding difficulties with pastor S. Keye   [1870?]

<u>45</u>: Akte.   ADS 2:601

BERKEMEIER, --. *See also* Berckemeyer [also Berckenmeyer], -- (ministerial candidate & thereafter pastor in New York)

von BERKES, Maria, born Kielborn. *See* Martini, Maria

BERNER, Emilie, born Zabel (in New York)

Inquiry by private person in Germany regarding her   1900

<u>39</u>: Reichs- u. auswaert.Ang.: N.Ib.1.6 [Nachforschungen]   ADS 2:310

BERNHARD, Heinrich Valentin (from Unkenbach, Duchy of Zweibruecken)

Division of property & details of his emigration to Pennsylvania   1739-1792

<u>201</u>: Archivabt. Herzogtum Zweibruecken: Akt Zweibruecken I, Nr. 1346: Unkenbach.   ADS 8:56

BERNSTEIN, Albert (from Moerfelden, Landkreis Gross-Gerau, HEL)

Lived in Moerfelden with three family members in 1933; emigrated   [1933]

<u>179</u>: Wiedergutmachungsakten: Ordner Nr. 061-00/05.   ADS 7:232

BERNZ, Johanne, born Krumme (from Bad Salzuflen)

Legacy to three citizens of Bad Salzuflen [She had emigrated to USA in 1853 & lived in Milwaukee]   n.d.

<u>104</u>   ADS 4:331

BERT, Sophia (from Rohrbach, Landkreis Gross-Gerau, HEL)

Emigration to America   1854

<u>173</u>: Gesinderegister [register of farmer workers] 1837-1843 [of Goddelau, Landkreis Gross-Gerau, HEL]: Nr. 110.   ADS 7:224

BERTHEAU, G. W.

San Francisco: Estate of   1884

39: Bestand Senat: Cl.VIII, No.X, Jg.1884 (Register) 38 [Erbschafts-Amt]   ADS 2:252

BERTLER, --

Estate of   1943
148: R XI.   ADS 6:260

BERTSCH [alias Barge] Jakob (died 1908 in Philadelphia)

Search for an alleged will   1909
152: Abt.405: Wiesbaden: Rep.405/I,2: Nr. 3446.
                                                   ADS 7:143

Estate of   1878-1928
148: R XI.   ADS 6:260

BETE, August (in Chicago)

Sentencing of Bete, formerly a one-year volunteer
   [military service?]   1913
227: Rep.MA 1921: A.V. V: Nr. 16012.   ADS 9:90

BETHAUSEN, J. E. Chr.

Philadelphia: Estate of   1899
39: Reichs- u. auswaert.Ang.: E.IV.73 [alt.Sig.]
   [Erbschaften]   ADS 2:301

BETZ, Fr. (in Baltimore)

Efforts of Anna Maria Eulenberg in Flammersfeld to
   clear the estate of her father; correspondence
   with relatives in Baltimore, contains letter
   from Betz to A. S. Eulenberg, dated 8 September
   1783, and an answer, dated 18 April 1784. Letters contain family news & information on conditions in America   1784
200: Abt.30: Reichsgrafschaft Sayn-Altenkirchen
   (u. Hachenburg): Nr. 4678.   ADS 8:10

BETZ, Jacob (from Trebur, Landkreis Gross-Gerau, HEL)

Emigrant to America   1854
173: Gesinderegister 1853-1894: Nr. 50. ADS 7:225

BEYER, --

Estate of   1943-1944
148: R XI.   ADS 6:260

BEYERLE, Jacob (in Pennsylvania?)

The conflict between the pastors in Pennsylvania
   & Beyerle   1764

500: Halle: Missionsbibliothek des Waisenhauses:
   Abteilung H IV (Nordamerika) Fach A, No. 16.
                                                 Not in ADS

BICKEL, F. A. (pastor; in Milwaukee)

Collection in Prussia regarding the founding of
   Lutheran churches & schools in America to be
   carried out by Bickel   1870
52: Han.Des.80 Hannover: A.II.B.II.4: Nr. 605.
                                                   ADS 4:83

BICKEL, Maria K. (widow)

San Francisco: Estate of   1887
39: Bestand Senat: Cl.VIII, No.X, Jg.1887 (Regis-
   ter) 4 [Erbschafts-Amt]   ADS 2:258

San Francisco: Estate of   1888
39: Bestand Senat: Cl.VIII, No.X, Jg.1888 (Regis-
   ter) 18 [Erbschafts-Amt]   ADS 2:260

San Francisco: Estate of   1889
39: Bestand Senat: Cl.VIII, No.X, Jg.1889 (Regis-
   ter) 18 [Erbschafts-Amt]   ADS 2:262

BIEGEMANN, Emil

Hoboken [New York]: Estate of   1902
39: Reichs- u. auswaert.Ang.: E.IV.2 [Erbschaften]
                                                   ADS 2:323

BIELINSKI, --

Estate of   1924-1927
148: R XI.   ADS 6:260

BIEMESDOERFER, Marie (from Neuburg; died 1896 in New
York)

Estate of   1898-1903
227: Rep.MA 1921: A.St.I: Nr. 228.   ADS 9:66

BIERBRAUER, Jakob Friedrich (in America)

Death of   1889
227: Rep.MA 1921: A.St.I: Nr. 663.   ADS 9:68

BIERINGER, Friedrich August (from Hoechst; died in
San Francisco)

Estate of   1885
152: Abt.228: Amt Hoechst: Nr. 262.   ADS 7:132

BIERMANN, John (in New York)

Information sought by private individual   1895
39: Reichs- u. auswaert.Ang.: N.Ib.8 [alt.Sig.]
   [Nachforschungen]   ADS 2:277

BIERWIRTH, Heinrich Carl (from Moisburg)

Biography in article entitled, "Von Schiffsjunge
zum Harvard-Professor" [From Cabin Boy to Har-
vard Professor] in *Harburger Jahrbuch*, Band V
(1955), 1-20.

ADS 2:614

BINKLI, -- (from Moersbach, Landkreis Homburg, Kurpfalz, RPL)

Emigration of   18th century

201: Archivabt. Ausfautheiakten [Kreis Homburg,
Moersbach]   ADS 8:63

BINNEWIES, Ludwig (laborer; from Varrigsen)

Emigration of delinquent to America   1860-1861

56: L Neu Abt.129A: Gr.11, Nr. 81.   ADS 4:191

BIRKEL, Marg[arethe](in Davenport [Iowa?])

File on [estate?]   1896

227: Rep.MA 1921: A.St.I: Nr. 203.   ADS 9:66

BISCHOFSHEIMER, Jette. *See* Pressburger, Karl

BLECKER, H. W. (in New Orleans)

Reply from Blecker regarding claims of -- Stoffer
[a tailor, presumably in Hamburg]   1838

39: Bestand Senat: Cl.VIII, No.X, Jg.1838 (Register) 1.b (Nordamerika)   ADS 2:136

BLEISTEIN, Clara (in New York)

Transmittal of money to Bleistein, former ward of
Hamburg orphanage   n.d.

39: Reichs- u. auswaert.Ang.: W.I.2b [alt.Sig.]
[Waisenhausang.]   ADS 2:279

BLEISTEIN, Frieda Aug. Joh. *See* Alves, Frieda Aug.
Joh., born Bleistein

BLEYLER, Valentin (from Wyhl, BWL; in America)

Transportation contract from LeHavre to New York
1888

288: Kommunalarchiv Wyhl.   ADS 10:150

BLOCH, Simon L. (in Philadelphia)

Made bequest to town of Sachsenhausen, HEL   1906
192   ADS 7:254

BLOHM, --

Estate of   1943

148: R XI.   ADS 6:260

BLUCH, Jonas Hirschel (in Germany?)

Inheritance [in America?]   1797-1804

500: Berlin-Dahlem: Preus.Geh.Staatsarchiv: Rep.
XI, 21a, Conv. 2; 3.   Not in ADS

BLUM, Eva Elisabetha. *See* Balbierer, Philipp Nickel

BLUM, Valentin. *See* Balbierer, Philipp Nickel

BLUME, Mathias (in Pittsburgh, [Pennsylvania?]).
See Hargienst, Jacob

BLUMENFELD, Natorp (born in Hamburg)

Draft reply to petition of American embassy to
have Blumenfeld, a naturalized American citizen,
exempted from German military service during his
residency in Berlin to study music   1914

39: Bestand Senat: Cl.VI, No.16p: Vol. 1, Fasc. 72
[Aeusserungsersuchen]   ADS 2:35

BLUMENFELD, --

Expulsion of Blumenfeld, an American citizen, for
avoidance of military service obligation   1912

39: Reichs- u. auswaert.Ang.: A.IV.3 [Ausweisungen]
ADS 2:372

BLUMENKRON, Abraham

Tax liability of Blumenkron in Hamburg; protest of
American consul   1849

39: Bestand Senat: Cl.VIII, No.X, Jg.1849 (Register)   ADS 2:155

BLUMENTHAL, -- (businessman in Baltimore; formerly
from Seesen)

Application for position as Braunschweig consul
1856

56: L Neu Abt.129A: Gr. 6: Nr. 63.   ADS 4:185

BLUMGARD, B. H. B. (in New York). *See* Rothstein, B.

BOAS, Emil J. (HAPAG [steamship line] agent in Washington)

Boas' citizenship   1897

39: Reichs- u. auswaert.Ang.: S.III.4 [alt.Sig.]
[Staatsangehoerigkeit]   ADS 2:292

BOBERG, Christine. *See* Deumeland, Christine, born
Boberg

BOCK, A.

Application of convict Bock for financial aid in
emigrating to America [from town of Dransfeld]

1865-1866

<u>52</u>: Han.Des.80: Hildesheim I: E.DD.35a: Nr. 1216.
ADS 4:89

BOCK, Emil (from Gernsbach, BWL)

His trip to America on the ship *Labrador*   1884

<u>260</u>: Rep.Abt. 233: Minist. des Innern: Fasz. 11411.
ADS 10:13

BOCK, --

Estate of   1941

<u>148</u>: R XI.   ADS 6:260

BOCKELMANN, Emil (pastor of Evang. Luth. St. John's Church, Waterloo, Ontario, Canada)

Letters   [1919-1922?]

<u>44</u>: Mappe Amerika.   ADS 2:600

BODE, Heinr. Chr. Aug. (from Einbeck)

Transport to America of Bode, sentenced to five months' imprisonment [Karrenstrafe]   1837

<u>52</u>: Han.Des.80: Hildesheim I: E.DD.37a: Nr. 1288.
ADS 4:89

BODENSCHATZ, Nikolaus

Applicability of the Bavarian-American treaty of 1868 in the case of Nikolaus Bodenschatz, a naturalized American citizen who returned to Bavaria & was fined for failing to comply with Bavarian military service obligation; fine was repaid to him   [late? 19th century]

<u>227</u>: Rep.MA 1921: A.V. II: Nr. 4131.   ADS 9:82

BODENSTAB, Leonhard

Transportation to America [from Neuenhaus?]   1863

<u>54</u>: Rep.122, II.1: Neuenhaus: Fach 96, Nr. 26.
ADS 4:133

BODENSTEIN, Wilhelm (from Jerze)

Emigration to America   1852

<u>56</u>: L Neu Abt.129A: Gr. 11, Nr. 53.   ADS 4:189

BOEHLE, Heinrich

Transmittal of report from German consulate in New York to Franz Ruhland, Hamburg, with a death certificate for Boehle   1901

<u>39</u>: Reichs- u. auswaert.Ang.: G.Ia.1.4 [Geschaefts-verkehr]   ADS 2:316

BOEHLER, Petrus (member Moravian Brethren; in South Carolina). *See* Moravian Brethren

BOEHM, Nikolaus. *See* Bommarius, Friedrich

BOEHME, A.

Consular report from Washington & New York regarding Boeme's release from prisoner of war status   1865

<u>39</u>: Bestand Senat: Cl.VIII, No.X, Jg.1865 (Register) II.   ADS 2:187

BOEHMER, -- (from Hunteburg)

Transportation of the vagabond Boehmer & Tewitz families to America   1839

<u>54</u>: Rep.122, X.B: Wittlage-Hunteburg: I. Fach 90, Nr. 31.   ADS 4:144

BOEHMKE, Heinrich

Request by Hamburg juvenile authorities for investigation by consulate in Cincinnati   1898

<u>39</u>: Reichs- u. auswaert.Ang.: V.III.3 [Vormund-schaftssachen]   ADS 2:299

BOERGER, -- (widow)

Letter on behalf of widow Boerger addressed to the Magistrate at Philadelphia   1785

<u>39</u>: Bestand Senat: Cl.VIII, No.X, Jg.1785 (Register) [Fuerschreiben]   ADS 2:103

BOERNER, -- (married woman)

Chicago: Estate of   1897

<u>39</u>: Reichs- u. auswaert.Ang.: E.IV.42 [alt.Sig.] [Erbschaften]   ADS 2:290

BOERSMANN, Martin (in New York 1867-1875)

Boersmann was born in 1851 in Elfershude bei Beverstedt & died in 1903 in Hannover. During his sojourn in New York he founded the first Plattdeutsch Verein [Low German Club]. He collected literature in the dialect & willed the collection to the Stadtbibliothek Hannover
ADS 4:267

BOESCH, --. *See* Friede, --

BOETCKER, L. Th. (married woman; in Hamburg)

Request by Mrs. Boetcker for assistance by the German Foreign Office in getting the birth certificate changed for her son, Wilhelm Boetcker, born in New Jersey   1900

<u>39</u>: Reichs- u. auswaert.Ang.: S.IV.13 [Standesamt-sachen]   ADS 2:312

BOETCKER, Wilhelm (born in New Jersey). *See* Boetcker, L. Th.

BOETTCHER, Georg E. R. (American citizen)

Negotiations regarding Boettcher's arrest at the
request of state's attorney in Breslau   1881
49: A.3.N.3: Nr. 22.   ADS 3:78

BOETTIGER, Adam II (of Biebesheim, Landkreis Gross-
Gerau, HEL)

Public aid for his emigration to America   n.d.
164: Abt. XI,4.   ADS 7:214

BOHN, Martin. *See* Cohn, Max.

BOLDT, Frederike, born Petersen (in Terre Haute,
Indiana)

Estate of   1906
39: Reichs- u. auswaert.Ang.: E.IV.38 [Erbschaften]
                                            ADS 2:343

BOLLANDER, Jacob (of Eidengesaess, Hesse-Hanau, HEL)

Emigration permit denied   1751
151: Bestand 80: Lit.A: Nr. 7-1/2.   ADS 7:78

BOLLER, Emanuel (former Secont-Jaeger [military rat-
ing]; in Town [?] North America)

Petition for renaturalization   1861
227: Rep.MA 1921: A.St.I: Nr. 1236.   ADS 9:71

BOLLING, --

General Consulate in New York: Case of Bolling *versus* Bolling   1898
39: Reichs- u. auswaert.Ang.: J.Ib.10.4 [alt.Sig.]
    [Justizsachen]   ADS 2:297

BOLSTER, Friedrich (teacher & choir director from
Hollfeld; in Reading [Pennsylvania])

Petition for re-employment in Bavaria   1849
227: Rep.MA 1898: A.St.: Nr. 5.   ADS 9:57

BOLTZIUM [BOLTZ? BOLTZIAN?] -- (pastor? in Ebenezer,
Georgia)

News of -- Boltzium & -- Gronau with accounting
records on funds forwarded them; correspondence
                                  1734; 1735-1758
500: Halle: Missionsbibliothek des Waisenhauses:
     Abteilung H IV (Nordamerika) Fach E, No. 2;
     Fach G, No. 3.   Not in ADS

Diary of   1733-1735
500: Halle: Missionsbibliothek des Waisenhauses:
     Abteilung H IV (Nordamerika) Fach J, No. 1b.
                                            Not in ADS

BOMANN, Hermann Otto

Emigration file contains letter from friend or
relative in Brooklyn [New York]   1860
55: Han.Des.74: Blumenthal: IX.J.1: Fach 301, Nr.
    10.   ADS 4:154

BOMMARIUS, Friedrich (from Kelsterbach, Landkreis
Gross-Gerau, HEL)

Cost of transportation to America of the following
persons: Johannes Jung I, Friedrich Bommarius,
Adam Schneider, Christian Renz, Philipp Rentz,
Georg Pflueger, Nikolaus Boehm, Philipp Laun II,
& Katharina Laun. (Costs detailed as 1880 gulden for ocean transportation, 15 gulden for
transportation from Kelsterbach to Mainz, 30
gulden pocket money, 100 gulden clothing money,
30 gulden travel to LeHavre   1852
175: Gemeinderechnung 1852, Seite 82.   ADS 7:228

BONHAG, Barbara

Measures to be taken against emigration agents in
LeHavre regarding their unlawful transportation
of Bonhag & Barbara Karl to America   1860
227: Rep.MA 1921: A.V. IV: Nr. 11863.   ADS 9:87

BONN, Anna Maria (from Leeheim, Landkreis Gross-
Gerau, HEL)

Emigration to America   1854
173: Gesinderegister [register of farm workers]
     1837-1843: Nr. 101.   ADS 7:224

BONNER, Anna, born Hofmann (widow; died 14 June 1908
in Brooklyn, New York)

Estate of   1909
152: Abt.405: Wiesbaden: Rep.405/I,2: Nr. 3448.
                                            ADS 7:143

BOOTMANN, -- (widow; in New York)

Repatriation of   1898
39: Reichs- u. auswaert.Ang.: S.Ie.1.6 [Schiffahrt]
                                            ADS 2:298

BOOTSMANN, Stienje (from Suedgeorgsfehn)

Emigration to America   1865
50: Rep.21a: Nr. 8938.   ADS 4:8

BORCHERS, Jacob (born 11 March 1840; from Ploenjes-
hausen)

Issuance of emigration permit   1858
62: A, Fach 6, Vol. 15.   ADS 4:233

BORCKS, --

Estate of   1942
<u>148</u>: R XI.   ADS 6:260

BORGAS, Johanne

Emigration to America   1850
<u>52</u>: Han.Des.80: Hildesheim I: E.DD.7i: Nr. 941.
                                             ADS 4:88

BORGES, Caspar Heinrich (bishop of Detroit, Michigan, 1872-1888)

Mentioned in "Bischoefe des Oldenburger Muensterland" [Bishops from the Archbishopric of Olden-
burg] in *Veroeffentlichungen aus dem Heimatskalender fuer das Oldenburger Muensterland*
(1952), 115.   ADS 4:248

BORGMANN, Albert (in Grenfell [state not given in
ADS description])

Estate of   1903
<u>39</u>: Reichs- u. auswaert.Ang.: E.IV.51 [Erbschaften]
                                             ADS 2:330

BORKMANN, Adolph (in Colusa, California)

Estate of   1910
<u>39</u>: Reichs- u. auswaert.Ang.: E.IV.2 [Erbschaften]
                                             ADS 2:362

BORNSCHEUER, W. (in New York)

Information regarding   1877
<u>39</u>: Bestand Senat: Cl.VIII, No.X, Jg.1877 (Regis-
ter) 7 [Nordamerika]   ADS 2:234

von BORSTEL, H.

San Francisco: Estate of   1910
<u>39</u>: Reichs- u. auswaert.Ang.: E.IV.13 [Erbschaften]
                                             ADS 2:362

von BOSEN, -- (officer, U.S. Army; died in Egg Harbor, 15 October 1778) [information seems unlikely]

Letters relating to inheritance   1785
<u>500</u>: Berlin-Dahlem: Preus. Geh. Staatsarchiv: Rep.
XI, 21a, Conv. 1,7(a).   Not in ADS

BOSSELMANN, Heinrich (on ship *John and Samuel*)

Death & estate of   1877
<u>39</u>: Bestand Senat: Cl.VIII, No.X, Jg.1877 (Regis-
ter) 35.   ADS 2:235

BOSSUT, George, *alias* Laymoine, George

Hamburg police search for Bossut who is sought by
New York police   1903
<u>39</u>: Reichs- u. auswaert.Ang.: P.III.1.1 [Polizei-
wesen]   ADS 2:332

BOWEN, John. *See* Jungbohn-Clement, Ale, *alias* Bowen,
John

BOYE, Charles (in Etna Mills [state not given in
ADS description])

Estate of   1905
<u>39</u>: Reichs- u. auswaert.Ang.: E.IV.85 [Erbschaften]
                                             ADS 2:339

BOYSEN, Peter M.

San Francisco: Estate of   1908
<u>39</u>: Reichs- u. auswaert.Ang.: E.IV.41 [Erbschaften]

BRAACK, J. M. (from Heide)

Claim for damages due to prohibited emigration to
America   1881
<u>39</u>: Bestand Senat: Cl.VIII, No.X, Jg.1881 (Regis-
ter) 7 [Nordamerika]   ADS 2:245

BRACHHAUSEN, --

Estate of   1941
<u>148</u>: R XI.   ADS 6:260

BRACHTHAEUSER, Daniel (of Suedershausen, HEL)

File on Johannes Mueller, Johann Jost Mueller's
widow, & Daniel Brachthaeuser, all of Suedershausen, regarding confiscation of the properties
of their brothers-in-law, Johann Jost Schmidts &
Johann Jost Hofheinz, both of Suedershausen, who
have emigrated to America   1789-1790
<u>152</u>: Abt.172: Landratsamt Dillenburg: Heft 3.
DLC has.   ADS 7:129

BRACHVOGEL, Jacob (in Prince's Bay, New York)

Estate of   1900
<u>39</u>: Reichs- u. auswaert.Ang.: E.IV.66 [Erbschaften]
                                             ADS 2:307

BRACK, Luise, born Rimbach (died in Baltimore)

Estate of   1902
<u>152</u>: Abt.405: Wiesbaden: Rep.405/I,2: Nr. 3450.
                                             ADS 7:143

BRADITSCH, -- (general in America)

File on   1841
229: Rep. Reg. Akten von Oberbayern, Kammer des
     Innern: 1157.130.   ADS 9:115

Confidential information regarding   1841
263: Rep.E 46-48: Minist. d. AA III: Polizei: Verz.
     2, 778/8.   ADS 10:45

Police surveillance of the alleged American general
                                             n.d.
227: MA 1898: A.St.: Nr. 1.   ADS 9:57

The incitement to emigration by the alleged North
   American militia general Braditsch   1841
227: Rep. Gesandtschaft Karlsruhe: Nr. 420.
                                   ADS 9:52

BRAMANN, Carl Theodor (paperhanger)

Expulsion from Hamburg as Social Democrat & emigration to America   1880-1881
39: Polizeibehoerde-Polit.Pol.: S 149/102: Bl. 1-9.
                                   ADS 2:414

BRAND, Claus. *See* Brandt, Claus

BRAND, Bernhard Heinrich (died in St. Louis)

Estate of   1878-1883
116: Minden: Praes.-regist.: I.L. Nr. 234. ADS 6:54

BRANDES, K. E.

Philadelphia: Estate of   1888
39: Bestand Senat: Cl.VIII, No.X, Jg.1888 (Regis-
    ter) 26 [Erbschafts-Amt]   ADS 2:260

BRANDHUBER, Josef (from Munich; in Utica [New York])

Repatriation of Brandhuber from an insane asylum in
   Utica   1876
227: Rep.MA 1921: A.V. II: Nr. 3915.   ADS 9:81

BRANDIS, Hermann (Doctor; in Hoboken, New Jersey)

Application to be Bremen consul in New Jersey 1856
48: B.13.b.0: Nr. 15.
    DLC has   ADS 3:10

BRANDIS, -- (Doctor; in Hoboken, New Jersey)

Seeks Hamburg consulate in New Jersey 1856-1857
39: Bestand Senat: Cl.VI, No.16p: Vol.4b, Fasc. 17.
                                   ADS 2:59

BRANDT, Ch. E. A. (pastor; in New York)

Pastor Brandt of Immigrants Church, New York, requests recommendation to emigrants from Hamburg
                                     1851
39: Bestand Senat: Cl.VIII, No.X, Jg.1851 (Regis-
    ter)   ADS 2:159

BRAND[T?] Claus (born 19 February 1837; from Seedorf)

Issuance of emigration permit   1857
62: A, Fach 6, vol. 13.   ADS 4:232

BRANDT, Detlef Christian

Deed to property in Franklin, near Milwaukee, Wisconsin   n.d.
47   ADS 2:608

BRANDT, Friedrich (born 29 November 1838; from Ober-
ochtenhausen)

Issuance of emigration permit   1857-1858
62: A, Fach 6, vol. 14.   ADS 4:233

BRANDT, G. F.

Death certificate for   1872
39: Bestand Senat: Cl.VIII, No.X, Jg.1872 (Regis-
    ter) 15 [Nordamerika]   ADS 2:220

BRANDT, Hans (in Seattle)

Estate of   1917
39: Reichs- u. auswaert.Ang.: E.IV.26 [Erbschaften]
                                   ADS 2:392

BRANDT, Hinrich (born in Fernau)

File on application for emigration, dated 21 February 1838   1856
62: A, Fach 6, vol. 10.   ADS 4:232

BRANDT, Hinrich (infantry soldier; from Seedorf)

Application for emigration certificate 1856-1857
62: A, Fach 6, vol. 12.   ADS 4:232

BRANDT, Karl Heinrich Wilhelm (from Hitzacker; died
in San Francisco)

Establishment of heirs [to Brandt's estate] and
   legal attestations thereto   1886
52: Han.Des.74: Dannenberg: Abt.II, G, Nr. 35.
                                   ADS 4:66

BRANDT, Wilhelm (from Einbeck)

Proposed transportation to America 1849-1852

52: Han.Des.80: Hildesheim I: E.DD.37a: Nr. 1298.
 ADS 4:89

BRASCH, J. G., *alias* Gottfried Brockmann

Military service obligation of 1876

39: Bestand Senat: Cl.VIII, No.X, Jg.1876 (Register) 8 [Nordamerika] ADS 2:232

BRAUER, August Ernst Friedrich (bank messenger)

Expulsion from Hamburg as Social Democrat & emigration to America 1880-1882

39: Polizeibehoerde-Polit.Pol.: S 149/109: Bl.
 218-224. ADS 2:414

BRAUER, Hans (in Lodi, California)

Estate of 1907

39: Reichs- u. auswaert.Ang.: E.IV.57 [Erbschaften]
 ADS 2:347

BRAUER, -- (*alias*). *See* Ramos, --

BRAUN, Anna. *See* Scharf, Anna

BRAUN, Christoph (American citizen; from Hassfurt)

Family of 1897

227: Rep.MA 1921: A.V. IV: Nr. 10252. ADS 9:83

BRAUN, Elisabeth

San Francisco: Estate of 1908

39: Reichs- u. auswaert.Ang.: E.IV.45 [Erbschaften]
 ADS 2:351

BRAUN, Emma

Exhumation of Mrs. Braun's body & its transmittal
 to New York 1899

39: Reichs- u. auswaert.Ang.: P.III.16.16 [Polizei-
 sachen] ADS 2:303

BRAUN, Georg (died in Cincinnati, Ohio)

Estate of 1910

152: Abt.405: Wiesbaden: Rep.405/I,2: Nr. 3451.
 ADS 7:143

BRAUN, Lorenz (from Texas; died 1 September 1890 in
Homburg v.d.H.)

Estate of 1900-1909

152: Abt.405: Wiesbaden: Rep.405/I,2: Nr. 3452.
 ADS 7:143

BRAUN, Robert (born 18 November 1872 in Hoboken, New
Jersey)

Personal file on him; expulsion 1897

39: Land. Bergedorf II: XI.C.53. ADS 2:467

BRAUN, -- (town clerk; in America)

Power of attorney from 1855-1857

227: Rep.MA 1898: A.St.: Nr. 1029. ADS 9:58

BRAUN. *See also* Brown

BRECHT, E.

New York: Estate of 1901

39: Reichs- u. auswaert.Ang.: E.IV.53 [Erbschaften]
 ADS 2:316

BREESE, Juergen Heinrich (from Niendorf)

Transmittal of information regarding death & estate of Breese in New Orleans 1832

52: Han.Des.74 Dannenberg: Abt.II, G, Nr. 9.
 ADS 4:65

BREIGER, W.

New York: Estate of 1901

39: Reichs- u. auswaert.Ang.: E.IV.52 [Erbschaften]
 ADS 2:316

BREITENBACH, Harriette. *See* Raffa, Harriette, divorced Breitenbach, born Rosenfeld

BREITUNG, J. W.

New York: Estate of 1895

39: Reichs- u. auswaert.Ang.: E.IV.55 [alt.Sig.]
 [Erbschaften] ADS 2:275

BREKELER, Margarete (died in North America)

Estate of 1927? [1827-1830?]*
*ADS description gives conflicting dates

116: Minden: Praes.-regist.: I.L.Nr. 232. ADS 6:54

BRENEK, Josef (an Austrian citizen)

Extradition of Brenek, wanted for robbery & theft,
 from the USA to Austria, via Hamburg 1906

39: Reichs- u. auswaert.Ang.: A.Ia.1.27 [Ausliefer-
 ungen] ADS 2:342

BRENNECKE, -- (pastor; from New York)

Permit to preach   1904

76: Repositur Specialia: Meine: 2.Conv.2, Aktenz.
203.   ADS 4:256

BREUEL, Ernst Carl Wilhelm (shopkeeper)

Expulsion from Hamburg as Social Democrat & emigration to America   1880-1893

39: Polizeibehoerde-Polit.Pol.: S 149/85: Bl. 1-
247.   ADS 2:413

BREUER-WIRTZ (WIRTH), --

Asserted estate in America worth millions [of dollars?]   1913

39: Reichs- u. auswaert.Ang.: D.Ih,46 [Erbschaften]
ADS 2:379

BREUER-WIRTH, --

Estate of   1929

148: R XI.   ADS 6:260

BREUNIG, Babette (widow; in New York)

File on   1894

227: Rep.MA 1921: A.St.I: Nr. 177.   ADS 9:66

BRICKENSTEIN, Eduard (naturalized American citizen)

Intercession of American consul in Bremen for
Brickenstein wanted for failure to fulfill Prussian military service obligation   1858

48: Dd.11.c.2.N.1.a.1.Nr. 68.   ADS 3:63

BRICKWEIDT, Wilhelm

Inquiry from Meppen as to whether Brickweidt already aboard ship bound for America   1846

39: Bestand Senat: Cl.VIII, No.X, Jg.1846 (Register).   ADS 2:149

BRIESKE, --

Estate of   n.d.

148: R XI.   ADS 6:260

BRINKMANN, E. (in New York)

German consular report from New York regarding
desertion of Brinkmann, steward aboard Hamburg
steamer *Holsatia*   1869

39: Bestand Senat: Cl.VIII, No.X, Jg.1869 (Register) 1 [Nordamerika]   ADS 2:200

BRINKMEYER, Joh[ann] Mathias (from Reimsloh)

Application for public assistance for transportation to America [Groenenberg-Melle]   1843

54: Rep.122, VII.B: Groenenberg-Melle: Regiminalia:
Fach 15, Nr. 7.   ADS 4:138

BRIX, Hermann (born in Luechow; died in America)

Estate of   1938-1940

95   ADS 4:289

BRIX, --

Search for Brix, said to be sought in American
newspapers   1902

39: Reichs- u. auswaert.Ang.: N.Ia.1.1 [Nachforsch-
ungen]   ADS 2:326

BROCKMANN, Gottfried. *See* Brasch, J. G.

BROCKMANN, H. (in Fort Stanton, New Mexico)

Estate of   1905

39: Reichs- u. auswaert.Ang.: E.IV.6 [Erbschaften]
ADS 2:339

BROCKMUELLER, J. *See* Rehagen, J. J. H.

BROCKSTEDT, W. (in Walla Walla, Washington)

Estate of   1905

39: Reichs- u. auswaert.Ang.: E.IV.50 [Erbschaften]
ADS 2:339

BRODA, Karl August Wilhelm (painter)

Expulsion from Hamburg as Social Democrat & emigration to America   1880-1881

39: Polizeibehoerde-Polit.Pol.: S 149/69: Bl. 1-10.
ADS 2:413

BRODA, Karl (in Dallas, Texas)

Estate of   1898

39: Reichs- u. auswaert.Ang.: E.IV.25 [alt.Sig.]
[Erbschaften]   ADS 2:296

BRODERSEN, -- (farmer; in Minnesota)

Estate of   1907

39: Reichs- u. auswaert.Ang.: E.IV.33 [Erbschaften]
ADS 2:347

BROEDERMANN, --. *See* Ahlers, --

BROEHAN, Peter Gerh.

New York: Estate of   1912
39: Reichs- u. auswaert.Ang.: E.IV.32 [Erbschaften]
                                               ADS 2:373

BROGLEY, Barbara (in Brooklyn, New York)

File on [estate?]   1895
227: Rep.MA 1921: A.St.I: Nr. 194.   ADS 9:66

BROMME, --

Emigration   1846
149: Abt.Hess.Hausarchiv: Konv. 653.   ADS 7:21

BRONIEC, --

Estate of   1940-1941
148: R XI.   ADS 6:260

BROWN, Frank (photographer)

Investigation by Hamburg police, at request of
    American police authorities, of Brown, wanted
    for breaking & entering   1906
39: Reichs- u. auswaert.Ang.: P.III.1.2 [Polizei-
    wesen]   ADS 2:344

BROWN, Johann (a Negro?)

Arrested in Wremen [Bremen?] & given clothing &
    food   1819-1820
52: Han.Des.74: Dorum: Fach 54, Nr. 4.   ADS 4:67

BROWN, --

Legality of American divorce decree before Hamburg
    Civil Registry   1895
39: Reichs- u. auswaert.Ang.: S.IV.1.6 [alt.Sig.]
    [Standesamtssachen]   ADS 2:279

BROWN, --, born Greuel (married woman; in New York)

Claim by Emily Diester, born Greuel [Hamburg?] for
    payment of loan   1873
39: Bestand Senat: Cl.VIII, No.X, Jg.1873 (Regis-
    ter) 17 [Nordamerika]   ADS 2:224

Claim by Emily Greuel [Hamburg?] for repayment of
    loan   1872
39: Bestand Senat: Cl.VIII, No.X, Jg.1872 (Regis-
    ter) 40 [Nordamerika]   ADS 2:221

BROWN. See also Braun

BROXTERMANN, Johann David (from Klosteroesede)

Transportation to America   n.d.

54: Rep.122: Iburg: Fach 311, Nr. 9.   ADS 4:135

BRUCHMUELLER, Heinrich Friedrich Wilhelm (vagabond
    iron-smelter laborer; from Kaierde)

Emigration to America   1850-1854
56: L Neu Abt.129A: Gr.11, Nr. 56.   ADS 4:189

BRUDER, Heinrich (journeyman carpenter; from Froes-
    chen; in New England)

Property [in Germany] of Bruder [see also Emmert,
    Peter]   1766
201: Archivabt. Grafschaft Hanau-Lichtenberg: Akt
    Nr. 1167.   ADS 8:67

BRUECK, --

Extradition to Saarbruecken from Philadelphia, via
    Hamburg   1911
39: Reichs- u. auswaert.Ang.: A.Ia.13 [Ausliefer-
    ungen]   ADS 2:367

BRUECKMANN, Engelbert (cigar maker)

Expulsion from Hamburg as Social Democrat & emi-
    gration to America   1880-1881
39: Polizeibehoerde-Polit.Pol.: S 149/28: Bl. 225-
    232.   ADS 2:411

BRUECKNER, Anna. See Stabel, Anna, born Brueckner

BRUEGGE, --

Hanseatic superior court seeks testimony of com-
    plainant (a woman) in New York in case of
    Bruegge versus Bruegge   1896
39: Reichs- u. auswaert.Ang.: J.Ib.12 [alt.Sig.]
    [Justizsachen]   ADS 2:283

BRUEGMANN, August

Estate of   1874
39: Bestand Senat: Cl.VIII, No.X, Jg.1874 (Regis-
    ter) 36 [Nordamerika]   ADS 2:228

BRUEGMANN, Claus

Chicago: Estate of   1880
39: Bestand Senat: Cl.VIII, No.X, Jg.1880 (Regis-
    ter) 4 [Nordamerika]   ADS 2:243

BRUELL, M. (in America)

Interrogation of Bruell as part of the investiga-
    tion of A. S. Feldheim & Co. [in Germany?] for
    fraud   1842-1843
227: Rep.MA 1898: A.St.: Nr. 6285.   ADS 9:59

BRUENNOW, --

Estate of   1942
<u>148</u>: R XI.   ADS 6:260

BRUMER, Maria Magdalena Fidelis. *See* Spinner, Maria
Magdalena Fidelis, born Brumer

BRUNDEL, --

Galveston [Texas]: Estate of   1912
<u>39</u>: Reichs- u. auswaert.Ang.: E.IV.92 [Erbschaften]
                                                    ADS 2:373

BRUNHOLTZEN, -- (pastor; died in Philadelphia)

His will contains bequests to brothers & sisters
   [in Germany]   1757
<u>500</u>: Halle: Missionsbibliothek des Waisenhauses:
        Abteilung H IV (Nordamerika) Fach A, Nr. 17.
                                                    Not in ADS

BRUNNER, Emil (in America)

Information regarding   1888
<u>227</u>: Rep.MA 1921: A.St.I: Nr. 1319.   ADS 9:72

BRUNNER, Joseph (journeyman butcher; from Lauingen)

Emigration to North America [apparently accompanying Georg Stengel]   1854
<u>247</u>: A 492.   ADS 9:158

BRUNNQUELL, Paul (of Milwaukee [Wisconsin])

Regarding his life   1880
<u>227</u>: Rep.MA 1921: A.St.I: Nr. 1298.   ADS 9:72

BRUNS, Bernhard (at Westphalia Settlement, Missouri)

Bruns was a medical doctor who emigrated to America in 1835. His wife Henriette & two of her
   brothers, Franz & Bernhard Geisberg, followed
   him to Missouri the following year   1835-1836
<u>147</u>   ADS 6:138

BRUNS, Henriette, born Geisberg (in Westphalia Settlement, Missouri)

Mrs. Bruns was the daughter of an educated family
   of officials in Muensterland. She married Dr.
   Bernhard Bruns in 1832; she emigrated to America
   with brothers Franz & Bernhard Geisberg in 1836
   to join husband. Much correspondence with family members in Muenster   1835-1865
<u>147</u>   ADS 6:138

BRYAN, F. H. (an American citizen)

Expulsion [from Germany]   1887
<u>39</u>: Bestand Senat: Cl.VIII, No.X, Jg. 1887 (Regis-
        ter) 15 [Nordamerika]   ADS 2:259

BUCHHAGE, Friedrich (farmhand; from Wenzen)

The dissolute way of life of Buchhage & his emigration to America   1867-1869
<u>56</u>: L Neu Abt.129A: Gr.11, Nr. 88.   ADS 4:192

BUCHHEIM, F.

Extradition to USA   1882
<u>39</u>: Bestand Senat: Cl.VIII, No.X, Jg.1882 (Regis-
        ter) 14 [Nordamerika]   ADS 2:249

BUCHHEIM, William

Extradition of Buchheim wanted in USA for counterfeiting   1881
<u>39</u>: Bestand Senat: Cl.VIII, No.X, Jg.1881 (Regis-
        ter) 13 [Nordamerika]   ADS 2:246

BUCHHOLTZ, J. H. M.

German consular report from San Francisco regarding   1871
<u>39</u>: Bestand Senat: Cl.VIII, No.X, Jg.1871 (Regis-
        ter) 32 [Nordamerika]   ADS 2:216

BUCHHOLTZ, -- (in Goodhull [no further geographic
description given in ADS])

Estate of   1912
<u>39</u>: Reichs- u. auswaert.Ang.: E.IV.87 [Erbschaften]
                                                    ADS 2:373

BUCHHOLZ, Chr.

Detroit, Michigan: Estate of   1910
<u>39</u>: Reichs- u. auswaert.Ang.: E.IV.87 [Erbschaften]
                                                    ADS 2:363

BUCHLI, Franz

Extradition to Switzerland from the USA, via Hamburg   1908
<u>39</u>: Reichs- u. auswaert.Ang.: A.Ia.17 [Ausliefer-
        ungen]   ADS 2:350

BUCK, --

Boston: Estate of   1912
<u>39</u>: Reichs- u. auswaert.Ang.: E.IV.88 [Erbschaften]
                                                    ADS 2:373

BUECHELBERGER, Bernhard (in New York)

File on [estate?]   1897
227: Rep.MA 1921: A.St.I: Nr. 206.   ADS 9:66

BUECHELCHEN, -- (in Texas)

Bookkeeping account? with emigrant [in Texas German colonies]   n.d.
154: CA 4d5: Nr. 9.   ADS 7:186

BUECHNER, Adolph (of Crumstaedt, Landkreis Gross-Gerau, HEL)

Emigration to North America   1849
167: Register ueber ausgestellte Heimatscheine
     1845-1849: Abt.XI,3.   ADS 7:217

BUEHRENHEIM, -- (widow)

Los Angeles: Estate of   1910
39: Reichs- u. auswaert.Ang.: E.IV.88 [Erbschaften]
                                          ADS 2:363

BUENGER, Heinrich (wagon maker; from Wenzen). See
Voss, August

BUERGER, Heinrich (died 1899? [1889?] in California)

Estate of   1891
227: Rep.MA 1921: A.St.I: Nr. 133.   ADS 9:66

BUESCH, Ludwig (a sailor)

San Francisco: Estate of   1916
39: Reichs- u. auswaert.Ang.: E.IV.2 [Erbschaften]
                                          ADS 2:390

BUEXLER, Friedrich (from Erfelden, Landkreis Gross-Gerau, HEL)

Applications by Buexler & Heinrich Wild for permit
     to emigrate to North America   1848
169: Abt.XI,4.   ADS 7:219

BUEXLER, Johann Friedrich (from Erfelden, Landkreis Gross-Gerau, HEL)

Emigration to America   before 1850
169: Ortsbuergerregister (first half 19th century)
                                          ADS 7:220

BUFFINGTON, Wilhelmine (widow; in New Rochelle, New York)

Estate of   1916
39: Reichs- u. auswaert.Ang.: E.IV.22 [Erbschaften]
                                          ADS 2:390

BULLMER, G. I. (from Iggersheim)

Emigration & exportation of personal property 1848
227: Rep.MA 1898: A.St.: Nr. 1026.   ADS 9:58

BUNGART, C. (in Brooklyn, New York)

Estate of   1901
39: Reichs- u. auswaert.Ang.: E.IV.64 [Erbschaften]
                                          ADS 2:316

BUNGER, H. (in New York)

Inquiry by private person in Germany regarding
                                                 1899
39: Reichs- u. auswaert.Ang.: N.Ib.1.6 [Nachforschungen]
                                          ADS 2:303

BURG, W.

New York: Information regarding   1878
39: Bestand Senat: Cl.VIII, No.X, Jg.1878 (Register) 25 [Nordamerika]   ADS 2:237

BURGHARDT, Leopold (died 16 April 1906 in Evansville, Indiana)

Estate of   1908
152: Abt.405: Wiesbaden: Rep.405/I,2: Nr. 3453.
                                          ADS 7:143

BURGSTEIN, --. See Kontrauth, Lea, born Heumann

BURIKE, Hermann (murdered in Kentucky)

Estate of   1887-1901
116: Muenster Reg.: Nr. 4269.   ADS 6:56

BURKARD, -- (pastor; in Baltimore, Maryland)

Correspondence   1887
120: Akte Frank Buchman (D II B).   ADS 6:75

BURMESTER, E. J. H.

New York: Death certificate of   1881
39: Bestand Senat: Cl.VIII, No.X, Jg.1881 (Register) 30 [Nordamerika]   ADS 2:247

BUSCHBECK, Adolf (American colonel; born in Koblenz)

Article regarding him in Koblenzer Heimatblatt,
     5. Jahrgang, Nr. 47 (25 November 1928)
200: Abt.701: Nachlaesse, etc.: Nr. 968. ADS 8:53

BUTENSCHOEN, C. (died in New York)

Inquiry from Flensburg regarding Butenschoen's estate 1844

39: Bestand Senat: Cl.VIII, No.X, Jg.1844 (Register). ADS 2:146

BUTTERHOF, W. (died in St. Louis [Missouri])

Property claim of 1858-1859

227: Rep.MA 1898: A.St.: Nr. 1033. ADS 9:58

BUTTNER, -- (in New York)

Claim of Buttner family through Meyer & Co. 1863

39: Bestand Senat: Cl.VIII, No.X, Jg.1863 (Register) II. ADS 2:183

BUTZ, -- (in America). *See* Bedassem, J.

CAESAR, Carl (American citizen)

Permit to operate business n.d.

48: B.13.c.1.b: Nr. 17.
DLC has. ADS 3:42

CAHNBLEY, Johann Hermann

Request by American consulate for birth certificate for 1908

39: Reichs- u. auswaert.Ang.: S.IV.3.1 [Standesamtssachen] ADS 2:354

CAMMERHOF, -- (member Moravian Brethren; in America)

Letters from 1747-1749

500: Herrnhut: Archiv: Rep.14.a: Nr. 28.
Not in ADS

CANNAM, -- (widow; in Berlin Township [county not given in ADS description] Michigan)

Estate of 1907

39: Reichs- u. auswaert.Ang.: E.IV.83 [Erbschaften] ADS 2:347

CAPLÉ [Caplé] Wilhlem (boy; from Langelsheim)

Emigration, with Caplé family, to America 1852-1853

56: L Neu Abt.129A: Gr.11, Nr. 60. ADS 4:190

CAPPES, -- (commissioner; Indian Point, Texas)

Correspondence with Texas Verein [in Braunfels, Texas] 1846-1852

154: CA 4b3: Nr. 5.
DLC has. ADS 7:167

CARL, Francis

St. Louis: Estate of 1910

39: Reichs- u. auswaert.Ang.: E.IV.69 [Erbschaften] ADS 2:362

CARLÉ [Carlé] Friedrich (died in New Jersey)

Information regarding estate of 1896

152: Abt.405: Wiesbaden: Rep.405/I,2: Nr. 3544. ADS 7:143

CARLSEN, Georg (in California)

Estate of 1911

39: Reichs- u. auswaert.Ang.: E.IV.42 [Erbschaften] ADS 2:368

CARLSON, Carl Rudolph (bartender)

Expulsion from Hamburg as Social Democrat & emigration to America 1880-1885

39: Polizeibehoerde-Polit.Pol.: S 149/35: Bl. 279-304. ADS 2:412

CARN, --

Attempt of U.S. consul to interven in case of Carn, a Mormon 1852

39: Bestand Senat: Cl.VIII, No.X, Jg.1852 (Register) I.2 [Nordamerika] ADS 2:161

Expulsion of Mormon, said to be a relative of the U.S. Consul 1852-1853

39: Bestand Senat: Cl.VII, Lit. H^{f}: Nr. 4, Vol. 14. ADS 2:68

Letter from Hamburg consul Bromberg regarding case of 1853

39: Bestand Senat: Cl.VIII, No.X, Jg.1853 (Register) I.1 [Nordamerika] ADS 2:162

CARSTENS, Christian Boje Heinrich (cigar worker)

Expulsion from Hamburg as Social Democrat & emigration to America 1881

39: Polizeibehoerde-Polit.Pol.: S 149/303: Bl. 305-312. ADS 2:416

CARSTENSEN, Niels Andersen (shoemaker)

Expulsion from Hamburg as Social Democrat & emigration to America 1883-1884

39: Polizeibehoerde-Polit.Pol.: S 149/439: Bl. 313-325. ADS 2:418

CASPELMANN, Caroline

Chicago: Search for 1877

39: Bestand Senat: Cl.VIII, No.X, Jg.1877 (Register) 54 [Nordamerika] ADS 2:235

CASPERSEN, Claus Christian (from Zennhusen, SHL)

Petition to be released from military service
obligation for purpose of emigrating to America
1867
13: Alphabetischen Sammlung.　　　　ADS 1:49

CASSENS, John H. (in St. Louis, Missouri)

Information regarding Cassens sought by private
individual　　　　1894
39: Reichs- u. auswaert.Ang.: N.Ib.22 [alt.Sig.]
[Nachforschungen]　　　　ADS 2:271

CASTEN, Moritz (in Richmond County, Georgia)

Claim of right to inheritance　　　　1822
48: B.13: Nr. 3 [Erbschafts-Anspruch]　　ADS 3:1

CASTON, Sophie

Cincinnati, Ohio: Estate of　　　　1910
39: Reichs- u. auswaert.Ang.: E.IV.90 [Erbschaften]
ADS 2:363

von CHOMINSKI, Theodor (painter; in America)

Claim for support by Anna Quild in Munich against
von Chominski　　　　1887
227: Rep.MA 1921: A.St.I: Nr. 1252.　　ADS 9:72

CHRISTGAU, John (died in Tonavanda [Tonawanda?])

File on [estate?]　　　　1891
227: Rep.MA 1921: A.St.I: Nr. 251.　　ADS 9:66

CHRISTIAN, -- (chairmaker; from Aurich)

Transporation to America　　　　1865
50: Rep.21a: Nr. 8885.　　　　ADS 4:7

CHRISTENSEN, --. See Soerensen, --

CHRISTOFFERS, Johann

Emigration file contains letter from friend or
relative in New York　　　　1854
55: Han.Des.74: Blumenthal: IX.J.1: Fach 301, Nr.
10.　　　　ADS 4:154

CLAPP, B. (in New York)

Letters to C. A. Offensandt　　　　1832-1834
48: Bestand 7,76: C. A. Offensandt: Nr. 48-50.
ADS 3:75

CLAUSEN, Hans Reimer (Senator; Iowa)

Biography in *Schlewswig-Holsteinischer Kunstkalen-
dar 1923*, p. 66
24: Drucksachen.　　　　ADS 1:58

CLAUSSEN, H. A. (farmer; in California)

Estate of　　　　1902
39: Reich- u. auswaert.Ang.: E.IV.11 [Erbschaften]
ADS 2:323

CLAUSSEN, -- (from Kiel, SHL)

Banquet for Claussen, a lawyer, before his emigration to America　　　　1851
2: Abt.106: Cc: Fasc.1, Nr. 6.　　　ADS 1:17

CLAUSSEN, -- (in Davenport [Iowa?])

Letter from Landdrostei in Heide regarding Claussen's affairs　　　　1834
39: Bestand Senat: Cl.VIII, No.X, Jg.1834 (Register)　　　　ADS 2:132

CLAUSSEN, --. See Paulsen, --

CLEMENT, Ale (Olde) Jungbohm. See Jungbohm-Clement,
Ale

CLEMENT, Alfred (in Virginia)

Correspondence　　　　1864
2: Nachlass Knud Jungbohm-Clement.　　ADS 1:37

CLEMENT, ERmin (in New York & Boston)

Corrrespondence　　　　1867-1871
2: Nachlass Knud Jungbohn-Clement.　　ADS 1:37

CLEVE, Johann Heinrich (shepherd; from Klein Rhueden?)

Emigration to America with Friedrich Haus, Friedrich Greune, Johann Heinrich Appuhn, Wilhelm
Wolze, Wilhelm Greune, Wilhelmine Bartum, Sophie
Hoebel, Heinrich Christian Steinmann, Marie Bartum　　　　1852
56: L Neu Abt.129A: Gr.11, Nr. 78.　　ADS 4:191

COHEN, Sa[muel?]. See Copelmann, Hirsch

COHN, Max (in Philadelphia)

Transmittal of funds swindled by Cohn (*alias?* Martin Bohn) from M. M. Strupp in Meiningen　1876
39: Bestand Senat: Cl.VIII, No.X, Jg.1876 (Register) 32 [Nordamerika]　　　　ADS 2:233

von COLL, [H.] (bookkeeper in German colony in Texas)

Service contract with German colonies in Texas
1844

154: CA 4c1: Nr. 40.
DLC has.   ADS 7:177

COMING, J. H. (from Hildesheim)

Coming's death & estate in America   1873

39: Bestand Senat: Cl.VIII, No.X, Jg.1873 (Register) 11 [Nordamerika]   ADS 2:224

CONRATH, -- (died in Tiffin, Ohio)

File on [estate?]   1899-1901

227: Rep.MA 1921: A.St.I: Nr. 310.   ADS 9:66

CONSTANTINE, Frank J.

Hamburg police investigation, at request of American police authorities, of Constantine, wanted
for murder   1906

39: Reichs- u. auswaert.Ang.: P.III.1.1 [Polizei-
wesen]   ADS 2:344

CONZE, Alexander

Born in Hamburg; short time in New York in 1844;
then in Milwaukee, where he began a school for
one year; killed on 23 February 1847 at the battle of Buena Vista, near Saltillo, Mexico [an
American soldier]

66: 33.a. Nr. 3, 14 (Fach XII, 5)[Aufzeichnungen]
ADS 4:243

COPELMANN, Hirsch

Investigation in Hamburg, at request of American
police authorities, of Copelmann & Sa. Cohen
1898

39: Reichs- u. auswaert.Ang.: P.III.1.2 [Polizei-
wesen]   ADS 2:298

CORDES, A. (in Philadelphia)

Correspondence [Mary J. Drexel Home]   1890

120: Akte Frank Buchman (D II B).   ADS 6:75

CORDES, E.

Expulsion of Cordes from Hamburg. Cordes emigrated
to America before completing his military service
obligation & was naturalized there   1886

39: Bestand Senat: Cl.VIII, No.X, Jg.1886 (Register) 5 [Nordamerika]   ADS 2:257

CORDES, Friedrich Theodor (in America)

Letter from mayor of Amarin [not otherwise described in ADS description] regarding condition
& reputation of Cordes, then in America   1840

39: Bestand Senat: Cl.VIII, No.X, Jg.1840 (Register).   ADS 2:139

CORDES, H.

Detroit: Estate of   1897

39: Reichs- u. auswaert.Ang.: E.IV.46 [alt.Sig.]
[Erbschaften]   ADS 2:290

CORDES, Johann Heinrich

Information requested by German consulate in New
York regarding Cordes   1870

55: Han.Des.80: Reg.Stade: N.Tit.1: Nr. 8, RR 748.
ADS 4:155

CORDES, Joh. Hinr.

New York: Estate of   1878

39: Bestand Senat: Cl.VIII, No.X, Jg.1878 (Register) 39 [Nordamerika]   ADS 2:238

CORDES, --

New York: Estate of   1880

39: Bestand Senat: Cl.VIII, No.X, Jg.1880 (Register) 33 [Nordamerika]   ADS 2:244

CORDES, -- (pastor; in Philadelphia)

Correspondence   1892

120: Akte Frank Buchman (D II B).   ADS 6:75

CORDSEN, --

Estate of   1941

148: R XI.   ADS 6:261

CORDUA, Christoph August (in California)

Genealogy of Cordua family

21   ADS 1:56

CRILL, John Hanns. *See* Weber, Hattie A. A.

CRENDE, W. R. (doctor; from New York)

Regarding passport of alleged Dr. Crende   n.d.

151: Bestand 82: Rep.III: Gef.248: Nr. 23.
ADS 7:86

CRIPPEN, Hawley Harvey (doctor; an American citizen)

Hamburg police search for Dr. Crippen & Ethel
Clara LeNeve   1910
39: Reichs- u. auswaert.Ang.: P.III.2.6 [Polizei-
wesen]   ADS 2:365

CSUKA, --. *See* Kovacs, --

CULMANN, Carl (practicing engineer; from Bergzabern)

Trip [emigration?] to England, France, & North
America   1849
227: Rep.MA 1895: A.V.: Nr. 1123.   ADS 9:57

DAHL, John (in Brooklyn, New York)

Estate of   1907
39: Reichs- u. auswaert.Ang.: E.IV.67 [Erbschaften]
ADS 2:347

DAHLMANN, Jacobus (of Duesseldorf)

Letter to Prime Minister Andreas Gottlieb von
Bernstorff regarding emigration of Palatines
to Pennsylvania   1719
52: Calenberger Brief-Archiv Des.24: Pfalz: Nr.
138.   ADS 4:56

DAHM, Amds. [Amadeus?] Detlef Ferdinand

Hamburg Consulate in Philadelphia: Death certificate for Dahm who was killed while on journey
from Boston to Philadelphia   1866-1867
39: Bestand Senat: Cl.VI, No.16p: Vol.4b: Fasc.
12g.   ADS 2:57

DAMBMANN, Christian (from Ruesselsheim, HEL)

Secret [illicit] emigration to North America 1850
Application for emigration to America   1852
182: Abt. XI,4.   ADS 7:235

DAMBMANN, Wilhelm (from Ruesselsheim, HEL)

Release from Hessian citizenship for the purpose
of emigration to America   1893
182: Abt. XI,4.   ADS 7:236

DAMMEL, Philipp (from Nauheim, Landkreis Gross-Gerau,
HEL)

Emigration   1880?
180: Ortsbuergerregister 1761-1886.   ADS 7:233

DAMROSCH, --, born Gerras

Request made to American consulate in Hamburg for
an American death certificate for Mrs. Damrosch
1910

39: Reichs- u. auswaert.Ang.: S.IV.3.1 [Standes-
amtssachen]   ADS 2:367

DANIELSEN, -- (widow; in Langenhorn)

American pension for   1874-1879
16: Nr. 239.   ADS 1:50

DANKER, Christ[ian]

Chicago: Estate of   1906
39: Reichs- u. auswaert.Ang.: E.IV.11 [Erbschaften]
ADS 2:343

DANNER, Josef (in America)

Transmittal of proceeds from sale of property [in
Germany?]   1850-1851
227: Rep.MA 1898: A.St.: Nr. 6279.   ADS 9:59

DANZ, -- (member Moravian Brethren; in North Carolina)

Travel diaries of -- Danz, -- Wagner, -- Schreyer,
-- Mueller, -- Krause, -- Wallis (brothers) from
London to North Carolina   1774
500: Herrnhut: Archiv: Rep.14.Ba: Nr. 2e.
Not in ADS

DANZIGER, -- (in Brooklyn, New York)

Estate of   1917
39: Reichs- u. auswaert.Ang.: E.IV.45 [Erbschaften]
ADS 2:392

DANZIGER, --. *See also* Tiedemann, Johann H. A.

DAUER, Jakob (from Wisper, HEL)

Emigration to North America   1853
152: Abt.231: Langenschwalbach: Nr. 1151.
ADS 7:133

DAUM, Philipp (from Ruesselsheim, HEL)

Emigration to America   1849
182: Abt. XI,4.   ADS 7:235

DAVID, -- (in America)

Application for inheritance tax exemption   1833
39: Bestand Senat: Cl.VIII, No.X, Jg.1833 (Register) 1.c [Zehnten-Amt]   ADS 2:132

DAVIS, Julius. *See* Allardt, B. H.

DECK, Elisabeth J. M. (in Marcy, New York? [or New Jersey?])

Estate of   1909

39: Reichs- u. auswaert.Ang.: E.IV.14 [Erbschaften]
ADS 2:356

DEETJEN, -- (from Bockel, NL)

Sale of farm in order to emigrate to America 1875
59   ADS 4:227

DEHN, Carl B.

Consular report fron San Francisco regarding death
certificate for Dehn   1869

39: Bestand Senat: Cl.VIII, No.X, Jg.1869 (Register) 9 [Nordamerika]   ADS 2:201

DEHNKE, Emil

New York: Estate of   1906

39: Reichs- u. auswaert.Ang.: E.IV.60 [Erbschaften]
ADS 2:343

DELIUS, L. E. (an American citizen)

Guardianship for   1900

39: Reichs- u. auswaert.Ang.: V.III.3 [Vormund-
schaftssachen]   ADS 2:312

DEINERT, C. A. P.

Chicago: Application for release from German citizenship for Deinert & his family   1875

39: Bestand Senat: Cl.VIII, No.X, Jg.1875 (Register) 29 [Nordamerika]   ADS 2:231

Application for release from German citizenship
1876

39: Bestand Senat: Cl.VIII, No.X, Jg.1876 (Register) 5 [Nordamerika]   ADS 2:232

DEINLEIN, Barbara (died in Baltimore, Maryland)

File on [estate?]   1889-1899

227: Rep.MA 1921: A.St.I: Nr. 356.   ADS 9:67

DEINLEIN, Philipp (died in Baltimore, Maryland)

File on [estate?]   1889-1899

227: Rep.MA 1921: A.St.I: Nr. 356.   ADS 9:67

DELBIG, Karl (from Philadelphia; died in Philadelphia)

File one [estate?]   1824

227: Rep. Bayerische Gesandtschaft London: Nr. 868.
ADS 9:47

DELBIG, Kasimir (from Zweibruecken; died in Philadelphia)

File on [estate?]   1824

227: Rep. Bayerische Gesandtschaft London: Nr. 868.
ADS 9:47

DELONG, --. See de Long, --

DENECKE, Joh. Konrad (from Emden)

Transport of criminal to America   1863

50: Rep.21a: Nr. 9016.   ADS 4:7

DENECKE, J. (miner from Clausthal; died in California)

Estate of   1858

68: VI.2.b: Fach 162, Nr. 43.   ADS 4:247

DENECKE, W. (Union soldier in U.S. Civil War; died
at Gettysburg, Pennsylvania [apparently not as a
consequence of the war, however])

Estate of   1905

39: Reichs- u. auswaert.Ang.: E.IV.62 [Erbschaften]
ADS 2:339

DENKER, Emilie Karoline. See Wulf, Emilie Karoline,
born Denker

DE ROY, Peter. See de Roy, Peter

DESSAU, Georg (in Indianapolis, Indiana)

Inquiry by private persons in Germany regarding
1901

39: Reichs- u. auswaert.Ang.: N.Ib.1.9 [Nachforsch-
ungen]   ADS 2:319

DETERING family (of Einbeck, NL)

Genealogical studies regarding

72: Harland's Nachlass: Manuscripte Nr.9: Vol.
VIII.   ADS 4:250

DETLEF, Peter

Death & estate of   1876

39: Bestand Senat: Cl.VIII, No.X, Jg.1876 (Register) 28 [Nordamerika]   ADS 2:233

Estate of   1877

39: Bestand Senat: Cl.VIII, No.X, Jg.1877 (Register) 4 [Nordamerika]   ADS 2:233

Estate of   1878

39: Bestand Senat: Cl.VIII, No.X, Jg.1878 (Register) 12 [Nordamerika]   ADS 2:237

DETTMER, --. *See* Leiding, M. C. L., born Dettmer

DEUMELAND, Christine, born Boberg (in Wilmington, North Carolina)

Estate of   1900
39: Reichs- u. auswaert.Ang.: E.IV.42 [Erbschaften]
ADS 2:307

DEUTSCH, Emil (in New York)

Inquiry of Reichskanzler regarding   1917
39: Reichs- u. auswaert.Ang.: N.Ia.3.1 [Nachforsch-
ungen]   ADS 2:393

DHIEL, Johannes. *See* Diehl, Johannes

DHIEL, Jacob. *See* Diehl, Johannes

DIBBERN, H. J.

San Francisco: Estate of   1903
39: Reichs- u. auswaert.Ang.: E.IV.95 [Erbschaften]
ADS 2:330

DICKMANN, Johann

Statement & complaint regarding his imprisonment
in an American insane asylum   1901-1919
116: Muenster Reg.: Nr. 4101.   ADS 6:56

DIEFFENBACH, Georg (from San Antonio)

File on [estate?]   1895-1900
227: Rep.MA 1921: A.St.I: Nr. 389.   ADS 9:67

DIEFFENBACKER, John (died in San Francisco)

File on [estate?]   1890
227: Rep.MA 1921: A.St.I: Nr. 362.   ADS 9:67

DIEHL, Georg Peter (from Ruesselsheim, HEL)

Emigration to North America   1832
182: Abt. XI,4.   ADS 7:235

DIEHL, Johann (from Geinsheim, Landkreis Gross-Gerau,
HEL)

Application for assistance from church funds for
emigration to America   1852
170: Abt. XI,4.   ADS 7:221

DIEHL, Johannes (from city of Zweibruecken; in Reading, Berks County, Pennsylvania)

Proceedings leading to the partial confiscation of

his property because of illicit emigration &
transferring [property title] to his son Jacob
Dhiel [Diehl]   1788-1803
201: Zweibruecken III: Akt Nr. 2027 I.   ADS 8:59

DIEHL, Johannes J., I (from Goddelau, Landkreis Gross-
Gerau, HEL)

Emigration to America   1868
173: Ortsbuergerregister.   ADS 7:224

DIERKS, H. C. (in Chicago)

Inquiry from private person regarding   1896
39: Reichs- u. auswaert.Ang.: N.Ib.10 [alt.Sig.]
[Nachforschungen]   ADS 2:284

DIESTER, Emily, born Greuel. *See* Brown, --, born
Greuel

DIETHER, Adam (musketeer; from Geinsheim, Landkreis
Gross-Gerau, HEL)

Application for emigration permit to America 1850
170: Abt. XI,4.   ADS 7:221

DIETL, Jakob (in America)

Information regarding   1886
227: Rep.MA 1921: A.St.I: Nr. 1424.   ADS 9:72

DIETMANN, I. (from Neufra, BWL)

Application for emigration permit   1842
262: Ho 193: Gammertingen: I 13863.   ADS 10:35

DIETRICH, Carl August Siegmund (chemist)

Expulsion from Hamburg as Social Democrat & emigration to America   1880-1881; 1896
39: Polizeibehoerde-Polit.Pol.: S 149/104: Bl. 1-
136.   ADS 2:414

DIETRICH, --

Estate of   1944
148: R XI.   ADS 6:261

DIETZ, Emanuel (died in St. Louis, Missouri)

File on [estate?]   1900
227: Rep.MA 1921: A.St.I: Nr. 1437.   ADS 9:72

DIETZMANN, Johann Andreas (from Uslar)

Proposed transportation to America   1844-1847
52: Han.Des.80: Hildesheim I: E.DD.44a: Nr. 1716.
ADS 4:91

DILDEY, Maria. *See* White, Georg Christoph

DINS, Fritz (in Sheboygan, Wisconsin)

Estate of   1911
39: Reichs- u. auswaert.Ang.: E.IV.53 [Erbschaften]
                                           ADS 2:369

DIPPEL, --

Mentioned in letter from Roxbury [Massachusetts?]
                                                1724
119: Lit. K: Nr. 36.   ADS 6:74

DOBEL, Moritz (in Tulpehocken, Pennsylvania). *See*
Bartholomaeus, Carl

DOELL, Friedrich (of Rossbach)

Confiscation of property of soldier deserting in
   America   1786
151: Bestand 82: Rep.E: 55.Amt Bieber X: Nr. 7.
                                           ADS 7:83

DOERING, Otto (pastor, St. John's Evangelical Lutheran Church, Napoleon, Indiana)

Letter   [1919-1922?]
44: Mappe Amerika.   ADS 2:599

DOERR, Wilhelm (from Alsfeld; died in San Francisco
on 2 February 1892)

Estate of; inheritance rights of Johann Conrad
   Doerr & wife in Offenbach & of Ludwig Doerr &
   wife in Gruenberg   1896-1897
149: Rep.49a: Nr. 563/2.   ADS 7:15

DOERRIES, Christian (laborer; from Vehrssen)

Emigration to America   n.d.
52: Han.Des.74: Coppenbruegge: VIII, H,7, Nr. 6.
                                           ADS 4:65

DOERRIES, M.

New York: Estate of   1890
39: Bestand Senat: Cl.VIII, No.X, Jg.1890 (Register) 25 [Erbschafts-Ang.]   ADS 2:264

DOERSAM, Johann Adam (from HEL)

Emigration to America   1871
149: Rep.51 V: Fuerth: Nr. 128 D.   ADS 7:2

DOHNE, Carl Friedrich (cigar worker)

Expulsion from Hamburg as Social Democrat & emigration to America   1880-1881

39: Polizeibehoerde-Polit.Pol.: S 149/245: Bl.
   265-269.   ADS 2:415

DOHSE, I. W.

San Francisco: Estate of   1888
39: Bestand Senat: Cl.VIII, No.X, Jg.1888 (Register) 18 [Erbschafts-Amt]   ADS 2:260

DOHSE, Peter Friedrich (of Zennhusen, SHL)

Petition for relase from citizenship   1867
13: Alphabetischen Sammlung.   ADS 1:49

DOLL, John (died in Brooklyn, New York)

File on [estate?]   1895
227: Rep.MA 1921: A.St.I: Nr. 391.   ADS 9:67

DOHRN, John. *See* Dorn, John

DOMINICUS, *alias* Schrader, Friedr[ich] Eduard (in
Goettingen)

Transportation to America   1836
52: Han.Des.80: Hildesheim I: E.DD.38b: Nr. 1426.
                                           ADS 4:90

DOMINIQUE, Karl Ludwig (businessman; from Zweibrueck-
en; died in Philadelphia)

File on   1833
227: Rep. Bayerische Gesandtschaft London: Nr. 891.
                                           ADS 9:47

DONINGER, Josef (died in America)

File on [estate?]   1887
227: Rep.MA 1921: A.St.I: Nr. 340.   ADS 9:67

DONNER, Wilhelm (in Arkansas)

Investigation regarding his [Wuerttemberg] military service obligation   1900-1913
263: Rep.E. 46-48: Minist. d. AA. III: Verz.1:
   148/17.   ADS 10:43

DONOP, August Wilhelm (an American citizen)

Permit to do business in Bremen   1856
48: B.13.C.1.b: Nr. 12.
   DLC has.   ADS 3:42

DORMANN, -- (from Worringen; in Chicago)

Minor children of a couple who died during trip to
   America   1854-1858
115: Rep.D9: Koeln: Abt.I.e: Nr. 287.   ADS 6:32

DORN, Georg (from Neusses; American citizen)

Petition for reversion to Bavarian citizenship
1899

227: Rep.MA 1921: A.St.I: Nr. 1244.   ADS 9:72

DORN (DOHRN), John

New York: Estate of   1914

39: Reichs- u. auswaert.Ang.: E.IV.61 [Erbschaften]
ADS 2:384

DOSE, Fritz (in Davenport, Iowa)

Estate of   1906

39: Reichs- u. auswaert.Ang.: E.IV.3 [Erbschaften]
ADS 2:343

DOTT, J. (in Chicago)

Information sought by private person   1895

39: Reichs- u. auswaert.Ang.: N.Ib.14 [alt.Sig.]
[Nachforschungen]   ADS 2:277

DOULEY, L. J. (in Texas)

Report regarding estates of -- Kleeman & -- Adler
in Bavaria sent to Douley   1901

227: Rep.MA 1921: A.V. V: Nr. 15509.   ADS 9:90

DRAPPELDREI, -- (died in New York)

File on [estate?]   1900

227: Rep.MA 1921: A.St.I: Nr. 415.   ADS 9:67

DRESEL, Eugen (a German citizen)

Extradition to Berlin from USA, via Hamburg   1914

39: Reichs- u. auswaert.Ang.: A.Ia.11 [Ausliefer-
ungen]   ADS 2:383

DREWES, Melchior Jacob (farmhand)

Released on 28 September 1840 from prison in Osnabrueck; to be transported to America   1840

100: Kirchspielsgericht Westerende-Ott.: Loc.17,
Nr. 5.   ADS 4:307

DREWING, Wilhelm

New York: Information regarding   1879

39: Bestand Senat: Cl.VIII, No.X, Jg.1879 (Register) 8 [Nordamerika]   ADS 2:240

DREWITZ, Anna Elisabeth. *See* Drewitz, Johann Jacob

DREWITZ, Conrad. *See* Drewitz, Johann Jacob

DREWITZ, Johann Jacob (of Uckersdorf, HEL)

Application of Anna Elisabeth, Conrad, & J. J.
Drewitz for permission to emigrate to America;
also regarding properties to be left behind
1773-1796

152: Abt.172: Dillenburg LR: Heft 1.
DLC has.   ADS 7:129

DREYER, Henriette (spinster; from Lutter a. B.)

Incarceration in the house of corrections at
Bevern & emigration to America   1856

56: L Neu Abt.129A: Gr.11, Nr. 62.   ADS 4:190

DREYFUSS, Heinrich (businessman; in New York)

Court decision in case of Leon & Moses Cerf [a
firm in Landau] *versus* Heinrich Dreyfuss   1870

227: Rep.MA 1921: A.St.I: Nr. 248.   ADS 9:66

DRIESSLER, -- (pastor; in Pennsylvania?)

Correspondence with Driessler in Frederica; & correspondence regarding him with others on subject
of pastoral call   1748

500: Halle: Missionsbibliothek des Waisenhauses:
Abteilung H IV (Nordamerika) Fach B, Nr. 2.
Not in ADS

DROEGE, Heinrich (salesman of pots & pans; from
Helmscherode)

Punishment for fraud & emigration to America
1856-1859

56: L Neu Abt.129A: Gr.11, Nr. 52.   ADS 4:189

DRUIVENGA, J.

Extradition of Druivenga from the USA to Weener,
via Hamburg   1902

39: Reichs- u. auswaert.Ang.: D.Ic,51 [Ausliefer-
ungen]   ADS 2:328

DUDERSTADT, Ludwig

San Francisco: Estate of   1911

39: Reichs- u. auswaert.Ang.: E.IV.98 [Erbschaften]
ADS 2:369

DUEHRKOP, Friedr[ich] Wilh[elm]

New York: Estate of   1907

39: Reichs- u. auwaert.Ang.: E.IV.1 [Erbschaften]
ADS 2:346

DUEHRKOP, J. H. W. (in Chicago)

Inquiry by private person regarding   1896

39: Reichs- u. auswaert.Ang.: N.Ib.18 [alt.Sig.]
[Nachforschungen]   ADS 2:284

DUELSEN, Heinrich (in Michigan)

Estate of   1917
39: Reichs- u. auswaert.Ang.: E.IV.17 [Erbschaften]
                                           ADS 2:392

DUENSING, H. Ph.

Search for   1875
39: Bestand Senat: Cl.VIII, No.X, Jg.1875 (Register) 15 [Nordamerika]   ADS 2:230

DUET, --

Estate of   1944
148: R XI.   ADS 6:261

DUN(G)STAETTER, Ludwig (from Gross-Gerau, HEL)

Emigration to America, with appendices   1860
174: XI,4.   ADS 7:227

DUTTENHOEFER, Chr. (died in America)

File on [estate?]   1886
227: Rep.MA 1921: A.St.I: Nr. 337.   ADS 9:66

DUVALL, Henry (from New York)

Investigation of   1896
39: Reichs- u. auswaert.Ang.: N.Ia.7 [alt.Sig.]
[Nachforschungen]   ADS 2:284

EARLE, -- (American citizen)

Extradition of -- Earle & -- Hermann to France
from Norway, via Sassnitz [both American citi-
zens]   1914
39: Reichs- u. auswaert.Ang.: A.Ia.3 [Ausliefer-
ungen]   ADS 2:383

EBEL, Konrad (in New York)

Konrad Ebel's right of inheritance to the estate
of Magdalena Ebel of Boehl   1884-1886
227: Rep.MA 1921: A.St.I: Nr. 454.   ADS 9:67

EBELING, Heinrich (day laborer; from Kirchberg)

Emigration to America   1858
56: L Neu Abt.129A: Gr.11, Nr. 57.   ADS 4:189

EBENSTEIN, Kurt (in Chicago)

Inquiry by private person in Germany regarding
                                           1900

39: Reichs- u. auswaert.Ang.: N.Ib.1.4 [Nachforsch-
ungen]   ADS 2:309

**EBERHARD, Friedrich (from Karlsbrunn, Principality
of Nassau-Saarbruecken, RPL)**

Inventory of the possessions of the widow of Martin Eberhard, smelter owner in Karlsbrunn, of
Margarete [Eberhard?] born Reppert, & of the
estate of Friedrich Eberhard   1770
200: Abt.22: Fuerstentum Nassau-Saarbruecken: Nr.
3656.   ADS 8:9

EBERHARD, Margarete, born Reppert. *See* Eberhard,
Friedrich

EBERHARD, -- (widow of Martin). *See* Eberhard, Friedrich

**EBERHARD, -- (from Karlsbrunn, Principality of
Nassau-Saarbruecken, RPL)**

Debts of -- Eberhard & -- Kramer   1779-1781
200: Abt.22: Fuerstentum Nassau-Saarbruecken: Nr.
3507.   ADS 8:10

**EBERLING, Balthasar (from Erfelden, Landkreis Gross-
Gerau, HEL)**

Application for public assistance for emigration
to New York   1867
Payment therefore, also noted   1867
169: Urkundenband zur Gemeinderechnung 1867.
                                    ADS 7:220

EBNER, Ch. (seaman)

Baltimore: Estate of   1902
39: Reichs- u. auswaert.Ang.: E.IV.42 [Erbschaften]
                                    ADS 2:323

EBSEN, Hans Heinrich (pastor)

Article on his emigration to America at the time
of the uprising in Schleswig-Holstein in *Jahrbuch
fuer die Schleswigsche Geest*, 4.Jg. (1956)
                                    ADS 1:67

ECKARDT, Ch. Fr.

San Francisco: Estate of   1896
39: Reichs- u. auswaert.Ang.: E.IV.35a [alt.Sig.]
[Erbschaften]   ADS 2:282

ECKHOFF, Henry (in California)

Estate of   1915
39: Reichs- u. auswaert.Ang.: E.IV.45 [Erbschaften]
                                    ADS 2:388

EDEL, Matth[aeus?] (died in Jefferson City [Missouri])

File on [estate?]   1870

227: Rep.MA 1921: A.St.I: Nr. 421.   ADS 9:67

EDELMAN, Johann Christ (founder of anti-Christian
sect).   *See* Wied (Duchy of)

EDENS, Claus Heinrich

Expulsion of   1911

39: Reichs- u. auswaert.Ang.: A.IV.1 [Ausweisungen]
                                             ADS 2:368

EDSEN, H. E.

New York: Information regarding   1877

39: Bestand Senat: Cl.VIII, No.X, Jg.1877 (Regis-
    ter) 11 [Nordamerika]   ADS 2:234

EDWARD, E. (in Akron, Ohio)

Estate of   1885

39: Bestand Senat: Cl.VIII, No.X, Jg.1885 (Regis-
    ter) 20 [Nordamerika]   ADS 2:255

EDWARDS, -- (in America)

Estate of   1908

39: Reichs- u. auswaert.Ang.: E.IV.4 [Erbschaften]
                                             ADS 2:351

EEGEBERG, Christ.

New York: Estate of   1909

39: Reichs- u. auswaert.Ang.: E.IV.38 [Erbschaften]
                                             ADS 2:357

EGG, Kaspar (from Steinbach)

Return of soldier who had been indentured in
America [after American Revolution]   1786

151: Bestand 10: a: Verz.2: Acc.1906/20 II: Nr. 6
     [Rueckkehr]   ADS 7:46

EGGERING, Heinrich (from Raunheim, Landkreis Gross-
Gerau, HEL)

Emigration to America with wife & her sister 1880

181: Raunheimer Evangelischen Pfarrei Familien-
     register (Lehrer Johannes Buxbaum). ADS 7:234

EGING, Chr. (died in San Francisco)

File on [estate?]   1895-1896

227: Rep.MA 1921: A.St.I: Nr. 487.   ADS 9:67

EHARDT (EHRHARDT), -- (in Streator, Illinois)

Estate of   1913

39: Reichs- u. auswaert.Ang.: E.IV.71 [Erbschaften]
                                             ADS 2:379

EHINGER, August (in Belleville, Illinois)

Letter   1848

216: Bestand 191-1: in Nr. 221.   ADS 8:85

EHLERS, Theodor Karl Heinrich (chairman of the Union
of Hamburg Citizens)

Trips to New York   1880; 1889-1896

39: Polizeibehoerde-Polit.Pol.: S 149/188: Bl. 1-7.
                                             ADS 2:415

EHRENBERG? Diedrich.   *See* Wehrenberg, Diedrich

EHRHARD, Peter (from Munich; died in Texas)

Death certificate   1843-1846

227: Rep. Bayerische Gesandtschaft London: Nr. 448.
                                             ADS 9:46

EHRHARDT, --.   *See* Ehardt, --

EHRICH (EHRIC), E. (in Sacramento, California)

Estate of   1890

39: Bestand Senat: Cl.VIII, No.X, Jg.1890 (Regis-
    ter) 9 [Erbschafts-Ang.]   ADS 2:264

EICHENMUELLER, Elisabeth

Return to America of Elisabeth Eichenmueller &
Katharina Puehl   1914

227: Rep.MA 1936: Deutsches Reich: 2494/380.
                                             ADS 9:103

EICKE, Betty (born 1873 in Bremen; emigrated to USA
in 1887; was nurse & supervisor in Lawrence, Massachusetts)

Material on the life of this distinguished German-
American nurse has been collected by Amtsrat in
R. Lucé, Kreisarchivpfleger, Eschershausen, NL

                                             ADS 4:255

EISKIRCH, Thomas (from Amt Gemuenden, RPL)

Sale of land by Eiskirch & Michel Reichel who
intend to emigrate to the French "new land"
[contains accounts only]   1764

200: Abt.53 C16: Amt Gemuenden: Nr. 80.   ADS 8:12

EISNER, -- (draftsman)

Extradition to Prenzlau from USA, via Hamburg 1912

39: Reichs- u. auswaert.Ang.: A.Ia.9 [Ausliefer-
ungen]   ADS 2:372

EISENHAUER, --

Estate of   1941

148: R XI.   ADS 6:261

ELLING, -- (assumed to have been in Philadelphia)

Estate of   1816

227: Rep. Bayer. & Pfaelz. Gesandtschaft Paris:
Nr. 8377.   ADS 9:50

ELSASSER, Theodor (in Buffalo [New York?])

Complaint in Reichstag   1904

227: Rep.MA 1936: Deutsches Reich: Nr. 507.
                                         ADS 9:102

ELSEL, Paul (pastor, Lutheran Church, Mapleton, Minnesota)

Letter   [1919-1922?]

44: Mappe Amerika.   ADS 2:599

ELTER, Margarethe. *See* Jorel, Marie

d'ELTOUR, Enrique (from America)

Settlement in Carlshafen & exemption from all
[taxes] for seven years   1786

151: Bestand 5: Nr. 1816.   ADS 7:38

ELVER(S), Chr.

Chicago: Estate of   1882

39: Bestand Senat: Cl.VIII, No.X, Jg.1882 (Register) 10 [Nordamerika]   ADS 2:248

EMIG, -- (of Rosdorff)

Sohn Emig & nine young men, all present at the inn
in the city of Wuerzburg, subjects of Hesse-
Darmstadt, who plan to emigrate to America 1773

158: Hessen: Ugb D42: Nr. 46 [Sohn Emig] ADS 7:193

EMMERICH, Friedrich (died in San Francisco)

Estate of   1890-1891

116: Minden: Praes-regist.: I.L.Nr. 236. ADS 6:54

EMMERICH, M. P.

Hamburg police search for Emmerich, wanted in

America   1909

39: Reichs- u. auswaert.Ang.: P.III.1.2. [Polizei-
wesen]   ADS 2:358

EMMERICH-ASTOR, --

Estate of   1930-1941

148: R XI.   ADS 6:261

EMMERT, Peter (from Froeschen, RPL; in New England)

Property of [in Germany]   1770

201: Archivabt. Grafschaft Hanau-Lichtenberg: Akt
Nr. 1164.   ADS 8:67

ENDER, --

Estate of   1942-1944

148: R XI.   ADS 6:261

ENGEL, Matheiss (from Amt Birkenfeld, Duchy of Zweibruecken). *See* Wittib, Jacob Mausen

ENGLER (or HAEFNER), --

Inquiry by Pastor -- Pohle in Philadelphia   1889

227: Rep.MA 1921: A.St.I: Nr. 1441.   ADS 9:73

ENNEN, R. J. (from Marx)

Emigration of convict to America   1863-1864

50: Rep.21a: Nr. 8947.   ADS 4:8

ENSMENGER, Abraham (in America)

File on   1886-1898

227: Rep.MA 1921: A.St.I: Nr. 456.   ADS 9:67

EPPLE, Adam (from Emmendingen, BWL; in America)

Information regarding the family of Adam Epple
                                               1884

288: Kommunalarchiv Emmendingen.   ADS 10:110

EPPLE, E.

Complaint made from New York by E. Epple & Conrad
Hoffmann regarding agent Kohn of Hamburg   1873

39: Bestand Senat: Cl.VIII, No.X, Jg.1873 (Register) 33 [Nordamerika]   ADS 2:225

ERB, A (in Blue Bell [state not given in ADS descrip-
tion])

Matter of inheritance   1892

227: Rep.MA 1921: A.St.I: Nr. 471.   ADS 9:67

ERBE-KOSERITZ, Ernst Ludwig (in America)

Supporting details for the statement prepared by Bankiers Finanzrat [banking councillor] Federer regarding the Erbe-Koseritz affair   1835

263: Rep.E 10: Minist.des Innern: Nr. 118.
                                                ADS 10:53

ERBEN, Bohumil

Internal correspondence regarding assistance to Erben, a refugee monk from Austria, who has become a protestant, in his emigration to Texas
                                                      1858

39: Geistl.Minis.: III,B: Fasc.20: Lose Akten 1858.
                                                ADS 2:483

ERICH, E.   *See* Ehrich, E.

ERIKSON, K. Ch. (in New York).   *See* Wideen, K. Ch.

ERNST, A. (in San Francisco)

Accusation by Ernst against Hamburg firm   1871

39: Bestand Senat: Cl.VIII, No.X, Jg.1871 (Register) 40 [Nordamerika]   ADS 2:217

ERNST, Marie

New York: Estate of   1903

39: Reichs- u. auswaert.Ang.: E.IV.93 [Erbschaften]
                                                ADS 2:330

ERPENSTEIN, Anton.   *See* Fenstermacher, G.

ESCHENFELDER, Barbara.   *See* Stoppel, Barbara, born Eschenfelder

ESCHERICH, Josefine (in Richmond [state not given in ADS description])

Legal affairs   1895

227: Rep.MA 1921: A.St.I: Nr. 1446.   ADS 9:73

ETTWEIN, -- (member Moravian Brethren; in America)

Letters & memoranda from   1771-1787

500: Herrnhut: Archiv: Rep.14.A: Nr. 47.
                                                Not in ADS

EULENBERG family.   *See* Betz, Fr.

EWERT, Bertha.   *See* Ewert, F. E.

EWERT, F. E. (in Providence [Rhode Island?])

Investigation of Ewert in Providence. He had requested that his daughter Bertha, left a public

charge in Hamburg, be sent to join him   1894

39: Reichs- u. auswaert.Ang.: W.I.4 [alt.Sig.] [Waisenhaussachen]   ADS 2:272

FABER, Agnes (in New York).   *See* von Melle, Emil

FABER, Catharina (in New York).   *See* von Melle, Emil

FABER, Constanze (in New York).   *See* von Melle, Emil

FABER, J. H. (in New York).   *See* von Melle, Emil

FACH, Barbara (spinster; in Dieberg)

Had heirs in America   1912

149: Rep.49a: Nr. 565/17.   ADS 7:16

FAHL, --

Estate of   1943-1944

148: R XI.   ADS 6:261

FAHRENKAMP, --

Estate of   1941

148: R XI.   ADS 6:261

FALK, Austrian.   *See* Austrian, Falk

FALK, Michael (in Lincoln [state not given in ADS description])

Estate of   1902

39: Reichs- u. auswaert.Ang.: E.IV.15 [Erbschaften]
                                                ADS 2:323

von FALKENHAUSEN, Hugo (in America)

Inquiry regarding   1887

227: Rep.MA 1921: A.St.I: Nr. 1459.   ADS 9:73

FARKAS, -- (Hungarian citizen)

Extradition to Hungary from the USA, via Hamburg
                                                      1910

39: Reichs- u. auswaert.Ang.: A.Ia.26 [Auslieferungen]   ADS 2:360

FARNY, Johann Jakob (from Bad Duerkheim; in Philadelphia)

Power of attorney given to his friend Georg Ulrich, returning to Germany, to institute a court case
                                                      1798

204: II.R.19.   ADS 8:70

FAULHABER, Magdalene (from Kimbach, Landkreis Gross-Gerau, HEL)

Emigrant to America   1888

173: Gesinderegister 1853-1894: Nr. 251. ADS 7:225

FAULSTROH, Heinrich (shot & killed on 8 July 1909 near Spokane, Washington)

Estate of   1909-1911

152: Abt.405: Wiesbaden: Rep.405/I,2: Nr. 3458.
                                        ADS 7:143

FEDDERSEN, Theod[or]

San Francisco: Estate of   1911

39: Reichs- u. auswaert.Ang.: E.IV.46 [Erbschaften]
                                        ADS 2:369

FEDER, Johann (died 1892 in St. Louis)

File on [estate?]   1895-1897

227: Rep.MA 1921: A.St.I: Nr. 498.   ADS 9:67

FEHLING, Friedrich Johann Karl (cigar worker)

Expulsion from Hamburg as Social Democrat & emigration to America   1883

39: Polizeibehoerde-Polit.Pol.: S 149/436: Bl.
    409-424.   ADS 2:418

FEIERABEND, Max (in St. Paul [Minnesota?])

File on [estate?]   1896-1898

227: Rep.MA 1921: A.St.I: Nr. 505.   ADS 9:67

FEIERABEND, Rosa (in Bridgeport, Connecticut)

Estate of   1915

39: Reichs- u. auswaert.Ang.: E.IV.37 [Erbschaften]
                                        ADS 2:388

FEIL, Anton

Residence with North American missions   1842

229: Rep. Reg. Akten von Oberbayern, Kammer des
     Innern: 647.6.   ADS 9:115

FEINER, Joh[ann]

Regarding a certain Joh. Feiner, allegedly from
   Bavaria   1856

227: Rep.MA 1898: A.St.: Nr. 10.   ADS 9:58

FELDHUSEN, J.

Alleged citizenship of   1862

39: Bestand Senat: Cl.VIII, No.X, J.1862 (Register)[Amerika]   ADS 2:180

FELDHUSEN, --

Consular report on Feldhusen's military service
   obligation [ADS description does not state which
   consulate]   1861

39: Bestand Senat: Cl.VIII, No.X, Jg.1861 (Register) I [Amerika]   ADS 2:179

FELDTMANN, J. (in Douglas, Washington)

Estate of   1902

39: Reichs- u. auswaert.Ang.: E.IV.29 [Erbschaften]
                                        ADS 2:323

FELDTMANN, Peter August (died in New York)

Inquiry by private persons regarding survivors
   [heirs?] of   1896

39: Reichs- u. auswaert.Ang.: N.Ib.16 [alt.Sig.]
    [Nachforschungen]   ADS 2:284

FELDTMANN, P. E.

Report to judge in Hamburg from German consulate
   in New York regarding   1901

39: Reichs- u. auswaert.Ang.: G.Ia.1.11 [Geschaefts-
    verkehr]   ADS 2:316

FELKNER, Peter (died 29 February 1904 at Bath, near
   Corning, New York)

Estate of   1908-1911

152: Abt.405: Wiesbaden: Rep.405/I,2: Nr. 3533.
                                        ADS 7:143

FELLMER, Wilhelm (a Union soldier [during Civil War])

Death & estate of Fellmer   1874

39: Bestand Senat: Cl.VIII, No.X, Jg.1874 (Register) 33 [Nordamerika]   ADS 2:228

FENSTERMACHER, G. (in Hazleton, Pennsylvania?)

Fenstermacher seeks information regarding Anton
   Erpenstein   1855

39: Bestand Senat: Cl.VIII, No.X, Jg.1855 (Register)   ADS 2:169

FERBER, T.

German consular report from New York regarding
   Ferber's accusation against Meyer & Co.   1869

39: Bestand Senat: Cl.VIII, No.X, Jg.1869 (Register) 40 [Nordamerika]   ADS 2:203

FERRIS, Mose

Hamburg police search for Ferris, wanted for murder in Roanoke [Virginia]   1907

39: Bestand Senat: Cl.VIII, No.X, Jg.1871 (Register) 23 [Nordamerika]   ADS 2:215

FITZNER, Carl (in Oelse, Kreis Striegau, Silesia)

His inquiry regarding emigration to America   1850

39: Bestand Senat: Cl.VIII, No.X, Jg.1850 (Register)   ADS 2:158

FLAD, F. (in Potosi, Buffalo, & Niagara Falls, New York)

Correspondence [F. Flad was a woman.]   1856
216: Bestand 191-1: in Nr. 85.   ADS 8:85

FLADO, Robert (in Puerto Plata [state or country not given in ADS description])

Estate of   1897
225: Rep.168: Verz.1: Fasz.19: Nr. 1-2: 1.
                                             ADS 9:15

FLASSNER, Eltje (from Popens; unmarried)

Emigration to America   1866
50: Rep.21a: Nr. 8942.   ADS 4:8

FLEISCHMANN, Henry J.

Hamburg police search for Fleischmann, wanted by American police authorities   1902

39: Reichs- u. auswaert.Ang.: P.III.1.1 [Polizeiwesen]   ADS 2:326

FLEISSNER-METZLER, --

Estate of   1943-1944
148: R XI.   ADS 6:261

FLOTO, E. (in New York)

Correspondence   1857-1858
39: Bestand Firmenarchive: H. J. Merck & Co.:
    A.V.35.   ADS 2:484

FOERSTMANN, Karl (from Bodenburg)

Emigration to America   1860
56: L Neu Abt.129A: Gr.11: Nr. 27.   ADS 4:187

FOLKERS, W. (journeyman carpenter; of Leer)

Return from America   1874-1875
50: Rep.21a: Nr. 8123.   ADS 4:7

FORGÉ [Forgé] Karl (cigar maker; from Gandersheim)

Emigration to America   1863

56: L Neu Abt.129A: Gr.11: Nr. 38.   ADS 4:188

FORQUIGNON, Grayson. *See* Forquignon, William

FORQUIGNON, Joseph. *See* Forquignon, William

FORQUIGNON, William

Transmittal of a San Francisco consular report to William Forquignon [in Hamburg] regarding Joseph & Grayson Forquignon [in America?]   1901

39: Reichs- u. auswaert.Ang.: G.Ia.1.18 [Geschaeftsverkehr]   ADS 2:317

FORSCHNER, August Justus (factor)

Expulsion from Hamburg as Social Democrat & emigration to America   1880-1881

39: Polizeibehoerde-Polit.Pol.: S 149/27: Bl.
    483-488.   ADS 2:411

FORSTER, Georg (died of accident on 15 April 1909 in Yakima County, Washington, while working on irrigation)

Estate of   1910
152: Abt.405: Wiesbaden: Rep.405/I,2: Nr. 3460.
                                      ADS 7:143

FORSTER, -- (in Yakima, Washington)

Estate of   1910
39: Reichs- u. auswaert.Ang.: E.IV.9 [Erbschaften]
                                      ADS 2:362

FORTMANN [FORDMANN] John (died in Silver Reef, Utah)

Estate of Fortmann, said to have been from Province of Hannover, & search for heirs [file destroyed]   1886

52: Han.Des.74: Dannenberg: Abt.II.G: Nr. 34.
                                      ADS 5:66

FORTNER, M. Dwight

Hamburg police search for Fortner, wanted in America   1909

39: Reichs- u. auswaert.Ang.: P.III.1.1 [Polizeiwesen]   ADS 2:358

FOSTER, W. R.

Investigation in Hamburg, at request of American police authorities, regarding   1898

39: Reichs- u. auswaert.Ang.: P.III.1.4 [Polizeiwesen]
                                      ADS 2:298

FRAEN(C)KEL, David (from Sarstedt)

Proposed transportation to America of Fraenkel who is destined to enter the prison labor camp at Hameln   1841

52: Han.Des.80: Hildesheim I: E.DD.9a: Nr. 999.
ADS 4:88

FRAEN(C)KEL, David. *See also* Fraen(c)kel, Simon

FRAENKEL, Jacob

Hamburg police search requested by American police authorities for Fraenkel accused of theft   1904

39: Reichs- u. auswaert.Ang.: P.III.1.1 [Polizei-wesen]   ADS 2:336

FRAEN(C)KEL, Seelig. *See* Fraen(c)kel, Simon

FRAEN(C)KEL, Simon

Letter from Amt Ruthe, Hannover, regarding remittance of funds by Simon & Seelig Fraen(c)kel to transport brother David Fraen(c)kel, detained in workhouse at Hameln   1841

39: Bestand Senat: Cl.VIII, No.X, Jg.1841 (Regis-ter)   ADS 2:141

Second letter from Amt Ruthe, Hannover, regarding remittance of funds by Simon & Seelig Fraenckel for their brother David Fraenckel, "a vagabond & criminal," to emigrate to America   1845

39: Bestand Senat: Cl.VIII, No.X, Jg.1845 (Regis-ter)   ADS 2:148

Further letter from Amt Ruthe, Hannover, regarding funds from Simon & Seelig Fraenckel to transport David Fraenckel to America   1846

39: Bestand Senat: Cl.VIII, No.X, Jg.1846 (Regis-ter)   ADS 2:150

FRANĈIĈ, Franz

Transmittal [of extradited prisoner] from the USA to Austria, via Hamburg   1901

39: Reichs- u. auswaert.Ang.: A.Ia.38 [Ausliefer-ungen]   ADS 2:313

FRANCIS, Josef (engineer; in New York)

Metal plates for ship construction   1859
227: Rep.MA 1895: A.V: Nr. 461.   ADS 9:56

FRANCKE, August Hermann

Mentioned in letter from Roxbury [Massachusetts?]
1724

119: Lit.K, Nr. 36.   ADS 6:74

FRANK, Moses Josef (in Chicago)

Estate of   1901

152: Abt.405: Wiesbaden: Rep.405/I,2: Nr. 3461.
ADS 7:143

FRANTZ, Anna Margareta (miller's daughter; from Kar-rach)

Emigration to America   1854
231: Rep.212/15: Rothenburg o.d.T.: Nr. 995.
ADS 9:129

FRANZEN, Anna. *See* Teichwitz, Anna

FREDERICA, Oscar

Philadelphia: Estate of   1913
39: Reichs- u. auswaert.Ang.: E.IV.17 [Erbschaften]
ADS 2:379

FREIBERG, Eduard (in Albany [New York?] prison)

Inquiry by private person in Germany regarding
1904

39: Reichs- u. auswaert.Ang.: N.Ib.1.1 [Nachforsch-ungen]   ADS 2:336

FREISE, Carl (in Marquette, Michigan)

Estate of   1913
39: Reichs- u. auswaert.Ang.: E.IV.5 [Erbschaften]
ADS 2:379

FREITAG, -- (lawyer)

Return to Osterode of the wife of Dr. -- Freitag who went to America   1840

52: Han.Des.80: Hildesheim I: E.DD.43a: Nr. 1668.
ADS 4:90

FRENSKY, Oskar August (from Ratibor [Upper Silesia]; in New York)

Investigation for breach of §171 of *Reichsstraf-gesetz* [criminal law code]   1881

149: Rep.49a: Nr. 566/10.   ADS 7:18

FRENSSEN, Gustav

Letters from America [printed]   1923
24: Drucksachen.   ADS 1:57

FRENZEL, -- (from Kelsterbach, Landkreis Gross-Gerau, HEL)

Applications for emigration permits to North America for -- Frenzel, -- Hirsch, & -- Weber   1845

175: XI,4.   ADS 7:228

FRENZEL, --

Estate of   1944
148: R XI.   ADS 6:261

FRETZ, Friedrich (in San Francisco)

File on [estate?]   1897-1899
227: Rep.MA 1921: A.St.I: Nr. 512.   ADS 9:67

von FREUDEN, Julius Nicolaus (American citizen)

Guardianship for   1888
39: Ritzebuettel I: Gericht u. Recht: Vormund.:
    Nr. 147.   ADS 2:465

FREYDANK, Louis G. A.

New York: Statements regarding death of Freydank
    aboard the *Kalliope*   1880
39: Bestand Senat: Cl.VIII, No.X, Jg.1880 (Regis-
    ter) 46 [Nordamerika]   ADS 2:245

FRICK, H.

St. Louis: Death certificate of   1881
39: Bestand Senat: Cl.VIII, No.X, Jg.1881 (Regis-
    ter) 34 [Nordamerika]   ADS 2:247

FRICK, Karl

Investigation in Hamburg at the request of Amer-
    ican police authorities, regarding   1898
39: Reichs- u. auswaert.Ang.: P.III.1.1 [Polizei-
    wesen]   ADS 2:298

FRICKE, Heinrich August (imprisoned in Pittsburgh,
Pennsylvania)

Fricke's former financial condition [in Bockenem]
                                           1865
52: Han.Des.80: Hildesheim I: E.DD.2a: Nr. 672.
                                        ADS 4:87

FRICKE, Matthias (itinerant laborer; from Astfeld)

Emigration to America   1854-1858
56: L Neu Abt.129A: Gr.11: Nr. 22.   ADS 4:187

FRIED, W. (died in Greensburgh, Westmoreland County,
Pennsylvania)

Receipt of estate of   1824
200: Abt.441: Regierung Koblenz: Nr. 4652.
                                        ADS 8:32

FRIEDE, --

Lodging of a legal action in Wilmington, North

Carolina, in case of Friede *versus* Boesch   1905
39: Reichs- u. auswaert.Ang.: J.Ib.57 [Justiz-
    sachen]   ADS 2:340

FRIEDEL, C. (in Washington [D.C.?])

Interrogation of Friedel & the woman pastor* --
    Beck in Cleveland [Ohio?] in the case of Eden-
    feld *versus* Freiherr von Seckendorf's Relict
                                       1850-1852
    *Could refer to the wife of a Pastor -- Beck.
227: Rep.MA 1898: A.St.: Nr. 6283.   ADS 9:59

FRIEDLAENDER, Daniel Julius

New Orleans: Death certificate for   1879
39: Bestand Senat: Cl.VIII, No.X, Jg.1879 (Regis-
    ter) 59 [Nordamerika]   ADS 2:242

New Orleans: Death certificate for   1880
39: Bestand Senat: Cl.VIII, No.X, Jg.1880 (Regis-
    ter) 2 [Nordamerika]   ADS 2:243

FRIEDLAENDER, Demetrix [Demetrius?] Petrowich

Hamburg police search, at request of American pol-
    ice authorities, for   1900
39: Reichs- u. auswaert.Ang.: P.III.1.1 [Polizei-
    wesen]   ADS 2:310

FRIEDMANN, Henry (in New York)

Determination of facts regarding   1898
227: Rep.MA 1921: A.St.I: Nr. 1476.   ADS 9:73

FRIEDRICH, Adam

Complaint of emigrant Friedrich against Hamburg
    firm   1896
39: Reichs- u. auswaert.Ang.: A.III.3 [alt.Sig.]
    [Auswanderung]   ADS 2:280

FRIEDRICH, Johannes (from Goddelau, Landkreis Gross-
Gerau, HEL)

Emigrant to America   1821
173: Ortsbuergerregister 1777-1911.   ADS 7:224

FRIEDRICH, Magdalena. *See* Graffert, Magdalena, born
Friedrich

FRIEDRICH, Rosette

New York: Estate of   1896
39: Reichs- u. auswaert.Ang.: E.IV.10 [alt.Sig.]
    [Erbschaften]   ADS 2:282

FRIEDRICHS, Johann Georg (pastor in South Carolina)

Letter to Ministry [of Religion]   1755

<u>39</u>: Geistl.Minis.: III,A 1 t: Gebundene Akten; Nr. XXI, p.55; *see also* III,B, Fasc.1: Lose Akten 1760 [two letters]   ADS 2:478

FRIES, Anna, born Mundi

Repatriation from St. Louis [Missouri] to Hamburg
1907
<u>39</u>: Reichs- u. auswaert.Ang.: H.IIb.3 [Heimschaff-ungsverkehr]   ADS 2:348

FRIES, Wilhelm Heinrich (died in America)

Estate of   1901
<u>152</u>: Abt.405: Wiesbaden: Rep.405/I,2: Nr. 3462.
ADS 7:143

FRISCH, P. R.

New York: Estate of   1900
<u>39</u>: Reichs- u. auswaert.Ang.: E.IV.40 [Erbschaften]
ADS 2:307

FRISCHKORN, Andreas (died in San Francisco)

Estate of   1910
<u>152</u>: Abt.405: Wiesbaden: Rep.405/I,2: Nr. 3463.
ADS 7:143

FRITZ, Heinrich (of Berkach, Landkreis Gross-Gerau, HEL)

Application for emigration permit for Fritz & his family to go to America to seek a better life
1851-1852
Also shown in an emigrant list with wife & four children   1852
<u>163</u>: Abt. XI/4.   ADS 7:213

FRITZ, Josef (died 1899 in San Francisco)

File on [estate?]   1900
<u>227</u>: Rep.MA 1921: A.St.I: Nr. 522.   ADS 9:67

FRITZ, Wilhelm Heinrich (died in Cleveland, Ohio)

Information regarding estate of   1895
<u>152</u>: Abt.405: Wiesbaden: Rep.405/I,2: Nr. 3546.
ADS 7:143

FRITZ, -- (in Philadelphia area)

Payment of [Wuerttemberg] military pension to the former *Ulan* [lancer] Fritz, through the consulate in Philadelphia   1903-1909
<u>263</u>: Rep.E. 46-48: Minist. d. AA III: Verz.1: Nr. 147/11.   ADS 10:43

FRITZKE, Herm[ann]

Boston: Estate of   1905
<u>39</u>: Reichs- u. auswaert.Ang.: E.IV.11 [Erbschaften]
ADS 2:339

FROEBEL, Julius (in USA)

Correspondence   1850
<u>150</u>: Froebel.   ADS 7:27

FROH, J. Ch. C. (in Rolla, Rolette County, North Dakota)

Inquiry by Hamburg Orphanage Commission regarding
1900
<u>39</u>: Reichs- u. auswaert.Ang.: W.I.6 [Waisenhaus-Ang.]   ADS 2:312

FUCHS, William N. (an American citizen)

Estate of   1916
<u>39</u>: Reichs- u. auswaert.Ang.: E.IV.31 [Erbschaften]
ADS 2:391

FUEGMANN, --

Estate of   1942
<u>148</u>: R XI.   ADS 6:261

FUELLGRABE, Carl Heinrich (forest laborer; from Lonauer Hammerhuette)

Emigration to North America   1853
<u>68</u>: VI.2.b: Fach 162, Nr. 37.   ADS 4:247

FUES family. *See* Bergisch Gladbach

FUHRHOPE, G. H. (in Harriesburg,* Pennsylvania)

Estate of   1876
*So spelled.
<u>39</u>: Bestand Senat: Cl.VIII, No.X, Jg.1876 (Regis-ter) 9 [Nordamerika]   ADS 2:232

FULLERTON, George Stuart (Ph.D.; university professor in New York)

Application for acceptance or recognition [of academic title]; disapproved   1908-1915
<u>226</u>: Rep.MK 3: Nr. 11344.   ADS 9:36

FUNK, H., born Nachtigal (widow; in California)

Estate of   1884
<u>39</u>: Bestand Senat: Cl.VIII, No.X, Jg.1884 (Regis-ter) 25 [Erbschafts-Amt]   ADS 2:252

FUNK, H.

San Francisco: Estate of   1887

39: Bestand Senat: Cl.VIII, No.X, Jg. 1887 (Register) 10 [Erbschafts-Amt]   ADS 2:258

GACK, F. (American citizen; died at Hamburg)

Estate of   1894

39: Reichs- u. auswaert.Ang.: E.IV.89 [alt.Sig.]
[Erbschaften]   ADS 2:269

GAEBLER, C. F.

Letter from Bretwig bei Dresden regarding imprisonment & extradition for fraud of Gaebler who had
emigrated to America   1839

39: Bestand Senat: Cl.VIII, No.X, Jg.1839 (Register)   ADS 2:138

GAERTNER, Elisabetha. *See* Gaertner, Johann

GAERTNER, Johann (master miller; from Einberg)

Issuance of passport to him & his daughter Elisabetha for purpose of emigration to America   1839

224: Landesregierungs-Archiv: Tit.III.c.8.40.
                                            ADS 9:10

GAETH, -- (in New Orleans)

File on missing person   n.d.

30: IV.B: Nr. 136.   ADS 1:62

GAHL, --

Estate of   1943

148: R XI.   ADS 6:261

GAMWEG, Hermann (from Wregen, HEL)

Emigration of   n.d.

151: Bestand 180: Arolsen: Acc.1937/76,5: Nr. 393.
                                            ADS 7:113

GARBERS, Hugo (in Oregon)

Estate of   1914

39: Reichs- u. auswaert.Ang.: E.IV.64 [Erbschaften]
                                            ADS 2:384

GARBIE, --

File on the cost of transporting convict Garbie
"aus dem Grimm" to America   1846

96: B.5.Nr. 14: Archiv des Klosters St. Michaelis.
                                            ADS 4:289

GARDTHAUSEN, Johann Friedrich Heinrich (basketmaker)

Expulsion from Hamburg as Social Democrat & emigration to America   1881; 1897

39: Polizeibehoerde-Polit.Pol.: S 149/361: Bl. 1-6.
                                            ADS 2:417

GAREIS, --

Estate of   1941

148: R XI.   ADS 6:261

GARRET, Henriette

Estate of   1933-1944

148: R XI: 6 vols.   ADS 6:261

GARRET, Oesterreich? [Austria]

Estate of   1933-1944?

148: R XI.   ADS 6:261

GARVE, Carl Heinrich Christian (tobacco worker)

Expulsion from Hamburg as Social Democrat & emigration to America   1880-1890

39: Polizeibehoerde-Polit.Pol.: S 149/23: Bl. 1-60.
                                            ADS 2:411

GAS, -- (ship's engineer; died in prisoner of war
camp in Hot Springs, North Carolina)

Estate of   1917

39: Reichs- u. auswaert.Ang.: E.IV.49 [Erbschaften]
                                            ADS 2:392

GASPARY, --

Fee due New York consulate in case of Gaspary *versus* Kallmann   1902

39: Reichs- u. auswaert.Ang.: J.Ib.10.5 [Justiz-
sachen]   ADS 2:325

GASS, --. *See* Gas, --

GASSLER, Thomas (American citizen)

Extradition to Austria from Hamburg   1913

39: Reichs- u. auswaert.Ang.: A.Ia.1.18 [Ausliefer-
ungen]   ADS 2:377

GATERMANN, J. G. (in Springfield, Illinois)

Estate of   1905

39: Reichs- u. auswaert.Ang.: E.IV.31 [Erbschaften]
                                            ADS 2:339

GAULLIER, Henry (American citizen)

Complaint of   1897
<u>227</u>: Rep.MA 1921: A.V. V: Nr. 15775.   ADS 9:90

GAUSMANN, --. *See* Walter, --

GEBHARDT, Mathilde

New York: Estate of   1905
<u>39</u>: Reichs- u. auswaert.Ang.: E.IV.36 [Erbschaften]
                           ADS 2:339

GEERTZ, Johann Hermann (Hemme, SHL)

Petition for release from citizenship   1858
<u>13</u>: Alphabetischen Sammlung.   ADS 1:49

GEERTZ, Th. (widow; in Hamburg)

Transmittal of a report & enclosures to Mrs.
  Geertz from German consulate in New York   1901
<u>39</u>: Reichs- u. auswaert.Ang.: G.Ia.1.9 [Geschaefts-
  verkehr]   ADS 2:316

GEGGLE, -- (from Dettlingen, BWL)

Letters from an emigrant   1765
<u>262</u>: Ho 201: Glatt: I 4249.   ADS 10:37

GEHRCKE, -- (brewer; from Stargard)

Inquiry regarding son in America   1799-1800
<u>500</u>: Berlin-Dahlem: Preus. Geh. Staatsarchiv: Rep.
  XI, 21a, Conv. 2,8.   Not in ADS

GEISBERG, Bernhard (in Westphalia Settlement, Missouri). *See* Bruns, Bernhard; Bruns, Henriette

GEISBERG, Franz (in Westphalia Settlement, Missouri).
See Bruns, Bernhard; Bruns, Henriette

GEISBERG, Henriette. *See* Bruns, Henriette

GEISS, Franziska. *See* Perl, Franziska, born Geiss

GEORG, Heinrich (in Texas)

Information sought by private person regarding
                             1897
<u>39</u>: Reichs- u. auswaert.Ang.: N.Ib.1.11 [alt.Sig.]
  [Nachforschungen]   ADS 2:291

GEORGE, A. Frieda Eva (in Hamburg)

Claims for support against P. A. Welch in Indiana
  to be transmitted by German consulate in

Cincinnati   1895
<u>39</u>: Reichs- u. auswaert.Ang.: J.Id.4 [alt.Sig.]
  [Justizsachen]   ADS 2:276

GERETSHUBER, Xaver (American citizen)

File on [estate?]   1893-1894
<u>227</u>: Rep.MA 1921: A.St.I: Nr. 541.   ADS 9:67

GERHARD, August Albert Friedrich (cigar worker)

Expulsion from Hamburg as Social Democrat & emigration to America   1880-1881
<u>39</u>: Polizeibehoerde-Polit.Pol.: S 149/248: Bl. 7-
  12.   ADS 2:415

GERHARD, E. (painter; in Newark, New Jersey)

Hamburg Orphanage Commission seeks information regarding   1897
<u>39</u>: Reichs- u. auswaert.Ang.: W.I.4 [alt.Sig.]
  [Waisenhaus-Ang.]   ADS 2:293

GERHARDT, Philipp V? (from Gross-Gerau, HEL)

Emigration to North America   1852
<u>174</u>: Abt.XI,4.   ADS 7:226

GERISCH, Paul (in Wisconsin)

Estate of   1909
<u>39</u>: Reichs- u. auswaert.Ang.: E.IV.5 [Erbschaften]
                           ADS 2:356

GERKEN, Claus H. (in Davenport, Iowa)

Estate of   1907
<u>39</u>: Reichs- u. auswaert.Ang.: E.IV.62 [Erbschaften]
                           ADS 2:347

GERRAS, --. *See* Damrosch, --, born Gerras

GERSTMAIER, -- (child; in New York)

Property of   1842
<u>227</u>: Rep.MA 1898: A.St.: Nr. 6288.   ADS 9:59

GESENER, --

Estate of   1941-1942
<u>148</u>: R XI.   ADS 6:261

GETZMANN, --. *See* Warcham, E? born Getzmann

GEVEKOTH, H. (doctor; died in Brooklyn, New York)

File on [estate?]   1897
<u>227</u>: Rep.MA 1921: A.St.I: Nr. 561.   ADS 9:67

GIESE, --

Request of Hamburg court for sworn testimony by witness in Talmage, Nebraska, to be obtained by Chicago consulate   1901

<u>39</u>: Reichs- u. auswaert.Ang.: J.Ib.7 [Justizwesen]
                                            ADS 2:318

GIESEMANN, Clemens (of Ruedigheim, HEL)

Emigration [to Carolinas?]   1741

<u>151</u>: Bestand 80: Lit.B: Nr. 1. [Unterthanen]
                                            ADS 7:79

GIESSLER, Marie, born Schur

Report of German consulate in New York regarding
                                            1901

<u>39</u>: Reichs- u. auswaert.Ang.: G.Ia.25 [Geschaefts-
verkehr]   ADS 2:317

GILLER. *See* Pope & Giller Brewery

GLAEFKE, Andreas (in New York)

Transfer of his property to the control of his aunt, Margaretha Rauert of Burg [auf Fehmarn]
                                            1834

<u>5</u>: Rubr.XIV: Nr. 74 [Uebertragung]   ADS 1:43

GLAESER, A. H. (American citizen)

A. H. Glaeser *versus* L. Glaeser (Hamburg citizen)
                                            1840

<u>39</u>: Bestand Senat: Cl.VII, Lit.J^b, No.12, vol.18, gg.5 [Forderung]   ADS 2:74

GLAESER, Jacob Heinrich

Request by U.S. consul for Senat [Hamburg] intervention in matter of one Glaeser (born in Hamburg but living in America) against J. H. Glaeser
                                            1840

<u>39</u>: Bestand Senat: Cl.VIII, No.X, Jg.1840 (Register) I.1 [Nordamerika]   ADS 2:139

GLAESSL, Johann Michael

Investigation of Hamburg emigration agency which apparently failed to honor its transportation contract with Glaessl & Anna Zeitler who arrived late in Hamburg; financial aid to the emigrants   1857-1858

<u>227</u>: Rep.MA 1921: A.V. IV: Nr. 11799.   ADS 9:67

GLASER, Karol (from Neustadt a. H.; died in Chicago)

File on [estate?]   1894-1899

<u>227</u>: Rep.MA 1921: A.St.I: Nr. 542.   ADS 9:67

GLASER, -- (German citizen)

Extradition from the USA   1910

<u>39</u>: Reichs- u. auswaert.Ang.: A.Ia.11 [Auslieferungen]   ADS 2:361

GLIESSMANN, Claus (Shoemaker)

Expulsion from Hamburg as Social Democrat & emigration to America   1880-1881

<u>39</u>: Polizeibehoerde-Polit.Pol.: S 149/256: Bl. 15-19.   ADS 2:415

GLITTKOWSKI, --

Estate of   1941?

<u>148</u>: R XI.   ADS 6:261

GLOCK, Edm[und?] (died 1899 in New York)

File on [estate?]   1900

<u>227</u>: Rep.MA 1921: A.St.I: Nr. 577.   ADS 9:68

GLOOTZ, Meta (in San Mateo County [California?])

Estate of   1907

<u>39</u>: Reichs- u. auswaert.Ang.: E.IV.35 [Erbschaften]   ADS 2:347

GLOYSSTEIN, Georg

Emigration file contains letter from friend or relative in New York   1860

<u>55</u>: Han.Des.74: Blumenthal: IX.J.1: Fach 301, Nr. 10.   ADS 4:154

GLOYSSTEIN, Heinrich. *See* Gloysstein, Georg

GLUECK, Anna (of Brooklyn [New York])

File on [estate?]   1896

<u>227</u>: Rep.MA 1921: A.St.I: Nr. 557.   ADS 9:67

GOEBEL, A. (in America). *See* Bedassem, J.

GOEBEL, Heinrich (born 20 April 1818 in Springe; died 16 December 1893 in New York; inventor of electric light bulb)

Information regarding   n.d.

<u>107</u>   ADS 4:333

GOEBEL, Johann Jacob Christian (machine builder)

Expulsion from Hamburg as Social Democrat & emigration to America   1881

<u>39</u>: Polizeibehoerde-Polit.Pol.: S 149/346: Bl. 206-233.   ADS 2:417

GOEDE, --

Chicago: Estate of Goede brothers 1894

39: Reichs- u. auswaert.Ang.: E.IV.88 [alt.Sig.] [Erbschaften] ADS 2:269

GOEGG, -- (German revolutionary; in America). *See* German Political Refugees [subject index]

GOELLER. Coelestin. *See* Goeller, Kilian

GOELLER, Kilian (of Hofaschenbach, HEL)

The brothers Kilian & Coelestin Goeller accused of enticing citizens to emigrate 1805

151: Bestand 98: Fulda X: Acc.1875/27: Nr. 46. ADS 7:91

GOERDELER, R. (an American citizen)

Release from the insane asylum in Schwetz 1894

39: Reichs- u. auswaert.Ang.: M.Ib.3 [alt.Sig.] [Medizinalsachen] ADS 2:270

GOERKE, Joh[anna] Chatharina Maximilliane, formerly Schellhorn. *See* Schellhorn, D. C.

GOHR, A., born Kaehler

Repatriation of Mrs. Gohr from America to Hamburg (not carried out) 1901

39: Reichs- u. auswaert.Ang.: H.IIb.9 [Heimschaffungsverkehr] ADS 2:318

GOLDSCHMIDT, Leopold (died 1899 in New York)

File on [estate?] 1899

227: Rep.MA 1921: A.St.I: Nr. 572. ADS 9:68

GOLDSCHMIDT, Loeb (from Trebur, Landkreis Gross-Gerau, HEL)

Emigration of Jewish person who had been registered in the community in 1830 n.d.

183: Abt.XI,4. ADS 7:237

GOLDSTEIN, Lippa (an American citizen; died at Hamburg)

Estate of 1895

39: Reichs- u. auswaert.Ang.: E.IV.49 [alt.Sig.] [Erbschaften] ADS 2:275

GOLDSTEIN, Samuel (died in San Francisco)

File on [estate?] 1898-1899

227: Rep.MA 1921: A.St.I: Nr. 566. ADS 9:68

GOLLER, --

Inheritance of millions [of dollars?] in America 1912

227: Rep.MA 1936: A.St.: Ma 98073-98074. ADS 9:98

GOLTZHEIM, Moritz. *See* Stach von Goltzheim, Moritz

GOODMAN, Mayer Wolf (died 1895 in Chicago)

File on [estate?] 1898

227: Rep.MA 1921: A.St.I: Nr. 564. ADS 9:68

GORISSEN, Friedrich Vivian Guenther (born in New Orleans)

Request for information regarding Gorissen's domicile, due to petition of American minister seeking to free him from German military service 1873

39: Bestand Senat: Cl.I, Lit.T, No.14, vol.4, Fasc. 5b, Invol.10 [Auskunftersuchen] ADS 2:5

GORISSEN, Friedrich Vivian Guenther. *See also* Gorrissen, V. F. G.

GORDON, Andrew (died in Muskegon)

File on [estate?] 1896-1902

227: Rep.MA 1921: A.St.I: Nr. 555. ADS 9:67

GORMLEY, John

Hamburg police search for Gormley, wanted in New York 1908

39: Reichs- u. auswaert.Ang.: P.III.1.1 [Polizeiwesen] ADS 2:353

GORRISSEN, V. F. G.

Exemption of Gorrissen from military service obligation 1873

39: Bestand Senat: Cl.VIII, No.X, Jg.1873 (Journal) 4 [Nordamerika] ADS 2:224

GORRISSEN, V. F. G. *See also* Gorissen, Friedrich Vivian Guenther

GOSSLER, J. H. *See* Midgeon, --, widow of Gossler

GOTTHARDT, -- (linen weaver; from Muenden)

Emigration to America 1840

52: Han.Des.80: Hildesheim I: E.DD.41a: Nr. 1562. ADS 4:90

GOTTLIEB, A.

New York: Death certificate for   1879

39: Bestand Senat: Cl.VIII, No.X, Jg.1879 (Register) 18 [Nordamerika]   ADS 2:240

GOTTRON, Johann (owner of restaurant "Zum Loewen";
from Raunheim, Landkreis Gross-Gerau, HEL)

Emigration to America with wife & 4 children 1854

181: Raunheimer Evangelischen Pfarrei Familienregister (Lehrer Johannes Buxbaum) ADS 7:234

GOTTRON, Philipp (owner of lime kiln & brick oven;
from Raunheim, Landkreis Gross-Gerau, HEL)

Emigration to America with wife & 9 children (brother of Johann Gottron)   1854

181: Raunheimer Evangelischen Pfarrei Familienregister (Lehrer Johannes Buxbaum) ADS 7:234

GRABBE, Ludolph (in Davenport, Iowa)

Estate of   1910

39: Reichs- u. auswaert.Ang.: D.Ih,35 [Erbschaften]
ADS 2:362

GRABSCH, Wilhelm (in California)

Estate of   1909

39: Reichs- u. auswaert.Ang.: E.IV.77 [Erbschaften]
ADS 2:357

GRADER, --

Collection of fees due German consulate in New
York from [lawyers] Gottschalck & Goldenberg
in the divorce proceedings of Grader *versus*
Grader   1908

39: Reichs- u. auswaert.Ang.: J.Id.18 [Justiz-Ang.]
ADS 2:352

von GRAETZ, Henriette Graetzel. *See* Graetzel, Henriette

GRAETZEL VON GRAETZ, Henriette (of Goettingen)

Papers regarding her inheritance in Charleston
1794

52: Han.Des.92: London Kanz.: LXXXV, Nr. G,8.
ADS 4:95

GRAF, John (in Iowa)

Estate of   1899

39: Reichs- u. auswaert.Ang.: E.IV.18 [alt.Sig.]
[Erbschaften]   ADS 2:301

GRAFFERT, Georg (died 1886 in Belleville, Illinois)

Estate of   1901-1902

152: Abt.405: Wiesbaden: Rep.405?I,2: Nr. 3465.
ADS 7:143

GRAFFERT, Magdalena, born Friedrich (died February
1898 in Belleville? Illinois)

Estate of   1901-1902

152: Abt.405: Wiesbaden: Rep.405/I,2: Nr. 3465.
ADS 7:143

GRAH, Max (journalist; in Los Angeles)

Correspondence   n.d.

118: Unregistered: Zimmer V, Aktenraum, Schrank 5.
ADS 6:73

GRANDT, Johann Friedrich (shoemaker)

Expulsion from Hamburg as Social Democrat & emigration to America   1883

39: Polizeibehoerde-Polit.Pol.: S 149/438: Bl. 26-
29.   ADS 2:418

GRASSEL, J. B. (died? 1897 in Buffalo [New York?])

File on [estate?]   1900-1901

227: Rep.MA 1921: A.St.I: Nr. 574.   ADS 9:68

GRATWOHL, --

Estate of   1944

148: R XI.   ADS 6:261

GRAY, Augustus (in Jacksonburg, Butler? County, Ohio)

Report to Grand Duke Ludwig II   1848

149: Abt.Hess.Hausarchiv: Konv. 662.   ADS 7:21

GREEF, Emil (manufacturer in New York)

Residency in Norden [Germany]   1864

50: Rep.21a: Nr. 8115.   ADS 4:7

GREENWOLDT, D. *See* Gruenwoldt, D.

GREHL, H. Th. (woman; in New York)

Inquiry by private person in Germany regarding
1900

39: Reichs- u. auswaert.Ang.: N.Ib.1.8 [Nachforsch-
ungen]   ADS 2:310

GREIF, J. Chr. (married couple; living in Sommers-
dorf, BYL)

Inquiry regarding claim of inheritance from the
North American states   1851

231: Rep.270/II: Reg.Mittelfranken: Kammer des
Innern: Abgabe 1932: Nr. 633.   ADS 9:134

GREIFENBERG, Carl (typesetter)

Expulsion from Hamburg as Social Democrat & emigration to America   1880-1881

39: Polizeibehoerde-Polit.Pol.: S 14/89: Bl. 297-
313.   ADS 2:413

GREIL, J.

New York: Estate of   1895

39: Reichs- u. auswaert.Ang.: E.IV.58 [alt.Sig.]
[Erbschaften]   ADS 2:275

GREINER, Peter (butcher; from Bischofsgruen; died in
South Carolina)

File on [estate?]   1822-1825

227: Rep. Bayerische Gesandtschaft London: Nr. 866.
ADS 9:47

GRENTZSCH, --

Inheritance of   1882

39: Bestand Senat: Cl.VIII, No.X, Jg.1882 (Register) 5 [Nordamerika]   ADS 2:248

GREUEL, --. See Brown, --, born Greuel

GREULICH, Josef (from Dronheim, Landkreis Gross-
Gerau, HEL)

Emigration to America   1922

168: Abmelderegister.   ADS 7:218

GROCH, W. (in America). See Bedassem, J.

GRUELING, Jakob (from Gross-Gerau, HEL)

Emigration of Jakob Grueling & Adam Kolb to North
America & the lease of land by the community
1846

174: Abt.XI,4.   ADS 7:226

GREUNE, Friedrich (farmhand; from Klein Rhueden).
See Cleve, Johann Heinrich

GREUNE, Wilhelm (from Klein Rhueden). See Cleve,
Johann Heinrich

GRIESE, Heinrich Detlef (cigar worker)

Expulsion from Hamburg as Social Democrat & emigration to America   1881

39: Polizeibehoerde-Polit.Pol.: S 149/338: Bl. 339-
347.   ADS 2:416

GRIESEDIECK, --

St. Louis: Estate of   1901

39: Reichs- u. auswaert.Ang.: E.IV.23 [Erbschaften]
ADS 2:315

GRIESGE, P. (from Westen)

Transportation to America   1850

52: Han.Des.74: Westen-Thedinghusen: I.56: Nr. 45.
ADS 4:80

GRILL, H. H. F. (an American citizen)

Attempt by U.S. Consul to have Grill declared no
longer subject to military service. Request
denied because Grill had not renounced Hamburg
citizenship   1852

39: Bestand Senat: Cl.VIII, No.X, Jg.1852 (Register) I.1 [Nordamerika]   ADS 2:161

GROENING, --

Estate of   1941

148: R XI.   ADS 6:261

GROENWOLD(T), Daniel

San Francisco: Estate of   1877

39: Bestand Senat: Cl.VIII, No.X, Jg.1877 (Register) 51 [Nordamerika]   ADS 2:235

GROLL, Adam (born 5 September 1864 in New York)

Negotiations regarding military service obligation
of former American citizen   1888-1889

152: Abt.415: Ruedesheim: Nr. 38.   ADS 7:156

GROMANN, -- (from Bad Salzuflen)

Emigrant to America sent money in 1869 to erect a
large tombstone in the Salzufler Alleefriedhof
[Salzuflen Allee Cemetery] for his father, Franz
Gromann, a former laborer in the saltwork & who
had lived in the Schiesshofstrasse   1869

104   ADS 4:331

GRONAU, -- (pastor; in Ebenezer, Georgia). See
Boltzium, --

GRONEWALDT, --

German consulate in New York in the case of Gronewaldt *versus* Wallace   1898

<u>39</u>: Reichs- u. auswaert.Ang.: J.Ib.10.6 [alt.Sig.]
[Justizsachen]   ADS 2:297

GROSS, Adolph (in Ohio)

Correspondence   1860-1861

<u>2</u>: Nachlass Knud Jungbohn-Clement.   ADS 1:39

GROSS, Franz (died 1899 in Buffalo [New York?])

File on [estate?]   1899-1902

<u>227</u>: Rep.MA 1921: A.St.I: Nr. 569.   ADS 9:68

GROSS? Jakob Joachim. *See* Joachim, Jakob

GROSS, Johann Carl Juergen (restaurant owner)

Expulsion from Hamburg as Social Democrat & emigration to America   1880-1885

<u>39</u>: Polizeibehoerde-Polit.Pol.: S 149/327: Bl. 404-
463.   ADS 2:416

GROSS, Johann (in California)

Estate of   1908

<u>39</u>: Reichs- u. auswaert.Ang.: E.IV.97 [Erbschaften]
                          ADS 2:352

GROSS, --

Estate of   1941
<u>148</u>: R XI.   ADS 6:261

GROTH, Claus

Trip to America (printed in Low German dialect)
                          n.d.

<u>24</u>: Drucksachen.   ADS 1:57

GROTKE (GROTHE) H. (from Hagenow)

Estate of   1872

<u>39</u>: Bestand Senat: Cl.VIII, No.X, Jg.1872 (Regis-
ter) 11 [Nordamerika]   ADS 2:220

GROTTER, W. (died in New York)

File on [estate?]   1894

<u>227</u>: Rep.MA 1921: A.St.I: Nr. 545.   ADS 9:67

GRUEBEL, --. *See* Schmidt, --, born Gruebel

GRUENBERG, M.

Request of the Hamburg court to have consulate in
New York seize funds & negotiable instruments in
possession of Gruenberg, a tailor, who fled to
America   1900

<u>39</u>: Reichs- u. auswaert.Ang.: J.Ib.3 [Justizwesen]
                          ADS 2:309

GRUENDER, Johann Jakob (died of self-inflicted accident on 5 August 1907 at Pittsburgh on Pennsylvania Railroad)

Estate of   1909

<u>152</u>: Abt.405: Wiesbaden: Rep.405/I,2: Nr. 3467.
                          ADS 7:143

GRUENDERT, Karl

Extradition of Gruendert, a prisoner, from the USA
to Saxony-Altenburg, via Hamburg   1902

<u>39</u>: Reichs- u. auswaert.Ang.: A.Ia.18 [Ausliefer-
ungen]   ADS 2:322

GRUENEBERG, Max

New York: Information regarding   1879

<u>39</u>: Bestand Senat: Cl.VIII, No.X, Jg.1879 (Regis-
ter) 56 [Nordamerika]   ADS 2:242

GRUENEISEL, Josef (died in Morristown [state not
given])

File on [estate?]   1900-1901

<u>227</u>: Rep.MA 1921: A.St.I: Nr. 579.   ADS 9:68

GRUENWALDT, Daniel

Report on heirs of Gruenwaldt, *alias* Greenwold, in
San Francisco & Hamburg   1876

<u>39</u>: Bestand Senat: Cl.VIII, No.X, Jg.1876 (Regis-
ter) 1 [Nordamerika]   ADS 2:232

Estate of   1875

<u>39</u>: Bestand Senat: Cl.VIII, No.X, Jg.1875 (Regis-
ter) 10 [Nordamerika]   ADS 2:230

GRUND, --. *See* Jaentsch, --

GRUNDMANN, Friedrich (American citizen)

Complaint of arrest   1898

<u>227</u>: Rep.MA 1921: A.V.V: Nr. 15786.   ADS 9:90

GRUNDNER, Karl (apprentice cigarmaker; from Bodenstein)

Convicted for misappropriation, theft, fraud, &
counterfeiting; emigration to America   1855-1864

<u>56</u>: L Neu Abt.129A: Gr.11: Nr.29.   ADS 4:187

GRUPEN, Auguste (in Charleston, South Carolina)

Transmittal of a will [inheritance?] to Grupen, unmarried   1862

52: Calenberger Brief-Archiv Des.15: Privatakten Band I: G, Nr. 241.   ADS 4:55

GUEMPEL, Justine. *See* Marquis, Justine, born Guempel

GUENDEL, -- (died 1876 in America)

File on [estate?]   1894

227: Rep.MA 1921: A.St.I: Nr. 546.   ADS 9:67

GUENTHER, Barbara (widow; from Hetzlos)

Financial aid for her transportation to New York   1886

227: Rep.MA 1921: A.V.II: Nr. 3953.   ADS 9:81

GUENTHER, E. E. (from Bledeln)

Approval of funds for Guenther's emigration to America   1867

52: Han.Des.80: Hildesheim I: E.DD.26a: Nr. 1138.   ADS 4:89

GUENTHER, H. (from Sorsum)

Approval of funds for transportation of convict Guenther to America   1866

52: Han.Des.80: Hildesheim I: E.DD.26a: Nr.1135.   ADS 4:88

GUENTHER, Julius. *See* Hinze, Christoph

GUER, D. (in New Orleans)

Letter to C. A. Offensandt   1832

48: Bestand 7,76: C. A. Offensandt: Nr. 56.   ADS 3:75

GUETH, Johann Friedrich (Pioneer [military rating]; from Geinsheim, Landkreis Gross-Gerau, HEL)

Application for release from military [unit?] for purpose of emigration to America   1850

170: Abt.XI,4.   ADS 7:221

von GUMPPENBERG-PEUERBACH, Julius *Freiherr* [baron] (died 1928 in America)

File on [estate?]   [20th century]

252: Nr.331.   ADS 9:170

GUNDLACH, H. C.

St. Louis: Estate of   1890

39: Bestand Senat: Cl.VIII, No.X, Jg.1890 (Register) 3 [Erbschafts-Ang.]   ADS 2:264

GUNDLACH, Johann

Transmittal of money to Gundlach by German consulate in New York   1909

39: Reichs- u. auswaert.Ang.: G.Ia.1.3 [Geschaeftsverkehr]   ADS 2:357

GUNTZEL, F. W.

Letter from Sachsen-Coburg-Gotha probate[?] court in Wiesenfeld bei Coburg regarding Guntzel, a prisoner in Hamburg; court to pay costs of transporting him to America upon completion of sentence   1841

39: Bestand Senat: Cl.VIII, No.X, Jg.1841 (Register)   ADS 2:142

GUTH, Susanne (from Niederwoellstadt; died in Philadelphia)

Administration of estate   1869-1870

149: Rep.49a: Nr.566/7 [Curatel]   ADS 7:18

GUTHMANN, Emanuel (widow of; in Gernsheim)

Application of Joseph Seligmann in Chicago regarding widow's will   1898

149: Rep.49a: Nr. 565/14.   ADS 7:17

GUTMANN, -- (in New York)

Interrogation of Gutmann in the case of I. Hiller of Frankfurt *versus* A. Ott of Nuernberg   1852

227: Rep.MA 1898: A.St.: Nr. 6294.   ADS 9:59

HAABEN, --

Investigation of alleged estate in America   1894

152: Abt.405: Wiesbaden: Rep.405/I,2: Nr. 3548.   ADS 7:144

HAAGE, Barbara (in America)

Letter from   1867

218: Nr. 132 (Umschlag)   ADS 8:88

HAAGE, Franz (in America)

Letter from   1865

218: Nr. 132 (Umschlag)   ADS 8:88

HAAK, Harm Friedrich (from Wittmund)

Emigration of convict to America   1856-1859
<u>50</u>: Rep.21a: Nr. 8946.   ADS 4:8

HAAK, Peter (in Davenport [Iowa?])

Estate of   1911
<u>39</u>: Reichs- u. auswaert.Ang.: E.IV.59 [Erbschaften]
                                            ADS 2:369

HAAS, Christian (tailor)

Expulsion from Hamburg as Social Democrat & emi-
    gration to America   1880-1881
<u>39</u>: Polizeibehoerde-Polit.Pol.: S 149/106: Bl. 168-
    173.   ADS 2:414

HAAS, Friedrich

New Orleans: Estate of   1899
<u>39</u>: Reichs- u. auswaert.Ang.: E.IV.81 [alt.Sig.]
    [Erbschaften]   ADS 2:301

HAAS, Friedrich Augustin (shoemaker; from Zwei-
    bruecken; in Philadelphia)

Estate of   1831
<u>227</u>: Rep. Bayer. & Pfaelz. Gesandtschaft Paris:
    Nr. 8450.   ADS 9:50

HAAS, -- (from Muelheim)

Care of Haas children & their transportation to
    America   1886
<u>227</u>: Rep.MA 1921: A.V.II: Nr. 3950.   ADS 9:81

HAASE, -- (widow of -- Medefind, a wheelwright; from
    Hannover)

Estate of widow -- Medefind, who later married
    Haase & emigrated to America   1828
<u>52</u>: Han.Des.92: London Kanzlei LXXXV: Nr. M, 7.
                                            ADS 4:95

HABENSTEIN, Manfred (from Crumstaedt, Landkreis
    Gross-Gerau, HEL)

Emigration to New York of Jewish person with wife
    & child   1938
<u>167</u>: Abmelderegister 1938-1957.   ADS 7:217

HABERKORN, Franz (in Cleveland [Ohio])

File on [estate?]   1894-1901
<u>227</u>: Rep.MA 1921: A.St.I: Nr. 612.   ADS 9:68

HACHFELD, Wilhelm Karl August (from Gandersheim)

Sentenced for failure to fulfill military obliga-
    tion & emigration to America   1864-1865
<u>56</u>: L Neu Abt.129A: Gr.11, Nr. 39.   ADS 4:188

HACKER, Ernst Ludwig August (shoemaker)

Expulsion from Hamburg as Social Democrat & emi-
    gration to America   1880-1881
<u>39</u>: Polizeibehoerde-Polit.Pol.: S 149/243: Bl. 200-
    204.   ADS 2:415

HAECKL, Georg (from Gressenwoehr, Oberpfalz; in
    America)

Fund set up to have a mass said annually in the
    parish church at Schlicht for Georg & Anna Wind,
    married couple, farmers, from Koedrig, who died
    in America   1858
<u>222</u>: Rep.123: Amberg: Nr. 220.   ADS 9:2

HAEFNER, --. *See* Engler, --

HAEUSER, -- (pastor; in Sechshelden & Dillenburg;
    from America)

Expulsion of alleged Reformed Church minister
                                            1876-1877
<u>152</u>: Abt.405: Wiesbaden: Rep.405/I,5: Nr. 126.
                                            ADS 7:148

HAEUSINGER, H. A. (in Galveston, Texas). *See*
    Andler, F.

HAEUSLER, --

Emigration of the Haeusler family to America, 2
    Conv.   1810-1814
<u>229</u>: Rep.G.R.: General-Registratur: Fasz. 419.41.
                                            ADS 9:110

von HAGEN, Erich Carl (Captain)

Application for 7000-Reichsthaler loan made by von
    Hagen, just returned from America [colonization
    proposal?]   1785
<u>52</u>: Calenberger Brief-Archiv Des.15: Privatakten
    Band I: H, Nr. 37.   ADS 4:55

HAGEN, M. D.

Boston: Estate of   1897
<u>39</u>: Reichs- u. auswaert.Ang.: E.IV.26 [alt.Sig.]
    [Erbschaften]   ADS 2:290

HAGER, Georg (in Denver, Colorado)

File on [estate?]   1898

227: Rep.MA 1921: A.St.I: Nr. 617.   ADS 9:68

HAHN, Charles (died in Philadelphia)

File on [estate?]   1900
227: Rep.MA 1921: A.St.I: Nr. 638.   ADS 9:68

HAHN, Josef (from Weltenburg)

Trip to America   1885
227: Rep.MA 1921: A.St.I: Nr. 1524.   ADS 9:73

HALL, Chr. (in America). *See* Bedassem, J.

von HALL, Dietrich

Trial of von Hall for adultery with the Dutchess
Jacobe; defendant was freed under the condition
that he go into exile in America or India
1599-1601
115: Hauptgericht Juelich: Nr. 64.   ADS 6:35

HALLER, Arnold (died in Chicago)

Estate of   1898-1901
152: Abt.405: Wiesbaden: Rep.405/I,2: Nr. 3468.
ADS 7:144

HAMER, Hans Heinrich (shoemaker)

Expulsion from Hamburg as Social Democrat & emigration to America   1881
39: Polizeibehoerde-Polit.Pol.: S 149/342: Bl. 40-
47.   ADS 2:417

HAMM, Scholte (captain)

San Francisco: Estate of   1912
39: Reichs- u. auswaert.Ang.: E.IV.99 [Erbschaften]
ADS 2:373

HAMMANN, Christoph (from Geinsheim, Landkreis Gross-
Gerau, HEL)

Emigration applications of Hammann & Ph[ilipp]
André's wife   1852
170: Abt.XI,4.   ADS 7:221

HAMMER, Anna Katherina. *See* Wiesen, Joh. Georg

HAMMER, Ernst

New York: Estate of   1902
39: Reichs- u. auswaert.Ang.: E.IV.26 [Erbschaften]
ADS 2:323

HAMMER, Moses (from Ruesselsheim, HEL)

Petition for community financial assistance for
the purpose of emigration   1855
182: Abt.XI,4.   ADS 7:235

HAMMERSCHMIDT, -- (in Naperville, Illinois). *See*
von Oven family

HAMMERSTEIN, --

Investigation of German Consul in Chicago regarding estate of   1895
152: Abt.405: Wiesbaden: Rep.405/I,2: Nr. 3549.
ADS 7:144

HANCHEN, -- (from Zweibruecken arrondisement, Kurpfalz, RPL)

Emigration to Philadelphia   1812
201: Archivabt. Departement Donnersberg I: Akt Nr.
86.   ADS 8:63

HANCKE [HAUCKE] -- (born 1852 in Otterndorf)

Estate of (in New York bank)   1929
100: Kirchspielsgericht Wester-Ilienworth: V.C.
Nr. 2.   ADS 4:311

HANDSCHUH, Johann Friedrich (pastor; in Pennsylvania)

File on his pastoral call to Pennsylvania   1747
500: Halle: Missionsbibliothek des Waisenhauses:
Abteilung H IV (Nordamerika) Fach A, Nr. 5.
Not in ADS

HANFMANN, Johann (died in San Jose, California)

File on [estate?]   1897-1900
227: Rep.MA 1921: A.St.I: Nr. 590.   ADS 9:68

HANS, Friedrich (servant; from Klein Rhueden?). *See*
Cleve, Johann Heinrich

HANSEN, C. F. (in Chicago)

Information sought by private person   1895
39: Reichs- u. auswaert.Ang.: N.Ib.9 [alt.Sig.]
[Nachforschungen]   ADS 2:277

HANSEN, H.

San Francisco: Estate of   1905
39: Reichs- u. auswaert.Ang.: E.IV.84 [Erbschaften]
ADS 2:339

HANSEN, Peter Christian

Memoirs in *Deutsche Guttempler*, Nr. 8 & 9 [not
   further described in ADS]

17   ADS 1:52

HANSEN, Theodor L. (in Milwaukee, Wisconsin)

Estate of   1913

39: Reichs- u. auswaert.Ang.: E.IV.12 [Erbschaften]
                                                       ADS 2:379

HANSEN, -- (brewery owner)

San Francisco: Estate of   1908

39: Reichs- u. auswaert.Ang.: E.IV.60 [Erbschaften]
                                                       ADS 2:352

HANSEN, --

Chicago: Estate of   1913

39: Reichs- u. auswaert.Ang.: E.IV.112 [Erbschaft-
   en]   ADS 2:379

HANSMANN, Julius (of Hagen)

Application of pastor -- Hansmann for issuance of
   an emigration certificate for his son Julius
   [presumably born in 1837] who is subject to
   military service   1857-1858

52: Han.Des.80: Woelpe: B.ee.VI.4: Nr. 235.
                                                       ADS 4:85

HANSTEIN, P. (in New York)

Estate of   1898

39: Reichs- u. auswaert.Ang.: E.IV.12 [alt.Sig.]
   [Erbschaften]   ADS 2:296

HARDENBERG family (of Einbeck, NL)

Genealogical studies regarding

72: Harlands Nachlass: Manuscripte Nr.9: vol.VIII.
                                                       ADS 4:250

HARDORF, Franz (in America)

Estate of   1907

39: Reichs- u. auswaert.Ang.: E.IV.76 [Erbschaften]
                                                       ADS 2:347

HARDT, Josef (from Motten, Landgericht Brueckenau;
   in Springfield [state not given in ADS descrip-
   tion])

Paper machines & fabrication methods

227: Rep.MA 1895: A.V.: Nr. 498.   ADS 9:56

HARDT, -- (married couple; in Rochester [state not
   given in ADS description])

Inquiry by Hamburg Orphanage Commission regarding
                                                       1902

39: Reichs- u. auswaert.Ang.: W.I.3 [Waisenhaus-
   Ang.]   ADS 2:328

HARFFEN, Philip. *See* Wittib, Jacob Mausen

HARGIENST, Jacob (from Bischweiler; died at Fort
   Liegenaur [Ligonier, Pennsylvania?]

Estate of [file contains letters from Stephan
   Barrer, born in Ansbach, a Reformed Church minister at Pittsburgh, & from Mathias Blume]   1790

201: Archivabt. Grafschaft Hanau-Lichtenburg: Nr.
   3492.   ADS 8:67

HARKAVY, Carpel (an American citizen)

Investigation of Harkavy by American consul in
   Hamburg   1895

39: Reichs- u. auswaert.Ang.: N.Ia.3 [alt.Sig.]
   [Nachforschungen]   ADS 2:277

HARMES, J. N. (from Bremen?)

Bremen Senat proceedings regarding Harmes' claim
   for repayment of special emigration tax levy
   [Abschoss]   1823

48: B.13.C.1.a: Nr. 20.
   DLC has.   ADS 3:38

HARMS, C. F. H.

Cincinnati: Information regarding   1879

39: Bestand Senat: C1.VIII, No.X, Jg.1879 (Register) 46 [Nordamerika]   ADS 2:241

HARPER, Otto. *See* Tauber-Harper, Otto

HARRING, Harro (a Mormon?)

Liberation of   1854

39: Bestand Senat: C1.VI, No.16p: vol.3b, Fasc. 2f
   [Freilassung]   ADS 2:37

HARTH, Johann N., II (from Goddelau, Landkreis Gross-
   Gerau, HEL)

Emigrant to America   1861

173: Ortsbuergerregister.   ADS 7:224

HARTH, Nikolaus (from Goddelau, Landkreis Gross-
   Gerau, HEL)

Emigrant to New York   1883

<u>173</u>: Gesinderegister 1853-1894: Deckblatt, rueck-
seite.   ADS 7:225

HARTIG, G. (died 1898 in Armstrong County, Pennsylvania)

Estate of   1902-1903
<u>152</u>: Abt.405: Wiesbaden: Rep.405/I,2: Nr. 3469.
                                            ADS 7:144

HARTJE, --

Application by Mrs. -- Hartje for the transportation of her husband to America   1840
<u>52</u>: Han.Des.80: Hildesheim I: E.DD.7i: Nr. 923.
                                            ADS 4:88

HARTMANN, Friedrich (in Chicago)

Inquiry by private person in Germany regarding
                                            1899
<u>39</u>: Reichs- u. auswaert.Ang.: N.Ib.1.12 [Nachforsch-
ungen]   ADS 2:303

HARTMANN, Jakob. *See* Schreiner, Elisabeth, widow of
Hartmann, born Lang

HARTMANN, Julie. *See* Andersen, Julie, born Hartmann

HARTMANN, -- (family; from Wisper, HEL). *See* Assus
family

HARTMANN, -- (former shift foreman)

Emigration of the family of   1853
<u>68</u>: VI.2.b: Fach 162, Nr. 38.   ADS 4:247

HARTMANN, -- (day laborer; from Dransfeld)

Releases from citizenship of minor children of
                                            1883
<u>71</u>: Mappe 20.F: 051-20.   ADS 4:250

HARTUNG, Albert (cigar worker)

Expulsion from Hamburg as Social Democrat & emigration to America   1881
<u>39</u>: Polizeibehoerde-Polit.Pol.: S 149/340: Bl. 106-
115.   ADS 2:417

HARTUNG, Heinrich (farmhand; from Mahlum)

Dispensation from military obligation & emigration
to America   1866
<u>56</u>: L Neu Abt.129A: Gr.11, Nr. 68. ADS 4:190

HARTUNG, Hugo (commercial agent; from Volkersheim)

Emigration to America   1859
<u>56</u>: L Neu Abt.129A: Gr.11, Nr. 85. ADS 4:192

HARVES, Johann Diedrich (from Goedestorf)

Release from citizenship   1851
<u>52</u>: Han.Des.80: Syke: B.aa.VI.4: Nr. 461. ADS 4:85

HARTWIG, Alexander (an American citizen)

Compulsory education of Alexander & Gertrud Hartwig   1913
<u>39</u>: Reichs- u. auswaert.Ang.: V.III.2 [Vormund-
schaftswesen]   ADS 2:382

HARTWIG, Gertrud. *See* Hartwig, Alexander

HARTWIG, -- (theological candidate)

Declaration of availability as pastor of the Rhynbeck & Auf den Camp congregations in America
                                    6 August 1745
<u>39</u>: Geistl.Minis.: II,6: Protokoll lit.A, 1720-
1745: p.349 (VII).   ADS 2:471

HARVEY, John

Request of Harvey of Padington [not further described in ADS] that the American consulate for-
ward his pension to him   1914
<u>39</u>: Reichs- u. auswaert.Ang.: P.I.2 [Pensionier-
ungen]   ADS 2:385

HASENCLEVER, Franz Caspar (iron & steel fabricator;
from Juelich?)

Accused of having encouraged citizens to emigrate
secretly [illicitly] to New York   1764
<u>115</u>: Rep.B.2: Juel.-Berg. Landesarchiv II: Nr. 391.
                                            ADS 6:17

HASEY, --. *See* Reinhart, --

HASHAGEN, Friedrich

Emigration file contains letter from relative or
friend in Wilmington [state not given in ADS
description]   1853
<u>55</u>: Han.Des.74: Blumenthal: IX.J.1: Fach 301, Nr.
10.   ADS 4:154

HASSEL, -- (non-commissioned officer in Hessian
Jaeger Corps during American Revolution)

Interrogation of Hassel during investigation of
Trenton [New Jersey] debacle   1778 (1782)
<u>151</u>: Bestand 4: Kriegssachen: 328 Nr. 141; 142.
DLC has.   ADS 7:37

HASSHAGEN, Gerhard Wilhelm

Emigration file contains letters from friends or
relatives in Brooklyn [New York]   1861

55: Han.Des.74: Blumenthal: IX,J.1: Fach 301, Nr.
10.   ADS 4:154

HASSHAGEN, Gerhard Wilhelm. *See also* HASHAGEN,
Friedrich

HATJE, Ludwig (in Columbia, South Carolina)

Hamburg Consulate in Charleston: Inquiry regarding
Hamburg-born man in asylum & his death   1878

39: Bestand Senat: Cl.VI, No.16p: vol.4b: Fasc. 5,
Invol. 3.   ADS 2:53

HATTENBERG, Heinrich Johann Franz (painter)

Expulsion from Hamburg as Social Democrat & emigration to America   1880-1881

39: Polizeibehoerde-Polit.Pol.: S 149/224: Bl. 222-
232.   ADS 2:415

HAUCKE, --. *See* Hancke, --

HAUF, Heinrich (from Leeheim, Landkreis Gross-
Gerau, HEL)

Emigration   ca. 1850

178: Ortsbuergerregister 1803-1871, p. 23.
                                           ADS 7:231

HAUF, Jacob (of Berkach, Landkreis Gross-Gerau, HEL)

Granted certificate of release from Hessian citizenship for purpose of emigration to America
                                             1849

163: Abt.XI/4.   ADS 7:213

von HAUFF, Walther

Emigrant?   n.d.

48: M.6.e.9 [g].   ADS 3:48

HAUPTMANN, -- (pastor; died 1891 in Brooklyn [New
York])

File on [estate?]   1898-1899

227: Rep.MA 1921: A.St.I: Nr. 618.   ADS 9:68

HAUSCHILD, William

Philadelphia: Estate of   1906

39: Reichs- u. auswaert.Ang.: E.IV.2 [Erbschaften]
                                           ADS 2:343

HAUSELT, Charles (leather manufacturer; in New York)

File on [estate?]   1895

227: Rep.MA 1921: A.St.I: Nr. 614.   ADS 9:68

HAUSMANN, -- (from Neuwied; in Washington [state?])

Estate of   1838-1859

200: Abt.403: Oberpraesidium Koblenz (Rheinprovinz): Nr. 4047.   ADS 8:18

HAUX, Charles (administrator; in Hoboken)

File on [estate?]   1898

227: Rep.MA 1921: A.St.I: Nr. 624.   ADS 9:68

HAYN, Mathilde

New York: Estate of   1917

39: Reichs- u. auswaert.Ang.: E.IV.30 [Erbschaften]
                                           ADS 2:392

HEBEL, Karoline. *See* Hebel, Klemens

HEBEL, Klemens (in New York)

Care of the children, in Bosenbach, of Klemens &
Karoline Hebel, both in New York   1906

227: Rep.MA 1935: A.St.: MA 98041.   ADS 9:93

HECHENLEITNER, -- (in Arizona)

Transmittal of support money for his illegitimate
child by Marie Fuchs, a maid   1917

227: Rep.MA 1936: A.St.: MA 98052.   ADS 9:94

HECK, Luise Christine

New York: Estate of   1916

39: Reichs- u. auswaert.Ang.: E.IV.20 [Erbschaften]
                                           ADS 2:390

HECKMANN, Karl (from Einbeck)

Transportation to America   1845

52: Han.Des.80: Hildesheim I: E.DD.37a: Nr. 1294.
                                           ADS 4:89

HECKSCHER, --

Inquiry by Dr. -- Golette, M.D., New York, regarding the sending of Heckscher (a lottery collect-
or) to Hamburg   1872

39: Bestand Senat: Cl.VIII, No.X, Jg.1872 (Register) 56 [Nordamerika]   ADS 2:222

HECTEL, Anton (from Muendlfeld; in Illinois)

    File on   1857-1883

    <u>227</u>: Rep.MA 1921: A.St.I: Nr. 580.   ADS 9:68

HEER, Samuel (from Aschenhausen, Thuringia)

    Emigration of   1842; 1846

    <u>500</u>: Weimar: Thueringisches Staatsarchiv: Dermbach
        II: Gb.2: Nr. 63-92; 93-114; 144-183; 186;
        188.   Not in ADS

HEERLEIN, Johann Heinrich Friedrich (basket weaver)

    Expulsion from Hamburg as Social Democrat & emi-
        gration to America   1881

    <u>39</u>: Polizeibehoerde-Polit.Pol.: S 149/335: Bl. 233-
        242.   ADS 2:416

HEESCH, John

    San Francisco: Estate of   1897

    <u>39</u>: Reichs- u. auswaert.Ang.: E.IV.42 [alt.Sig.]
        [Erbschaften]   ADS 2:290

HEGGER, Johann (military pensioner; died 1863 in
America)

    Estate & pension   1898-1900

    <u>227</u>: Rep.MA 1921: A.St.I: Nr. 634.   ADS 9:68

HEGGER, John H. (in South Ottawa [state not given in
ADS description])

    Estate of   1899

    <u>225</u>: Rep.168: Verz.1: Fasz.19: Nr. 1-3. ADS 9:15

HEGGER, John H. (in Shreveport, Louisiana)

    Estate of   1899

    <u>39</u>: Reichs- u. auswaert.Ang.: E.IV.56 [alt.Sig.]
        [Erbschaften]   ADS 2:301

HEIBER, Jul. (a German citizen)

    Prosecution for forgery of certificate of Heiber
        expelled from America   1911

    <u>39</u>: Reichs- u. auswaert.Ang.: J.Ic.3 [Justizsachen]
                                   ADS 2:369

HEIDE, Heinrich (Henry)(from Obermarsberg, RWL)

    Emigrated to North America; became wealthy & aided
        his native village most generously 1914-1931

    <u>136</u>   ADS 6:118

HEIDER, -- (in India). *See* Klingenberg, Elisabeth,
born Pomann

HEIDORN, --. *See* Salomon, --

HEIDTMANN, Heinrich Joachim (in Chicago)

    Inquiry by private person regarding   1896

    <u>39</u>: Reichs- u. auswaert.Ang.: N.Ib.20 [alt.Sig.]
        [Nachforschungen]   ADS 2:284

HEILBECKER, Lisette (from Wallrabenstein, HEL)

    Emigration to America   1884

    <u>152</u>: Abt.229: Amt Idstein: Nr. 355. ADS 7:132

HEILBRONN, --

    EState of   1940-1941

    <u>148</u>: R XI.   ADS 6:261

HEILMANN, Margaretha. *See* Becker, Margaretha, born
Heilmann

HEILMANN, --

    Estate of   1941

    <u>148</u>: R XI.   ADS 6:261

HEIN, Bertha (in Philadelphia)

    Application of -- Wegmann [in Hamburg] for a death
        certificate for   1918

    <u>39</u>: Reichs- u. auswaert.Ang.: S.IV.30 [Standesamts-
        sachen]   ADS 2:396

HEIN, H.

    St. Louis: Estate of   1894

    <u>39</u>: Reichs- u. auswaert.Ang.: E.IV.60 [alt.Sig.]
        [Erbschaften]   ADS 2:269

HEINEMANN, Paul

    New York: Estate of   1907

    <u>39</u>: Reichs- u. auswaert.Ang.: E.IV.13 [Erbschaften]
                                    ADS 2:346

HEINEMANN, -- (an American citizen)

    Residency permit for   1911

    <u>39</u>: Reichs- u. auswaert.Ang.: A.IV.2 [Ausweis-
        ungen]   ADS 2:368

HEINEMEIER, Anna (in Philadelphia)

    Hamburg Orphanage Commission seeks information re-
        garding   1897

    <u>39</u>: Reichs- u. auswaert.Ang.: W.I.1.1 [alt.Sig.]
        [Waisenhaus-Ang.]   ADS 2:293

HEINRICHS, Oswald (in Alma, Colorado)

Estate of   1914

39: Reichs- u. auswaert.Ang.: E.IV.36 [Erbschaften]
                                                        ADS 2:384

HEINZ, Georg Michael (from Kapellen; died in America)

File on [estate?]   1886

227: Rep.MA 1921: A.St.I: Nr. 1529a.   ADS 9:73

HEINZE, Paul (born in Kansas City)

Request by Hamburg Orphanage Commission for birth
   & christening certificates for   1902

39: Reichs- u. auswaert.Ang.: W.I.8 [Waisenhaus-
   Ang.]   ADS 2:328

HEINZELMANN, H? (pastor? in Pennsylvania)

Assignment to Pennsylvania with H. Schulzen
                                               1751-1752

500: Halle: Missionsbibliothek des Waisenhauses:
   Abteilung H IV (Nordamerika) Fach A, Nr. 6.
                                               Not in ADS

HEINZEN, -- (German revolutionary; in New York). *See*
German Political Refugees

HEISCHMANN, John J. (pastor, St. Peter's Evang.Luth.
Church, Brooklyn, New York)

Letter   [1919-1922]

44: Mappe Amerika.   ADS 2:599

HEISE, Heinrich

Hamburg Consulate in San Francisco: Recommendation
   for Heise   1869-1893?

39: Bestand Senat: Cl.VI, No.16p: Fasc.10k, invol.
   1 [Empfehlung]   ADS 2:56

HEISS, Michael

Residence with North American missions   1842

229: Rep.Reg. Akten von Oberbayern, Kammer des
   Innern: 647.6.   ADS 9:115

HEITHAUS, --

New Orleans: Estate of   1913

39: Reichs- u. auswaert.Ang.: E.IV.60 [Erbschaften]
                                               ADS 2:379

HELFERICH, Adam (from Zotzenbach)

Emigration   before 1847

149: Rep.51V: Fuerth: Nr. 25 S.   ADS 7:1

HELLER, Adam (in America)

Estate of Peter Laubenstein [in Germany?], inquiry
   made by Heller   1887

227: Rep.MA 1921: A.St.I: Nr. 707.   ADS 9:69

HELLERBACH, Peter Josef (died 1894 in Bombai, South
America [*sic*])

Alleged estate of $50 million in San Francisco
                                               1901-1903

152: Abt.405: Wiesbaden: Rep.405/I,2: Nr. 3470.
                                               ADS 7:144

HELM, J. A. (in Weissenburg Township, Pennsylvania
[county not given in ADS description])

Letter from emigrant Helm regarding claim against
   shipping agent   1840

39: Bestand Senat: Cl.VIII, No.X, Jg.1840 (Register) 95 [Polizeiverwaltung]   ADS 2:140

HELMBRECHT, Georg Heinr[ich] Chr. (in Dransfeld)

Transportation of convict [Karrengefangenen] Helmbrecht to America   1837

52: Han.Des.80: Hildesheim I: E.DD.35a: Nr. 1213.
                                               ADS 4:89

HELLMANN, Louis (in Chicago)

Inquiry by private person in Germany regarding
                                               1902

39: Reichs- u. auswaert.Ang.: N.Ib.1.5 [Nachforsch-
   ungen]   ADS 2:326

HELLWEG, Adolph (pastor; Lankenau Hospital, Philadelphia)

Letter   [1919-1922?]

44: Mappe Amerika.   ADS 2:600

HELMEL, Georg (in Chicago)

Granting of power of attorney   1898

227: Rep.MA 1921: A.St.I: Nr. 627.   ADS 9:68

HELMUTH, -- (pastor; in Pennsylvania)

File on pastoral calls for -- Helmuth & -- Schmidt
                                               1767-1769

500: Halle: Missionsbibliothek des Waisenhauses:
   Abteilung H IV (Nordamerika) Fach A, Nr. 10.
                                               Not in ADS

HENNE, -- (captain, 12th Regiment, Missouri Volun-
teers; of Davenport, Iowa)

Newspaper clipping regarding [veteran from Schles-
   wig-Holstein]   n.d.

47   ADS 2:608

HENNING, H. E. (pastor, St. John's Lutheran Church, Cullman, Alabama)

Letter   [1919-1922?]
44: Mappe Amerika.   ADS 2:599

HENRICI, --

Estate of   1925-1927
148: R XI.   ADS 6:261

HENSCHEL, C. (in New York). *See* von Melle, Emil

HENSEL, Carl (in Sheboygan, Wisconsin)

Estate of   1913
39: Reichs- u. auswaert.Ang.: E.IV.21 [Erbschaften]
                                               ADS 2:379

HENSEL, Charles. *See* Henze, K.

HENSEL, F. A. (in Salem, Oregon)

Estate of   1911
39: Reichs- u. auswaert.Ang.: E.IV.74 [Erbschaften]
                                               ADS 2:369

HENZE, Heinrich (day laborer; from Ahlshausen)

Emigration to America   1862
56: L Neu Abt.129A: Gr.11: Nr. 20.   ADS 4:186

HENZE, K., *alias* Charles Hensel (in Ohio)

Estate of   1896
39: Reichs- u. auswaert.Ang.: E.IV.18 [alt.Sig.]
    [Erbschaften]   ADS 2:282

HEPPENHEIMER, Fr. (doctor; from New York)

Use of the Archive for historical & literary pur-
    poses   1892
227: Rep.MA 1921: I.V.: Nr. 2294.   ADS 9:77

HERBST, Johann (secretary of the Directorate of the Lutheran Seminary & pastor of the Lutheran congregation at Gettysburg, Pennsylvania)

Signer of memorial addressed to Hamburg Spiritual
    Ministry   1828
39: Reistl.Minis.: III.B Fasc.14: Lose Akten 1828.
                                               ADS 2:480

HERF, -- (of Kreuznach; American citizen)

Administration of estate   1918-1922
200: Abt.441: Regierung Koblenz: Nr. 25636.
                                               ADS 8:26

HERING, Th. F. (Oakland, California)

Estate of   1894
39: Reichs- u. auswaert.Ang.: E.IV.78 [alt.Sig.]
    [Erbschaften]   ADS 2:269

HERING, -- (in New York)

Issuance of certificate of residency [German citi-
    zenship?]   1918
39: Reichs- u. auswaert.Ang.: S.III.2.20 [Staats-
    angehoerigkeit]   ADS 2:396

HERMANN, Johann Carl Friedrich (basket weaver)

Expulsion from Hamburg as Social Democrat & emi-
    gration to America   1880-1881
39: Polizeibehoerde-Polit.Pol.: S 149/93: Bl. 324-
    331.   ADS 2:413

HERMANN, J. F. W. *See* Strassen, H.

HERMANN, Wilhelm

Inquiry [from, or to, New London] regarding where-
    abouts of   1875
39: Bestand Senat: Cl.VIII, No.X, Jg.1875 (Regis-
    ter) 14 [Nordamerika]   ADS 2:230

HERMANN, --. *See* Earle, --

HEROLD, Eva (died in San Francisco)

File on [estate?]   1896
227: Rep.MA 1921: A.St.I: Nr. 582.   ADS 9:68

HERPPICH, Kath[arina] (died 1896 in Brooklyn, New York])

File on [estate?]   1897-1898
227: Rep.MA 1921: A.St.I: Nr. 589.   ADS 9:68

HERRLICH, Johannes (in San Francisco)

Letter to Georg Friedrich Buek in Hamburg   1851
39: Bestand Familienarchiv: Fam. Buek: I.Konv. H.
                                               ADS 2:487

HERRMANNS, Nicklaus. *See* Wittib, Jacob Mausens

HERRMANN, -- (lived in St. Louis [Missouri])

Request of the forester -- Seel, in Hahnstaetten,
for investigation regarding his mother-in-law's
alleged estate   1900

152: Abt.405: Wiesbaden: Rep.405/I,2: Nr. 3471.
                                                 ADS 7:144

HERTEL, Frederick (in Chicago)

Application to be Bremen consul in Chicago   1862

48: B.13.6.0: Nr. 17.   ADS 3:10
    DLC has.

HERTLE, Daniel (in Chicago)

Letter   1858

216: Bestand 191-1: in Nr. 221.   ADS 8:85

HERZ, Magalena, born Maier (in Philadelphia)

Information regarding   1900

227: Rep.MA 1921: A.St.I: Nr. 1562.   ADS 9:73

HERZHEIM, -- (American citizens)

Residence of Mr. & Mrs. -- Herzheim [of Hannover?]
    at the court of Lippe in Detmold   1917

124: D.Akten: Fach 23, Nr. 3.   ADS 6:84

HESER, August. See Heyer, August

HESS, A. (from Hesse)

Emigration to Ohio, then California   before 1857

149: Rep.51 V: Fuerth: Nr. 122 K.   ADS 7:1

HESSE, August (vagrant handworker; from Langelsheim)

Emigration to America   1858-1860

56: L Neu Abt.129A: Gr.11: Nr. 61.   ADS 4:190

HESSE, Gottfried Wilhelm Heinrich (mason)

Expulsion from Hamburg as Social Democrat & emi-
    gration to America   1880-1881

39: Polizeibehoerde-Polit.Pol.: S 149/48: Bl. 254-
    268.   ADS 2:412

HESSEL, Gottfried (doctor; in St. Paul, Minnesota)

Inquiry by private person in Germany regarding
                                                1901

39: Reichs- u. auswaert.Ang.: N.Ib.1.5 [Nachforsch-
    ungen]   ADS 2:319

HESSLING, -- (an American citizen)

Expulsion [from Hamburg?]   1894

39: Reichs- u. auswaert.A.: A.IV.1 [alt.Sig.]
    [Ausweisungen]   ADS 2:268

HEUBER, Lorenz (from Zeugleben; in America)

Bankruptcy of   1848

227: Rep.MA 1898: A.St.: Nr. 6291.   ADS 9:59

HEUCK, F.

San Francisco: Estate of   1906

39: Reichs- u. auswaert.Ang.: E.IV.53 [Erbschaften]
                                             ADS 2:343

HEUCKE, E. (married woman; in New York)

Inquiry by private person regarding   1896

39: Reichs- u. auswaert.Ang.: N.Ib.15 [alt.Sig.]
    [Nachforschungen]   ADS 2:284

HEUMANN, Lea. See Kontrauth, Lea, born Heumann

HEYER, August

Chicago: Information regarding   1879

39: Bestand Senat: Cl.VIII, No.X, Jg.1879 (Regis-
    ter) 4 [Nordamerika]   ADS 2:240

HEYER, Carl Christoph Ludwig (plumber)

Expulsion from Hamburg as Social Democrat & emi-
    gration to America   1880-1881

39: Polizeibehoerde-Polit.Pol.: S 149/57: Bl. 332-
    346.   ADS 2:412

HEYES, Emma (teacher)

Complaint by Heye, of Philadelphia, regarding her
    detention in a Hamburg insane asylum   1910

39: Reichs- u. auswaert.Ang.: M.Ib.5 [Medizinal-
    sachen]   ADS 2:364

HIESLAND, Henry (pastor; American citizen)

Complaint of American consul in Hamburg regarding
    handling of Hiesland's travel papers in the
    Zeven office   1839-1840

52: Han.Des.9: sog. Geheime Reg.: A., Nr. 4.
                                             ADS 4:100

HILB, Esther, born Wormser (in Little Rock, Arkansas)

Estate of   1908

39: Reichs- u. auswaert.Ang.: E.IV.2 [Erbschaften]
                                             ADS 2:351

HILKE, --

Philadelphia: Estate of   1902
39: Reichs- u. auswaert.Ang.: E.IV.37 [Erbschaften]
                                            ADS 2:323

HILLE, C. (machinist)

New York: Citizenship of   1882
39: Bestand Senat: Cl.VIII, No.X, Jg.1882 (Regis-
    ter) 17.   ADS 2:249

HILLE, Fritz. See Hille, Wilhelm

HILLE, Wilhelm (audit clerk; from Clausthal)

Emigration with brother Fritz Hille   1857-1858
68: VI.2.b: Fach 162, Nr. 42.   ADS 4:247

HILSLAND, Henry (an American citizen)

Rejection of Hilsland's passport issued by American
    consul in Hamburg   1839-1840
52: Calemberger Brief-Archiv Des.15: Privatakten
    Band I: H, Nr. 96.   ADS 4:55

HILTON, Frank

Hamburg police search for Hilton, wanted in America
                                                           1909
39: Reichs- u. auswaert.Ang.: P.III.1.5 [Polizei-
    wesen]   ADS 2:359

HINDELANG, Franz (reported to have been in America)

Investigation of Hindelang for theft   1848-1849
227: Rep.MA 1898: A.St.: Nr. 6292.   ADS 9:59

HINRICHS, -- (border guard; from Rechtebe)

Emigration to America   1865
52: Han.Des.74: Hagen: Regiminalia Fach 270: Nr. 4
    [Steuersachen]   ADS 4:70

HINSCH, Ch.

American military pension for   1898
39: Reichs- u. auswaert.Ang.: P.I.1 [Pensionier-
    ungen]   ADS 2:298

HINZE, Christoph (a musketeer from Gross Dahlum who
did not return from America)

File on wife's dealings with Julius Guenther re-
    garding interest on loan & mortgage on farm at
    Sollingen   1785
56: L Alt Abt.8, Bd.2: Jerxheim: Nr. 178.
                                            ADS 4:166

HIRSCH, Johanna (from Klein-Gerau, Landkreis Gross-
Gerau, HEL)

Emigration of Jewish person to USA   1937
176: Abmelderegister 1934-1960.   ADS 7:229

HIRSCH, Raphael. See Angermann, Betty, born Rendsburg

HIRSCH, Richard

Imprisonment of [in USA?]   1864
39: Bestand Senat: Cl.VIII, No.X, Jg.1864 (Regis-
    ter) I [Amerikanische V.S.]   ADS 2:185

HIRSCH, -- (from Kelsterbach, Landkreis Gross-Gerau,
HEL). See Frenzel, --

HIRSCH, -- (a child)

Commencement of guardianship proceedings. Mother
    resident of Hamburg; father, an American citizen,
    died in America   1895
39: Reichs- u. auswaert.Ang.: V.III.2 [alt.Sig.]
    [Vormundschaftswesen]   ADS 2:279

HIRSCHFELD, --

Estate   n.d.
148: R XI.   ADS 6:261

HIRSCHKORN, Leiser (called Ludwig)(from Gernsheim,
Landkreis Gross-Gerau, HEL)

Emigration to America, via London   1939
171: Judenkartei [Jewish card file]   ADS 7:222

HIRSCHMANN, Franziska (in New York)

Sworn testimony of Mrs. Hirschmann in the matter
    of the bankruptcy proceedings of her husband L.
    Hirschmann of Ansbach   1845-1846
227: Rep.MA 1898: A.St.: Nr. 6290.   ADS 9:59

HIRSCHMANN, H. See Jahn, Fr.

HIRT, Karl (in Chicago)

Inquiry by private party in Germany regarding 1898
39: Reichs- u. auswaert.Ang.: N.Ib.5 [Justizsachen]
                                            ADS 2:297

HIRTH, Jakob (died in Chicago)

File on [estate?]   1899-1900
227: Rep.MA 1921: A.St.I: Nr. 636.   ADS 9:68

HOCHHEIMER, -- (from Dornheim, Landkreis Gross-Gerau, HEL)

Emigration of family   ca. 1860
168: Schulgeschichte.   ADS 7:218

HOCKENDORFF, H. (naturalized American citizen)

Punishment of Hockendorff in Hamburg for failure to fulfill [German] military service obligation   1894
39: Reichs- u. auswaert.Ang.: M.II.13 [alt.Sig.] [Militaerang.]   ADS 2:270

HOEBEL, Sophie (spinster; from Klein Rheuden). *See* Cleve, Johann Heinrich

HOECK, -- (of Goettingen)

Proposed transportation to America of -- Hoeck, son of Hofrat [councillor] -- Hoeck   1863
52: Han.Des.80: Hildesheim I: E.DD.38b: Nr. 1477.   ADS 4:90

HOEFER, P. (from Luebeck)

Report from Washington regarding death & estate of   1873
39: Bestand Senat: Cl.VIII, No.X, Jg.1873 (Register) 35 [Nordamerika]   ADS 2:225

HOEFLER, Georg (in New York)

Hoefler offers to sell plans to the fortifications at Ingolstadt   1882
227: Rep.MA 1921: A.V. V: Nr. 16077.   ADS 9:91

HOEGES, A. A. J. M.

San Francisco: Estate of   1908
39: Reichs- u. auswaert.Ang.: E.IV.46 [Erbschaften]   ADS 2:351

HOEHN, Franz

Extradition to Darmstadt from USA, via Hamburg   1914
39: Reichs- u. auswaert.Ang.: A.Ia.6 [Auslieferungen]   ADS 2:383

HOELLDORFER, -- (in America)

Possible repatriation of the insane married couple Hoelldorfer   1905
227: Rep.MA 1936: A.St.: MA 98038.   ADS 9:93

HOEPPNER, Friedrich (in Burlington, Iowa)

Estate of   1894
39: Reichs- u. auswaert.Ang.: D.Ih, 35 [Erb-

schaften]   ADS 2:269

HOERMERTE, Appolonia (from Gernsheim, Landkreis Gross-Gerau, HEL)

Emigration to Canada with 2 children   1913
171: Abmelderegister 1912-1924.   ADS 7:222

HOESLER, Valentin, Junior (from Unteralba, Thuringia)

Application for emigration   1839
500: Weimar: Thueringisches Staatsarchiv: Dermbach II: Gb.2: Nr. 63-92; 93-114; 144-183; 186; 188.   Not in ADS

HOETZEL, Leonhard (in America)

Inquiry regarding   1888
227: Rep.MA 1921: A.St.I: Nr. 1532.   ADS 9:73

van HOFEN, --

Estate of   1922-1927
148: R XI.   ADS 6:261

HOFFGUTH, -- (alleged pastor in America)

Letter from Stuttgart regarding   4 November 1746
39: Geistl. Minis.: II,7: Protokoll lit.B, 1746-1758: pp.44 f (V); p.46 (IV); p.48; p.49 (I); p.89 (III).   ADS 2:471

HOFFGUTH, -- (allged pastor in New York). *See* Knoll, -- (pastor in New York)

HOFFMANN, Albert H. G. (businessman; from Bad Salzuflen)

Information regarding the son of founder of Hoffmann'schen Staerkefabrik [Hoffmann's Starch Factory] [see also Hoffmann, Reinhold, Wilhelm] n.d.
104   ADS 4:331

HOFFMAN(N), B. M.

Inquiry by U.S. consulate regarding   1913
39: Reichs- u. auswaert.Ang.: N.Ia.3.5 [Nachforschungen]   ADS 2:381

HOFFMANN, Conrad. *See* Epple, E.

HOFFMANN, F. A. (in Milwaukee? Wisconsin)

Seeks Hamburg consulship   1857-1858
39: Bestand Senat: Cl.VI, No.16p: vol.4b: Fasc. 18.   ADS 2:59

HOFFMANN, Jeremias (paymaster)

Extradition of   1853-1856

<u>151</u>: Bestand 5: Gesandtschaften: Berlin: b.: Nr.
150a.   ADS 7:45

HOFFMANN, J. F. C.

St. Louis: Estate of   1905

<u>39</u>: Reichs- u. auswaert.Ang.: E.IV.27 [Erbschaften]
                                                  ADS 2:339

HOFFMANN, J. N. (pastor, First Lutheran Church,
Chambersburg, Franklin County, Pennsylvania)

Letter to Hamburg Spiritual Ministry [Ministry for
Religion] regarding the will of -- Schrader 1841

<u>39</u>: Geistl. Minis.: III,B: Fasc.17: Lose Akten
1841 (4).   ADS 2:482

HOFFMANN, Reinhold (seaman; from Bad Salzuflen)

Information regarding the son of the founder of
Hoffmann'schen Staerkefabrik [Hoffmann's Starch
Factory] [*see also* Hoffmann, Albert H. G.]

<u>104</u>   ADS 4:331

HOFFMANN, Wilhelm (of Hoffmanns Staerkefabrik [Hoff-
mann's Starch Factory] Salzuflen)

Preparation for a study trip to the USA   1899

<u>114</u>: Lipp.Reg.: Ministerium: IV.Abt.Reg.: 21c, Nr.
6 [in vol. III]   ADS 6:12

HOFFMEISTER, E.

New York: Estate of   1888

<u>39</u>: Bestand Senat: Cl.VIII, No.X, Jg.1888 (Regis-
ter) 23 [Erbschafts-Amt]   ADS 2:260

HOFFMEISTER, Sophie (spinster; from Ortshausen)

Punishment for theft & emigration to America
                                                  1856-1857

<u>56</u>: L Neu Abt.129A: Gr.11, Nr. 75.   ADS 4:191

HOFHEINZ, Johann Jost (from Suedershausen, HEL).
See Brachthaeuser, Daniel

HOFMANN, Anna. *See* Bonner, Anna, born Hofmann

HOHL, --

Extradition from USA to Hamburg   1894

<u>39</u>: Reichs- u. auswaert.Ang.: D.Ic.59.   ADS 2:267

HOHMANN, Bertha

Transmittal of a report from the Chicago consulate

to the Hamburg juvenile [guardianship] authorities   1900

<u>39</u>: Reichs- u. auswaert.Ang.: V.III.2 [Vormund-
schaftssachen]   ADS 2:312

HOHMANN, Caspar Adam (in Thuringia)

Application for emigration   1840

<u>500</u>: Weimar: Thueringisches Staatsarchiv: Dermbach
II: Gb.2: Nr. 63-92; 93-114; 144-183; 186;
188.   Not in ADS

HOLLERBACH, -- (missionary)

Report from Chicago consulate regarding the heirs
of   1901

<u>39</u>: Reichs- u. auswaert.Ang.: G.Ia.1.24 [Geschaefts-
verkehr]   ADS 2:317

HOLTEGEL, A. D. (a minor; American citizen)

Appointment of guardian for   1914

<u>39</u>: Reichs- u. auswaert.Ang.: V.III.2.2 [Vormund-
schaftswesen]   ADS 2:387

HOLZHAUSEN, Oscar Robert (in Chicago)

Inquiry by private person in Germany regarding
                                                  1901

<u>39</u>: Reichs- u. auswaert.Ang.: N.Ib.10 [Nachforsch-
ungen]   ADS 2:319

HOLZLEITNER, Leopold

Reclamation of his personal property left in America   1902

<u>227</u>: Rep.MA 1921: A.St.I: Nr. 1261.   ADS 9:72

HOMANN, Karl (a seaman)

Boston: Estate of   1879

<u>39</u>: Bestand Senat: Cl.VIII, No.X, Jg.1879 (Regis-
ter) 45 [Nordamerika]   ADS 2:241

HOMRINGHAUSEN, Anna (died 14 May [1910?] in St.
Louis [Missouri])

Estate of   1910

<u>152</u>: Abt.405: Wiesbaden: Rep.405/I,2: Nr. 3473.
                                                  ADS 7:144

HONISCH, --

Estate of   1943-1944

<u>148</u>: R XI.   ADS 6:261

HONNEF, H.

New York: Death certificate for   1880

39: Bestand Senat: Cl.VIII, No.X, Jg.1880 (Register) 24 [Nordamerika]   ADS 2:244

HOOSE, Julie (in New York). *See* von Melle, Emil

HOOVE, T.

San Francisco: Estate of   1890

39: Bestand Senat: Cl.VIII, No.X, Jg.1890 (Register) 26 [Erbschafts-Ang.]   ADS 2:264

HOPPE, Charles (in San Jose [California?])

Estate of   1915

39: Reichs- u. auswaert.Ang.: E.IV.17 [Erbschaften]   ADS 2:388

HORNEBURG, L. P. (an American citizen)

Complaint of Horneburg against police judge Dr. Homann   1874

39: Bestand Senat: Cl.VIII, No.X, Jg.1874 (Register) 27 [Nordamerika]   ADS 2:228

HORNFLECK, Georg

Request for financial aid to emigrate to America   1854? 1881?

152: Abt.228: Amt Hoechst: Nr. 331.   ADS 7:132

HORNUNG, Salomon

Extradition from Hamburg to Hungary of Hornung who was excluded by U.S. Immigration Service in New York   1912

39: Reichs- u. auswaert.Ang.: A.Ia.1.10 [Auslieferungen]   ADS 2:371

HOTTELMANN, Hugo (in Brooklyn, New York)

Hamburg Military Commission seeks information regarding   1896

39: Reichs- u. auswaert.Ang.: N.Ia.13 [alt.Sig.] [Nachforschungen]   ADS 2:284

HOTZEN, Anna Margarethe. *See* Kramer, Joh. Diedrich

HOVEY, --. *See* Love, --, born Hovey

HOWITZ, F. (in New York)

Inquiry by private person in Germany regarding   1899

39: Reichs- u. auswaert.Ang.: N.Ib.1.2 [Nachforschungen]   ADS 2:303

HOWOLD, Christian

Chicago: Estate of   1915

39: Reichs- u. auswaert.Ang.: E.IV.41 [Erbschaften]   ADS 2:388

HOYER, Emma (in New York). *See* von Melle, Emil

HUBER, E. W. (married in Port Gibson, Mississippi)

File on   1900-1901

227: Rep.MA 1921: A.St.I: Nr. 641.   ADS 9:68

HUBER, Josef (innkeeper in Chicago)

Collection of gift tax from   1908

227: Rep.MA 1936: A.St.: MA 98068.   ADS 9:97

HUCK, C. Ludwig (died in St. Louis)

Estate of   1888-1891

116: Minden: Praes.-regist.: I.L.: Nr. 238.   ADS 6:54

HUEDEPOLL, H. E. *See* Pitzing, Rud.

HUELSMANN, --

Estate of   1941

148: R XI.   ADS 6:261

HUETTMANN, Fritz (a prisoner at Stillwater, Minnesota, convicted of murder)

Repatriation of   1901

39: Reichs- u. auswaert.Ang.: H.IIb.3 [Heimschaffungsverkehr]   ADS 2:318

HUISMANN, Hero (in Brooklyn, New York)

Estate of   1911

39: Reichs- u. auswaert.Ang.: E.IV.48 [Erbschaften]   ADS 2:369

HULL, Joseph A.

Hamburg police investigation, at the request of American police authorities, regarding Hull, wanted for defalcation of funds   1906

39: Reichs- u. auswaert.Ang.: P.III.1.3 [Polizeiwesen]   ADS 2:344

HUMMEL, Adam II (of Bauschheim, Landkreis Gross-Gerau, HEL)

Register of Removals [Abmelderegister] contains notation of his emigration to America   1890

162   ADS 7:212

HUMMERT, Fred (farmer; from Schuettorf)

Emigrated in 1914 & owned farm near Charlotte,
   North Carolina; founded German colony? & aided
   his old hometown after 1918

106   ADS 4:332

HUNGER, --

Estate of   1941
148: R XI.   ADS 6:261

HUNT, --

Chicago consulate: case of Hunt *versus* Jaekel 1896
39: Reichs- u. auswaert.Ang.: J.Ib.10.11 [alt.Sig.]
   [Justizsachen]   ADS 2:283

HUNTEMANN, Gottfried (in Kansas City)

Estate of   1908
39: Reichs- u. auswaert.Ang.: E.IV.57 [Erbschaften]
                                               ADS 2:351

HUPFELD family

Emigration tax levied on   1795
151: Bestand 19(b): Nr. 11.   ADS 7:74

HURLEMANN, Konrad (shepherd; from Bodenstein)

Emigration to America   1852
56: L Neu Abt.129A: Gr.11: Nr. 28.   ADS 4:187

HUSTEDT, Ehler (from Blender? [Blendern?])

Transportation to America   1831?
52: Han.Des.74: Westen-Thedinghusen: I.56: Nr. 31.
                                               ADS 4:80

HUWALD, Wilhelm Johannes (Hemme, SHL)

Emigration to America   1862
13: Alphabetischen Sammlung.   ADS 1:49

IGELHAUT, Joh[ann] (in Fresno, California, in 1889)

File on   1897-1902
227: Rep.MA 1921: A.St.I: Nr. 676.   ADS 9:68

IGNATIUS, -- (miner)

Extradition from USA to Bochum, via Hamburg   1909
39: Reichs- u. auswaert.Ang.: A.Ia.5 [Ausliefer-
   ungen]   ADS 2:355

IHDE, Marie C. (widow; died in Baltimore, Maryland)

Estate of   1910
152: Abt.405: Wiesbaden: Rep.405/I,2: Nr. 3475.
                                               ADS 7:144

IHLE, Magdalena (from Pirmasens; "presently" in
   Philadelphia)

Inheritance matters   1885-1886
227: Rep.MA 1921: A.St.I: Nr. 656.   ADS 9:68

IHMSEN, --

Estate of   1942
148: R XI.   ADS 6:261

ILLGE, Johann Gottlieb (in America)

Estate of   1906
39: Reichs- u. auswaert.Ang.: E.IV.13 [Erbschaften]
                                               ADS 2:343

ILLIGER, -- (theological candidate; in Rochester [New
   York?] formerly of Hamburg)

Decision to go to America in the hope of becoming
   a pastor of a German Evangelical congregation
   there   1845
39: Geistl. Minis.: II,11: Konventsprotokolle: pp.
   149; 157.   ADS 2:476

IMBSWEILER, Anna Elisabetha. *See* Lindemann, Anna
   Elisabetha, born Imbsweiler

IRION, J. (from Jacobs-Vale [Jakobsthal?] Kurpfalz)

Emigration permit   1766
201: Archivabt. Historischer Verein: Abt.BI: Nr.
   769.   ADS 8:63

ISENECKER, Wilhelm Johann Christian (cigar worker)

Expulsion from Hamburg as Social Democrat & emigration to America   1880-1881
39: Polizeibehoerde-Polit.Pol.: S 149/145: Bl. 470-
   479.   ADS 2:414

ISSLER, --

Washington: Planned assassination of the Crown
   Prince by Issler & -- Wiegand   1878
39: Bestand Senat: Cl.VIII, No.X, Jg.1878 (Register) 62 [Nordamerika]   ADS 2:239

JACKE, *alias* WOECKENER, August Bernhard (footman
   from Wallenstedt, Flecken Gronau)

Proposed approval of funds for transportation of

Jacke to America   1868

52: Han.Des.80: Hildesheim I: E.DD.24a: Nr. 1120.
                                        ADS 4:88

JACOB, --

Estate of   1943
148: R XI.   ADS 6:261

JACOBI, Amalie (in Milwaukee, Wisconsin)

Letter to brother-in-law -- Brandt in Hamburg 1865
47   ADS 2:608

JACOBI, Leonhard W. A. (in Milwaukee, Wisconsin)

Death certificate for   1878
47   ADS 2:608

JACOBS, Karoline (spinster; from Ellierode)

Convicted of fraud & misappropriation of funds;
   confined to house of corrections in Bevern; emigrated to America   1859-1860
56: L Neu Abt.129A: Gr.11: Nr. 32.   ADS 4:187

JACOBSEN, Johann Peter (brushmaker)

Expulsion from Hamburg as Social Democrat & emigration to America   1881
39: Polizeibehoerde-Polit.Pol.: S 149/341: Bl.
   393-404.   ADS 2:417

JACOBSEN, Ludwig Wilhelm (cigar worker)

Expulsion from Hamburg as Social Democrat & emigration to America   1881
39: Polizeibehoerde-Polit.Pol.: S 149/259: Bl.
   347-353.   ADS 2:416

JACOBSON, Annie W. (in Denver, Colorado)

Estate of   1916
39: Reichs- u. auswaert.Ang.: E.IV.7 [Erbschaften]
                                        ADS 2:390

JACOBY, Johann (died in Columbus, Mississippi)

Estate of   1895
152: Abt.405: Wiesbaden: Rep.405/I,2: Nr. 3551.
                                        ADS 7:144

JACOBY, Margarethe (in New Rochelle [New York])

File on [estate?]   1898
227: Rep.MA 1921: A.St.I: Nr. 678.   ADS 9:68

JACOBY, -- (American Methodist minister)

File on his activities in Bremen   1857-1858
48: 4,48.A.1.c.1.c.   ADS 3:71

JAEGER, Bendix (in Brooklyn, New York)

Estate of   1909
39: Reichs- u. auswaert.Ang.: E.IV.2 [Erbschaften]
                                        ADS 2:356

JAEGLY, Josef (died 1893 in San Antonio, Texas)

File on [estate?]   1900
227: Rep.MA 1921: A.St.I: Nr. 682.   ADS 9:68

JAEGLY, Kunigunde (died 1893 in San Antonio, Texas)

File on [estate?]   1900
227: Rep.MA 1921: A.St.I: Nr. 682.   ADS 9:68

JAEKEL, Theodor

Naturalization of the first secretary of the German General Consulate in New York   1907
39: Reichs- u. auswaert.Ang.: S.III.1 [Staats-
   angehoerigkeit]   ADS 2:349

JAEKEL, --. See Hunt, --

JAENTSCH, --

German General Consulate in New York in case of
   Jaentsch versus Grund   1898
39: Reichs- u. auswaert.Ang.: J.Ib.10.6 [alt.Sig.]
   [Justizsachen]   ADS 2:297

JAEP, Karl (day laborer; from Einbeck)

Proposed transportation to America   1846
52: Han.Des.80: Hildesheim I: E.DD.37a: Nr. 1296.
                                        ADS 4:89

JAHN, A. F. (mason; in New York)

Complaint of Jahn against Mayer & Co.   1873
39: Bestand Senat: Cl.VIII, No.X, Jg.1873 (Register) 21a [Auswanderer-Deputation]   ADS 2:223

Report of New York consulate regarding Jahn's complaint against Boehling of Meyer & Co., Hamburg
                                        1874
39: Bestand Senat: Cl.VIII, No.X, Jg.1874 (Register) 6 [Nordamerika]   ADS 2:227

JAHN, Fr.

Accusation of Hamburg Consul in New York that

H. Hirschmann has defrauded Jahn   1856

39: Bestand Senat: Cl.VIII, No.X, Jg.1856 (Register) 7.   ADS 2:169

JAHN, H. (in America)

Letter from   1840

218: Nr. 132 (Umschlag)   ADS 8:88

JAHNKE, J. K.

New York: Estate of   1889

39: Bestand Senat: Cl.VIII, No.X, Jg.1889 (Register) 14 [Erbschafts-Ang.]   ADS 2:262

JAKOBSEN, Hermann (in Oregon)

Estate of   1913

39: Reichs- u. auswaert.Ang.: E.IV.45 [Erbschaften]   ADS 2:379

JAMROZY, J. (in Chicago)

Permit for his son to marry [in Hamburg?]   1896

39: Reichs- u. auswaert.Ang.: S.IV.4 [alt.Sig.] [Standesamtssachen]   ADS 2:287

JANIS, Wilhelm (from Wregen, HEL)

Emigration of   n.d.

151: Bestand 180: Arolsen: Acc.1937/76,5: Nr. 392.   ADS 7:113

JANSEN, H. W. (professor; in Rapid City, South Dakota)

Estate of   1899

39: Reichs- u. auswaert.Ang.: E.IV.50 [alt.Sig.] [Erbschaften]   ADS 2:301

JANSON, Just. *See* Kief, Chr.

JANSKI, --

Complaint of emigrant Janski regarding Hamburg firm   1911

39: Reichs- u. auswaert.Ang.: A.III.3 [Auswanderung]   ADS 2:368

JANSS, -- (in New York)

Inquiry by Hamburg Orphanage Commission regarding alleged married couple Janss   1900-1901

39: Reichs- u. auswaert.Ang.: W.I.3 [Waisenhaus-Ang.]   ADS 2:312

JANSSEN, Reemt (from Emden)

Transportation of prisoner Janssen to America 1855

50: Rep.6: Zuchthaus-Sachen, Spec.: Nr. 15.
                                               ADS 4:5

JANSSEN, -- (seaman; in New York)

Repatriation of   1898

39: Reichs- u. auswaert.Ang.: S.Ie.1.9 [Schiffahrt]   ADS 2:298

JARCHOW, Fritz (in Toledo, Ohio)

Estate of   1914

39: Reichs- u. auswaert.Ang.: E.IV.37 [Erbschaften]   ADS 2:384

JARMULOWSKY, --

New York: Seized letter from firm Jarmulowsky & Weber in New York addressed to a Jarmulowsky in Hamburg contained stolen Breslau-Schweidnitz-Freiburger Railroad shares   1880

39: Bestand Senat: Cl.VIII, No.X, Jg.1880 (Register) 22 [Nordamerika]   ADS 2:243

JARMULOWSKY, -- (an American citizen)

Extradition from The Netherlands to the USA   1918

39: Reichs- u. auswaert.Ang.: A.Ib.3 [Auslieferungen]   ADS 2:394

JARZ, -- (of Bergedorf, HH)

Banishment to America for chronic drunkenness 1868

39: Bergedorf II: Nr. 41, vol.7, Nr. 5. [Abscheibung]   ADS 2:467

JASSER, --

Estate of   1943

148: R XI.   ADS 6:261

JASTER, Charles F.

Hamburg police search for Jaster, wanted in Wheeling [West Virginia?]   1908

39: Reichs- u. auswaert.Ang.: P.III.1.2 [Polizeiwesen]   ADS 2:353

JAYME, Elisabethe (from Rohrbach, Landkreis Gross-Gerau, HEL)

Emigration to America   1854

173: Gesinderegister 1837-1843 [of Goddelau, Landkreis Gross-Gerau, HEL]: Nr. 86.   ADS 7:224

JEBENS, Hans Johann (of Hemmerwurth, SHL)

Petition for release from citizenship   1867

13: Alphabetischen Sammlung.   ADS 1:49

JEEP, Gr. Christ. Ludwig (baker; from Dransfeld)

Release from Prussian citizenship   1880
71: Mappe 20.E: 051-20.   ADS 4:250

JENNFELD, H. (in Galveston, Texas)

Estate of   1904
39: Reichs- u. auswaert.Ang.: E.IV.30 [Erbschaften]
                                       ADS 2:335

JENSEN, Anna

New York: Estate of   1900
39: Reichs- u. auswaert.Ang.: E.IV.4 [Erbschaften]
                                       ADS 2:307

JEVE, --

New York: Information regarding   1877
39: Bestand Senat: Cl.VIII, No.X, Jg.1877 (Register) 46 [Nordamerika]   ADS 2:235

JOACHIM, Jakob (in St. Louis, Missouri)

Petition of widow -- Gross of Hanhofen [Germany]
for data on the property ownership of Jakob
Joachim   1863-1865
227: Rep. MA 1921: A.St.I: Nr. 1565.   ADS 9:73

JOB, P.

Philadelphia: Estate of   1902
39: Reichs- u. auswaert.Ang.: E.IV.30 [Erbschaften]
                                       ADS 2:323

JOCHUMSEN, Johann Carl Theodor (cigar worker)

Expulsion from Hamburg as Social Democrat & emigration to America   1880-1881
39: Polizeibehoerde-Polit.Pol.: S 149/107: Bl.
416-423.   ADS 2:414

JOERG, Philipp (died in New York)

File on [estate?]   1900
227: Rep.MA 1921: A.St.I: Nr. 681.   ADS 9:68

JOHANNES, Heinrich Friedrich Johann (cigar worker)

Expulsion from Hamburg as Social Democrat & emigration to America   1881
39: Polizeibehoerde-Polit.Pol.: S 149/339: Bl.
462-469.   ADS 2:417

JOHANNSEN, J. (an emigrant)

Counterfeit money received by Johannsen from Petersen in Hamburg   1872

39: Bestand Senat: Cl.VIII, No.X, Jg.1872 (Register) 9.   ADS 2:218

JOHN, -- (in America)

Search for Mrs. -- John and children   1888
227: Rep.MA 1921: A.St.I: Nr. 1572.   ADS 9:73

JOKEL, Jakob (in New York)

File on   1879
227: MA 1921: A.St.I: Nr. 651.   ADS 9:68

JONAS, Alexander (editor of New York Volkszeitung
[newspaper])

Hamburg police surveillance over Jonas
                          1880; 1881; 1912
39: Polizeibehoerde-Polit.Pol.: S 220: Bl. 1-14.
                                       ADS 2:418

JOPP, B. F. A. (in New York)

Information on Jopp requested by private person
[in Germany?]   1897
39: Reichs- u. auswaert.Ang.: N.Ib.1.2 [alt.Sig.]
                                       ADS 2:291

JORDAN, Anna Martha

Emigration to America   n.d.
253: Fach 24, Fasz. 37.   ADS 9:173

JORDAN, Caroline (daughter of peasant; from Klein
Rhueden)

Emigration to America   1852
56: L Neu Abt. 129A: Gr.11, Nr. 77.   ADS 4:191

JOREL, Marie (of Omaha [Nebraska])

Estate of her mother Margarethe Elter   1874
227: Rep.MA 1921: A.St.I: Nr. 646.   ADS 9:68

JOST, Peter (died in America)

Request for a death certificate on   1885
227: Rep.MA 1921: A.St.I: Nr. 1570.   ADS 9:73

JUEHLF, H. F. (from Jever; died in St. Louis)

Estate of   n.d.
53: Bestand 31: Oldenburg: 31-15-23-23. ADS 4:114

JUENGST. See Juengsten

JUENGSTEN? Johann Friedrich

Reception of wife, of Herborn [probably returning
from America] & the education of her children
1790-1793

152: Dillenburg LR: Heft 2.   ADS 7:129

JUERGENS, Fred D. (in Chicago)

Inquiry by private party in Germany regarding 1901

39: Reichs- u. auswaert.Ang.: N.Ib.1.2 [Nachforsch-
ungen]   ADS 2:319

JUERGENS, --

Estate of   1942
148: R XI.   ADS 6:261

JUERRIERS, Karl (mason; from Delligsen)

Emigration to America   1869

56: L Neu Abt.129A: Gr.11, Nr. 31.   ADS 4:187

JUHL, -- (in Michigan)

New York: Claim of married couple Juhl regarding
false $100-bill received in Hamburg   1877

39: Bestand Senat: Cl.VIII, No.X, Jg.1877 (Register) 17 [Nordamerika]   ADS 2:234

New York: Claim of married couple Juhl regarding
false $100-bill received in Hamburg   1878

39: Bestand Senat: Cl.VIII, No.X, Jg.1878 (Register) 2 [Nordamerika]   ADS 2:236

JUNG, Heinrich (from Frankelbach; died in Salem
[state not given])

File on [estate?]   1887-1889

227: Rep.MA 1921: A.St.I: Nr. 658.   ADS 9:68

JUNG, Johannes, [Senior]. See Bommarius, Friedrich

JUNG, Philipp (Ph.D.; from Ruesselsheim, Hesse)

Emigration to America   1851
182: Abt. XI,4.   ADS 7:235

JUNG, Philipp, Junior (of Nied, Hesse)

Emigration   1881-1882
152: Abt.228: Amt Hoechst: Nr. 1248.   ADS 7:131

JUNGBOHN-CLEMENT, Ale (Olde) alias John Bowen (in
Portland, Boston, West Philadelphia)

Correspondence   1842; 1844; 1856; 1859; 1869
2: Nachlass Kund Jungbohn-Clement.   ADS 1:36

JUNGE, P. K. D.

Galveston, Texas: Estate of   1910
39: Reichs- u. auswaert.Ang.: E.IV.58 [Erbschaften]
                                  ADS 2:362

JUNGE, -- (an American citizen)

Arrest of Junge for failure to fulfill [German]
military service obligation   1894

39: Reichs- u. auswaert.Ang.: M.II.14 [alt.Sig.]
[Militaerang.]   ADS 2:270

JUNGE, --

Emigration of?   1930
115: Rep.D9 III: Abt.E: Nr. 147.   ADS 6:34

JUNGEL, Anna, born Erforth (in New York)

File on [estate?]   1896
227: Rep.MA 1921: A.St.I: Nr. 670.   ADS 9:68

JUNGHANS, Friedrich (laborer; from Wolperode)

Regarding his debts for food & emigration to
America with family   1865-1866

56: L Neu Abt.129A: Gr.11, Nr. 92.   ADS 4:192

JUNGWIRTH, --

Extradition to Hungary from USA, via Hamburg 1912

39: Reichs- u. auswaert.Ang.: A.Ia.1.28 [Ausliefer-
ungen]   ADS 2:372

JUST, Karl (from Helmershausen, Thuringia)

Application for emigration   1840

500: Weimar: Thueringisches Staatsarchiv: Dermbach II: Gb.2: Nr. 63-92; 93-114; 144-183;
186; 188.   Not in ADS

KAEHLER, A. See Gohr, A., born Kaehler

KAEMMERER, Andreas

New York: Death certificate for   1877
39: Bestand Senat: Cl.VIII, No.X, Jg.1877 (Register) 22 [Nordamerika]   ADS 2:234

KAEMMERER, G. H.

Sons of Kaemmerer seek intervention of German
government in a long-delayed trial in Boston
1884

39: Bestand Senat: Cl.VIII, No.X, Jg.1884 (Register) 3 [Nordamerika]   ADS 2:253

KAERKER, L.

New York: Death certificate for 1878

39: Bestand Senat: Cl.VIII, No.X, Jg.1878 (Register) 46 [Nordamerika] ADS 2:238

KAESEBIER, --

Mentioned in letter from Roxbury 1724

119: Litt.K: Nr. 36. ADS 6:74

KAGEL, Hans (in Newark, New Jersey)

Estate of 1907

39: Reichs- u. auswaert.Ang.: E.IV.29 [Erbschaften] ADS 2:347

KAHL, M. (in Cleveland, Ohio)

Estate of 1913

39: Reichs- u. auswaert.Ang.: E.IV.87 [Erbschaften] ADS 2:379

KAHLE, Heinrich (miller's apprentice; from Kirchberg)

Punished for theft & emigrated to America 1859-1861

56: L Neu Abt.129A: Gr.11, Nr. 59. ADS 4:190

KAHN, Berta (from Worfelden, Landkreis Gross-Gerau, HEL; in Milwaukee [Wisconsin])

Claim for indemnification & data on her emigration 1937

186: Wiedergutmachungsakten: Ordner Nr. 000-30. ADS 7:240

KAHN, Erna (of Bischofsheim, Landkreis Gross-Gerau, HEL)

Emigration to Evansville [state not given in ADS description] 1939

165: Judenkartei. ADS 7:215

KAHN, Ferdinand (of Bischofsheim, Landkreis Gross-Gerau, HEL)

Emigration to New York 1938

165: Judenkartei. ADS 7:215

KAHN, Helmut (of Bischofsheim, Landkreis Gross-Gerau, HEL)

Emigration to New York 1939

165: Judenkartei. ADS 7:215

KAI, Fal

Transmittal of [extradited prisoner] from New York to Austria, via Hamburg 1901

39: Reichs- u. auswaert.Ang.: A.Ia.32 [Auslieferungen] ADS 2:313

KAINER, --

Request of civil court in Hamburg for information regarding American law in case of Wantzelius versus Kainer's heirs 1905

39: Reichs- u. auswaert.Ang.: J.Ib.53 [Justizsachen] ADS 2:340

KAISER, Auguste (in Dubuque, Iowa)

Estate of 1899

39: Reichs- u. auswaert.Ang.: E.IV.35 [Alt.Sig.] [Erbschaften] ADS 2:301

KAISER, Carl. See Caesar, Carl

KAISER, Max (in Chicago)

Report of Kaiser's address to Hamburg information office [credit bureau] 1904

39: Reichs- u. auswaert.Ang.: H.Id.1.11 [Handel u. Verkehr] ADS 2:335

KALLMANN, --. See Gaspary, --

KAMPMEIER, -- (in Chicago)

Inquiry by private person in Germany regarding 1908

39: Reichs- u. auswaert.Ang.: N.Ib.1.1 [Nachforschungen] ADS 2:353

KANITZ, -- (in Chicago)

Inquiry by Mrs. -- Kanitz regarding Dr. Franz Th. Mueller in Hamburg 1896

39: Reichs- u. auswaert.Ang.: N.Ib.14 [alt.Sig.] [Nachforschungen] ADS 2:284

KARCK, Ferdinand (in New York. See von Melle, Emil

KARL, Barbara. See Bonhag, Barbara

KARL, Jos[ef] (died in Somerville [state not given in ADS description])

File on [estate?] 1900

227: Rep.MA 1921: A.St.I: Nr. 320. ADS 9:66

KARL, -- (in Baltimore [Maryland])

Notes on a letter from Neustadt a.d. Aisch to businessman -- Karl 1821

39: Bestand Senat: Cl.VIII, No.X, Jg. 1821 (Register) ADS 2:122

KAUCK, Balthasar (from Gauersheim, Grafschaft Falkenstein, RPL)

Division of the property of Kauck, an emigrant,
among his relatives remaining in Germany   1769

201: Archivabt. Grafschaft Falkenstein: Akt Nr.
126.   ADS 8:61

KAUFHOLD, Georg (died 26 December 1911? in St. Louis,
Missouri)

Estate of   1911-1912
152: Abt.405: Wiesbaden: Rep.405/I,2: Nr. 3478.
                                        ADS 7:144

KAUFHOLD, Heinrich (died in New York)

Estate of   1900
152: Abt.405: Wiesbaden: Rep.405/I,2: Nr. 3477.
                                        ADS 7:144

KAUFFMANN, Anton

Complaint by Kauffmann & colleagues against Williams & Co.   1856

39: Bestand Senat: Cl.VIII, No.X, Jg.1856 (Register) 16 [Auswanderer-Verhaeltnisse] ADS 2:170

KAUFMANN, Christian (servant; from Hallensen). See
Voss, August

KAUL, Heinrich [Senior] (from Naheim, Landkreis
Gross-Gerau, HEL)

Emigration with financial aid from community 1849
180: XI,4.   ADS 7:233

KAUL, Maria. See Kaul, Philipp

KAUL, Philipp (of Berkach, Landkreis Gross-Gerau,
HEL)

Emigration to America   1861
163: Abt.XI,4.   ADS 7:213

Census of 1861 contains notation of Philipp &
Maria Kaul's emigration to America   1861
163: Abt.XI,4.   ADS 7:213

KAUSERS, --. See Meyer, --, born Kausers

KAYSER, Jeanette (in Norden [Germany])

Transportation of Jeanette Kayser, having no fixed
abode, & her illegitimate children to America
                                            1867
50: Rep.21a: Nr. 8997.   ADS 4:8

KAYSER, --

New York: Information regarding   1877
39: Bestand Senat: Cl.VIII, No.X, Jg.1877 (Register) 44 [Nordamerika]   ADS 2:235

KAYSER, --

Estate of   1943-1944
148: R XI.   ADS 6:261

KECSKEMETHY, Victor (a prisoner)

Extradition to Austria-Hungary from New York or
Buenos Aires, via Hamburg   1902
39: Reichs- u. auswaert.Ang.: A.Ia.1.22 [Ausliefer-
ungen]   ADS 2:322

KEELSON (also Koelsen), C. (in Travemuende)

Request by U.S. Consul in Hamburg for his extradition   1871
39: Bestand Senat: Cl.VIII, No.X, Jg.1871 (Register) 3 [Nordamerika]   ADS 2;214

KEHL, Karl Ludwig August (lived in Minneapolis,
Minnesota)

Inquiry regarding estate   1895
152: Abt.405: Wiesbaden: Rep.405/I,2: Nr. 3553.
                                        ADS 7:144

KEHLMEIER, -- (in New York)

Appointment of guardian for   1917
39: Reichs- u. auswaert.Ang.: V.III.2.5 [Vormund-
schaftswesen]   ADS 2:393

KEHN, Jacob. See Rehn, Jacob

KEIL, -- (married woman)

American court case regarding inheritance of 1911
39: Reichs- u. auswaert.Ang.: E.IV.5 [Erbschaften]
                                        ADS 2:368

KEITEL, Karl Friedrich August (machine builder)

Expulsion from Hamburg as Social Democrat & emigration to America   1880-1881
39: Polizeibehoerde-Polit.Pol.: S 149/70: Bl. 1-
18.   ADS 2:413

KELKENINK, Friederike (in Morton Grove, Illinois)

Estate of   1915
39: Reichs- u. auswaert.Ang.: E.IV.40 [Erbschaften]
                                        ADS 2:388

KELLER, Auguste (from Worfelden, Landkreis Gross-
Gerau, HEL; in Milwaukee [Wisconsin])

Letter requesting birth certificate   1957

186: Wiedergutmachungsakten, Ordner Nr. 000-30.
ADS 7:240

KELLER, Benjamin (member of directorate of Lutheran
Seminary & pastor of Lutheran congregation in Car-
lisle, Pennsylvania)

Signer of memorial addressed to Hamburg Spiritual
Ministry [Ministry of Religion]   1828

39: Geistl.Minis.: III,B Fasc.14: Lose Akten 1828.
ADS 2:480

KELLER, Matthaeus

Emigration to America   1837

161: Abt.XI, Abschnitt 4.   ADS 7:211

KELLERMANN, August (died in San Francisco)

File on [estate?]   1899-1900

227: Rep.MA 1921: A.St.I: Nr. 315.   ADS 9:66

KELLNER, Pauline Mary (an American citizen)

Appointment of guardianship for   1915

39: Reichs- u. auswaert.Ang.: V.III.2.3 [Vormund-
schaftswesen]   ADS 2:390

KELLY, Paul

Hamburg police search for Kelly, wanted for murder
in New York   1907

39: Reichs- u. auswaert.Ang.: P.III.1.1 [Polizei-
wesen]   ADS 2:348

KELP, Heinrich (handworker; from Badenhausen)

Emigration [of person committing a misdemeanor] to
America   1860-1861

56: L Neu Abt.129A: Gr.11, Nr. 24.   ADS 4:187

KEMMERLING, J. A. (in America). *See* Bedassem, J.

KEMP, A. J., *alias* Henderson

Hamburg police search for Kemp, wanted [in America]
for defalcation of funds   1907

39: Reichs- u. auswaert.Ang.: P.III.1.3 [Polizei-
wesen]   ADS 2:348

KERLING, Eduard

New York: Estate of   1912

39: Reichs- u. auswaert.Ang.: E.IV.14 [Erbschaften]
ADS 2:373

KERN, Johann Adam (from Raunheim, Landkreis Gross-
Gerau, HEL)

Emigration to America with wife & 8 children 1867

181: Raunheimer Evangelischen Pfarrei Familienregister (Lehrer Johannes Buxbaum). ADS 7:234

KERTZ, Lina (widow)

New York: Letter addressed to Friedrich Muehlfenz
in Hamburg regarding affairs of Mrs. Kertz 1878

39: Bestand Senat: Cl.VIII, No.X, Jg.1878 (Register) 5 [Nordamerika]   ADS 2:236

KESSLER, -- (from Heerda [Germany])

Emigration of Kessler family to America 1847-1854

500: Weimar: Thueringisches Staatsarchiv: XIX.C.c.
55 [Film]   Not in ADS

KESTING, --

Estate of   1942

148: R XI.   ADS 6:261

KESTINGER, --

Estate of   1921-1927

148: R XI.   ADS 6:261

KEYL, S. (Lutheran minister, New York & Hamburg)

Files of   [1873-1874]

45: Akte.   ADS 2:601

KIEF, Chr. (from Flecken Herzberg)

Proposed transportation of Just Janson & Chr.
Kief to America   1854

52: Han.Des.80: Hildesheim I: E.DD.47a: Nr. 1751.
ADS 4:91

KIEFER, Daniel (from Goddelau, Landkreis Gross-
Gerau, HEL)

Emigration to New York   1882

173: Gesinderegister 1853-1894: Deckblatt, rueckseite.   ADS 7:225

KIEFER, Jettche (from Goddelau, Landkreis Gross-
Gerau, HEL)

Emigration to America   1883

173: Gesinderegister 1853-1894: Deckblatt, rueckseite.   ADS 7:225

KIEFER, Rosine (from Goddelau, Landkreis Gross-
Gerau, HEL)
Emigration to America   1883

KLEMMER, --

New York: Estate of   1906
39: Reichs- u. auswaert.Ang.: E.IV.78 [Erbschaften]
                                         ADS 2:343

KLENNOW, F. W. F. (in Kenockee, Michigan)

Estate of   1895
39: Reichs- u. auswaert.Ang.: E.IV.64 [alt.Sig.]
    [Erbschaften]   ADS 2:275

KLINGE, Wilhelm (day laborer; from Seesen)

File on emigration to North America   1844
56: Akten der Stadt Seesen [uncatalogued]
                                         ADS 4:193

KLINGENBERG, Elisabeth, born Pomann (in Brookville
[state not given in ADS description])

Inquiry of the Oberaltenburg Landgericht [court]
by Mrs. Klingenberg regarding the estate of an
alleged Nabob in India, -- Heider   1848-1850
227: Rep.MA 1921: A.St.I: Nr. 1514.   ADS 9:73

KLINTWORTH, Anna (in Orange, California)

Letter to Sophie Maibohn in Ahrenswolde bei Harse-
    feld   1883
63   ADS 4:239

KLOCKE, F. H. W. (in Columbus, Ohio)

Inquiry by Hamburg Orphanage Commission regarding
                                                     1900
39: Reichs- u. auswaert.Ang.: W.I.4 [Waisenhaus-
    Ang.]   ADS 2:312

KLOPPENBURG, M. (married woman; Davenport, Iowa)

Estate of   1905
39: Reichs- u. auswaert.Ang.: E.IV.3 [Erbschaften]
                                         ADS 2:339

KLOTZBACH, Jacob (from Lenders, Thuringia)

Application for emigration   1840
500: Weimar: Thueringisches Staatsarchiv: Dermbach
     II: Gb.2: Nr. 63-92; 93-114; 144-183; 186;
     188.   Not in ADS

KNAAK, Louis (Landpoint, Idaho)

Estate of   1906
39: Reichs- u. auswaert.Ang.: E.IV.37 [Erbschaften]
                                         ADS 2:343

KNABSCHNEIDER, Johannes (from Durlach, Baden; in
Pennsylvania)

Inheritance of   1769
260: Nr. 477.   ADS 10:1

KNIERER, --

Official inquiry, via German consulate in Cincinnati, regarding Knierer brothers; information
sought by a family member in Hamburg   1913
39: Reichs- u. auswaert.Ang.: N.Ia.1.1 [Nachforsch-
    ungen]   ADS 2:380

KNIESCH, Mrs. A. C. C., born Krawaak, widow of
Ockens (in New York)

Inquiry by General Poorhouse & Orphanage Commission [in Germany] regarding   1901
39: Reichs- u. auswaert.Ang.: W.I.6 [Waisenhaus-
    Ang.]   ADS 2:321

KNIGHT, Edward A.

Investigation in Hamburg, at request of American
    police authorities, regarding   1898
39: Reichs- u. auswaert.Ang.: P.III.1.3 [Polizei-
    wesen]   ADS 2:298

KNITTEL, Wilhelm

New York: Estate of   1916
39: Reichs- u. auswaert.Ang.: E.IV.5 [Erbschaften]
                                         ADS 2:390

KNOBLAUCH, Joh[ann] (from Odernheim; in Cincinnati)

Complaint of   1852
227: Rep.MA 1898: A.St.: Nr. 6274.   ADS 9:59

KNOCKE, J. F.

Sentenced to lengthy penitentiary sentence in
    Frankfurt/Main for political activities; expul-
    sion to America   1838
200: Abt.403: Oberpraesidium Koblenz (Rheinpro-
     vinz): Nr. 2528.   ADS 8:21

KNOEBEL, -- (a German citizen)

Extradition to Freiburg/Breisgau from the USA, via
    Hamburg   1910
39: Reichs- u. auswaert.Ang.: A.Ia.12 [Ausliefer-
    ungen]   ADS 2:361

KNOEDLER, Beatrice, born Bachmann (in Mosinee, Wisconsin)

Letter to [cousin] George Bachmann   1935
<u>63</u>   ADS 4:238

KNOEDLER, Beatrice, born Bachmann (in Mosinee, Wisconsin). *See also* Bachmann, August

KNOEPFEL, Johann (soldier; from Oberbeisheim; said to have been drowned in America)

Petition of widow for support   1779
<u>151</u>: Bestand 5: Nr. 16775.   ADS 7:39

KNOLL, -- (pastor; in America)

Correspondence from Spiritual Ministry [Ministry of Religion] Hamburg   7 May 1734
<u>39</u>: II,6: Protokoll lit.A, 1720-1745: p.76 (4).
     ADS 2:469

Letter from Knoll regarding congregation
    20 May 1735
<u>39</u>: Geistl. Minis.: II,6: Protokoll lit.A, 1720-1745: p.79.   ADS 2:469

Letter from Knoll regarding -- Hoffguth who asserts himself to be a pastor   26 August 1746
<u>39</u>: Geistl.Minis.: II,7: Protokoll lit.B, 1746-1758: pp.30 f (IV).   ADS 2:471

KNOPP, Johann (died April 1909 in Fredericksburg, Texas)

Estate of   1910
<u>152</u>: Abt.405: Wiesbaden: Rep. 405/I,2: Nr. 3480.
     ADS 7:144

KNUDTEN [Knudtsen?] Riewert C. (in San Jose, California)

Death certificate, dated 12.7.1881   1881
<u>38</u>: Verz. Nr.2.3 (Osterlandfoehr): Nr. 1980.
     ADS 1:70

KOBURGER, -- (from Buxtehude)

Emigrants to America [picture of Mr. & Mrs. Koburger in American costume]   ca. 1840
<u>65</u>   ADS 4:242

KOCH, Carl (in Lawrence, Massachusetts)

Transmittal of money deposited with Hamburg Orphanage Commission to Carl Koch, formerly a ward thereof   1901
<u>39</u>: Reichs- u. auswaert.Ang.: W.I.2 [Waisenhaus-Ang.]   ADS 2:321

KOCH, Christian (wagon-maker's apprentice; from Kaierde). *See* Voss, August

KOCH, Heinrich (IX)* (in Alsfeld)

He & his wife had heirs in America   1902
*[Ninth person of this name having an estate file in this court]
<u>149</u>: Rep.49a: Nr. 565/12.   ADS 7:17

KOCH, I. (in America). *See* Bedassem, J.

KOCH, Otto (in Hoboken, New Jersey)

Estate of   1914
<u>39</u>: Reichs- u. auswaert.Ang.: E.IV.33 [Erbschaften]
     ADS 2:384

KOCH family. *See* Bergisch-Gladbach [geographic index]

KOCHSCHUETZKE, -- (married couple; in Stillson, Hancock County, Iowa)

Inquiry by Hamburg Orphanage Commission regarding Mr. & Mrs. Kochschuetzke in matter of Wilhelm Peters   1898
<u>39</u>: Reichs- u. auswaert.Ang.: W.I.2 [Waisenhaus-Ang.]   ADS 2:299

KOCK, -- (married couple; in New York)

Information requested by private person [in Germany] regarding   1897
<u>39</u>: Reichs- u. auswaert.Ang.: N.Ib.1.8 [alt.Sig.] [Nachforschungen]   ADS 2:291

KOCK, -- (in Portland, Oregon)

Estate of   1908
<u>39</u>: Reichs- u. auswaert.Ang.: E.IV.7 [Erbschaften]
     ADS 2:351

KOEHLER, Georg (of Philadelphia)

Report on his estate   1872
<u>227</u>: Rep. MA 1936: A.St.: MA 98071.   ADS 9:97

KOEKER, C. H. (in Montague, Michigan)

Correspondence [Zoar, Sanatorium for the Epileptic & Nervous Diseases]   1901
<u>120</u>: Akte Frank Buchman (D II B).   ADS 6:75

KOEHLER, Carl (laborer; from Schoeningen)

Emigration with family   1853
<u>105</u>: Ratsakten: Nr. 25:4.   ADS 4:332

KOEHN, Karl (in Detroit, Michigan)

Estate of   1901
39: Reichs- u. auswaert.Ang.: E.IV.13 [Erbschaften]
                                        ADS 2:315

KOEHNE, Auguste

New York: Estate of   1896
39: Reichs- u. auswaert.Ang.: E.IV.75 [alt.Sig.]
    [Erbschaften]   ADS 2:282

KOELLMANN, -- (bookseller; from Augspurg)

Release from citizenship   1827
96: B.26.Nr.9: Archiv des Klosters St. Michaelis.
                                        ADS 4:289

KOELSCH, Marie Ottilia (from Erfelden, Landkreis
   Gross-Gerau, HEL)

Emigration to America   1847-1852
169: Abt. XI,4.   ADS 7:219

KOELSEN, C. See Keelson, C.

KOENIG, Bertha. See Behrens, August

KOENNECKE, Wilhelm (book printer)

Expulsion from Hamburg as Social Democrat & emigration to America   1880-1883
39: Polizeibehoerde-Polit.Pol.: S 149/78: Bl. 264-
    303.   ADS 2:413

KOEPPLINGER, G. (from Landersdorf, Bavaria)

Emigration to North America   [19th century]
231: Rep. 270/II: Reg. Mittelfranken: Kammer des
     Innern: Abgabe 1932: Nr. 724.   ADS 9:134

KOERNER, Louise. See Spangenberg, Louise, born
   Koerner

KOESTER, Hubertus Bernhard (died 22 July 1910 in St.
   Louis, Missouri)

Estate of   1910
152: Abt.405: Wiesbaden: Rep.405/I,2: Nr. 3482.
                                        ADS 7:144

KOESTER, H. P. (in Denver, Colorado)

Estate of   1904
39: Reichs- u. auswaert.Ang.: E.IV.7 [Erbschaften]
                                        ADS 2:335

KOESTER, -- (colony doctor; in Texas)

Service contract with doctor for German colonies
   in Texas   1844
154: CA 4c 1: Nr. 37.
     DLC has.   ADS 7:177

KOESTNER, Joseph

Hamburg police search for Koestner, wanted for murder in America   1903
39: Reichs- u. auswaert.Ang.: P.III.1.2 [Polizei-
    wesen]   ADS 2:333

KOETZL, Josef (died in 1881 in Dyersville [state not
   given in ADS description])

File on [estate?]   1899
227: Rep. MA 1921: A.St.I: Nr. 308.   ADS 9:66

KOHL, Adam (died in West Fray? [state not given in
   ADS description])

File on [estate?]   1894
227: Rep. MA 1921: A.St.I: Nr. 279.   ADS 9:66

KOHLBERG, Herm.

San Francisco: Estate of   1916
39: Reichs- u. auswaert.Ang.: E.IV.24 [Erbschaften]
                                        ADS 2:391

KOHLE, -- (pastor; from Destedt)

"Wie der Pastor Kohle aus Destedt Feldprediger
   bei den braunschweigischen Truppen in Amerika
   wurde" [How Pastor Kohle from Destedt Became
   Chaplain with the Brunswick Troops in America]
   Braunschweigisches Magazin (1927) Nr. 2, pp.
   29-31   1776
                                        ADS 4:340

KOHNERT, Wilhelmine (in Hoboken, New Jersey)

Estate of   1912
39: Reichs- u. auswaert.Ang.: E.IV.93 [Erbschaften]
                                        ADS 2:373

KOLB, Adam. See Grueling, Jakob

KOLLATH, --

Extradition to Hungary from New York, via Hamburg
                                             1910
39: Reichs- u. auswaert.Ang.: A.Ia.1.25 [Ausliefer-
    ungen]   ADS 2:360

KOLLIGS, Gustav (died in New Orleans)

Estate of   1900-1902
152: Abt.405: Wiesbaden: Rep.405/I,2: Nr. 3483.
                                             ADS 7:144

KONCIKOWSKI, --

Estate of   1940-1941
148: R XI.   ADS 6:261

KONOPKY, Adam (in Brooklyn, New York)

Prosecution for bigamy   1895
39: Reichs- u. auswaert.Ang.: S.IV.1.5 [alt.Sig.]
   [Standesamtsachen]   ADS 2:279

KONTRAUTH, Lea, born Heumann (in New York)

Sworn testiomy of Mrs. Kontrauth in her paternity
   case against -- Burgstein   1843
227: Rep.MA 1898: A.St.: Nr. 6272.   ADS 9:58

KOOB, Johann Georg (in Houston [Texas?])

Estate of   1894
227: Rep. MA 1921: A.St.I: Nr. 285.   ADS 9:66

KOOP, Georg (died in America)

Death certificate for   1891
227: Rep. MA 1921: A.St.I: Nr. 254.   ADS 9:66

KOOP, Johann (in Houston, Texas)

Estate of   1881
39: Bestand Senat: Cl.VIII, No.X, Jg.1881 (Register) 7 [Nordamerika]   ADS 2:246

KOOP, M. (widow; in Darlington, Wisconsin)

Estate of   1903
39: Reichs- u. auswaert.Ang.: E.IV.3 [Erbschaften]
                                             ADS 2:330

KOPISCH, Adolph (in America)

Letter from Prussian court of law in Schmiedeberg
   regarding life or death of Adolph Kopisch,
   father of Emanuel Adolphus Benedictus Kopisch,
   a minor   1841
39: Bestand Senat: Cl.VIII, No.X, Jg.1841 (Register) 1.   ADS 2:142

KOPISCH, Emanuel Adolphus Benedictus. *See* Kopisch,
Adolph

KOPPITZ, K.

Report from Boston regarding death of   1873
39: Bestand Senat: Cl.VIII, No.X, Jg.1873 (Register) 28 [Nordamerika]   ADS 2:225

KORDENAT, Metta Catherine. *See* Rendtorff, Metta
Catherine

KORDER, Georg (died in Cochrane [state not given in
ADS description])

File on [estate?]   1899
227: Rep. MA 1921: A.St.I: Nr. 306.   ADS 9:66

KORELL, Anna Maria (from Strebendorf)

Emigration to, & disgraceful treatment in, New
   York   1855-1856
149: Rep.39/1: Konv.135, Fasz.11.   ADS 7:10

KOSERITZ, Ernst Ludwig. *See* Erbe-Koseritz, Ernst
Ludwig

KOSSUTH, -- (Hungarian national)

Report on the whereabouts & [political] agitation
   of   1852
227: Rep. MA 1898: A.St.: Nr. 7.   ADS 9:57

KOSZTAS, Martin (in New York?)

Letter to Austrian chargé d'affaires regarding
                                             1853
48: B.13.a: Nr. 42.   ADS 3:4

KOVACS (Csuka), -- (Hungarian national)

Extradition to Hungary from USA, via Bremerhaven
                                             1913
39: Reichs- u. auswaert.Ang.: A.Ia.1.9 [Ausliefer-
   ungen]   ADS 2:382

KOWALSKI, Richard (in Illinois)

Estate of   1915
39: Reichs- u. auswaert.Ang.: E.IV.22 [Erbschaften]
                                             ADS 2:388

KRAEH, Johann (journeyman mason; from Lauingen)

Emigration to North America   1866-1867
247: A 495.   ADS 9:158

KRAEMER, Michael (died in Cleveland, Ohio)

Estate of   1896
152: Abt.405: Wiesbaden: Rep.405/I,2: Nr. 3555.
                                             ADS 7:144

KRAEMER, --

Baptismal entry for child of Kraemer family who
emigrated to Pennsylvania via Duisburg-Ruhrort
1750

126: Taufregister der Evang.Gem. Ruhrort. ADS 6:91

KRAETZER, Karl (died in Buffalo)

File on [estate?]   1894-1895
227: Rep. MA 1921: A.St.I: Nr. 277.   ADS 9:66

KRAFFT, Hugo Victor (peddler)

Expulsion from Hamburg as Social Democrat & emigration to America   1880-1881
39: Polizeibehoerde-Polit.Pol.: S 149/83: Bl. 1-11.
                                        ADS 2:413

KRAFT, J. M. (from Coburg; in America?)

Improved American kiln   1847
227: Rep. MA 1895: A.V.: Nr. 382.   ADS 9:56

KRALL, Gustav (died 29 September 1909 in San Antonio,
Texas)

Estate of   1911-1914
152: Abt.405: Wiesbaden: Rep.405/I,2: Nr. 3485.
                                        ADS 7:145

KRAMER, Christian (died in New York)

File on [estate?]   1873
55: Han.Des.80: Reg.Stade: N. Tit.1: Nr. 8, RR 748.
                                        ADS 4:155

KRAMER, Joh[ann] Diedrich

Request to marry Anna Margarethe Hotzen & permit
to emigrate   1840
54: Depositum 50b: Stadtarchiv Quakenbrueck: IV:
Nr. 1597.   ADS 4:148

KRAMER, Johann Matthias

Kramer wrote book entitled *Neueste und richtigste
Nachricht von der Landschaft Georgia in den
Engellaendischen Amerika* [Latest and most accur-
ate information on the Georgia region of English
America] (Goettingen: Verlag Johann Peter Schmidt,
1746). Kramer was lektor of Italian 1746-1753 at
the University of Goettingen & then emigrated to
Georgia in 1754 or 1755

43: H A VI 104.   ADS 2:589

KRAMER, -- (from Karlsbrunn, Principality of Nassau-
Saarbruecken, RPL). *See* Eberhard, --

KRAPP, Johann (from Gross-Gerau, HEL)

Petition of his wife for his release from penitentiary at Marienschloss for the purpose of emigration to North America   1846
174: Abt. XI,4.   ADS 7:226

KRATZ, Charles

Hamburg police search for C. Kratz & John K. Murrell, both wanted by American police authorities
1902
39: Reichs- u. auswaert.Ang.: P.III.1.2 [Polizei-
wesen]   ADS 2:326

KRATZKE, --

Estate of   1942
148: R XI.   ADS 6:261

KRAUSE, -- (pastor; in America)

Report regarding activities in America & as founder
of Old Lutheran congregation in Hamburg   1842
39: Geistl.Minis.: II,9: Protokolle 1795-1850:
p.527; *see also* pp. 522; 548; 561.

KRAUSE, -- (Member Moravian Brethren; in North Carolina). *See* Danz, --

KRAWAAK, --. *See* Kniesch, Mrs. A. C. C., born Krawaak, widow of Ockens

KREGER, --

Documentation needed for Kreger's marriage [to
Kruederer?] requested by U.S. consul   1890-1891
39: Bestand Senat: Cl.VI, No.16p: vol.3b, fasc.
10r [Beschaffung]   ADS 2:41

KREISSLER, C. (carpenter; in New York)

Kreissler's complaint against agent August Behrens
1873
39: Bestand Senat: Cl.VIII, No.X, Jg.1873 (Register) 22 [Auswanderer-Deputation]   ADS 2:223

Kreissler's complaint against agent August Behrens
1874
39: Bestand Senat: Cl.VIII, No.X, Jg.1874 (Register) 3 [Auswanderer-Deputation]   ADS 2:226

KRESS von KRESSENSTEIN, -- (in America)

Material on a member of this family who emigrated
to America   n.d.

[Apply to the Kress von Kressenstein'sches Archiv,
Hofen ueber Aalen, Wuerttemberg]

KRETSCHMAR, --

Estate of   1941
148: R XI.   ADS 6:261

KRETZSCHMER, August (in Kiester, Minnesota)

Estate of   1910
39: Reichs- u. auswaert.Ang.: E.IV.48 [Erbschaften]
                                             ADS 2:362

KREUTZ, Hans (cigar worker)

Expulsion from Hamburg as Social Democrat & emi-
    gration to America   1881
39: Polizeibehoerde-Polit.Pol.: S 149/378: Bl. 12-
    16.   ADS 2:417

KREUTZBERGER, Jacob (from Weilau, Thuringia)

Application for emigration   1836
500: Weimar: Thueringisches Staatsarchiv: Dermbach
    II: Gb.2: Nr. 63-92; 93-114; 144-183; 186;
    188.   Not in ADS

KREZ, Konrad (born 1828 in Landau, Pfalz; in America)

Large portion of the literary remains of Krez who
    was condemned to death for his membership in the
    Studentenlegion [Student Legion]; he was able
    to flee to North America where he became a poet,
    lawyer, & general in the American Civil War.
    The literary remains include letters, poems,
    memoirs, & fragments of a description of America
                                         19th century
208   ADS 8:75

KRICKELSDORF, W.

Hamburg consul in Baltimore: Military pension for
    Krickelsdorf   1868
39: Bestand Senat: Cl.VIII, No.X, Jg.1868 (Regis-
    ter) C.24 [Nordamerika]   ADS 2:198

KRIETE, Johann (died at Mobile, Alabama)

Report of death   1848 or 1849
52: Calenberger Brief-Archiv Des.15: Privatakten,
    Band I: K, Nr. 280.   ADS 4:55

KRITTMANN, J[oseph]

Efforts of American consul to repossess personal
    possessions taken from Krittmann by the mayor of
    Hohenfelde in 1811   1820
39: Bestand Senat: Cl.VII, Lit.J^b: Nr. 20 u [Ver-
    wendung]   ADS 2:71

Investigation of damage to personal liberty &
    property in Hamburg by the French occupants
    [mairie]   1820

39: Polizeibehoerde-Krim.: Jg.1820: Nr. 343.
                                             ADS 2:409

KRKOVIC, Johann (an Austrian national)

Extradition of Krkovic, wanted for forgery of bank
    notes, from the USA to Austria, via Hamburg 1906
39: Reichs- u. auswaert.Ang.: A.Ia.1.34 [Ausliefer-
    ungen]   ADS 2:342

KROECKEL, J. C. A. See Meyer, Amalie

KROEGER, Hans

San Francisco: Estate of   1914
39: Reichs- u. auswaert.Ang.: E.IV.50 [Erbschaften]
                                             ADS 2:384

KROEGER, Johann (in Davenport, Iowa)

Estate of   1905
39: Reichs- u. auswaert.Ang.: E.IV.63 [Erbschaften]
                                             ADS 2:339

KROGMANN, Charles

Chicago: Estate of   1915
39: Reichs- u. auswaert.Ang.: E.IV.15 [Erbschaften]
                                             ADS 2:388

KRONSHAGE, Bernard (from Bad Salzuflen)

Information regarding [son of founder of Lippische
    Zuckerfabrik]
104   ADS 4:331

KRONSHAGE, Ernest H. (writer & critic in Wisconsin)

Information regarding [brother of Theodore Krons-
    hage]
104   ADS 4:331

KRONSHAGE, Theodor (from Bad Salzuflen)

Information regarding [son of founder of Lippische
    Zuckerfabrik; emigrated to Wisconsin & was prom-
    inent in Madison & member of board of regents]
104   ADS 4:331

KRUEDERER, --. See Kreger, --

KRUEGER, Gustav (in Loup County, Nebraska)

Estate of   1900
39: Reichs- u. auswaert.Ang.: E.IV.6 [Erbschaften]
                                             ADS 2:307

KRUEGER, Mathilde A. W. E.   *See* Mueller, John

KRUEGER, -- (widow)

Washington [D.C.?]: Estate of   1894
39: Reichs- u. auswaert.Ang.: E.IV.61 [alt.Sig.]
[Erbschaften]   ADS 2:269

KRUEGER, --

Estate of   1941
148: R XI.   ADS 6:261

KRUEMMEL, -- (in Ackly, Iowa)

Estate of   1913
39: Reichs- u. auswaert.Ang.: E.IV.33 [Erbschaften]
                                       ADS 2:379

KRUG, Adam (died in Eureka, California)

File on [estate?]   1890
227: Rep. MA 1921: A.St.I: Nr. 250.   ADS 9:66

KRUG, -- (pastor; in Pennsylvania). *See* Voigt, --

KRUMB, -- (from Dornheim, Landkreis Gross-Gerau, HEL)

Emigration of family of 7 persons   1880
168: Abmelderegister.   ADS 7:218

KRUMM, --

Information regarding American law in the case of
Krumm *versus* Witt   1906
39: Reichs- u. auswaert.Ang.: J.Ib.65 [Justiz-
sachen]   ADS 2:344

KRUMME, Johanne. *See* Bernz, Johanne, born Krumme

KRUSE, Fritz (in Davenport, Iowa)

Estate of   1899
39: Reichs- u. auswaert.Ang.: E.IV.55 [alt.Sig.]
[Erbschaften]   ADS 2:301

KRUSE, -- (married woman; in Michigan)

Estate of   1913
39: Reichs- u. auswaert.Ang.: E.IV.10 [Erbschaften]
                                       ADS 2:379

KUEHL, Claus (in New York)

Inquiry by private person in Germany regarding
                                              1898

39: Reichs- u. auswaert.Ang.: N.Ib.1.8 [Nachforsch-
ungen]   ADS 2:297

KUEHL, Ernst

German consular report from New York regarding
Kuehl's accusation against Meyer & Co.   1869
39: Bestand Senat: Cl.VIII, No.X, Jg.1869 (Register) 41 [Nordamerika]   ADS 2:203

KUEHL, Simon (cabinetmaker)

Expulsion from Hamburg as Social Democrat & emigration to America   1881
39: Polizeibehoerde-Polit.Pol.: S 149/380: Bl.
110-113.   ADS 2:417

KUEHNE, R. (pastor of Evang. Luther. Friedens-Gemeinde
in Lincoln, Nebraska)

Letter   [1919-1922?]
44: Mappe Amerika.   ADS 2:599

KUENDEL, J. H.

Chicago: Estate of   1895
39: Reichs- u. auswaert.Ang.: E.IV.47 [alt.Sig.]
[Erbschaften]   ADS 2:275

KUEPPER, Friedrich Wilhelm (from Wermelskirchen)

German consular report from St Louis on trial
regarding [ownership?] of property of Kuepper,
wanted for fraud   1856
113: III. Hauptabt. Preuss. Min. des Auswaertigen:
Nr. 925a.2.   ADS 5:1

KUERVER, Catharina Maria. *See* Timm, Anna Catharina

KUFS, Emil (cigar worker)

Expulsion from Hamburg as Social Democrat & emigration to America   1881; 1885
39: Polizeibehoerde-Polit.Pol.: S 149/382: Bl.
136-139.   ADS 2:417

KUGELMANN, Willi (from Klein-Gerau, Landkreis Gross-
Gerau, HEL)

Emigration of Jewish person with wife & three sons
to the USA   1937
176: Abmelderegister 1937-1960.   ADS 7:229

KUMMER, Julius W. (died 27 May 1909 in Saginaw,
Michigan)

Estate of   1910
152: Abt.405: Wiesbaden: Rep.405/I,2: Nr. 3488.
                                       ADS 7:145

KUMPF, Peter A. (shoemaker; in Heppenheim)

Peter Kumpf & wife Katherina, born Silbermann, had
heirs in America   1909

149: Rep.49a: Nr. 565/8.   ADS 7:16

KURCH, Josef (died in Philadelphia, Pennsylvania)

Estate of   1901

152: Abt.405: Wiesbaden: Rep.405/I,2: Nr. 3487.
ADS 7:145

KURNICKI, L.

Seizure of money & negotiable paper by German consulate in New York from Kurnicki, sought by Hamburg police   1899

39: Reichs- u. auswaert.Ang.: P.III.6 [Polizei-
wesen]   ADS 2:303

KUERSCHNER, Gottfried Christian Carl (shoemaker)

Expulsion from Hamburg as Social Democrat & emigration to America   1880-1881

39: Polizeibehoerde-Polit.Pol.: S 149/31: Bl. 130-
135.   ADS 2:412

KUHN, Gustav (died in Atchison [Kansas?])

File on [estate?]   1900

227: Rep. MA 1921: A.St.I: Nr. 323.   ADS 9:66

KUNHARDT, George E. (in New York). See von Melle,
Emil

KUPFRIAN, -- (from Huetten)

Petition of Konrad Kupfrian & Eva Elisabeth Kupfrian, his sister, for title to the property of
their sister who has emigrated to America   1805

151: Bestand 81: Rep.E: 155.Amt Altengronau/Brand-
enstein IIIa: Nr. 25.   ADS 7:84

KURTZ, Benjamin (pastor; in Hagerstown, Maryland)

Report on Kurtz's arrival in Hamburg to raise
funds for a theological seminary in Gettysburg,
Pennsylvania   1826

39: Geistl. Minis.: II,9: Protokolle 1795-1850:
p.405 f.; see also pp. 521-522.   ADS 2:475

KURTZ, J. Daniel (first pastor of Lutheran congregation in Baltimore, Maryland)

Letter of introduction to Spiritual Ministry [Min-
istry of Religion] in Hamburg for his nephew
Benjamin Kurtz   1826

39: Geistl. Minis.: III,B, Fasc.14: Lose Akten
1826 (1).   ADS 2:479

KUSCHEL, C. R. (in Brooklyn, New York)

Prosecution for bigamy   1895

39: Reichs- u. auswaert.Ang.: S.IV.1.8 [alt.Sig.]
[Standesamtssachen]   ADS 2:279

KUSTNER, -- (corporal in Hessian? military unit in
America)

Interrogation of, during investigation of Trenton
[New Jersey] debacle   1778

151: Bestand 4: Kriegssachen; 328, Nr. 115.
DLC has.   ADS 7:37

KYRISS, Karl

Extradition of Kyriss from USA to Heilbronn, via
Hamburg   1900

39: Reichs- u. auswaert.Ang.: A.I.2 [Ausliefer-
ungen]   ADS 2:305

LACHER, Otto (died 1886 in America)

File on [estate?]   1887

227: Rep. MA 1921: A.St.I: Nr. 704.   ADS 9:69

LAGEMANN, August Eduard (mason)

Expulsion from Hamburg as Social Democrat & emigration to America   1881-1882

39: Polizeibehoerde-Polit.Pol.: S 149/260: Bl.
183-187.   ADS 2:416

LAGERSHAUSEN, August (wood dealer; from Gittelde)

Application by Lagershausen, guardian of the children of -- Pook (former toll-keeper) to obtain
funds due them as orphans for purpose of emigration to America   1857

56: L Neu Abt.129A: Gr.11, Nr. 45.   ADS 4:188

LAGERSHAUSEN, August (hand laborer; from Gittelde)

Punishment for fraud & misappropriation of funds &
his emigration to America   1866-1869

56: L Neu Abt.129A: Gr.11, Nr. 47.   ADS 4:189

LAHNSTEIN, Wilhelm (killed July 1916 in a tunnel ex-
plosion in Cleveland, Ohio; from Thalheim, Kreis
Limburg, HEL)

Estate of   1915-1916

152: Abt.405: Wiesbaden: Rep.405/I,2: Nr. 3489.
ADS 7:145

LAMBERT, Balthasar (in Skippack, near Philadelphia)

Power of attorney given by Lambert to Philipp Ulrich to enable him to collect debts owed him in
Moerzheim [RPL]   1739

211: Akt Nr. 11. ADS 8:79

LAMMERING, Albert, *alias* Schlachtmeyer (in Chicago)

Alleged murder of an innkeeper in Regensburg by
  Lammering 1913

227: Rep. MA 1936: A.St.: MA 98049. ADS 9:94

LAMPERT, Elise (in New York)

File on [estate?] 1886-1888
227: Rep. MA 1921: A.St.I: Nr. 697. ADS 9:69

LAMPRECHT, Johann (from Bamberg)

Measures taken for the mishandling of Lamprecht,
  who had represented himself to be an American
  citizen, by American ship officers 1854-1855

227: Rep. MA 1898: A.St.: Nr. 8. ADS 9:57

LANCK, Louis (Hamburg citizen)

Death certificate & estate of Lanck, who was
  drowned in the interior of California? 1850

42: Bestand H516: Hamburg Consulate in New York:
  1(a)(Anlage). ADS 2:521

LANDAUER, Emil (in Chicago)

Investigation regarding his military service obligation in Wuerttemberg 1900-1913

263: Rep. E 46-48: Minist.d.AA III: Verz.1: 148/17.
  ADS 10:43

LANDHINRICHS, -- (widow; in Hamburg)

American military pension for 1895

39: Reichs- u. auswaert.Ang.: P.I.3 [alt.Sig.]
  [Pensionen] ADS 2:277

LANDMANN, Seraphin (in New York)

Employment agency for maids in America 1883
227: Rep. MA 1921: A.V. IV: Nr. 11708. ADS 9:85

LANDMESSER, Jakob (from Durloch? Baden)

Emigration of 1770-1771
260: Nr. 217. ADS 10:1

LANDSBERG, Ludwig (sailor; from Zweibruecken; died
  in America)

File on [estate?] 1887
227: Rep. MA 1921: A.St.I: Nr. 703. ADS 9:69

LANG, Elisabeth. *See* Schreiner, Elisabeth, widow of
  Hartmann, born Lang

LANG, Eva Elisabetha, born Blum. *See* Balbierer,
  Philipp Nickel

LANG, Heinrich

Transmittal of a report from German consulate in
  New York to Mrs. Anna Lang, born Schmeil, regarding 1901

39: Reichs- u. auswaert.Ang.: G.Ia.1.16 [Geschaefts-
  verkehr] ADS 2:316

LANG, Johannes (from Erfelden, Landkreis Gross-
  Gerau, HEL)

Emigration to America 1850
169: Abt. XI,4. ADS 7:219

LANG, Nickel. *See* Balbierer, Philipp Nickel

LANGE, Carl Wilhelm Julius (mason)

Expulsion from Hamburg as Social Democrat & emigration to America 1880-1881

39: Polizeibehoerde-Polit.Pol.: S 149/58: Bl. 193-
  212. ADS 2:412

LANGE, Christian August (expeditor)

Expulsion from Hamburg as Social Democrat & emigration to America 1880-1881

39: Polizeibehoerde-Polit.Pol.: S 149/254: Bl.
  188-192. ADS 2:415

LANGE, F.

San Francisco: Death certificate of 1881
39: Bestand Senat: Cl.VIII, No.X, Jg.1881 (Register) 33 [Nordamerika] ADS 2:247

LANGE, Johann (from Kirchweyhe)

Issuance of emigration permit 1863
52: Han.Des.80: Syke: B.aa.VI.4: Nr. 463.
  ADS 4:85

LANGE, Julius (in New York)

File on [estate?] 1891
227: Rep. MA 1921: A.St.I: Nr. 720. ADS 9:69

LANGE, J. Fr. *See* Tiedemann, Johann H. A.

LANGE, M. W. E.

New Orleans: Death certificate of 1879
39: Bestand Senat: Cl.VIII, No.X, Jg.1879 (Register) 30 [Nordamerika] ADS 2:241

LANGEN, Elisabetha (from Gernsheim, Landkreis Gross-Gerau, HEL)

Emigration of Elisabetha & Margaretha Langen to North America   1924

<u>171</u>: Abmelderegister 1912-1924.   ADS 7:222

LANGEN, Margaretha. *See* Langen, Elisabetha

LANGER, --

Estate of   1942
<u>148</u>: R XI.   ADS 6:261

LAPPE, P. *See* Unkelbach, P., born Lappe

LASSON, Peter

Inquiry from lawyers Bullock & Ferguson of Montgomery City, Alabama, regarding heirs of Lasson   1871

<u>39</u>: Bestand Senat: Cl.VIII, No.X, Jg.1871 (Register) 12 [Nordamerika]   ADS 2:215

LAST, --

Extradition to Berlin from America, via Hamburg, of -- Last & -- Singer   1910

<u>39</u>: Reichs- u. auswaert.Ang.: A.Ia.20 [Auslieferungen]   ADS 2:361

LAUB, Gustav (in Shambaugh, Iowa)

Estate of   1907
<u>39</u>: Reichs- u. auswaert.Ang.: E.IV.3 [Erbschaften]   ADS 2:346

LAUDECKER, Heinrich (died 1900 in Detroit)

File on [estate?]   1900-1901
<u>227</u>: Rep. MA 1921: A.St.I: Nr. 762.   ADS 9:69

LAUMANN, Margarethe (from Gernsheim, Landkreis Gross-Gerau, HEL)

Emigration to USA with four children   1901
<u>171</u>: Abmelderegister 1900-1912.   ADS 7:222

LAUN, Amalie (from Ruesselsheim, HEL)

Application for emigration to America   1860
<u>182</u>: Abt. XI,4.   ADS 7:235

LAUN, Jacob, II [Junior](from Kelsterbach, Landkreis Gross-Gerau, HEL)

Emigration to America   1854
<u>175</u>: Abt. XI,4.   ADS 7:228

LAUN, Katharina. *See* Bommarius, Friedrich

LAUN, Philipp, II [Junior]. *See* Bommarius, Friedrich

de LAUNEY, Maria Ursula (died in New York)

Legality of death certificate for Maria Ursula de Launey, one of the heirs of -- Friedel, a surgeon in Bergen   1803-1812

<u>151</u>: Bestand 81: Rep.E: 21.Amt Bergen: I.a: Nr. 41.   ADS 7:82

LAUPPE, --

Estate of   1941
<u>148</u>: R XI.   ADS 6:261

LAUTENSCHLAEGER, Barbara (from Wembach, Landkreis Gross-Gerau, HEL)

Emigration to America   1854
<u>173</u>: Gesinderegister 1837-1843 [of Goddelau, Landkreis Gross-Gerau, HEL]: Nr. 113.   ADS 7:224

LAUTENSCHLAEGER, Barbara. *See also* Loenegens, Barbara, born Lautenschlaeger

LAUTENSCHLAEGER, Peter (of Biebesheim, Landkreis Gross-Gerau, HEL)

Emigrant to America [in list of emigrants . . .]   1859

<u>164</u>: Abt. XI,4.   ADS 7:214

LAUTERBACH, Andreas (died 1891 in Baltimore, Maryland)

Declaration of marital status   1899
<u>227</u>: Rep. MA 1921: A.St.I: Nr. 749.   ADS 9:69

LAYER, Eva (died in America)

File on [estate?]   1883-1884
<u>227</u>: Rep. MA 1921: A.St.I: Nr. 686.   ADS 9:69

LEDERHAUS, Ernst (of Flecken Salzdetfurth)

Proposed transportation of convict Lederhaus to America   1846-1854

<u>52</u>: Han.Des.80: Hildesheim I: E.DD.12a: Nr. 1069.   ADS 4:88

LEDERLE, Karl. *See* Astor, John Jacob

LEE, Edward

Hamburg police search for Lee, wanted in Niagara Falls [New York]   1908

<u>39</u>: Reichs- u. auswaert.Ang.: P.III.1.3   ADS 2:353

LEHMANN, Carl

Search for Lehmann's whereabouts in North America
1878

39: Bestand Senat: Cl.VIII, No.X, Jg.1878 (Register) 28 [Nordamerika]   ADS 2:237

LEHMANN, Ernst Heinr. Phil.

Petition of L. J. Lehmann to have his son Ernst H.
P. Lehmann exempted from American military service   1863

39: Bestand Senat: Cl.VIII, No.X, Jg.1863 (Register) 1 [Amerika]   ADS 2:182

LEHMANN, John

Request for information by Reichskanzler regarding
Lehmann's arrest in New York   1908

39: Reichs- u. auswaert.Ang.: P.III.13 [Polizei-
wesen]   ADS 2:353

LEHMANN, Wilhelm (from Nauroth, HEL)

Passage money for emigration to America 1853-1854

152: Abt.231: Langenschwalbach: Nr. 2873.
                               ADS 7:133

LEIBL, Josef Karl (first class seaman [Obermatrosse]
of the Reserve; in Chicago)

Personal file on   1913

227: Rep. MA 1921: A.V. V: Nr. 16387.   ADS 9:91

LEIDERSDORF, William (Wolf)

Alleged estate in America   1904

227: Rep. MA 1936: A.St.: MA 98054.   ADS 9:96

LEIDING, Mrs. M. C. L., born Dettmer

New York: Letter regarding the affairs of Mrs.
Leiding   1878

39: Bestand Senat: Cl.VIII, No.X, Jg.1878 (Register) 4 [Nordamerika]   ADS 2:236

LEIDING, --

New York: Letter from Dr. Carl Hartmann regarding
Leiding estate   1880

39: Bestand Senat: Cl.VIII, No.X, Jg.1880 (Register) 37 [Nordamerika]   ADS 2:244

LEIKAUF, Josef (from Katzdorf; died in Syracuse [New
York?])

Confession of murder committed by him   1890

227: Rep. MA 1921: A.St.I: Nr. 716.   ADS 9:69

LEIMBROCK, Jobst (from Wehringsdorf)

Transportation [to America] of   1842

54: Rep.122, VII.B: Groenenberg-Melle: Regiminalia:
Fach 15, Nr. 6.   ADS 4:138

LEINER, Julius (in America)

Search for   1887

227: Rep. MA 1921: A.St.I: Nr. 1589.   ADS 9:73

LEINER, Ludwig (journeyman maker of woolen material;
from Hornbach, Duchy of Zweibruecken; died in
Philadelphia)

Estate of   1784-1794

201: Zweibruecken III: Akt Nr. 2027 I.   ADS 8:59

LEITHMANN, Jakob (died in America)

File on [estate?]   1887

227: Rep. MA 1921: A.St.I: Nr. 702.   ADS 9:69

LEMKE, Oscar

Hamburg police search, at the request of American
police authorities, for Lemke [formerly] of Chicago   1900

39: Reichs- u. auswaert.Ang.: P.III.1.2 [Polizei-
wesen]   ADS 2:310

LEMKEN (LEMCKEN) Hermann Heinrich (pastor; in Ebenezer, Georgia)

Pastoral call to succeed Pastor -- Gronau; diary
of voyage from Halle to London   1745

500: Halle: Missionsbibliothek des Waisenhauses:
Abteilung H IV (Nordamerika) Fach J, Nr. 14;
Nr. 9.   Not in ADS

LENHARDT, -- (lived in Philadelphia)

Search for her   1898

227: Rep. MA 1921: A.St.I: Nr. 1606.   ADS 9:73

LENSCH, Heinrich Asmus (newspaper peddler)

Expulsion from Hamburg as Social Democrat & emigration to America   1880-1881

39: Polizeibehoerde-Polit.Pol.: S 149/30: Bl. 247-
250.   ADS 2:412

LENZ, Johann Christian (from Laubueschbach, HEL; in
America?)

Twenty-four letters from him to his brother Georg
Wilhelm Lenz [in Germany]   1849-1888

197: Abt.III. Familienarchiv: Nr. 17.   ADS 7:260

LIESER, Karl (died in Hartford [Connecticut])

File on [estate?]   1887

<u>227</u>: Rep. MA 1921: A.St.I: Nr. 705.   ADS 9:69

LIMPERT, Andreas (chronic alcoholic; from Oberweida, Thuringia)

Emigration of   1841

<u>500</u>: Weimar: Thueringisches Staatsarchiv: Dermbach
        II: Gb.2: Nr. 63-92; 93-114; 144-183; 186;
        188.   Not in ADS

LINCK, Louis (in Darmstadt)

Had heirs in America   1917

<u>149</u>: Rep. 49a: Nr. 565/10.   ADS 7:16

LINDEMANN, Anna Elisabetha, born Imbsweiler (from Unkenbach, Duchy of Zweibruecken)

Division of property & details of her emigration
    to Pennsylvania   1739-1792

<u>201</u>: Archivabt. Herzogtum Zweibruecken: Akt Zwei-
        bruecken I: Nr. 1346: Unkenbach.   ADS 8:56

LINDEMANN, August

Emigration? of   1900-1917

<u>115</u>: Rep. D9 III: Abt.E: Nr. 144.   ADS 6:34

LINDENBERG, Emilie

Chicago: Estate of   1904

<u>39</u>: Reichs- u. auswaert.Ang.: E.IV.25 [Erbschaften]
                                           ADS 2:335

LINDER, Carl (member Moravian Brethren; in Okak, Labrador)

Observations made by   1865

<u>500</u>: Herrnhut: Archiv: Rep.15.Ka: Nr. 9.
                                          Not in ADS

LINDHEIMER-SONNENBERG, Thilde (in Chicago)

Donation to community of Gross-Umstadt   n.d.

<u>149</u>: Rep. 48/1: Konv.8, Fasz.10.   ADS 7:14

LINDNER, Edward (civil engineer; in New York)

Claim for back pay? for Bavarian military service
                                                1867

<u>227</u>: Rep. MA 1921: A.V. V: Nr. 16156.   ADS 9:91

LINDNER, Michel (from Sondheim, Thuringia)

Passport for   1840

<u>500</u>: Weimar: Thueringisches Staatsarchiv: Dermbach
        II: Gb.2: Nr. 63-92; 93-114; 144-183; 186;
        188.   Not in ADS

LINDNER, Paul (a German citizen)

Extradition to Berlin from the USA, via Hamburg
                                                1914

<u>39</u>: Reichs- u. auswaert.Ang.: A.Ia.5 [Ausliefer-
        ungen]   ADS 2:383

LINGEN, Alexander (doctor; in West Alexandria, Preble County, Ohio)

Letter & newspaper clippings   1851

<u>151</u>: Verz.13: Unverzeichnet 16-18th centuries:
        Aktenpaket Nr. 4.   ADS 7:57

LINNERT, H. W. (an American citizen)

Planned repatriation of Linnert; request by Amer-
    ican consul in Hamburg for Linnert's release
    from insane asylum   1896

<u>39</u>: Reichs- u. auswaert.Ang.: H.IIa.8 [Heimschaff-
        ungsverkehr]   ADS 2:283

LINSENMEIER, A.

New York: Extradition to Baden of A. & K. Linsen-
    meier   1884

<u>39</u>: Bestand Senat: Cl.VIII, No.X, Jg.1884 (Regis-
        ter) 8 [Nordamerika]   ADS 2:253

LINZ, Else, born Mayer (from Nauheim, Landkreis Gross-Gerau, HEL)

Emigration of   1938?

<u>180</u>: Wiedergutmachungsakten.   ADS 7:233

LIPPMANN, Heinz

Emigration? of   1889-1905

<u>115</u>: Rep. D9 III: Abt.E: Nr. 143.   ADS 6:34

LIS, Fritz (baker)

Damage claim against City of Baltimore   1907

<u>39</u>: Reichs- u. auswaert.Ang.: G.Ia.7 [Geschaefts-
        verkehr]   ADS 2:347

LOCHMANN, Jacob (of Berkach, Landkreis Gross-Gerau, HEL)

Emigration to America with wife, three children, &
    one grandchild   1860

<u>163</u>: Abt. XI,4.   ADS 7:213

148: R XI.   ADS 6:261

LOVE, --, born Hovey (widow)

Washington: Plea for financial aid to educate her
two children   1887

39: Bestand Senat: Cl.VIII, No.X, Jg.1887 (Register) 18 [Nordamerika]   ADS 2:260

LUCKE, Christian (day laborer; from Muenchehof)

Emigration with family to America   1851

56: L Neu Abt.129A: Gr.11, Nr. 64.   ADS 4:190

LUEBKE, August

Washington: Estate of   1915

39: Reichs- u. auswaert.Ang.: E.IV.61 [Erbschaften]
ADS 2:388

LUECKE, Georg (in Hameln)

His attempts to promote emigration to South Carolina [missing?]   1753

52: Han.Des.74: Hameln: VIII.H.7: Nr. 1. ADS 4:71

LUEDDECKE, Carl (journeyman cabinetmaker; from
Volkersheim)

Emigration to America   1852

56: L Neu Abt.129A: Gr.11, Nr. 83.   ADS 4:191

LUEDEMANN, J. (in Staten Island, New York)

Estate of   1896

39: Reichs- u. auswaert.Ang.: E.IV.81 [alt.Sig.]
[Erbschaften]   ADS 2:282

LUEDERS, Charles

New York: Letter regarding Lueders, deceased 1878

39: Bestand Senat: Cl.VIII, No.X, Jg.1878 (Register) 43 [Nordamerika]   ADS 2:238

LUEGER, Thomas (day laborer in Ochsenfurt)

Claim against the Williamsburgh Savings Bank [state
not given in ADS description]   1897

227: Rep. MA 1921: A.St.I: Nr. 742.   ADS 9:69

LUEHMANN, L.

New York: Estate of   1900

39: Reichs- u. auswaert.Ang.: E.IV.80 [Erbschaften]
ADS 2:307

LUEHT, Rosalie. *See* Baumann, Rosalie, born Lueht

LUERS, Johann Heinrich (first bishop of Fort Wayne,
Indiana, 1858-1871)

Mentioned in "Bischofe des Oldenburger Muenster-
landes" [Bischops from the Oldenburger Bishopric]
*Veroeffentlichungen aus dem Heimatkalendar fuer
das Oldenburger Muensterland* (1952), p. 115

ADS 4:248

LUETGENS [*or* Luetjen] J. F. Ch. (in New York)

Inquiry by private person in Germany regarding
1899

39: Reichs- u. auswaert.Ang.: N.Ib.1.7 [Nachforsch-
ungen]   ADS 2:303

LUETJENS, -- (in Lincoln County, Oregon)

Estate of   1905

39: Reichs- u. auswaert.Ang.: E.IV.81 [Erbschaften]
ADS 2:339

LUGGE, Wilhelm (in New York)

Inquiry by private person in Germany regarding
1902

39: Reichs- u. auswaert.Ang.: N.Ib.1.4 [Nachforsch-
ungen]   ADS 2:326

LUM, Levin

Hamburg police search for Lum, wanted in America
1909

39: Reichs- u. auswaert.Ang.: P.III.1.4 [Polizei-
wesen]   ADS 2:358

LUND, -- (married couple; in Philadelphia)

Estate of   1898

39: Reichs- u. auswaert.Ang.: E.IV.35 [alt.Sig.]
[Erbschaften]   ADS 2:296

LUTZ, Georg (died in America)

Guardianship for   1888

227: Rep. MA 1921: A.St.I: Nr. 709.   ADS 9:69

LUTZ, -- (an American citizen)

Residency in Bavaria   n.d.

253: Fach 162, Fasz. 15.   ADS 9:173

MAASS, Carl Friedrich (typesetter)

Expulsion from Hamburg as Social Democrat & emigration to America   1880-1881

39: Polizeibehoerde-Polit.Pol.: S. 149/97: Bl. 1-
10.   ADS 2:413

<u>39</u>: Reichs- u. auswaert.Ang.: P.III.4 [alt.Sig.]
[Polizeisachen]   ADS 2:285

MARIKOFSKY, Roman (from Erfelden, Landkreis Gross-
Gerau, HEL)

Emigration to North America   1921
<u>169</u>: Abmelderegister 1920-1924.   ADS 7:220

MARKERT, M. A. (died in Charleston)

File on [estate?]   1898
<u>227</u>: Rep. MA 1921: A.St.I: Nr. 793.   ADS 9:69

MARKOWITZ, S. (an American citizen)

Arrest of   1918
<u>39</u>: Reichs- u. auswaert.Ang.: P.III.39 [Polizei-
wesen]   ADS 2:395

MARON, Peter

Complaint of emigrant Maron against Hamburg firm
1877
<u>39</u>: Bestand Senat: Cl.VIII, No.X, Jg.1877 (Regis-
ter) 11 [Auswanderer-Deputation]   ADS 2:233

MARONNE, --

Estate of   1941
<u>148</u>: R XI.   ADS 6:261

MARQUARDT, --. *See* Marckworth, --

MARQUIS, Justine, born Guempel

New York: Estate of   1911
<u>39</u>: Reichs- u. auswaert.Ang.: E.IV.28 [Erbschaften]
ADS 2:368

MARS, Herm. (a miller)

New York: Damages payable to Mars by agent Georg
Hirschmann in Hamburg   1881
<u>39</u>: Bestand Senat: Cl.VIII, No.X, Jg.1881 (Regis-
ter) 5 [Nordamerika]   ADS 2:246

von MARSCHALCK, Melusine (from Bockel, NL)

Emigrated to, & married in, America. Her father
was Carl *Freiherr* [Baron] von Marschalck & her
mother Louise von Dankwerth. Mother's letters
extant   [19th century]
<u>59</u>   ADS 4:227-228

MARSCHALCKEN, --

Baptismal entry for child of Marschalcken family

who emigrated to Pennsylvania, via Duisburg-
Ruhrort   1750
<u>126</u>: Taufregister der Evang.Gem. Ruhrort. ADS 6:91

von MARSCHALL, Friedrich (member Moravian Brethren;
in America?)

Reports from von Marschall & Nathanael Seidel
1763-1764
<u>500</u>: Herrnhut: Archiv: Rep.14.A: Nr. 45.
Not in ADS

MARSCHALL, Gottlieb (from Kaltennordheim, Thuringia)

Passport application for   1840
<u>500</u>: Weimar: Thueringisches Staatsarchiv: Dermbach
II: Gb.2: Nr. 63-92; 93-114; 144-183; 186;
188.   Not in ADS

MARTENS, August (in Salmon Falls, Eldorado County,
California)

Notice of death of   1854
<u>39</u>: Bestand Senat: Cl.VIII, No.X, Jg.1854 (Regis-
ter) III.2 [Nordamerika-Varia]   ADS 2:165

MARTENS, -- (tailor; from Walle)

Emigration to America   1863
<u>50</u>: Rep.21a: Nr. 8902.   ADS 4:8

MARTENS [*or* Marcus?] -- (in Mexico)

San Francisco: Estate of   1884
<u>39</u>: Bestand Senat: Cl.VIII, No.X, Jg.1884 (Regis-
ter) 40 [Erbschafts-Amt]   ADS 2:252

MARTIN, Joh. P. (from Traben, Grafschaft Sponheim,
RPL; in Philadelphia)

File on the expropriation of property belonging to
Martin who emigrated many years before to Phila-
delphia   1767-1773
<u>200</u>: Abt.33: Vordere Grafschaft Sponheim u. beide
Grafschaften gemeinschaftlich: Nr. 9328.
ADS 8:11

MARTIN, Karl (bookbinder; died in Chicago)

File on [estate?]   1895
<u>227</u>: Rep. MA 1921: A.St.I: Nr. 773.   ADS 9:69

MARTINI, Ch. A. (in Georgetown, Colorado)

Estate of   1901
<u>39</u>: Reichs- u. auswaert.Ang.: D.Ih, 35 [Erbschaften]
ADS 2:315

MARTINI, Maria, formerly von Berkes, born Kielborn

New York: Letter from Dr. C. Manz, notary, regarding inheritance due Mrs. Martini   1879

39: Bestand Senat: Cl.VIII, No.X, Jg.1879 (Register) 57 [Nordamerika]   ADS 2:242

MARX, Carl (Dr.; writer; from Trier)

Official papers regarding Marx's application for
permission to emigrate to America, dated 6 November 1845. [In ADS there is a lengthy analysis of
this application, which pertains to the writer of
Das Kapital. Marx was at the time in Brussels]
                                          1845

200: Abt.442: Regierung Trier: Nr. 6722. ADS 8:38

MARX, David (died in San Francisco)

File on [estate?]   1898-1899

227: Rep. MA 1921: A.St.I: Nr. 797.   ADS 9:69

MARX, E. W. *See* Mathes, E. W., born Marx

MASCHMANN, Georg Johann (in Savannah, Georgia)

Estate of   1917

39: Reichs- u. auswaert.Ang.: E.IV.10 [Erbschaften]
                                     ADS 2:392

MASSA, Wilhelm (laborer in shoe factory; in America)

Petition for separation submitted by Louise Massa,
wife [in Germany?]   1894-1895

227: Rep. MA 1921: A.St.I: Nr. 771.   ADS 9:69

MASTERSON, E. F.

Extradition to USA   1902

39: Reichs- u. auswaert.Ang.: A.Ia.21 [Ausliefer-
ungen]   ADS 2:328

MATES, --. *See* Meetz, --

MATHES, E. W., born Marx (in Lynn, Massachusetts)

Death & estate of   1876

39: Bestand Senat: Cl.VIII, No.X, Jg.1876 (Register) 17 [Nordamerika]   ADS 2:232

Boston: Estate of   1877

39: Bestand Senat: Cl.VIII, No.X, Jg.1877 (Register) 2 [Nordamerika]   ADS 2:233

Estate of widow -- Mathes who died in Lynn, Massachusetts   1878

39: Bestand Senat: Cl.VIII, No.X, Jg.1878 (Register) 14 [Nordamerika]   ADS 2:237

MATIZA, Elisabetha. *See* Schmelzeisen, Anton

MATTFELDT, -- (theological candidate)

Testimony of the Spiritual Ministry [Ministry of
Religion] regarding Mattfeldt who hopes to take
charge of a congregation in America   1852

39: Geistl. Minis.: II,10: Protokoll des Seniors
Strauch: p.8.   ADS 2:475

MATTFELDT, F. H. J. (pastor)

Application of former candidate Mattfeldt to be
re-accepted by Hamburg Spiritual Ministry [Min-
istry of Religion]. He was examined in 1845 &
approved; received call to pastorate of German
Evangelical Lutheran church at Pekin [Illinois?]
in 1852 & returned to Hamburg in 1855   1855

39: Geistl. Minis.: III,B Fasc.19: Lose Akten 1855
(1).   ADS 2:482

MATTHAEI, Carl A.

Petition of Joh. Diedr. Matthaei to have his son
Carl A. Matthaei liberated from American military service   1863

39: Bestand Senat: Cl.VIII, No.X, Jg.1863 (Register) I [Amerika]   ADS 2:182

Report from Hamburg consul in Washington [D.C.]
regarding   1863

39: Bestand Senat: Cl.VIII, No.X, Jg.1863 (Register) II [Amerika]   ADS 2:184

Reports regarding the liberation of Lieutenant
Matthaei   1864

39: Bestand Senat: Cl.VIII, No.X, Jg.1864 (Register) II.   ADS 2:185

MATTHIES, J. (in Durango, Colorado)

Estate of   1910

39: Reichs- u. auswaert.Ang.: E.IV.44 [Erbschaften]
                                     ADS 2:362

MAUER, Jakob Philipp (from Raunheim, Landkreis Gross-
Gerau, HEL)

Emigration to America with wife & ten children
                                          1854

181: Raunheimer Evangelischen Pfarrei Familienregister (Lehrer Johannes Buxbaum). ADS 7:234

MAUER, Johann Georg Wilhelm (from Raunheim, Landkreis
Gross-Gerau, HEL)

Emigration to America with wife & three children
                                          1854

181: Raunheimer Evangelischen Pfarrei Familienregister (Lehrer Johannes Buxbaum) ADS 7:234

MAUS, Peter (from Ibacherhof; formerly American soldier)

Inquiry regarding his death   1885-1886
227: Rep. MA 1921: A.St.I: Nr. 1624.   ADS 9:73

MAUSEN, Jacob (from Amt Birkenfeld, Duchy of Zweibruecken). *See* Wittib, Jacob Mausen

MAY, F. M.

Planned arrest of May, an American citizen   1896
39: Reichs- u. auswaert.Ang.: P.III.2.6 [alt.Sig.]
    [Polizeisachen]   ADS 2:285

MAY, Johann (of Hochstadt, Hesse-Hanau)

Emigration [to Carolina?] of   1741
151: Bestand 80: Lit.B: Nr. 1.   ADS 7:79

MAY, Martin. *See* May, Philipp

MAY, Philipp (from Kirschhausen, Kreis Heppenheim)

File on emigrants Philipp & Martin May, in New
    York   1844-1848
149: Rep.39/1: Konv.135, Fasz.2.   ADS 7:9

MAY, --

Transportation to America of May, a convict from
    the prison in Celle   1841
52: Han.Des.80: Hildesheim I: E.DD.7i: Nr. 930.
                                            ADS 4:88

MAYER, Else. *See* Linz, Else, born Mayer

MAYER, Emanuel (in New York)

File on [estate?]   1835
227: Rep. Bayerische Gesandtschaft London: Nr. 899.
                                            ADS 9:47

MAYER, Friedrich (died in San Andreas, California)

Estate of   1894-1895
152: Abt.405: Wiesbaden: Rep.405/I,2: Nr. 3558.
                                            ADS 7:145

MAYER, Josef (of Crumstaedt, Landkreis Gross-Gerau,
HEL)

Emigration to Minneapolis [Minnesota] of Jewish
    person   1936
167: Abmelderegister 1926-1938.   ADS 7:217

MAYER, Karl (doctor; in New York)

Mayer, former army surgeon [Markgrafl. Gardechirurg]
    & his son Karl Mayer in New York   1815-1825
227: Bayerische Gesandtschaft London: Nr. 847.
                                            ADS 9:47

MAYER, Levi (died 22 June 1908 in Philadelphia)

Estate of   1909-1911
152: Abt.405: Wiesbaden: Rep.405/I,2: Nr. 3492.
                                            ADS 7:145

MAYER, -- (in America)

Petition for information regarding the son of pensioned Oberleutnant [First Lieutenant] Georg
    Mayer   1858-1879
227: Rep. MA 1921: A.St.I: Nr. 1610.   ADS 9:73

MAYER, -- (from Lemfoerde; died in Texas)

Affairs of   1856-1859
52: Han.Des.74: Diepholz: Abt.II,N: Nr. 12.
                                            ADS 4:66

MAYERFELD, Ferdinand (of Crumstaedt, Landkreis Gross-
Gerau, HEL)

Emigration to New York of Jewish person with wife
                                             1938
167: Abmelderegister 1938-1957.   ADS 7:217

MAYERFELD, Sali (of Crumstaedt, Landkreis Gross-
Gerau, HEL)

Emigration to New York of Jewish person with wife
    & two children   1938
167: Abmelderegister 1938-1957.   ADS 7:217

MEDEFIND, --. *See* Haase, --

MEDOMAK, --

Letters from an emigrant   1849
In possession of Oberstudienrat i.R. Dr. Diedrich
    Mueller, Kehdinger Muehren 30, Stade, NL.
                                            ADS 4:336

MEETZ, Mathilde (widow; died on 24 August 1901 in
San Francisco)

Estate of   1902
152: Abt.405: Wiesbaden: Rep.405/I,2: Nr. 3493.
                                            ADS 7:145
San Francisco: Estate of   1902
39: Reichs- u. auswaert.Ang.: E.IV.8 [Erbschaften]
                                            ADS 2:323

MEHLING, Georg (in America)

Invalid pension for Mehling, a former corporal
1885

227: Rep. MA 1921: A.V. V: Nr. 16203.   ADS 9:91

MEIER, Diedrich Hermann (of Baltimore, Maryland)

Bremen passport issued in Washington   1855
48: B.13.b.1.a.1: Nr. 11.
    DLC has.   ADS 3:11

MEIER, Fr. F.

San Francisco: Estate of   1911
39: Reichs- u. auswaert.Ang.: E.IV.83 [Erbschaften]
                                          ADS 2:369

MEIER, Johann (from Bayreuth; died in Virginia)

File on [estate?]   1821-1826
227: Rep. Bayerische Gesandtschaft London: Nr. 863.
                                          ADS 9:47

MEIER, L. (in Hoboken, New Jersey)

Estate of   1916
39: Reichs- u. auswaert.Ang.: E.IV.44 [Erbschaften]
                                          ADS 2:391

MEINECKE, Sanncho* (in New York)

Search for   1901
*So spelled
227: Rep. MA 1921: A.St.I: Nr. 1648.   ADS 9:74

MEINTZEN, --. See Tiedemann, Johann H. A.

MEISNER, Engelbert. See Meixner, Engelbert

MEISTER, Gustav

Chicago: Estate of   1913
39: Reichs- u. auswaert.Ang.: E.IV.39 [Erbschaften]
                                          ADS 2:379

MEISTERKNECHT, H. S.

Transmittal of report from German consulate in
  Chicago to person holding power of attorney from
  Meisterknecht   1901
39: Reichs- u. auswaert.Ang.: G.Ia.1.10 [Geschaefts-
    verkehr]   ADS 2:316

MEIXNER [Meisner?] Engelbert (in America)

Search for   1891
227: Rep. MA 1921: A.St.I: Nr. 1632.   ADS 9:74

MELCHIOR, Christine (from Leeheim, Landkreis Gross-
  Gerau, HEL)

Emigration to America   1855
173: Gesinderegister 1853-1894: Nr. 75. ADS 7:225

MELINE, J. F. (in Cincinnati, Ohio)

Correspondence with Hamburg consulate in America
  regarding the forwarding of a letter to H. Bergmann   1858
39: Bestand Senat: Cl.VIII, No.X, Jg.1858 (Regis-
    ter) II.1.   ADS 2:174

von MELLE, Emil (in Hamburg)

Genealogical papers of Emil von Melle include the
  following names of persons living in New York in
  1846:

  Alfred Beneke, Julie Hoose, George E. Kunhardt,
  Marie A. Robinson, Therese Robinson, Jane E.
  Webster, Alfred Schlesigner, Emma Hoyer, Agnes
  Faber, Oskar & Elfriede Zoellikoffer, born Faber,
  J. H. Faber, Catharina Faber, Constanze Faber,
  Ferd[inand] Karck, C. Henschel   1846
39: Bestand Familien Archives: Fam. v. Melle.
                                         ADS 2:488

MELTES, --

Extradition to Hungary from USA, via Hamburg 1913
39: Reichs- u. auswaert.Ang.: A.Ia.38 [Ausliefer-
    ungen]   ADS 2:377

MELTZIAHN, H. See Melzian, H.

MELZIAN, H. (correct spelling Meltziahn)

New York: Estate of   1900
39: Reichs- u. auswaert.Ang.: E.IV.2 [Erbschaften]
                                         ADS 2:307

MENDHEIM, Kate Olga

San Francisco: Attempt to clear up the inheritance
  matters of Kate Olga Mendheim, a minor 1877
39: Bestand Senat: Cl.VIII, No.X, Jg.1877 (Regis-
    ter) 52 [Nordamerika]   ADS 2:235

MENG, Gertrud (widow of schoolteacher; from Dalheim
  [Grafschaft Falkenstein]). See Meng, --

MENG, -- (schoolteacher; from Hohensuelzen [Graf-
  schaft Falkenstein, RPL])

Petition for permit to emigrate to America by the
  schoolteacher -- Meng & the schoolteacher's
  widow Gertrud Meng of Dalheim with two children
  & her mother; to join five brothers & sisters
  already in America   1796

201: Archivabt. Grafschaft Falkenstein: Akt Nr. 139. ADS 8:61

MENGENSDORF, Julius

Hamburg police search for Mengensdorf, at request of American police authorities; Mengensdorf wanted in Cleveland [Ohio] for fraud n.d.

39: Reichs- u. auswaert.Ang.: P.III.1.2 [Polizei-wesen] ADS 2:336

MENGER, Anton (from Grebenroth, HEL)

Domicile of 1853-1864

152: Abt.231: Langenschwalbach: Nr. 1386. ADS 7:133

MENNINGER, Karoline (died in Cincinnati)

File on [estate?] 1895-1898

227: Rep. MA 1921: A.St.I: Nr. 781. ADS 9:69

MERESS, A.

Extradition from USA to Hungary, via Hamburg 1909

39: Reichs- u. auswaert.Ang.: A.Ia.1.25 [Ausliefer-ungen] ADS 2:355

MERKEL, Johannes (from Rockenhausen; died in America)

File on [estate?] 1890

227: Rep. MA 1921: A.St.I: Nr. 796. ADS 9:69

MERTENS, Friedrich

Hamburg consulate in Baltimore, Maryland: Estate of Mertens 1890

39: Bestand Senat: Cl.VI, No.16p: vol.4b, Fasc.4c: Invol. 9. ADS 2:52

MESTERN, Joh. See Moering, --

METHE, Heinrich (cigar worker)

Expulsion from Hamburg as Social Democrat & emi-gration to America 1881-1897

39: Polizeibehoerde-Polit.Pol.: S 149/324: Bl. 171-177. ADS 2:416

METZ, Gabriel (from Gansingen; died in America)

Request by Jakob Ackva of Kirn a. N. for entry into the records of the estate of Gabriel Metz 1877

227: Rep. MA 1921: I.V.: Nr. 2598. ADS 9:77

METZGER, Elise (died 1899 in Alameda [California])

File on [estate?] 1900

227: Rep. MA 1921: A.St.I: Nr. 802. ADS 9:69

METZGER, Heinrich, II [Junior?](from Goddelau, Land-kreis Gross-Gerau, HEL)

Emigrant to America 1868

173: Ortsbuergerregister. ADS 7:224

METZGER, Theodor (died 25 April 1901 in America)

Estate of 1901-1902

152: Abt.405: Wiesbaden: Rep.405/I,2: Nr. 3495. ADS 7:145

METZGER, Ulrich (from Zoeschingen)

Emigration to America 1873-1874

230: Rep. Akten der Bezirksaemter: Dillingen: Nr. 1927. ADS 9:122

METZLER, -- (born in Herborn). See Bartholomaeus, Carl

Metzler, --. See Fleissner-Metzler, --

MEYER, Adolph

German consul in New York regarding estate of 1869

39: Bestand Senat: Cl.VIII, No.X, Jg.1869 (Regis-ter) 15 [Nordamerika] ADS 2:201

MEYER, Albert

Emigration file contains letters from friends & relatives in San Francisco 1859

55: Han.Des.74: Blumenthal: IX.j.1: Fach 301, Nr. 10. ADS 4:154

MEYER, Albert (in Rock Rapid, Iowa)

Inquiry by Hamburg Orphanage Commission regarding 1902

39: Reichs- u. auswaert.Ang.: W.I.5 [Waisenhaus-Ang.] ADS 2:328

MEYER, Amalie (in Belleville [Illinois?])

Search for J. C. A. Kroeckel, heir of Amalie Meyer 1882

39: Bestand Senat: Cl.VIII, No.X, Jg.1882 (Regis-ter) 3 [Nordamerika] ADS 2:248

MEYER, A. (schoolteacher; from Malstadt)

Pardon under the condition that he immediately emigrate to America   1863

62: D i, Fach 102, vol. 16 [Regiminalia, Band II]
ADS 4:230

MEYER, A. E. A. (from Hamburg)

German consulate at Louisville, Kentucky: Seeks information regarding whereabouts of Meyer 1872

39: Bestand Senat: Cl.VI, No.16p: vol.4b: Fasc.21 [Ermittlung]   ADS 2:60

MEYER, Carl (waiter; from Schoeningen)

Emigration of   1853

105: Ratsakten, Nr. 25:4.   ADS 4:332

MEYER, Christian Reinhard (cigar dealer)

Expulsion from Hamburg as a Social Democrat & emigration to America   1876-1880

39: Polizeibehoerde-Polit.Pol.: S 149/65b: Bl. 221-247.   ADS 2:413

MEYER, Friedrich

New York: Estate of   1898

39: Reichs- u. auswaert.Ang.: D.Ih.35 [Erbschaften]
ADS 2:296

MEYER, Fr. (from Waetzum)

Approval of funds for Meyer's transportation to America [Hildesheim]   1867

52: Han.Des.80: Hildesheim I: E.DD.26a: Nr. 1137.
ADS 4:89

MEYER, G. F.

San Francisco: Estate of   1877

39: Bestand Senat: Cl.VIII, No.X, Jg.1877 (Register) 1 [Nordamerika]   ADS 2:233

MEYER, Hans

Expulsion of naturalized American citizen
1889-1890

39: Bestand Senat: Cl.I, Lit.T, No.20: vol.2b, Fasc. 4 [Ausweisung]   ADS 2:20

Complaint of Meyer, an American citizen, regarding treatment by Hamburg police authorities   1894

39: Reichs- u. auswaert.Ang.: P.III.16 [alt.Sig.] [Polizeisachen]   ADS 2:271

MEYER, Heinrich (from Bremen)

Inquiry regarding emigrant to New Orleans   1854

48: B.13.b.14: Nr. 4-6.
DLC has?   ADS 3:34

MEYER, Henry F. (in New York)

Estate of   1869

39: Bestand Senat: Cl.VIII, No.X, Jg.1869 (Register) 58 [Nordamerika]   ADS 2:204

MEYER, Johann Hinrich

Emigration file contains letter from friend or relative in New York   1859

55: Han.Des.74: Blumenthal: IX.J.1: Fach 301, Nr. 10.   ADS 4:154

MEYER, J. P. (in Chicago)

Application for release from Hamburg citizenship
1874

39: Bestand Senat: Cl.VIII, No.X, Jg.1874 (Register) 21 [Nordamerika]   ADS 2:228

MEYER, Riccardo

Meyer's transport from Tarnowitz to America   1850

39: Bestand Senat: Cl.VIII, No.X, Jg.1850 (Register)   ADS 2:158

MEYER, Wilhelm (peddler)

Extradition from New York to Kassel, via Hamburg
1906

39: Reichs- u. auswaert.Ang.: A.Ia.10 [Auslieferungen]   ADS 2:342

MEYER, -- (in Lemfoerde; died in Houston, Texas)

Inquiry by Mr. & Mrs. -- Meyer regarding the estate of their son who died in Houston, Texas
1844-1869

52: Han.Des.80: Diepholz: Miscellanea: Nr. 648.
ADS 4:84

MEYER, -- (married couple)

Cincinnati [Ohio]: Estate of   1901

39: Reichs- u. auswaert.Ang.: E.IV.85 [Erbschaften]
ADS 2:316

MEYER, --

Chicago: Information on Meyer brothers   1878

39: Bestand Senat: Cl.VIII, No.X, Jg.1878 (Register) 44 [Nordamerika]   ADS 2:238

MEYER, --, born Kausers (in Chicago)

Information on Mr. & Mrs. Meyer requested by a

governmental office   1894

39: Reichs- u. auswaert.Ang.: N.Ia.14 [alt.Sig.]
    [Nachforschungen]   ADS 2:271

MEYER, --

Estate of   1940-1941

148: R XI.   ADS 6:261

MEYERHOF, --

Testimony taken from Meyerhof at request of New
    York court   1915

39: Reichs- u. auswaert.Ang.: J.Ia.2.4 [Justiz-
    sachen]   ADS 2:389

MEYERLING, Engel (from Bokeloh)

Expulsion of   1836

54: Rep.122, VIII: Reg. Meppen: Nr. 851. ADS 4:141

MICHEL, Jakob. *See* Ruhland, Philipp

MICHEL, Johann H. (from Goddelau, Landkreis Gross-
    Gerau, HEL)

Emigrant to America   1850

173: Ortsbuergerregister.   ADS 7:224

MICHLER, Johannes (from Mittelbuchen, Hessen-Hanau).
    See Schwind, Peter

MIGEON, -- (widow of J. H. Gossler)

Request for information by widow of Gossler, now
    married to -- Migeon, regarding inheritance 1872

39: Bestand Senat: Cl.VIII, No.X, Jg.1872 (Regis-
    ter) 1 [Nordamerika]   ADS 2:219

MILLER, Michael (steward; an American citizen)

Arrest [in Hamburg] of   1918

39: Reichs- u. auswaert.Ang.: P.III.32 [Polizei-
    wesen]   ADS 2:395

MILLER, Philipp. *See* Mueller, Katharina

MILLER, --

Estate of   1943

148: R XI.   ADS 6:261

MINGE, W.

Complaint of   1866

39: Bestand Senat: Cl.VIII, No.X, Jg.1866 (Regis-
    ter) 14 [Auswanderer-Deputation]   ADS 2:190

von MIRBACH, -- (in America). *See* Bedassem, J.

MISCHLICH, Elisabetha (from Ruesselsheim, HEL)

Emigration to America   1851

182: Abt. XI,4.   ADS 7:235

MISHOFF family (of Einbeck, NL)

Genealogical studies regarding   n.d.

72: Harlands Nachlass: Manuscripte Nr.9: vol. VIII.
                                   ADS 4:250

MITTELBERGER, D.

Estate of   1877

39: Bestand Senat: Cl.VIII, No.X, Jg.1877 (Regis-
    ter) 23 [Nordamerika]   ADS 2:234

MOEHLE, Friedrich (day laborer, from Opperhausen)

Emigration of a delinquent to America   1858-1859

56: L Neu Abt.129A: Gr.11, Nr. 74.   ADS 4:191

MOEHLMANN, B.

Claim against the USA for a disability pension
                                   1878

49: A.3.N.3: Nr. 10a.   ADS 3:76

MOEHN, Martha Alwine (a German national; in New York)

Imposition of a guardianship for   1907

39: Reichs- u. auswaert.Ang.: V.III.2 [Vormund-
    schaftswesen]   ADS 2:349

MOELLER, Caspar

Transportation to America   n.d.

54: Rep.122, VII.B.: Groenenberg-Melle: Regimin-
    alia: Fach 15, Nr. 5.   ADS 4:138

MOELLER, Henry D. (in Arizona, Colorado)

Estate of   1917

39: Reichs- u. auswaert.Ang.: E.IV.24 [Erbschaften]
                                   ADS 2:392

MOELLER, Johann Heinrich (died in National Military
    Home in Ohio)

Estate of   1899-1903

152: Abt.405: Wiesbaden: Rep.405/I,2: Nr. 3497.
                                   ADS 7:145

MOELLER, -- (in New York & Williamsburg [state not
    given in ADS description])

Gift by Moeller brothers of 1200 Reichsmarke gold

to the congregation of Rhade for the construction
of an organ   1853

112: VII, Fach 18.V.13.   ADS 4:346

MOENNICH, Johanna

New York: Estate of   1902

39: Reichs- u. auswaert.Ang.: E.IV.34 [Erbschaften]
                                           ADS 2:323

MOERGEN, Simon (from Dahn; died in America)

Estate of   1854-1873

227: Rep. MA 1921: A.St.I: Nr. 1609.   ADS 9:73

MOERING, --

New York: M[arks] 4430 to be delivered to Joh.
Mestern, Hamburg, from Moering estate settlement
                                                     1879

39: Bestand Senat: Cl.VIII, No.X, Jg.1879 (Register) 53 [Nordamerika]   ADS 2:242

MOHR, Christoph (from Rodheim)

Returned from Pennsylvania & is accused of enticing
citizens to emigrate   1754

151: Bestand 86: Nr. 51975.   ADS 7:88

MOHR, Friedrich Wilhelm

An emigrant?   n.d.

48: M.6.e.9 (g).   ADS 3:48

MOHR, J. (in Chicago). See Mohr, M.

MOHR, M. (in Chicago)

Inquiry by private person in Germany regarding M.
& J. Mohr & Mrs. -- Sievers   1899

39: Reichs- u. auswaert.Ang.: N.Ib.1.4 [Nachforsch-
ungen]   ADS 2:303

MOLITOR, Anton (from Inneringen, BWL)

Debts of a would-be emigrant to America   1848

262: Ho 193: Gammertingen: II 8104.   ADS 10:36

MOLITOR, --

Estate of   1941-1942

148: R XI.   ADS 6:261

MOLKA, C. (an American citizen)

Molka's complaint regarding Hamburg police arrest
                                                     1901

39: Reichs- u. auswaert.Ang.: P.III.4 [Polizei-
wesen]   ADS 2:319

MOLKENBUHR, Heinrich (messenger for health insurance
office)

Expulsion from Hamburg as Social Democrat & emigration to America   1881-1882; 1896

39: Polizeibehoerde-Polit.Pol.: S 149/318: Bl.
266-283.   ADS 2:416

MOLL, --

Baptismal entry for child of Moll family emigrating
to Pennsylvania, via Duisburg-Ruhrort   1750

126: Taufregister der Evang. Gem. Ruhrort.
                                           ADS 6:91

MOLLENHAUER, Carsten (from Elbersdorf)

Application of defendant Mollenhauer for [public]
aid for journey to America   1846

62: A, Fach 6, vol. 6.   ADS 4:232

MONTAG, Chr. (in Pennsylvania)

Estate of   1904

39: Reichs- u. auswaert.Ang.: E.IV.40 [Erbschaften]
                                           ADS 2:335

MONTAG, Eduard (from Goddelau, Landkreis Gross-
Gerau, HEL)

Emigrant to Georgia   1882

173: Gesinderegister 1853-1894: Deckblatt, Rueckseite.   ADS 7:224

MONTAG, W. (died in Baltimore [Maryland])

Estate of   1929

116: Minden: Praes.-regist.: I.L.Nr. 239. ADS 6:54

MORÉ (Moré), H. F.

Expulsion to America; sentenced to lengthy penitentiary sentence in Frankfurt/Main for political activities   1838

200: Abt.403: Oberpraesidium Koblenz (Rheinprovinz): Nr. 2528.   ADS 8:21

MORRISON, Peter (from New Bedford [state not given
in ADS description])

Inquiry of American consul regarding Morrison,
said to have been shanghaied aboard Norwegian
ship Enigheden, Captain Peterson commanding, &
brought to Hamburg   1838

39: Bestand Senat: Cl.VIII, No.X, Jg.1838 (Register) 1.a [Nordamerika]   ADS 2:136

MOSEL, -- (from Hitzenhausen)

Transportation of the Mosel brothers & their families to America 1843

<u>54</u>: Rep.122,X.B: Wittlage-Hunteburg: I. Fach 46, Nr. 19.
DLC has. ADS 4:144

MOSES, Max (an American citizen)

Expulsion of 1913

<u>39</u>: Reichs- u. auswaert.Ang.: A.IV.4 [Ausweisungen]
ADS 2:378

MOSES, Peter W.

San Francisco: Estate of 1913

<u>39</u>: Reichs- u. auswaert.Ang.: E.IV.73 [Erbschaften]
ADS 2:379

MOSHER, -- (an *alias*). *See* Schlitz, --

MOSS, Leopold (in New York)

Inquiry by private person in Germany regarding 1900

<u>39</u>: Reichs- u. auswaert.Ang.: N.Ib.1.3 [Nachforschungen] ADS 2:309

MOST, Johannes (publisher of *Freiheit*; born in Augsburg in 1846; died in Cincinnati, Ohio, in 1906)

Hamburg police file on [entirely newspaper clippings] 1846-1906

<u>39</u>: Polizeibehoerde-Polit.Pol.: S 1015. ADS 2:418

MOTH, Louis (in New Orleans)

Questions his prosecution for exchange operations 1877

<u>39</u>: Bestand Senat: Cl.I: Lit.T, No.7, fol.6, Fasc. 6a [Anfrage] ADS 2:4

New Orleans: Inquiry from Moth regarding prosecution in the event of his return to Hamburg 1877

<u>39</u>: Bestand Senat: Cl.VIII, No.X, Jg.1877 (Register) 5 [Nordamerika] ADS 2:233

MOTT, Charles F.

Hamburg police search for Mott, wanted in America 1916

<u>39</u>: Reichs- u. auswaert.Ang.: P.III.1.1 [Polizeiwesen] ADS 2:391

MOTTE, Caspar (a notary?)

File of the Fulda administration regarding Motte, impressed into the Hessen-Hanau Freikorps 1781

<u>151</u>: Bestand 81: Rep.E: 1.Gen.IV.b: Nr. 8.
ADS 7:81

MOYER, G. H. Chr. (from Lemfoerde; died in Texas)

Correspondence with Hannoverian foreign minister regarding the estate of 1844-1847

<u>48</u>: C.26: Nr. 15.
DLC has. ADS 3:45

MUEGGE, Carl Friedrich August (from Gittelde)

Application for passport to America 1841

<u>56</u>: L Neu Abt.129A: Gr.11, Nr. 43. ADS 4:188

MUEHLENBERG, Heinrich Melchior (organizer & patriarch of the Lutheran church in North America; born in Einbeck in 1711, died in New Hannover, Pennsylvania in 1787)

Data regarding the father of n.d.

<u>72</u>: Schubmacher-Gildenbuch. ADS 4:251

Diaries 1742; 1748; 1753; 1759-1760
Letters & reports 1746-1747; 1753-1754

<u>500</u>: Halle: Missionsbibliothek des Waisenhauses: Abteilung H IV (Nordamerika) Fach D, Nrs. 1; 7b; 13-15; 17 [film] Not in ADS

[Heretofore] unlisted manuscripts 1, 4 n.d.
Diaries 1752; 1779-1781

<u>500</u>: Halle: Missionsbibliothek des Waisenhauses: Unverzeichnetes Stuecke [on reels with Fach D material above] Not in ADS

MUEHLENBERG, Heinrich Melchior. *See also* Rauss, Lucas

MUEHLENBERG, -- (son of Pastor Heinrich M. Muehlenberg)

Contract, termination thereof, & correspondence regarding, with businessman -- Niemeier in Luebeck [apparently apprenticeship agreement] 1765

<u>500</u>: Halle: Missionsbibliothek des Waisenhauses: Abteilung H IV (Nordamerika) Fach A, Nr. 18.
Not in ADS

MUEHLENBERG family (of Einbeck, NL)

Genealogical studies regarding n.d.

<u>72</u>: Harlands Nachlass: Manuscripte Nr.9: vol.VIII.
ADS 4:250

MUEHLIG, Friedrich (died in Philadelphia)

File on [estate?] 1900-1901

<u>227</u>: Rep. MA 1921: A.St.I: Nr. 803. ADS 9:69

MUELLER, Barbara (widow; from Oberschleichach)

File on   1891

227: Rep. MA 1921: A.V. IV: Nr. 10251.   ADS 9:82

MUELLER, Christian (from Trebur, Landkreis Gross-
Gerau, HEL)

Emigration to America of a person who had been
registered in the community in 1831   n.d.

183: Abt. XI,4.   ADS 7:237

MUELLER, Conrad

San Francisco: Estate of   1896

39: Reichs- u. auswaert.Ang.: E.IV.15 [alt.Sig.]
[Erbschaften]   ADS 2:282

MUELLER, Elisabeth Margaretha, born Mueller

Distribution of the property of Elisabeth Margaretha Mueller, born Mueller, the wife of Johann
Mueller, who has emigrated to America   1832

152: Abt.237: Rennerod: Nr. 2144.   ADS 7:135

MUELLER, Elisabeth (died in Pine Grove, Ohio)

Estate of   1902

152: Abt.405: Wiesbaden: Rep.405/I,2: Nr. 3496.
                                          ADS 7:145

MUELLER, Eva (died in Philadelphia)

Estate of   1900-1905

152: Abt.405: Wiesbaden: Rep.405/I,2: Nr. 3498.
                                          ADS 7:145

MUELLER, Franz (died in Westfal Township, Pike
County, Pennsylvania)

Estate of   1902

152: Abt.405: Wiesbaden: Rep.405/I,2: Nr. 3499.
                                          ADS 7:145

MUELLER, F. A.

Expulsion to America; sentenced to lengthy penitentiary sentence in Frankfurt/Main for political activities   1838

200: Abt.403: Oberpraesidium Koblenz (Rheinprovinz): Nr. 2528.   ADS 8:21

MUELLER, Georg (from Nauheim, Landkreis Gross-Gerau,
HEL)

Emigration of   1883

180: Ortsbuergerregister 1761-1886.   ADS 7:233

MUELLER, Georg Heinrich (from Duchy of Braunschweig;
murdered in New Orleans)

Correspondence regarding, & search for heirs of
                                          1853

48: B.13.b.4: Nr. 34.
DLC has?   ADS 3:21

MUELLER, Guntram (from Glatt, BWL; in New York)

Death certificate for   n.d.

262: Ho 202: Haigerloch: I 1459, 4500.   ADS 10:37

MUELLER, Heinrich (in Milwaukee, Wisconsin)

Estate of   1908

39: Reichs- u. auswaert.Ang.: E.IV.91 [Erbschaften]
                                          ADS 2:352

MUELLER, Heinrich Friedrich (from Hardegsen)

Transportation to America   1850

52: Han.Des.80: Hildesheim I: E.DD.39a: Nr. 1507.
                                          ADS 4:90

MUELLER, Heinrich. *See* Mueller, Philipp

MUELLER, Herm. (in Chicago)

Inquiry by Hamburg Orphanage Commission regarding
                                          1900

39: Reichs- u. auswaert.Ang.: W.I.5 [Waisenhaus-
Ang.]   ADS 2:312

MUELLER, H.

Refusal to permit the entry [into Hamburg] of
Mueller, a Wuerttemberg citizen, arriving from
New York   1894

39: Reichs- u. auswaert.Ang.: P.III.10 [alt.Sig.]
[Polizeisachen]   ADS 2:271

MUELLER, Jakob (died in Quincy, Illinois)

Estate of   1895

152: Abt.405: Wiesbaden: Rep.405/1,2: Nr. 3560.
                                          ADS 7:145

MUELLER, Johann (died at poor farm in Douglas County
[state not given in ADS description])

Estate of   1900

152: Abt.405: Wiesbaden: Rep.405/I,2: Nr. 3500.
                                          ADS 7:145

MUELLER, Johann. *See* Mueller, Elisabetha Margaretha,
born Mueller

MUELLER, Johann Jost (widow of; in Suedershausen, HEL). *See* Brachthaeuser, Daniel

MUELLER, Johannes (of Suedershausen, HEL). *See* Brachthaeuser, Daniel

MUELLER, John (in Burlington [state not given in ADS description])

Inquiry by Hamburg Orphanage Commission regarding Mathilde A. W. E. Krueger   1899

39: Reichs- u. auswaert.Ang.: W.I.2 [Waisenhaus-Ang.]   ADS 2:305

MUELLER, Joseph

New Orleans: Estate of   1909

39: Reichs- u. auswaert.Ang.: E.IV.29 [Erbschaften]
                                               ADS 2:357

MUELLER, J. R.

German consular report from New York regarding Mueller's accusation of fraud against J. R. Reinicke   1870

39: Bestand Senat: Cl.VIII, No.X, Jg.1870 (Register) 16 [Nordamerika]   ADS 2:210

MUELLER, Katharina (in America)

Inquiry regarding the estate of Philipp Miller*
*So spelled   1886

227: Rep. MA 1921: A.St.I: Nr. 1628.   ADS 9:73

MUELLER, Katherine (died in Cincinnati, Ohio)

Estate of   1909

151: Bestand 180: Homberg: Acc.1934/44: Nr. 33.
                                               ADS 7:118

MUELLER, Leo

Repatriation from Los Angeles to Hamburg   1911

39: Reichs- u. auswaert.Ang.: H.IIb.5 [Heimschaffungsverkehr]   ADS 2:369

MUELLER, Luise (from Roettau; in America?)

Power of attorney given to Johann Friedrich Stadel in Langenbach   n.d.

197: Abt.III: Familienarchiv: Nr. 1.   ADS 7:260

MUELLER, Michael (in America)

Investigation of Michael Mueller's marriage in America, as requested by Luise Mueller, his wife, in Munich   1890

227: Rep. MA 1921: A.St.I: Nr. 1629.   ADS 9:74

MUELLER, Nicolaus (in America)

Petition of Peter Mueller, brother, & Fr. Frichhorn, brother-in-law, for the property of Nicolaus Mueller, a soldier "who remained in America"
                                          1790-1791

151: Bestand 81: Rep.E: 130.Stadt Steinau II: Nr. 28.   ADS 7:83

MUELLER, Oswald (from Osnabrueck; killed in battle near Richmond [Virginia?])

Estate of   [1860s?]

53: Bestand 31: Oldenburg: 31-15-23-65. ADS 4:114

MUELLER, Ottmar (from Krumbach; soldier in U.S. Civil War)

Estate of Mueller who was killed during [Civil?] War; herein the notarization of two certifications   1866

227: Rep. MA 1921: A.St.I: Nr. 60.   ADS 9:64

MUELLER, Philipp (born 1884; from Klein-Rohrheim, Landkreis Gross-Gerau, HEL)

Emigration to America with Heinrich Mueller   n.d.

177: Familien- und Gemeindebuch der Realschule Gernsheim zur Einweihung des neuen Schulgebaeude 1911 [commemorative publication celebrating inauguration of new school building in Gernsheim 1911].   ADS 7:230

MUELLER, W. (in Louisiana)

Estate of   1902

39: Reichs- u. auswaert.Ang.: E.IV.25 [Erbschaften]
                                               ADS 2:323

MUELLER, -- (pastor; in Dayton, Ohio)

Correspondence [with Deaconess House]   1891-1895

120: Akte Frank Buchman (D II B)   ADS 6:75

MUELLER, -- (from Henigsen)

Voluntary collection of money for his emigration
                                               n.d.

52: Han.Des.74: Westen-Thedinghusen: I.96: Nr. 46.
                                               ADS 4:80

MUELLER, --

Estate of   1866

39: Bestand Senat: Cl.VIII, No.X, Jg.1866 (Register) I.1.   ADS 2:190

MUELLER, -- (member Moravian Brethren; in North Carolina). *See* Danz, --

MUND, H. (from Sievershausen)

Application for aid to transport Mund, a convict in chains, to America [at Einbeck]   1865

52: Han.Des.80: Hildesheim I: E.DD.71a: Nr. 1900.
                                            ADS 4:91

MUNDI, Anna. *See* Fries, Anna, born Mundi

MURRELL, John K. *See* Kratz, Charles

NAEGELE, Max (in America)

Family situation   1888-1892

227: Rep. MA 1921: A.St.I: Nr. 1654.   ADS 9:74

NALBANDIAN, Vahan

Hamburg police search for Nalbandian, wanted in America   1909

39: Reichs- u. auswaert.Ang.: P.III.1.6 [Polizei-
    wesen]   ADS 2:359

NANNE, Gebke (unmarried)

Transportation to America from Emden   1866

50: Rep.21a: Nr. 9017.   ADS 4:7

NAPP, Jacob (Bavarian national? in Buffalo [New York?])

Steam engine for small tugs   1862

227: Rep. MA 1895: A.V.: Nr. 599.   ADS 9:56

NATHAN, A. (in San Francisco)

Estate of   1884

39: Bestand Senat: Cl.VIII, No.X, Jg.1884 (Regis-
    ter) 22 [Erbschafts-Amt]   ADS 2:252

NATHAN, Hermann (in California)

Estate of   1918

39: Reichs- u. auswaert.Ang.: E.IV.121 [Erbschaften]
                                        ADS 2:394

NAUMANN, E. A.

San Francisco: Estate of   1908

39: Reichs- u. auswaert.Ang.: E.IV.63 [Erbschaften]
                                        ADS 2:352

NEB, H. G. C. *See* Weiss, H.

NEFF, Jacob

Emigration of   1786-1789

201: Archivabt. Leiningen-Hardenburg: Akt Nr. 135.
                                        ADS 8:62

NEHLSEN, Cl. Chr. (in Davenport, Iowa)

Estate of   1906

39: Reichs- u. auswaert.Ang.: E.IV.21 [Erbschaften]
                                        ADS 2:343

NEISSER, Georg (member Moravian Brethren; in Pennsylvania?). *See* Wiegner, Christ.

NEITZEL, C. A. (in Tranzfelde)

Inquiry by Neitzel regarding emigration to America via Hamburg   1848

39: Bestand Senat: Cl.VIII, No.X, Jg.1848 (Regis-
    ter) 2 [Auswanderer-Verhaeltnisse]   ADS 2:152

NEITZKE, George (in White River, Michigan)

Estate of   1908

39: Reichs- u. auswaert.Ang.: E.IV.15 [Erbschaften]
                                        ADS 2:351

NEPOMICENA, -- (an American citizen)

Nepomicena not to be granted [Hamburg] citizenship
                                        1831

39: Bestand Senat: Cl.VIII, No.X, Jg.1831 (Regis-
    ter) IV.4 [Buergerrecht]   ADS 2:131

von NESS, Jacob (in Chicago)

Death certificate for   1866

52: Callenberger Brief-Archiv Des.15: Privatakten
    Band 1: N, Nr. 9.   ADS 4:55

NESSE, Heinrich (from Ahnebergen)

Transportation to America   1849

52: Han.Des.74: Westen-Thedinghusen: I.56: Nr. 44.
                                        ADS 4:80

NETTÉ (Netté), -- (surgeon's helper; in Texas)

Service contract for German colonies in Texas 1844

154: CA 4cl: Nr. 39.
     DLC has.   ADS 7:177

NEU, Henry. *See* New, Henry

NEU, Hermann (from Moerfelden, Landkreis Gross-Gerau, HEL)

Lived in Moerfelden, with four family members, in 1933; emigrated to USA

179: Wiedergutmachungsakten, Ordner Nr. 061-00/05.
                                        ADS 7:232

NEUBASS, Max (U.S. soldier; died at Manila [Philip-
pine Islands])

Estate of   1900
152: Abt.405: Wiesbaden: Rep.405/I,2: Nr. 3503.
                                       ADS 7:145

NEUBAUER, Friedrich (in Lopez, Pennsylvania)

Transmittal of money to emigrant Neubauer   1895
39: Reichs- u. auswaert.Ang.: A.III.7 [alt.Sig.]
    [Auswanderung]   ADS 2:274

NEUENHAUS, --

Estate of   1943
148: R XI.   ADS 6:261

NEUMANN, Andreas (in St. Louis [Missouri])

Inquiry by private person [in Germany] regarding
                                               1896
39: Reichs- u. auswaert.Ang.: N.Ib.12 [alt.Sig.]
    [Nachforschung]   ADS 2:284

NEUMANN, Hans Heinrich (cabinet maker)

Expulsion from Hamburg as Social Democrat & emi-
    gration to America   1880-1881
39: Polizeibehoerde-Polit.Pol.: S 149/187: Bl.
    386-410.   ADS 2:414

NEUMANN, I. F. (merchant; in Charlestown [South Carolina?])

Claim of -- Sittmann, Berlin, widow   1794
500: Berlin-Dahlem: Preus.Geh. Staatsarchiv: Rep.
     XI, 21a, Conv.2; 2b.   Not in ADS

NEUMANN, Robert (pastor; in New York)

Warning regarding emigration propaganda made by
    Pastor Neumann, who has no pulpit, on behalf of
    Virginia, North & South Carolina   1881
115: Rep.G.29/8:Moenchen-Gl.: Nr. 609. ADS 6:40

NEUMANN, Siegfried Kurt (from Nauheim, Landkreis
Gross-Gerau, HEL)

Emigration of   1938?
180: Wiedergutmachungsakten.   ADS 7:233

NEUMANN, -- (in California)

Estate of   1917
39: Reichs- u. auswaert.Ang.: E.IV.41 [Erbschaften]
                                       ADS 2:392

NEUMAYER, Andreas (in New York)

File on [estate?]   1880
227: Rep. MA 1921: A.St.I: Nr. 1649.   ADS 9:74

NEUMAYR, Max (in Brooklyn)

Information regarding   1890
227: Rep. MA 1921: A.St.I: Nr. 1656.   ADS 9:74

le NEVE, Ethel Clara. See Crippen, Hawley Harvey

NEVE, J. L. (professor, Wittenberg College, Springfield, Ohio)

Letter   [1919-1922?]
44: Mappe Amerika.   ADS 2:599

NEW, Henny, born Schott (from Moerfelden, Landkreis
Gross-Gerau, HEL)

Emigration to Hoboken [New Jersey]   1938?
179: Wiedergutmachungsakten, Ordner Nr. 061-00/05.
                                       ADS 7:232

NICHOLAS, --

Nicholas' body sent from Wiesbaden to New York,
    via Hamburg   1899
39: Reichs- u. auswaert.Ang.: P.III.16.35 [Polizei-
    wesen]   ADS 2:303

NIEBERG, --

Consular report from New York regarding Nieberg's
    accusation against Falck & Co.   1869
39: Bestand Senat: Cl.VIII, No.X, Jg.1869 (Regis-
    ter) 93; 95 [Nordamerika]   ADS 2:207

NIEBERGALL, Ludwig (master baker; from Kaltennordheim, Thuringia)

Application for emigration   1839
500: Weimar: Thueringisches Staatsarchiv: Dermbach
     II: Gb.2: Nr. 63-92; 93-114; 144-183; 186;
     188.   Not in ADS

NIEDERRAD, R.

Expulsion to America; sentenced to lengthy peni-
    tentiary sentence in Frankfurt/Main for politi-
    cal activities   1838
200: Abt.403: Oberpraesidium Koblenz (Rheinprov-
     inz): Nr. 2528.   ADS 8:21

NIEHUUS, --

St. Louis: Estate of   1884

NOLD, Michael (in Michigan)

Estate of   1878

39: Bestand Senat: Cl.VIII, No.X, Jg.1878 (Regis-
    ter) 3 [Nordamerika]   ADS 2:236

NOLTE, Gerhard (in Einbeck)

Proposed transportation to America   1849

52: Han.Des.80: Hildesheim I: E.DD.37a: Nr. 1299.
                                        ADS 4:89

NOLTE, Hermann (in Mexico)

Letters to his brother Vincent Nolte   1850-1851

39: Bestand Familienarchive: Fam. Nolte: Nr. 11.
                                        ADS 2:490

NOLTE, Vincent (born 1779, died 1856; in Philadelphia)

Letter (25 pages) to his father in Hamburg, dated
    18 April 1808 recounting his shipwreck on the
    coast of Florida aboard the clipper *Carysford*
                                              1808

39: Bestand Familienarchive: Fam. Nolte: Nr. 11.
                                        ADS 2:489

NORDEN, H. H. (from Sottrum)

File on possible extradition of Norden, presently
    in Stade prison, to the authorities in Maryland
    & question of his citizenship   1842

52: Han.Des.9: sog. Geheime Reg.: A., Nr. 9.
    DLC has.   ADS 4:101

NORMAN, J. A.

Hamburg police search for Norman, wanted in Amer-
    ica   1909

39: Reichs- u. auswaert.Ang.: P.III.1.7 [Polizei-
    wesen]   ADS 2:359

NORTH, Jesse A. (an American citizen)

Correspondence with American consul regarding
    North, imprisoned for debt   1852-1853

48: Dd.11.c.2.N.1.a.1: Nr. 44; *and* Bestand 4,48:
    A.1.a.1.e.1.c.   ADS 3:60

von der Null, --

Inheritance tax   [1805?]

115: Berg. Apanagielreg.: A.: Nr. A27.   ADS 6:22

NUMERICH, Georg (from Erfelden, Landkreis Gross-
Gerau, HEL)

Emigration to America   1926

169: Ortsbuergerregister.   ADS 7:220

NYBOR, --. *See* Soerensen, --

OBERHOFER, Elise (in Neu Ulm [New Ulm, Minnesota])

File on [estate?]   1887

227: Rep. MA 1921: A.St.I: Nr. 845.   ADS 9:69

OBERLE, Walter Otto (an American citizen)

Naturalization [in Germany] of   1939-1942

288: Kommunalarchiv Sasbachwalden.   ADS 10:138

OCHS, Barbara (from Wolfskehlen, Landkreis Gross-
Gerau, HEL)

Emigration to America   1854

173: Gesinderegister 1837-1843 [of Goddelau, Land-
     kreis Gross-Gerau, HEL]: Nr. 124.   ADS 7:224

OCKELMANN, Heinrich Friedrich Johannes (peddler &
cigar dealer)

Expulsion from Hamburg as Social Democrat & emi-
    gration to America   1880-1881; 1902

39: Polizeibehoerde-Polit.Pol.: S 149/98: Bl.
    448-458.   ADS 2:414

OCKENS, --. *See* Kniesch, Mrs. A. C. C., born Krawaak, widow of Ockens

O'CONNEL, --, *alias* Ward, Hardy, Harding

Extradition from Prussia to the USA, via Hamburg
                                              1905

39: Reichs- u. auswaert.Ang.: A.Ia.24 [Erbschaften]
                                        ADS 2:338

OECHLER, Johann (died 21 January 1906 in Jersey
Shore, Lycoming County, Pennsylvania)

Estate of   1908-1911

152: Abt.405: Wiesbaden: Rep.405/I,2: Nr. 3504.
                                        ADS 7:145

OEHLENSCHLAEGER, Heinrich

New York: Information regarding   1879

39: Bestand Senat: Cl.VIII, No.X, Jg.1879 (Regis-
    ter) 12 [Nordamerika]   ADS 2:240

OEHLERT, Frederick (died in San Francisco)

Estate of   1928-1932

116: Minden: Praes.-regist.: I.L: Nr. 241.
                                        ADS 6:54

OELMANN, Karl (journeyman mechanic; from Gandersheim)

Confinement in house of corrections at Bevern &

emigration to America   1858-1861

56: L Neu Abt.129A: Gr.11, Nr. 36.   ADS 4:188

OESER, --. *See* Thiessen, L.

OEST family

A genealogical study of this Lower Saxon family is
in the possession of Rektor i.R. Richard Tiensch,
Scholienstrasse 50, Otterndorf, Niedersachsen.

OESTERREICHER, Falk. *See* Austrian, Falk

OESTREICHER, N. (in Cincinnati)

Military investigation of   1911

227: Rep. MA 1921: A.V. V: Nr. 16397.   ADS 9:91

OETTINGER, Reimund (died in America)

File on [estate?]   1888

227: Rep. MA 1921: A.St.I: Nr. 848.   ADS 9:69

OFFERMANN, Ferd[inand] (in Lancaster, Pennsylvania)

Estate of   1909

39: Reichs- u. auswaert.Ang.: E.IV.72 [Erbschaften]
                                        ADS 2:357

OHBEN, --

San Francisco: Estate of   1908

39: Reichs- u. auswaert.Ang.: E.IV.20 [Erbschaften]
                                        ADS 2:351

OHLERT, Thomas (in Calumet, Wisconsin)

Estate of   1910

39: Reichs- u. auswaert.Ang.: E.IV.12 [Erbschaften]
                                        ADS 2:362

OHLSEN, Eduard (cigar maker)

Expulsion from Hamburg as Social Democrat & emigration to America   1881-1882

39: Polizeibehoerde-Polit.Pol.: S 149/301: Bl.
459-472.   ADS 2:416

OHLSSON, C. (a tailor; in Chicago)

Information regarding Ohlsson sought by private
person [in Germany]   1894

39: Reichs- u. auswaert.Ang.: N.Ib.13 [alt.Sig.]
[Nachforschungen]   ADS 2:271

OLDAG, Carl (in Appleton, [Wisconsin?])

Estate of   1903

39: Reichs- u. auswaert.Ang.: E.IV.64 [Erbschaften]
                                        ADS 2:330

OLDENBURG, Henry

Chicago: Estate of   1915

39: Reichs- u. auswaert.Ang.: E.IV.13 [Erbschaften]
                                        ADS 2:388

von OLDERSHAUSEN, August Ferdinand Jobst Wilhelm
Otto *Freiherr* [Baron] (in Louisville, Kentucky)

Relationship to family in Germany   1844
[apparently returned to Germany upon assumption of
family seniority, 1847]

99: Nr. 309.   ADS 4:296

Son of Captain Carl Friedrich Burch. Detlef von
Oldershausen; asserts his title to Rittergut
[noble estate] Gebesee, Seegut, the Bohuslain-
schen inheritance estate, & the fishing rights
to Unstrut   1844

99: Nr. 311.   ADS 4:297

von OLDERSHAUSEN, Franz Ernst Otto Friedrich (in New
York)

Relationship to family in Germany   1844
[brother of August F. J. W. O. von Oldershausen &
son of Captain Carl F. B. D. von Oldershausen]

99: Nr. 309.   ADS 4:296

OLESKI, R. J. H.

Extradition from America to Flensburg, via Hamburg
                                        1904

39: Reichs- u. auswaert.Ang.: A.Ia.3 [Ausliefer-
ungen]   ADS 2:334

OLSEN, -- (in Leadville, Colorado)

Estate of   1883

39: Bestand Senat: Cl.VIII, No.X, Jg.1883 (Register) 3 [Nordamerika]   ADS 2:250

ONCKEN, J. G. *See* Baptists (Wiedertaeufer)[subject
index]

ONCKEN, --

Request for information by U.S. Consul regarding
Baptist minister Oncken   1840

39: Bestand Senat: Cl.VIII, No.X, Jg.1840 (Register) I.1 [Nordamerika]   ADS 2:139

OPITZ, -- (married woman; in Oregon)

Estate of   1911

39: Reichs- u. auswaert.Ang.: E.IV.60 [Erbschaften]
                                        ADS 2:369

OPP, Friedrich Wilhelm (died 1887 in Boston)

    File on [estate?]   1887

    <u>227</u>: Rep. MA 1921: A.St.I: Nr. 846.   ADS 9:69

OPPENHEIMER, Herz (in Baltimore [Maryland])

    File on [estate?]   1891

    <u>227</u>: Rep. MA 1921: A.St.I: Nr. 852.   ADS 9:69

OPPENHEIMER, Julius (from Moerfelden, Landkreis Gross-Gerau, HEL)

    Emigration to East Orange [New Jersey?]   1938?

    <u>179</u>: Wiedergutmachungsakten, Ordner Nr. 061-00/05.
                                ADS 7:232

OPPENHEIMER, Julius (of Nuernberg)

    Decision regarding the carrying out of a Bavarian
      civil case decision in the USA   1894

    <u>227</u>: Rep. MA 1921: A.St.I: Nr. 856.   ADS 9:69

OPPENHEIMER, Karl (died 1887? in New York)

    Inquiry regarding Karl & Wilhelm Oppenheimer 1899

    <u>227</u>: Rep. MA 1921: A.St.I: Nr. 1672.   ADS 9:74

OPPENHEIMER, Wilhelm. *See* Oppenheimer, Karl

ORDERDORK, -- (bishop; in New York)

    A scandalous trial [apparently of Orderdork] & his
      sentencing   1845

    <u>226</u>: Rep. MInn 7: 25127c.   ADS 9:18

ORTH, Henry (in New York)

    Information on Orth sought by private individual
                                    1894

    <u>39</u>: Reichs- u. auswaert.Ang.: N.Ib.10 [alt.Sig.]
      [Nachforschungen]   ADS 2:271

OST, -- (a minor)

    Guardianship in America   1913

    <u>39</u>: Reichs- u. auswaert.Ang.: V.III.1.43 [Vormund-
      schaftswesen]   ADS 2:382

OSTERHAUS, Peter Josef (U.S. general; born in Koblenz)

    Manuscript on his life; also published in *Koblenzer
      Heimatblatt*, 5 Jg, Nr. 33 (19 August 1928)

    <u>200</u>: Abt.701: Nachlaesse, etc.: Nr. 968. ADS 8:52

OSTERLOH, -- (married woman; in United States?)

    Consular note regarding her residency   1864

    <u>39</u>: Bestand Senat: Cl.VIII, No.X, Jg.1864 (Regis-
      ter) I.   ADS 2:185

OTT, Wolfgang (in America)

    Search for   1888

    <u>227</u>: Rep. MA 1921: A.St.I: Nr. 1670.   ADS 9:74

OTTE, H. Matth. (in Philadelphia)

    Letter to relatives in Germany   1725

    <u>77</u>: 4$^{\underline{o}}$, Bl. 19b-20.   ADS 4:259

OTTE, Johannes (murdered in Texas)

    Estate of   1894

    <u>39</u>: Reichs- u. auswaert.Ang.: E.IV.98 [alt.Sig.]
      [Erbschaften]   ADS 2:269

OTTEN, -- (an American citizen)

    Complaint of American consul [in Hamburg] regard-
      ing imprisonment of Otten for failure to fulfill
      [German] military service obligation   1894

    <u>39</u>: Reichs- u. auswaert.Ang.: M.II.1 [alt.Sig.]
      [Militaer-Ang.]   ADS 2:270

OTTEN, Peter (in California)

    Estate of   1897

    <u>39</u>: Reichs- u. auswaert.Ang.: E.IV.29 [alt.Sig.]
      [Erbschaften]   ADS 2:290

OTTENS, Peter (released convict)

    Transportation to America   1850-1851

    <u>100</u>: Gem. Oberndorf: XIII, 1. Nr. 3e. ADS 4:312

OTTENS, --

    Unsworn testimony in Boston in case of Ottens *ver-
      sus* Ottens   1910

    <u>39</u>: Reichs- u. auswaert.Ang.: J.Ib.35 [Justiz-
      sachen]   ADS 2:364

OTTMER, Wilhelm (from Moritzberg bei Hildesheim)

    Regarding letter sent from Annapolis, Maryland,
      containing money from Ottmer, serving in U.S.
      Army   1868-1869

    <u>52</u>: Han.Des.80: Hildesheim I: A.C.IV: Nr. 103.
                                ADS 4:86

OUTSTAD, --

    Estate of   1944
    <u>148</u>: R XI.   ADS 6:261

von OVEN family

    The family emigrated to Illinois in 1848 & descend-
    ants were later in New York, Milwaukee, Rochester,
    & Pittsburgh [file contains table of descendants]
    Mentioned in file is a Mrs. -- Hammerschmidt, a
    founder of St. John's Lutheran Church of Naper-
    ville, Illinois   n.d.
    <u>145</u>   ADS 6:131

OWEN, Andrew. *See* Williams, Edward

PAASCHE, --

    New York: Estate of   1879
    <u>39</u>: Bestand Senat: Cl.VIII, No.X, Jg.1879 (Regis-
    ter) 5 [Nordamerika]   ADS 2:240

PABST, Hans

    New York: Estate of   1915
    <u>39</u>: Reichs- u. auswaert.Ang.: E.IV.12 [Erbschaften]
                                 ADS 2:388

PACHMANN, Barbara. *See* Wurkler, Johann

PAETOW, -- (from Verden)

    Transportation of Paetow family to America   1845
    <u>52</u>: Han.Des.74: Westen-Thedinghusen: I.56: Nr. 57.
                                 ADS 4:80

PAETOW, -- (widow; in Lyons, Iowa)

    Inquiry by Hamburg Orphanage Commission regarding
                                 1902
    <u>39</u>: Reichs- u. auswaert.Ang.: W.I.4 [Waisenhaus-
    Ang.]   ADS 2:328

PAGEL, Annie (in Chicago)

    Information requested by private person regarding
                                 1897
    <u>39</u>: Reichs- u. auswaert.Ang.: N.Ib.1.4 [alt.Sig.]
    [Nachforschungen]   ADS 2:291

PAJONCZEK, --

    Estate of   [1940's?]
    <u>148</u>: R XI.   ADS 6:261

PALAND, Andreas (linen weaver; from Mahlum)

    Emigration to America   1852
    <u>56</u>: L Neu Abt.129A: Gr. 11, Nr. 63.   ADS 4:190

PALANDRINI, Francesco (an Italian national)

    Extradition to Philadelphia from Frankfurt/Main,
    via Hamburg   1914
    <u>39</u>: Reichs- u. auswaert.Ang.: A.Ia.7 [Ausliefer-
    ungen]   ADS 2:383

PANDITSCH, T. A. (in New York)

    Pricing of Oldenburg railroad shares for   1903
    <u>39</u>: Reichs- u. auswaert.Ang.: F.Ia.5 [Finanzwesen]
                                 ADS 2:330

PAPE, Heinrich (day laborer; from Astfeld)

    Sentenced for theft & emigration to America
                              1855-1859
    <u>56</u>: L Neu Abt.129A: Gr.11, Nr. 23.   ADS 4:187

PAPE, Peter Christoph

    St. Louis: Information regarding   1879
    <u>39</u>: Bestand Senat: Cl.VIII, No.X, Jg.1879 (Regis-
    ter) 54 [Nordamerika]   ADS 2:242

PAPENBERG, W. J. G. (employee aboard an American
ship; died at Tampico)

    Estate of   1897
    <u>39</u>: Reichs- u. auswaert.Ang.: E.IV.32 [alt.Sig.]
    [Erbschaften]   ADS 2:290

PAPENDIECK, C. H. H. (in Milwaukee, Wisconsin)

    Petition of B. Pearkes [in Bremen?] to have his
    stepson Papendieck named Bremen consul in Mil-
    waukee   1850
    <u>48</u>: B.13.b.O: Nr. 9.
    DLC has.   ADS 3:9

PAPENHEIMER. *See* Pappenheimer

PAPPEI, Michael (journeyman tanner; from Lauingen)

    Emigration to North America   1854
    <u>247</u>: A 493.   ADS 9:158

PAPPENHEIMER, Friedrich (from Dornheim, Landkreis
Gross-Gerau, HEL)

    Emigration to New York   1924
    <u>168</u>: Abmelderegister.   ADS 7:218

PAPPENHEIMER, Isaak (from Dornheim, Landkreis Gross-
Gerau, HEL)

Emigration to New York   1935
168: Abmelderegister.   ADS 7:218

PAPPENHEIMER, Karolina (from Dornheim, Landkreis
Gross-Gerau, HEL)

Emigration to New York   1920
168: Abmelderegister.   ADS 7:218

PAQUET, J. C. A.

Letter from City of Stuttgart regarding money for
Paquet to emigrate to America   1848
39: Bestand Senat: Cl.VIII, No.X, Jg.1848 (Register).   ADS 2:154

PARKER, --

Arrest of Parker, wanted in Chicago   1908
39: Reichs- u. auswaert.Ang.: P.III.2.10 [Polizei-
wesen]   ADS 2:353

PASOLD, Franz (in Redwood City, California)

File on [estate?]   1897
227: Rep. MA 1921: A.St.I: Nr. 210.   ADS 9:66

PASSAVANT, M. A. (pastor; in Pittsburgh)

Correspondence   1895
120: Akte Frank Buchmann (D II B).   ADS 6:75

PASSAVANT, W. A. (in Pittsburgh)

Correspondence   1895
120: Akte Frank Buchmann (D II B).   ADS 6:75

PAULHUFER, Franz H. (priest; in Milwaukee)

Application for re-employment in Bavaria   1856
227: Rep. MA 1898: A.St.: Nr. 1030.   ADS 9:58

PAULMANN, Carl. See Paulmann, Ludwig

PAULMANN, Christian. See Paulmann, Ludwig

PAULMANN, Heinrich. See Paulmann, Ludwig

PAULMANN, Ludwig (from Einbeck)

Court case: Carl & Christian Paulmann (Carl being
representative also of Ludwig Paulmann) versus
restaurant owner -- Wedemeyer, of Schuetzenhaus
near Einbeck, guardian of minors Carl & Heinrich

Paulmann, for recuperation of articles loaned
                                                 1838
72: Stadt Einbeck: Nr. 5.   ADS 4:251

Court case: The brothers Carl & Christian Paulmann,
of Hoppensen, for themselves & for Ludwig Paulmann, in America, versus Friedrich Schwerdtfeger
(Spruetzer) of Sievershausen, guardian of minor
August Paulmann, for recuperation of articles
loaned   1838
72: Sievershausen (Amt Hunnesrueck), Nr. 46.
                                         ADS 4:251

PAULSEN, --

Request of Hamburg court for foreclosure of third-
party debtor in Iowa, in the case of Paulsen
versus Claussen   1901
39: Reichs- u. auswaert.Ang.: J.Ib.60 [Justiz-
wesen]   ADS 2:318

PAULY, Matthias (from Langenschwalbach, HEL)

Emigration to America   1850
152: Abt.231: Langenschwalbach: Nr. 2861.
                                         ADS 7:133

PAZEY, --

Request of Prussian minister regarding estate of
Pazey who came from America to Hamburg [where
he died?]   1824
39: Bestand Senat: Cl.VIII, No.X, Jg.1824 (Register) 1.q [Antraege]   ADS 2:125

PEDALL, Georg. See Weiss, Margarethe

PEIN, H. H.

San Francisco: Estate of   1904
39: Reichs- u. auswaert.Ang.: E.IV.57 [Erbschaften]
                                         ADS 2:335

PEININGER, C. (in New York)

Complaint   1866
39: Bestand Senat: Cl.VIII, No.X, Jg.1866 (Register) 4 [Auswanderer-Deputation]   ADS 2:192

PEITMANN, Justus Julius Gottlieb (died in America)

Request for death certificate for Peitmann, formerly in the Scheither's Corps & then in London
with Elliot's Dragoon Regiment   1802
52: Han.Des.92: London Kanz.: LXXXV, Nr. P,2.
                                         ADS 4:95

PELITOWSKI, --, born Wedekind (widow; in New Orleans)

Death certificate for   1866

<u>52</u>: Calenberger Brief-Archiv: Des.15: Privatakten
Band I: P, Nr. 45.   ADS 4:56

PELS, Bartel C.

Hanseatic Superior Court seeks examination by consul of witness in Philadelphia in matter of Bartel C. Pels   1896

<u>39</u>: Reichs- u. auswaert.Ang.: J.Ib.50 [alt.Sig.]
[Justizsachen]   ADS 2:283

PELS, --

Hanseatic Superior Court seeks expert testimony in
Boston in case of Young & Kimball *versus* Pels
                                            1897

<u>39</u>: Reichs- u. auswaert.Ang.: J.Ib.7 [alt.Sig.]
[Justizsachen]   ADS 2:290

PELS, Eduard (in New York)

Transport of Pels, mentally disturbed, from New
York to Hamburg   1865

<u>39</u>: Bestand Senat: Cl.VIII, No.X, Jg.1865 (Regis-
ter) I.   ADS 2:187

PERL, Franziska, born Geiss (in Pennsylvania)

File on [estate?]   1899
<u>227</u>: Rep. MA 1921: A.St.I: Nr. 236.   ADS 9:66

PERNERSDORFER, --

Estate of   1941
<u>148</u>: R XI.   ADS 6:261

PERSEL, --

Estate of   1942
<u>148</u>: R XI.   ADS 6:261

PERTHES, Friedrich (in New York)

Letter from Franz Lieber, Hamburg, regarding Franz
Niebuhr   1838

<u>39</u>: Bestand Familienarchive: Fam. Perthes: Mappe
21a, Fol. 126-127.   ADS 2:490

PETER, Johannes (journeyman mason; from Unteralba,
Thuringia)

Application for emigration   1841
<u>500</u>: Weimar: Thueringisches Staatsarchiv: Dernbach
II: Gb.2.: Nr. 63-92; 93-114; 144-183; 186;
188.   Not in ADS

PETERS, Ernst Theodor

Philadelphia: Estate of   1903

<u>39</u>: Reichs- u. auswaert.Ang.: E.IV.73 [Erbschaften]
                                            ADS 2:330

PETERS, H.

Impending extradition of H. Peters, G. Treiber, &
G. L. Plath to the USA   1889

<u>39</u>: Bestand Senat: Cl.VIII, No.X, Jg.1889 (Regis-
ter) 3.   ADS 2:262

PETERS, Luise (in Thousand Island Park, New York)

Estate of   1905
<u>39</u>: Reichs- u. auswaert.Ang.: E.IV.53 [Erbschaften]
                                            ADS 2:339

PETERS, Wilhelm. *See* Kochschuetzke, Mr. & Mrs. --

PETERS, -- (died in Paris)

Inquiry by private person regarding Peters, a
German-American   1896

<u>39</u>: Reichs- u. auswaert.Ang.: N.Ib.2 [alt.Sig.]
[Nachforschungen]   ADS 2:284

PETERS, -- (an American citizen)

Expulsion of   1909
<u>39</u>: Reichs- u. auswaert.Ang.: A.IV.5 [Ausweisungen]
                                            ADS 2:356

PETERSEN, C.

Application for American military pension   1894
<u>39</u>: Reichs- u. auswaert.Ang.: P.I.1 [alt.Sig.]
[Pensionen]   ADS 2:271

PETERSEN, Frederike. *See* Boldt, Frederike, born
Petersen

PETERSEN, Jacob

New York: Estate of   1915
<u>39</u>: Reichs- u. auswaert.Ang.: E.IV.20 [Erbschaften]
                                            ADS 2:388

PETERSEN, Julius

Arrest for theft in New York   1909
<u>39</u>: Reichs- u. auswaert.Ang.: P.III.2.6 [Polizei-
wesen]   ADS 2:359

PETERSEN, Max

Chicago consulate loan to   1897
<u>39</u>: Reichs- u. auswaert.Ang.: U.I.1.4 [alt.Sig.]
[Unterstuetzungen]   ADS 2:292

PETERSEN, -- (seaman; in Galveston, Texas)

Repatriation of   1898

39: Reichs- u. auswaert.Ang.: S.Ie.1.6 [Schiffahrt]
ADS 2:298

PETERSEN, --

German consulate in New York: Divorce proceedings
concerning Petersen   1899

39: Reichs- u. auswaert.Ang.: J.Ib.10.4 [Justiz-
sachen]   ADS 2:303

PETERSEN, -- (businessman)

Extradition from USA to Halberstadt, via Hamburg
1907

39: Reichs- u. auswaert.Ang.: A.Ia.15 [Ausliefer-
ungen]   ADS 2:346

PETRI, Jacob (of Berkach, Landkreis Gross-Gerau, HEL)

Emigration to America   1868

163: Abt. XI/4.   ADS 7:213

PETRI (Petrie) Konrad (of Berkach, Landkreis Gross-
Gerau, HEL)

Census of 1861 contains notation of his emigration
to America   1861

163: Abt. XI/4.   ADS 7:213

PETROVSKY, Sigmund (master tailor; in Memmingen)

Petition for the recognition of his certification
[Zeugnisse] from the New York Institute of Sci-
ence in Rochester   1904

226: Rep. MK 3: Nr. 11795.   ADS 9:36

PEUERBACH. See von Gumppenberg-Peuerbach, Julius
Freiherr [Baron]

PFABE, Bernhard (in New York)

Inquiry by private person in Germany regarding
1899

39: Reichs- u. auswaert.Ang.: N.Ib.1.1 [Nachforsch-
ungen]   ADS 2:303

PFANNENSCHLAG, Therese (widow; in New York)

Estate of (mentions Karl Zipprich)   1897-1899
227: Rep. MA 1921: A.St.I: Nr. 1215.   ADS 9:71

PFANNEN-SCHMIDT, -- [should be Pfannenschmidt?]

Estate of Pfannen-Schmidt who died in American
military service, & pension for his widow in
Paderborn   1859-1886

116: Minden: Praes.-regist.: I.C. Nr. 13. ADS 6:53

PFEFFER, Johann (of Hamzelle)

Emigration of   1787

151: Bestand 98: Fulda X: Acc.1875/27: Nr. 10.
ADS 7:91

PFEFFER, -- (tailor; in Baltimore)

Inquiry from Hamburg consulate in Baltimore regarding Pfeffer who is suspected of a delict   1842

39: Bestand Senat: Cl.VIII, No.X, Jg.1842 (Register) 4 [Auswanderer-Verhaeltnisse] ADS 2:143

PFEIFER, J. L. (from Markt Nordheim)

Settlement in North America   1841-1842

227: Rep. MA 1898: A.St.: Nr. 1024.   ADS 9:58

PFEIFER family. See Bergisch-Gladbach [geographic
index]

PFEIFFER, Johann. See Piper, Johann

PFEIFFER, Simon (from Gochsheim, Hessen-Hanau)

Death certificate for Pfeiffer, a soldier in the
Freikorps who died in America, issued at the
request of -- Spangenberg, a farm laborer in
Bergen   1801-1802

151: Bestand 81: Rep.E: 21.Amt Bergen: Ia.: Nr. 38.
ADS 7:82

PFEIFFER, -- (a sailor; in New York)

Avoidance of military service obligation [deser-
tion?]   1912

39: Reichs- u. auswaert.Ang.: M.II.12 [Militaer-
wesen]   ADS 2:375

PFEIL, Anton. See Feil, Anton

PFIRRMENN, --

Estate of   1929-1933
148: R XI.   ADS 6:261

PFISTER, --

Estate of   1944
148: R XI.   ADS 6:261

PFLUEGER, Georg. See Bommarius, Friedrich

PFOERTNER, Frederike. See Poertner, Frederike

PFOERTNER, Wilhelmine (spinster; from Kirchberg)

Emigration to America   1858–1859
<u>56</u>: L Neu Abt.129A: Gr.11, Nr. 58.   ADS 4:190

PFORTNER, M. Dwight. *See* Fortner, M. Dwight

PHILIPP, Walb. (in New York)

Power of attorney   1894
<u>227</u>: Rep. MA 1921: A.St.I: Nr. 172.   ADS 9:66

PHILIPPS, Clarendon L. (an American citizen who died
in Germany)

Estate of   1879
<u>39</u>: Bestand Senat: Cl.VIII, No.X, Jg.1879 (Regis-
        ter) 24 [Nordamerika]   ADS 2:240

PHILLIPS Brothers (Junior)(in Aachen). *See* Asam, --

PIENING, Hinrich (cigar maker)

Expulsion from Hamburg as Social Democrat & emi-
        gration to America   1881–1882
<u>39</u>: Polizeibehoerde-Polit.Pol.: S 149/306: Bl.
        385-393.   ADS 2:416

PIEPER, -- (minors)

Request by Hamburg Orphanage Commission to have
        Pieper children, en route to America, delivered
        to their stepfather by the General Consulate in
        New York   1895
<u>39</u>: Reichs- u. auswaert.Ang.: W.I.1 [alt.Sig.]
        [Waisenhaus-Ang.]   ADS 2:279

PIPER, Johann (in Willow Springs, Illinois)

Estate of   1912
<u>39</u>: Reichs- u. auswaert.Ang.: E.IV.18 [Erbschaften]
                                                 ADS 2:373

PITTSCHNER, W.

San Francisco: Estate of   1914
<u>39</u>: Reichs- u. auswaert.Ang.: E.IV.7 [Erbschaften]
                                                 ADS 2:384

PITZING, Rud.

Report from German consulate in Chicago regarding
        Pitzing's rights to America citizenship through
        H. E. Huedepoll [presumably stepfather?]   1901
<u>39</u>: Reichs- u. auswaert.Ang.: G.Ia.26 [Geschaefts-
        verkehr]   ADS 2:317

PLAMBOECK, William (in Pennsylvania)

Estate of   1912
<u>39</u>: Reichs- u. auswaert.Ang.: E.IV.11 [Erbschaften]
                                                 ADS 2:373

PLATH, G. L.

Extradition of Plath to the USA   1887
<u>39</u>: Bestand Senat: Cl.VIII, No.X, Jg.1887 (Regis-
        ter) 4 [Nordamerika]   ADS 2:259

PLATH, G. L. *See also* Peters, H.

PLUEMER, Christian (soldier; from Schlewecke)

Emigration to America   1851
<u>56</u>: L Neu Abt.129A: Gr.11, Nr. 79.   ADS 4:191

PLUMEYER, Wilhelm Johann Martin (an American dentist)

Negotiations regarding his permit to practice in
        Ritzebuettel, Hamburg   1879
<u>39</u>: Ritzebuettel I: Gesundheitswesen: Heildiener:
        Nr. 305.   ADS 2:465

POEHLMANN, -- (in New York)

Claim of -- Poehlmann, businessman in Prussia, to
        estate of brother   1803–1804
<u>500</u>: Berlin-Dahlem: Geh. Staatsarchiv: Rep.XI, 21a,
        Conv. 3,5.   Not in ADS

POERTNER, Friederike (died in St. Louis)

Estate of   1890–1891
<u>116</u>: Minden: Praes.-regist.: I.L. Nr. 242.
                                                 ADS 6:54

POFLISCH, -- (married couple; in New York)

Information sought regarding them by private per-
        son [in Germany]   1895
<u>39</u>: Reichs- u. auswaert.Ang.: N.Ib.3 [alt.Sig.]
        [Nachforschungen]   ADS 2:277

POHLE, -- (pastor; in Philadelphia). *See* Engler, --

POLLACK, --

San Francisco: Estate of   1918
<u>39</u>: Reichs- u. auswaert.Ang.: E.IV.129 [Erbschaften]
                                                 ADS 2:394

POLLMANN, Johann (died in Texas)

Estate of   1926–1931
<u>116</u>: Minden: Praes.-regist.: I.L.Nr. 243-244.
                                                 ADS 6:54

POLLOCK, A. (of Washington)

Prize of 100,000 Franks for best equipment for saving lives at sea, offered by the heirs of Pollock who died in the accident on the *Bourgogne* 1899

39: Reichs- u. auswaert.Ang.: S.Ih.5 [Schiffahrt] ADS 2:304

POLLY, Martin (in Cincinnati [Ohio])

Interrogation of Martin Polly regarding the remarriage of widow Barbara Polly of Ebertshausen [Bavaria] 1843-1844

227: Rep. MA 1898: A.St.: Nr. 6262. ADS 9:58

POMANN, Elisabeth. *See* Klingenberg, Elisabeth, born Pomann

PONATH, Robert O. (former office clerk in Celle; later in Chicago)

Complaint against county superintendent[Kreishauptmann] Frank for the false arrest of complainant [Ponath] 1883-1884

52: Han.Des.80: Lueneburg II: Praesid.-Abt.: Regis. 1890: Hauptabt. 13: Nr. 282. [Beschwerde] ADS 4:93

POOK, --. *See* Lagershausen, August

POPE and GILLER BREWERY (in Warsaw, Illinois)

Claim of Bertha Koehler, heiress in Langsdorf [Hesse] against firm 1902

149: Rep.49a: Nr. 566/15. ADS 7:18

POPP, H. A. (in San Francisco)

Death & estate of 1872

39: Bestand Senat: Cl.VIII, No.X, Jg.1872 (Register) 16 [Nordamerika] ADS 2:220

Death & estate of 1873

39: Bestand Senat: Cl.VIII, No.X, Jg.1873 (Register) 19 [Nordamerika] ADS 2:224

POPPING, Friedrich (from Neuenkirchen)

Transportation to America 1837

54: Rep. 122, VII.B: Groenenberg-Melle: Regiminalia: Fach 15, Nr. 8. ADS 4:139

PORZELT, Chr. (died in San Francisco)

File on [estate?] 1891

227: Rep. MA 1921: A.St.I: Nr. 132. ADS 9:65

POST, -- (member Moravian Brethren; in Pennsylvania?). *See* Wade, John

POST, -- (in America). *See* Bedassem, J.

PRAAST, Rudolph Friedrich (expeditor)

Expulsion from Hamburg as Social Democrat & emigration to America 1880-1881; 1898

39: Polizeibehoerde-Polit.Pol.: S 149/100: Bl. 493-508. ADS 2:414

PRACHT, Georg (in America)

Information regarding 1888

227: Rep. MA 1921: A.St.I: Nr. 1317. ADS 9:72

PRAGEMANN, -- (pastor; in New York)

Pastoral call to New York 5 July 1720

39: Geistl. Minis.: II,6: Protokoll lit.A, 1720-1745: pp.4 und 4r (IV). ADS 2:468

Declaration of Palatine Germans in New York accepting Pragemann as pastor 1720

39: Geistl. Minis.: III,Alm: Gebundene Akten: Nr. CXCII, p. 494. ADS 2:477

PRECHT, Gottlieb Friedrich Eduard (an American citizen)

Deposition of 1857

48: B.13.C.1.b: Nr. 14. DLC has. ADS 3:42

PRELL, Gustav (in Newark, New Jersey)

Estate of 1905

39: Reichs- u. auswaert.Ang.: E.IV.55 [Erbschaften] ADS 2:339

PRENZLER, G. (in Burlington, Iowa)

Estate of 1915

39: Reichs- u. auswaert.Ang.: E.IV.36 [Erbschaften] ADS 2:388

PRESS, Heinrich Jakob (from Raunheim, Landkreis Gross-Gerau, HEL)

Emigration to America with wife & seven children 1854

181: Raunheimer Evangelischen Pfarrei Familienregister (Lehrer Johannes Buxbaum). ADS 7:234

PRESS, Johann Nicolaus (from Raunheim, Landkreis Gross-Gerau, HEL)

Emigration to America 1854

181: Raunheimer Evangelischen Pfarrei Familienregister (Lehrer Johannes Buxbaum) ADS 7:234

PRESS, Philipp Martin (soldier; in Raunheim, Landkreis Gross-Gerau, HEL)

Emigration to America   1856

181: Raunheimer Evangelischen Pfarrei Familienregister (Lehrer Johannes Buxbaum) ADS 7:234

There are also two letters extant from Press: In
the first letter he describes the boat trip from
Mainz to LeHavre; in the second he describes his
trip across the ocean to New York & reception by
the Gottron family which had emigrated two years
before.

There is also a letter from Press' wife, dated 27
January 1863, in which she tells the relatives
in Raunheim that he has been a cavalry corporal
for 18 months. He was killed in action in 1863
in the U.S. Civil War.

181   ADS 7:234

PRESSBURGER, Karl (in Chicago)

Request for information regarding Jette Bischofsheimer   1899

227: Rep. MA 1921: A.St.I: Nr. 1352.   ADS 9:72

PREUSS, Christoph

New York: Estate of   1918

39: Reichs- u. auswaert.Ang.: E.IV.125 [Erbschaft-
en]   ADS 2:394

PREUSSER(IN) Anna Maria (from Amt Friedwald)

Inheritance of a woman who had emigrated to America
twelve years before & from whom nothing had been
heard thereafter [the "(in)" in surname above
merely denotes a female person; the modern surname would be Preusser]   1775

200: Abt.30: Reichsgrafschaft Sayn-Altenkirchen
(u. Hachenburg): Nr. 3410.   ADS 8:10

PRICKER, Tr. (from Frepsum)

Arrest & emigration to America   1836-1843

50: Rep.28: 6: Nr. 2577.   ADS 4:12

PRIES, H. F.

Cincinnati [Ohio]: Estate of   1903

39: Reichs- u. auswaert.Ang.: E.IV.52 [Erbschaften]
                                         ADS 2:330

PRINZ, Georg (an American citizen)

Arrest in Cuxhaven   1909

39: Reichs- u. auswaert.Ang.: P.III.20 [Polizei-
wesen]   ADS 2:359

PROBANDT, --

Estate of   1929-1931

148: R XI.   ADS 6:261

PROBST, August (vagabond laborer; from Gandersheim)

Emigration to America   1858-1860

56: L Neu Abt.129A: Gr.11, Nr. 35.   ADS 4:188

PRUNHUBER, Franz (died 1900 in New York)

Information regarding   1900

227: Rep. MA 1921: A.St.I: Nr. 1355.   ADS 9:72

PUDLACK, W.

San Francisco consulate: Cost of support for 1915

39: Reichs- u. auswaert.Ang.: U.Ia.1.4 [Unter-
stuetzungen]   ADS 2:390

PUEHL, Katharina. See Eichenmueller, Elisabeth

PUHLMANN, Martha (in Chicago)

Inquiry by private person in Germany regarding
                                                1898

39: Reichs- u. auswaert.Ang.: N.Ib.1.4 [Nachforsch-
ungen]   ADS 2:297

PULSSKI [Pulaski?] --

Estate of   1899-1913

148: R XI.   ADS 6:261

PUNDT, H. (from Omaha, Nebraska)

American consul in Hamburg seeks information regarding   1896

39: Reichs- u. auswaert.Ang.: N.Ia.5 [alt.Sig.]
[Nachforschungen]   ADS 2:284

PURTZ, --

Extradition to Zwickau from USA, via Hamburg 1912

39: Reichs- u. auswaert.Ang.: A.Ia.5 [Ausliefer-
ungen]   ADS 2:372

QUERLING, Joh. (died aboard brig *Georgia* on trip to
San Jago [San Diego? Santiago?])

Estate of   1872

39: Bestand Senat: Cl.VIII, No.X, Jg.1872 (Register) 50 [Nordamerika]   ADS 2:222

QUILD, Anna. See von Chominski, Theodor

RAAB, Johann (in Springfield [state not given in ADS description])

File on [estate?]   1896-1898
227: Rep. MA 1921: A.St.I: Nr. 870.   ADS 9:70

RAABE, --

The descendants of Wilhelm Raabe, a German poet of
Eschershausen, live in America. Material on the
family has been colleced by Amtsrat i.R. Lucé,
Kreisarchivpfleger, Eschershausen, Niedersachsen
                                           ADS 4:255

RABENAU, --

Hamburg police search for Rabenau, wanted in USA
                                           1912
39: Reichs- u. auswaert.Ang.: P.III.2.2 [Polizei-
wesen]   ADS 2:375

RAFE, Borchert

Emigration file contains letters from relatives &
friends in New York   1853
55: Han.Des.74: Blumenthal: IX.J.1: Fach 301, Nr.
10.   ADS 4:154

RAFFA, Harriette, divorced Breitenbach, born Rosenfeld (in Chicago)

Information regarding her marriage   1900
227: Rep. MA 1921: A.St.I: Nr. 1704.   ADS 9:74

von RAGUÉ (Ragué), Louis (pastor; in Quincy, Illinois)

Correspondence   1887
120: Akte Frank Buchman (D II B).   ADS 6:75

RAHN, Detlef (in Colfax, Washington)

Estate of   1913
39: Reichs- u. auswaert.Ang.: E.IV.91 [Erbschaften]
                                           ADS 2:379

RAHTGENS, P. (in Texas)

Estate of   1905
39: Reichs- u. auswaert.Ang.: E.IV.18 [Erbschaften]
                                           ADS 2:339

RAIBLE, Andreas (from Weildorf; in St. Louis)

Death certificate for   n.d.
262: Ho 202: Haigerloch: I 1459, 4500.   ADS 10:37

RAISS, David (from Moerfelden, Landkreis Gross-
Gerau, HEL)

Emigration, via Antwerpen?   1910
179: Abt. XI,4.   ADS 7:232

RALF [or Ralfs] Christ.

St. Louis, Estate of   1895
39: Reichs- u. auswaert.Ang.: E.IV.11 [alt.Sig.]
[Erbschaften]   ADS 2:275

RAMM, --

Fee due consulate in New York in case of Ramm versus Ramm   1902
39: Reichs- u. auswaert.Ang.: J.Ib.10.6 [Justiz-
sachen]   ADS 2:325

RAMOS, --, alias Brauer & Steiner

American police search for   1896
39: Reichs- u. auswaert.Ang.: P.III.1.1 [alt.Sig.]
[Polizeisachen]   ADS 2:285

RAMSDORF, -- (an American citizen)

Expulsion of   1909
39: Reichs- u. auswaert.Ang.: A.IV.4 [Ausweisungen]
                                           ADS 2:356

RAMSTORF, C. (in Alameda County, California)

Estate of   1902
39: Reichs- u. auswaert.Ang.: E.IV.14 [Erbschaften]
                                           ADS 2:323

RANG, J. N. (tailor; from Nuernberg; died during
voyage to North America)

Estate of   [1849?]
231: Rep. 270/II: Reg. Mittelfranken: Kammer des
Innern: Abgabe 1932: Nr. 604.   ADS 9:134

RAPP, Georg (Economy Colony, Pennsylvania)

Petition of Eligius Leinweber et al in Offenbach
for information regarding the Rapp inheritance
case   1903-1925
149: Rep.49a: Nr. 565/4.   ADS 7:16

RAPPEL, -- (book printer)

Extradition from USA to Munich, via Hamburg 1906
39: Reichs- u. auswaert.Ang.: A.Ia.16 [Ausliefer-
ungen]   ADS 2:342

RASCH, --. *See* Resch, --

RATHE, Heinr. Christ. (from Stockhausen)

  Residency of Rathe, now in America   1872-1873
  50: Rep. 21a: Nr. 8116.   ADS 4:7

RATHJEN, Johann Heinrich August (journeyman cabinet
  maker)

  Expulsion from Hamburg as Social Democrat & emigration to America   1880-1881
  39: Polizeibehoerde-Polit.Pol.: S 149/39: Bl. 44-
  51.   ADS 2:412

RAUCH, Georg (in St. Louis)

  Inquiry by private person in Germany regarding
  1900
  39: Reichs- u. auswaert.Ang.: N.Ib.1.5 [Nachforsch-
  ungen]   ADS 2:309

RAUCH, Ludwig David (drummer; from Langstadt)

  Petition of Susanna Margarethe Engelhard, born
  Rauch, Wilhelmine Ernestine Rauch, & Friedericke Henriette Rauch for donation of the property of their brother Ludwig David Rauch who
  deserted in America & is now deceased   1784
  151: Bestand 81: Rep.E: 2.Altstadt Hanau III: Nr.
  3.   ADS 7:81

RAUSCHENBUSCH, August (born 13 February 1816 in Altena, Westphalia; died 5 December 1899 in Wandsbek
  bei Hamburg; professor of theology in Rochester
  1858-1888)

  Estate of   [1899]
  46   ADS 2:604

RAUSS, Lucas (in Pennsylvania?)

  Pastor -- Muehlenberg's defense against [accusa-
  tions] of Rauss   1761
  500: Halle: Missionsbibliothek des Waisenhauses:
  Abteilung H IV (Nordamerika) Fach A, Nr. 15.
  Not in ADS

RAVENHORST, -- (died in Ebenezer, Georgia)

  Estate of   1784
  500: Berlin-Dahlem: Preus.Geh. Staats-Archiv: Rep.
  XI, 21,a, Conv.1,6.   Not in ADS

REBSTOCK, Mathias (died 1906 in Braziel, Indiana)

  Estate of   1907-1908
  152: Abt.405: Wiesbaden: Rep.405/I,2: Nr. 3509.
  ADS 7:146

RECK, -- (sculptor; died in America)

  Estate of   1896
  39: Reichs- u. auswaert.Ang.: E.IV.14 [alt.Sig.]
  [Erbschaften]   ADS 2:282

REEMERS, -- (convict; from Osteel)

  Emigration to America   1863-1865
  50: Rep.21a: Nr. 8968.   ADS 4:8

  Transportation from Norden to America   1863
  50: Rep.32a.: Nr. 2431.   ADS 4:13

REESE, George (in Pittsburgh, Pennsylvania)

  Estate of   1899
  39: Reichs- u. auswaert.Ang.: E.IV.20 [alt.Sig.]
  [Erbschaften]   ADS 2:301

REESE (Ries) Michael (Jew; died in San Francisco)

  Personal file on   1878
  226: Rep. MInn 16: Nr. 52644.   ADS 9:19

REHAGEN, J. J. H., *alias* J. Brockmueller (of Neu-
  Guelze, Mecklenburg)

  Expulsion of [naturalized] American citizen who
  emigrated [from Germany] before fulfilling his
  military service obligation   1886
  39: Bestand Senat: Cl.VIII, No.X, Jg.1886 (Register) 13 [Nordamerika]   ADS 2:257

REHBOCK, Valentin (of Bieber, Hessen-Hanau)

  Fined for encouraging other persons to emigrate
  1765
  151: Bestand 80: Lit.A, Nr. 9.   ADS 7:78

REHFELD, J. F.

  New York: Estate of   1882
  39: Bestand Senat: Cl.VIII, No.X, Jg.1882 (Register) 18 [Nordamerika]   ADS 2:249

REHN, Jacob (master shoemaker; in Altenmuhr bei
  Gunzenhausen, Bavaria)

  Letter from Rehn requesting information on the
  plight of certain emigrants, via Hamburg, to
  America   1837
  39: Bestand Senat: Cl.VIII, No.X, Jg.1837 (Register).   ADS 2:134

REICH, *alias* Rieck, Lina (wife of American citizen)

  Interned at Friedrichsberg [asylum]   1895
  39: Reichs- u. auswaert.Ang.: M.Ib.2 [alt.Sig.]
  [Medizinalsachen]   ADS 2:276

REICHARDT, Franz (died 16 October 1908 in Los Angeles, California)

Estate of   1910
152: Abt.405: Wiesbaden: Rep.405/I,2: Nr. 3510.
                                                    ADS 7:146

REICHERT, -- (vice corporal, Third English-Waldeck Regiment)

Courtmartial for his enticement [of others] to
    desertion   1782
151: Bestand 118: Nr. 986.
     DLC has.   ADS 7:101

REIDER, Ludwig (in Darmstadt)

Had heirs in America   1908
149: Rep.49a: Nr. 565/9.   ADS 7:16

REIF, --

Estate of   1941
148: R XI.   ADS 6:262

REIMER, Georg Otto (cigar manufacturer)

Expulsion from Hamburg & emigration to America
    [Social Democrat]   1880-1892
39: Polizeibehoerde-Polit.Pol.: S 149/25: Bl. 158-
    277.   ADS 2:411

REIMER, Hinrich (in Zennhusen, SHL)

Petition to be released from military service
    obligation for purpose of emigrating to America
                                                         1867
13: Alphabetischen Sammlung.   ADS 1:49

REIMERS, John (in Seattle, Washington)

Estate of   1913
39: Reichs- u. auswaert.Ang.: E.IV.113 [Erbschaften]
                                                    ADS 2:379

REIMERS, J. P. N.

San Francisco: Estate of   1888
39: Bestand Senat: Cl.VIII, No.X, Jg.1888 (Regis-
    ter) 18 [Erbschafts-Amt]   ADS 2:260

REINECKE, Wilh.

San Francisco: Estate of   1897
39: Reichs- u. auswaert.Ang.: E.IV.4 [alt.Sig.]
    [Erbschaften]   ADS 2:289

REINERT, Ernst August (from Naensen)

Emigration to America   1852
56: L Neu Abt.129A: Gr.11, Nr. 69.   ADS 4:190

REINHARD, --

Estate of   1941
148: R XI.   ADS 6:262

REINHART, -- (worker)

Trial against -- Hasey for mortal injury to Rein-
    hart near Seattle [Washington]   1907
39: Reichs- u. auswaert.Ang.: J.Ic.9 [Justiz-Ang.]
                                                    ADS 2:348

REINHEIMER, Isidor (from Goddelau, Landkreis Gross-
    Gerau, HEL)

Emigration of Jewish person to America   1920
173: Abmelderegister 1916-1924: Nr. 338. ADS 7:224

REINHOLD, --

San Francisco: case of Reinhold versus Ruether 1896
39: Reichs- u. auswaert.Ang.: J.Ib.10.2 [alt.Sig.]
    [Justizsachen]   ADS 2:283

REINICKE, J. R. See Mueller, J. R.

REINISCH, Andreas Josef (paperhanger; in America)

Inquiry regarding him   1885-1886
227: Rep. MA 1921: A.St.I: Nr. 1688.   ADS 9:74

REINKING, Anna Caroline (from Leer)

Situation of Anna Carolina Reinking & her possible
    emigration to America   1869-1870
50: Rep.21a: Nr. 8948.   ADS 4:8

REIS. See Raiss, David

von REISCHACH, Hans Friedrich Count (from Riet, BWL)

Family archive contains notes on his emigration to
    USA around 1880

[Apply to Archiv der Grafen von Reischach, Riet,
Baden-Wuerttemberg]

REISINGER, Hugo (industrialist & Bavarian Privy Coun-
    cilor; in New York)

Estate of   1914-1936
227: Rep. MA 1936: A.V.: Nr. 298.   ADS 9:100

RENDTORFF, Wanda Lucile (in Reedsburg, Wisconsin, & Maywood, Illinois)

Photograph of grave   1906
Photograph   n.d.

2: Abt.399 (Fram.Archiv Rendtorff): Nr. 305, 309.
            ADS 1:35

RENNER, Anton (in Chicago)

Report of his address   1900

227: Rep. MA 1921: A.St.I: Nr. 1703.   ADS 9:74

RENNER, Johann (in America)

Report regarding   1872-1874
227: Rep. MA 1921: A.St.I: Nr. 1673.   ADS 9:74

RENTZ, Philipp. *See* Bommarius, Friedrich

RENVENUTI, Kuni (in New York)

Complaint against the German consul in New York
               1905

227: Rep. MA 1921: I.V.: Nr. 1862.   ADS 9:77

RENZ, Christian. *See* Bommarius, Friedrich

REPP, Anna Margaretha (from Kaltensundheim, Thuringia)

Emigration of   1857

500: Weimar: Thueringisches Staatsarchiv: Dermbach II: Gb.2: Nr. 63-92; 93-114; 144-183; 186; 188.   Not in ADS

REPPERT, Margarete. *See* Eberhard, Margarete, born Reppert

RESCH? [*or* Rasch] -- (from Frankfurt/Main)

Intercession of City of Frankfurt/Main on behalf of fifteen-year-old boy being impressed for military duty in America   1777

151: Bestand 81: Rep.E: 1.Gen.III, Nr. 8. ADS 7:80

RESTORFF, Adolph (in Brillion, Wisconsin)

Estate of   1917

39: Reichs- u. auswaert.Ang.: E.IV.22 [Erbschaften]
           ADS 2:392

RETBERG, Friedrich (from Bremen)

File regarding attempted transmittal, via Baltimore consulate, of unclaimed deposit; later paid out to F. W. Retberg, father of Friedrich Retberg, a cigar manufacturer in Bremen   1868-1869

48: B.13.b.2: Nr. 26.   ADS 3:19

RETTSTADT, W. (from Elze?)

Approval of funds covering cost of emigration for convict Rettstadt   1865-1866

52: Han.Des.80: Hildesheim I: E.DD.4a: Nr. 687.
           ADS 4:87

REUSS, Hermann (small cottage owner; from Brunsen)

Sale of his farm (No.Ass.8) to the farmer & master carriage maker Ludwig Riemenschneider (No.Ass.36) of Brunsen, & Reuss' emigration to America   1865

56: L Neu Abt.129A: Gr.32, Nr. 211.   ADS 4:193

REYNOLDS family. *See* Bergisch-Gladbach [geographical index]

RHEIN, -- (an emigrant to America)

Complaint of   1856

39: Bestand Senat: Cl.VIII, No.X, Jg.1856 (Register) 12 [Auswanderer-Verhaeltnisse] ADS 2:170

RICHTER, Carl August Hermann (pick & shovel laborer; from Clausthal-Zellerfeld). *See* Bauer, Albert

RICHTER, Friedrich Albert (cigar worker)

Expulsion from Hamburg as Social Democrat & emigration to America   1881

39: Polizeibehoerde-Polit.Pol.: S 149/302: Bl. 278-290.   ADS 2:416

RICHTER, Heinrich Joseph (first bishop of Grand Rapids, Michigan, 1883-1916)

Mentioned in "Bischoefe des Oldenburger Muensterland" [Bishops of the Oldenburg Bishopric] *Veroeffentlichungen aus dem Heimatskalender fuer das Oldenburger Muensterland* (1952), p. 115.

           ADS 4:248

RICHTER, Max (furnace man; in Brooklyn)

Estate of   1912

39: Reichs- u. auswaert.Ang.: E.IV.100 [Erbschaften]
           ADS 2:373

RICHTER, Wilhelm (musician; from Gandersheim)

Emigration to America   1859
56: L Neu Abt.129A: Gr.11, Nr. 37.   ADS 4:188

RICHTER, --

Estate of   1942
148: R XI.   ADS 6:262

RICKLEF, -- (bookkeeper; from Jever; died as a soldier in America)

Claim by widow for payment of back wages [1860's]
<u>53</u>: Bestand 31: Oldenburg; 31-15-23-83.   ADS 4:114

RIECK, Lina. *See* Reich, Mrs. Lina

RIECKE, Carl (in Chicago)

Inquiry by private person in Germany regarding
1898
<u>39</u>: Reichs- u. auswaert.Ang.: N.Ib.1.12 [Nachforschungen]   ADS 2:297

RIEDELN, -- (married woman; in Berlin)

Claim of inheritance [in America]   1794
<u>500</u>: Berlin-Dahlem: Preus.Geh.Staatsarchiv: Rep.
XI, 21a: Conv. 2; 2c.   Not in ADS

RIEDMUELLER, Karl (convict in penitentiary at Jackson, Michigan)

Pardon petition   1900
<u>227</u>: Rep. MA 1921: A.St.I: Nr. 883.   ADS 9:70

RIEGE, Georg (in Pittsburgh, Pennsylvania)

Estate of   1915
<u>39</u>: Reichs- u. auswaert.Ang.: E.IV.56 [Erbschaften]
                                           ADS 2:388

RIES, Michael. *See* Reese, Michael

RIESS, Joachim (in New York)

Administration of possessions upon his having left
wife & four minor children behind; he had been a
master shoemaker & former citizen of Burg auf
Fehmarn   1859
<u>5</u>: Rubr.XIV: Nr. 117 [Vermoegens-Angelegenheit]
                                           ADS 1:43

RIGBY, --. *See* Lockent, --

RIMBACH, Luise. *See* Brach, Luise, born Rimbach

RINGS, Fritz (in Detroit, Michigan)

Application for extension of his foreign vacation
1910
<u>227</u>: Rep. MA 1921: A.V.V: Nr. 16402.   ADS 9:91

RINTELMANN, Gerhard

[Gift?] to the church at Rowan, North Carolina
1772

<u>39</u>: Bestand Senat: Cl.VIII, No.X, Jg.1772 (Register)   ADS 2:98

RIPCKE, Johannes

New York: Information regarding   1879
<u>39</u>: Bestand Senat: Cl.VIII, No.X, Jg.1879 (Register) 52. [Nordamerika]   ADS 2:242

RIPPE, -- (a convict)

Transportation to America   1839
<u>52</u>: Han.Des.74: Westen-Thedinghusen: I.56: Nr. 29.
                                           ADS 4:80

RIRK, Albertine (in Ontonagon, Michigan)

Estate of   1896
<u>39</u>: Reichs- u. auswaert.Ang.: E.IV.78 [alt.Sig.]
[Erbschaften]   ADS 2:282

RISER, -- (a Mormon)

Liberation of   1854
<u>39</u>: Bestand Senat: Cl.VI, No.16p: vol.3b, Fasc. 2f
[Freilassung]   ADS 2:37

RISSE, --

St. Louis consulate: Risse *versus* Risse   1896
<u>39</u>: Reichs- u. auswaert.Ang.: J.Ib.10.5 [alt.Sig.]
[Justizsachen]   ADS 2:283

RITTER, Georg J. N.

Galveston, Texas: Estate of   1904
<u>39</u>: Reichs- u. auswaert.Ang.: E.IV.58 [Erbschaften]
                                           ADS 2:335

RITTER, J. (from Aschaffenburg)

Release [from Bavarian citizenship]   1808
<u>232</u>: M. P. A. V.: Nr. 2640.   ADS 9:136

RITTER, Karl (from Hildesheim)

Transportation of convict to America   1840
<u>52</u>: Han.Des.80: Hildesheim I: E.DD.7i: Nr. 920.
                                           ADS 4:88

RITTGERODT, Friederike (servant girl; from Dannhausen)

Trial for theft & emigration to America 1857-1859
<u>56</u>: L Neu Abt.129A: Gr.11, Nr. 30.   ADS 4:187

RITZ, -- (a minor; in USA)

Guardianship matters   1918
39: Reichs- u. auswaert.Ang.: V.III.2.2 [Vormund-
    schaftssachen]   ADS 2:396

RITZMANN, M.

New York: Estate of   1901
39: Reichs- u. auswaert.Ang.: E.IV.56 [Erbschaften]
                                             ADS 2:316

RIX, Ch. Friedr.

Transmittal of prisoner Rix from Altona to Chicago
    where he is wanted for falsification of documents
                                                  1896
39: Reichs- u. auswaert.Ang.: D.Ic.38 [Ausliefer-
    ungen]   ADS 2:280

ROBBE, Gerd (of Leer; journeyman carpenter)

Return from America   1877
50: Rep.21a: Nr. 8124.   ADS 4:7

ROBINSON, Marie A. (in New York). *See* von Melle,
Emil

ROBINSON, Therese (in New York). *See* von Melle,
Emil

ROCH, Peter (from Niederlustadt [Herrschaft])

Data on his emigration   1741-1758
201: Archivabt. Johanniterorden: Akt Nr. 58.
                                             ADS 8:61

ROCHELLE (ROSENKILDE), Emil

New York: Estate of   1907
39: Reichs- u. auswaert.Ang.: E.IV.70 [Erbschaften]
                                             ADS 2:347

RODEWALD, -- (in New York)

Report from   1851
48: B.13a: Nr. 35-37 [Schreiben]   ADS 3:3

ROEMMICH, Philipp Dietrich (of Edenkoben, Pfalz
[Palatinate])

Petitions for information regarding an inherit-
    ance in Philadelphia   1830
227: Rep. Bayerische Gesandtschaft London: Nr. 883.
                                             ADS 9:47

ROEPER, William (in West Bend, Wisconsin)

Estate of   1915
39: Reichs- u. auswaert.Ang.: E.IV.38 [Erbschaften]
                                             ADS 2:388

Estate of   1917
39: Reichs- u. auswaert.Ang.: E.IV.23 [Erbschaften]
                                             ADS 2:392

ROESCH, --

Extradition of -- Roesch & -- Specht to Rosslau
    from the USA   1913
39: Reichs- u. auswaert.Ang.: A.Ia.19 [Ausliefer-
    ungen]   ADS 2:377

ROETHER, Johann Joachim (mason)

Expulsion from Hamburg as Social Democrat & emi-
    gration to America   1880-1881
39: Polizeibehoerde-Polit.Pol.: S 149/326: Bl.
    340-354.   ADS 2:416

ROEVER, Carl (miner; from Clausthal)

Emigration of   1862
68: IV.2.b: Fach 162, Nr. 45.   ADS 4:247

ROGGE, Fritz

New York: Estate of   1908
39: Reichs- u. auswaert.Ang.: E.IV.90 [Erbschaften]
                                             ADS 2:352

ROHDE, Friedr. Joh. Christ. (disappeared in North
America)

Estate of   1897
39: Reichs- u. auswaert.Ang.: E.IV.51 [alt.Sig.]
    [Erbschaften]   ADS 2:290

ROHDE, Herm.

San Francisco: Estate of   1908
39: Reichs- u. auswaert.Ang.: E.IV.12 [Erbschaften]
                                             ADS 2:351

ROHDE, Irmgard (in New York)

Appointment of guardianship for   1917
39: Reichs- u. auswaert.Ang.:V.III.2.4 [Vormund-
    schaftswesen]   ADS 2:393

ROHDE, Karl (from Osterode)

Proposed aid for emigration to America   1866
52: Han.Des.80: Hildesheim I: E.DD.43a: Nr. 1679.
                                             ADS 4:91

ROHDE, Minna, born Toede (in Chicago)

Inquiry by private person in Germany regarding
1898

39: Reichs- u. auswaert.Ang.: N.Ib.1.2 [Nachforsch-
ungen]   ADS 2:297

ROHDE, -- (in Hundeskopf bei Tempelberg, Pomerania)

Inquiry about emigration to America   1848

39: Bestand Senat: Cl.VIII, No.X, Jg.1848 (Register)   ADS 2:153

ROHLFS, Hennig (resident of Bergedorf, HH)

Banishment to America   1864

39: Bergedorf II: Nr. 41, vol.6, Nr. 11 [Abschieb-
ung]   ADS 2:467

ROLF, Gesche (married woman)

Chicago: Letter regarding her   1878

39: Bestand Senat: Cl.VIII, No.X, Jg.1878 (Register) 8 [Nordamerika]   ADS 2:236

ROLLE, Franz (notary public? in Chicago)

Complaint against Bavarian consulate in Chicago
1870

227: Rep. Gesandtschaft Karlsruhe: Nr. 206.
ADS 9:52

ROMANOWSKI, Willi

Boston: Estate of   1915

39: Reichs- u. auswaert.Ang.: E.IV.28 [Erbschaften]
ADS 2:388

ROSE, Eugen G. (in Arkansas)

Estate of [file missing]   1914

39: Reichs- u. auswaert.Ang.: E.IV.31 [Erbschaften]
ADS 2:384

ROSE, Johann Carl Heinrich (or John Rose)

German consulate in Pittsburgh: Request for ex-
tract from Rose's testament   1876

39: Bestand Senat: Cl.VI, No.16p: vol.4b: Fasc. 25,
Invol. 1.   ADS 2:60

ROSE [or Rosse] --

Estate of   1880

39: Bestand Senat: Cl.VIII, No.X, Jg.1880 (Register) 3 [Nordamerika]   ADS 2:243

ROSENBAUM, --

San Francisco: Estate of   1912

39: Reichs- u. auswaert.Ang.: E.IV.77 [Erbschaften]
ADS 2:373

ROSENBLATT, Ascher (in New York)

Case of Rosenblatt versus Herz Steinberger, of
Angerod   1851

149: Rep.49a: Nr. 566/1.   ADS 7:17

ROSENFELD, Harriette. See Raffa, Harriette, divorced
Breitenbach, born Rosenfeld

ROSENKILDE, Emil. See Rochelle, Emil

ROSENTHAL, Amalie (from Moerfelden, Landkreis Gross-
Gerau, HEL)

Lived in Moerfelden with three family members; emigrated to USA   1933?

179: Wiedergutmachungsakten, Ordner Nr. 061-00/05.
ADS 7:232

ROSENTHAL, Ferdinand (cable maker; from Osterode)

Proposed aid to transport Rosenthal, a convict,
to America   1865

52: Han.Des.80: Hildesheim I: E.DD.43a: Nr. 1678.
ADS 4:90

ROSENTVAIG, R. (an American citizen)

Expulsion from Hamburg   1897

39: Reichs- u. auswaert.Ang.: A.IV.1 [alt.Sig.]
[Ausweisungen]   ADS 2:288

ROSKOP, Johann (from Miselohe)

Emigration tax   1805

115: Berg. Apanagialreg.: A. Nr. A29.   ADS 6:22

ROSS, -- (an American citizen)

Expulsion from Bavaria   1850-1851

227: Rep. MA 1898: A.St.: Nr. 6.   ADS 9:57

ROSS, -- (in New York)

Complaint of emigrant Ross against agent Falck
1864

39: Bestand Senat: Cl.VIII, No.X, Jg.1864 (Register) III.   ADS 2:185

ROSSE. --. See Rose, --

ROTH, Christian (in America)

Claim of inheritance [in Germany]   1854? 1881?
152: Abt.228: Amt Hoechst: Nr. 331.   ADS 7:132

ROTH, Johann (from Holzhausen)

Expropriation of the property of Roth, of the first
[Hessian?] Battalion, who deserted in America
1784-1800
151: Bestand 10:a: Nr. 55.   ADS 7:46

ROTH, Philipp I (from Erfelden, Landkreis Gross-
Gerau, HEL). See Rothmann, Philipp

ROTH, R. F.

Transmittal [of extradited prisoner] from the USA
to Ulm, via Hamburg   1901
39: Reichs- u. auswaert.Ang.: A.Ia.1 [Ausliefer-
ungen]   ADS 2:312

ROTH, -- (married couple; in New York)

Inquiry by Orphanage Commission regarding the financial standing of the Roths   1895
39: Reichs- u. auswaert.Ang.: W.I.5 [alt.Sig.]
[Waisenhausang.]   ADS 2:279

ROTHGIESSER, J. (in Hamburg)

Transmittal of a report from German consulate in
New York   1901
39: Reichs- u. auswaert.Ang.: G.Ia.1.3 [Geschaefts-
verkehr]   ADS 2:316

ROTH[MANN] Adam (from Erfelden, Landkreis Gross-
Gerau, HEL; in Kansas)

Letter to mayor [of Erfelden] & brother Philipp
regarding arrival & conditions in Kansas   1884
169: Abt. XI,4.   ADS 7:219

ROTH[MANN] Philipp I (from Erfelden, Landkreis Gross-
Gerau, HEL)

Emigration to America   1880
169: Ortsbuergerregister.   ADS 7:220

ROTHMANN, Philipp (from Erfelden, Landkreis Gross-
Gerau, HEL; in Kansas)

Correspondence with Mayor --Schaefer [in Erfelden]
regarding his life   1878-1884
169: Abt. XI,4.   ADS 7:219

ROTHMANN, Philipp (in Kansas). See also Schmenger,
Johannes

ROTHSTEIN, B. (in New York)

Inquiry by private person in Germany regarding B.
Rothstein & B. H. B. Blumgard   1898
39: Reichs- u. auswaert.Ang.: N.Ib.1.10 [Nachforsch-
ungen]   ADS 2:297

ROTTMEIER, Bernhard (former corporal of 4th Infantry
Regiment; lived in America)

Application for granting of a medal   1910
227: Rep. MA 1921: A.V. V: Nr. 16403.   ADS 9:91

ROWOLD, Karoline (from Wolfshagen)

Vagrancy & emigration to America   1850-1851
56: L Neu Abt.129A: Gr.11, Nr. 90; 91 (Band 2)
ADS 4:192

de ROY, Peter (in Wisconsin)

Estate of   1910
39: Reichs- u. auswaert.Ang.: E.IV.75 [Erbschaften]
ADS 2:362

ROYCE, George Monroe (American pastor)

Criminal case against   1903
227: Rep. MA 1921: A.V. V: Nr. 15813.   ADS 9:90

ROYER, Franz

New Orleans: Estate of   1907
39: Reichs- u. auswaert.Ang.: E.IV.112 [Erbschaften]
ADS 2:347

RUDOLPH, August

New York: Apparent taking advantage of the emigrant
Rudolph by agent B. Karlsberg, of Hamburg   1878
39: Bestand Senat: C1.VIII, No.X, Jg.1878 (Register) 63 [Nordamerika]   ADS 2:239

RUDOLPH, Balthasar (from Ruesselsheim, HEL)

Emigration to America   1851
182: Abt. XI,4.   ADS 7:235

RUDOLPH, Philipp II (from Ruesselsheim, HEL)

Emigration to America   1851
182: Abt. XI,4.   ADS 7:235

RUDOLPH, -- (Doctor of Divinity & former pastor; in
America)

Emigration of   1849-1853
500: Weimar: Thueringisches Staatsarchiv: XIX.C.c.
56; 57.   Not in ADS

RUPPERT, -- (an American citizen)

    Trial of Ruppert for customs violation   1894

    <u>39</u>: Reichs- u. auswaert.Ang.: J.Ic.4 [alt.Sig.]
    [Justizsachen]   ADS 2:270

RUPUIK? [Rupnik?] --

    Estate of   1942

    <u>148</u>: R XI.   ADS 6:262

RUSCH (or Schwenke) Johann

    San Francisco: Estate of   1906

    <u>39</u>: Reichs- u. auswaert.Ang.: E.IV.89 [Erbschaften]
    ADS 2:343

RYER, W. (at Ellis Island, New York)

    Estate of   1918

    <u>39</u>: Reichs- u. auswaert.Ang.: E.IV.161 [Erbschaften]
    ADS 2:394

SACHS, W.

    Complaint by Sachs regarding mayor & city officials
    of Everett, Washington   1910

    <u>39</u>: Reichs- u. auswaert.Ang.: G.Ia.1.5 [Geschaefts-
    verkehr]   ADS 2:363

SAEVECKE, Carl Wilhelm Gustav (bookbinder)

    Expulsion from Hamburg as Social Democrat & emi-
    gration to America   1879-1881; 1899
    [includes file of Carl Woelky]

    <u>39</u>: Polizeibehoerde-Polit.Pol.: S 149/507: Bl.
    427-460.   ADS 2:418

SAFFT, Wilh.

    San Francisco: Death certificate for   1877

    <u>39</u>: Bestand Senat: Cl.VIII, No.X, Jg.1877 (Regis-
    ter) 38 [Nordamerika]   ADS 2:235

SALOMON, --

    Exclusion by U.S. Immigration Service of -- Salo-
    mon & -- Heidorn [and return to Hamburg]   1912

    <u>39</u>: Reichs- u. auswaert.Ang.: A.Ib.2 [Ausliefer-
    ungen]   ADS 2:372

SALOMONSKI, Louis (in New York)

    Information requested by private person regarding
    1897

    <u>39</u>: Reichs- u. auswaert.Ang.: N.Ib.1.1 [alt.Sig.]
    [Nachforschungen]   ADS 2:291

SAMMIS, -- (an American)

    Residence in Detmold   1917

    <u>124</u>: D. Akten: Fach 23, Nr. 3.   ADS 6:84

SAND, Louis (in America)

    Release of Sand, a Bavarian citizen, from U.S.
    Army   1862

    <u>227</u>: Rep. MA 1921: A.St.I: Nr. 1237.   ADS 9:71

SANDEL, -- (in Chicago)

    Information regarding Sandel sought by private
    person [in Germany]   1894

    <u>39</u>: Reichs- u. auswaert.Ang.: N.Ib.14 [alt.Sig.]
    [Nachforschungen]   ADS 2:271

SANDER, Adam (from Blaubach; in New York)

    Data regarding the donation of $1000 to the com-
    munity of Blaubach [Kreis Kusel] by Sander, an
    emigrant   1875

    <u>207</u>: Nr. 950-01.   ADS 8:74

    Charitable gift to the community of Blaubach [Bav-
    aria]   1875

    <u>227</u>: Rep. MA 1921: A.V. V: Nr. 15139.   ADS 9:89

SANDERS, Henry C. (an American citizen)

    Estate of   1913

    <u>39</u>: Reichs- u. auswaert.Ang.: E.IV.58 [Erbschaften]
    ADS 2:379

SANDROCK, Adam (tanner; from Vacha, Thuringia)

    Emigration application   1833

    <u>500</u>: Weimar: Thueringisches Staatsarchiv: Dermbach
    II: Gb.2: Nr. 63-92; 93-114; 144-183; 186;
    188   Not in ADS

SASS, B.

    San Francisco: Estate of   1887

    <u>39</u>: Bestand Senat: Cl.VIII, No.X, Jg.1887 (Regis-
    ter) 16 [Erbschafts-Amt]   ADS 2:258

SASS, Hans Detlef (cigar worker)

    Expulsion from Hamburg as Social Democrat & emi-
    gration to America   1887-1888

    <u>39</u>: Polizeibehoerde-Polit.Pol.: S. 149/495: Bl.
    461-475.   ADS 2:418

SATTLER, Jost (Swiss citizen)

    Extradition to Switzerland from the USA, via Ham-
    burg   1908

39: Reichs- u. auswaert.Ang.: A.Ia.19 [Auslieferungen]
ADS 2:350

SAUER, Wilh[elm] (in Lancaster, Pennsylvania)

Death certificate   1877
39: Bestand Senat: Cl.VIII, No.X, Jg.1877 (Register) 55 [Nordamerika]   ADS 2:236

Death certificate   1878
39: Bestand Senat: Cl.VIII, No.X, Jg.1878 (Register) 49 [Nordamerika]   ADS 2:238

SCHAAF, M. F. (an American citizen)

Application to be permitted to remain in Hamburg
1900
39: Reichs- u. auswaert.Ang.: A.IV.3 [Ausweisung]
ADS 2:306

SCHABIROWSKI, --

Delivery of a summons & legal complaint of the County Court of Adams County, Colorado, in the divorce case, Schabirowski *versus* Schabirowski
1909
39: Reichs- u. auswaert.Ang.: J.Ia.4.9 [Justizsachen]   ADS 2:358

SCHACHTEL, Dorothea, born Nocker (widow; died 9 April 1906 in New York)

Estate of   1902-1909
152: Abt.405: Wiesbaden: Rep.405/I,2: Nr. 3516.
ADS 7:146

SCHAD, Daniel, III (from Gross-Gerau, HEL)

Travel costs for emigration   1854
174: Abt. XI,4.   ADS 7:227

SCHADT, Daniel (from Gross-Gerau, HEL)

Transportation contract from Mainz to New York
1854
174: Abt. XI,4.   ADS 7:227

SCHAEBER, Aug. Am. (in San Francisco)

Consular report on Schaeber's military service obligation   1861
39: Bestand Senat: Cl.VIII, No.X, Jg.1861 (Register) I [Amerika]   ADS 2:178

SCHAEFER, Carl (died in Guatemala)

Estate of [Schaefer was in the Hamburg project to colonize Guatemala]   1844; 1860
52: Han.Des.9: sog.Geheime Reg.: A., Nr. 13. DLC has.   ADS 4:101

SCHAEFER, Friedrich (from Baltimore)

Correspondence with Bremen consul regarding his treatment in Bremen   after 1800
48: Dd.11.c.N.1.a.1: Nr. 4-7; 9 a-c; 10-12.
ADS 3:57

SCHAEFER, Georg (in Georgetown [District of Columbia?])

File on [estate?]   1893
227: Rep. MA 1921: A.St.I: Nr. 965.   ADS 9:70

SCHAEFER, Johann Christoph (from Raunheim, Landkreis Gross-Gerau, HEL)

Emigration to America with wife and four children
1854
181: Raunheimer Evangelischen Pfarrei Familienregsiter (Lehrer Johannes Buxbaum) ADS 7:234

SCHAEFER, John E. (in Riverdale, Illinois)

Statement of A.Schumacher of Hamburg in support of the claim of the widow J. Schmidt against Schaefer   1895
39: Reichs- u. auswaert.Ang.: H.Id.26 [alt.Sig.] [Handel]   ADS 2:276

SCHAEFER, Wilhelm (last maker; from Gandersheim)

Emigration to America   1867
56: L Neu Abt.129A: Gr.11, Nr. 40.   ADS 4:188

SCHAEFER, -- (from Dornheim, Landkreis Gross-Gerau, HEL)

Emigration of family   ca. 1860
168: Schulgeschichte.   ADS 7:218

SCHAEFER, --

Estate of   1942
148: R XI.   ADS 6:262

SCHAEFER, -- (from Lohr)

Emigration file on   1788
232: Rep.77: Aschaffenburger Archivreste: M.P.A. 12464.   ADS 9:137

SCHAEFER (or Schaeffer) --

Investigation to determine whether person of this name left will; heirs in America   1901
149: Rep. 49a: Nr. 565/13.   ADS 7:17

SCHAEFFER, Friedericke (in Brooklyn)

File on   1894-1897

227: Rep. MA 1921: A.St.I: Nr. 1009.   ADS 9:70

SCHAEFFER, Paul Wilhelm (in America)

Letters from   1776

500: Hanau: Geschichtsverein: Mss. Nr. 559; 559a-f.
                                               Not in ADS

SCHAEUFFLER, Friedr[ich]

New York: Estate of   1911

39: Reichs- u. auswaert.Ang.: E.IV.64 [Erbschaften]
                                               ADS 2:369

SCHAFFER, -- (in Germantown, Pennsylvania)

Inheritance of Mrs. Philipp Schaffer to be paid to
    children   1898

149: Rep.49a: Nr. 565/15.   ADS 7:17

SCHAFFNER, Dora

San Francisco: Estate of   1904

39: Reichs- u. auswaert.Ang.: E.IV.15 [Erbschaften]
                                               ADS 2:335

SCHAFFNER, Elisabeth (from Goddelau, Landkreis Gross-
Gerau, HEL)

Emigrant to Brooklyn   1937

173: Abmelderegister 1930-1943: Nr. 96. ADS 7:224

SCHAFFNER, Stadian (from Goddelau, Landkreis Gross-
Gerau, HEL)

Emigrant to America   1882

173: Gesinderegister 1853-1894: Deckblatt, Rueck-
     seite.   ADS 7:225

SCHALTER, Franz Balthasar (from Assenheim)

Emigration permit for   1767-1768

201: Archivabt. Leiningen-Hardenburg: Akt. Nr. 133.
                                               ADS 8:62

SCHAPER, John (from Hamburg)

Hanseatic Ministry in Berlin: Statement regarding
    Schaper's expulsion from the USA   1899

39: Hanseat. u. Hambg.dipl.Vertretungen: HG.VII.
    Z.5.   ADS 2:399

Schaper's complaint regarding his expulsion from
    the USA   1899

39: Reichs- u. auswaert.Ang.: D.Id, 16 [Ausweis-
    ungen]   ADS 2:300

SCHARF, Anna

Arrest on board *Bohemia* at Cuxhaven as suspected
    Socialist Anna Braun, also wanted for theft 1885

39: Bestand Senat: Cl.I, Lit.T., No.7, Fol.6, Fasc.
    11a, Invol. 19 [Verhaftung]   ADS 2:4

SCHARFENBERG, H. (in Empire, Oregon)

Estate of   1908

39: Reichs- u. auswaert.Ang.: E.IV.24 [Erbschaften]
                                               ADS 2:351

SCHATZ, Wilhelm (day laborer; from Schlewecke)

Punishment & emigration to America   1853-1856

56: L Neu Abt.129A: Gr.11, Nr. 80.   ADS 4:191

SCHAUB, W.

Pursuit of Schaub who fled to America   1892

227: Rep. MA 1921: A.V. V: Nr. 15753.   ADS 9:90

SCHAUM, Elisabeth (died 1891 in Philadelphia)

File on [estate?]   1892-1893

227: Rep. MA 1921: A.St.I: Nr. 945.   ADS 9:70

SCHAUMBURGER, Elise (widow; in Cincinnati)

File on [estate?]   1899

227: Rep. MA 1921: A.St.I: Nr. 1054.   ADS 9:71

SCHAUS, Karl Jakob Christian (in New York)

Application of Karl Schaus, a teacher & father of
    Karl Jakob Christian Schaus, for son's release
    from [German] citizenship due to his domicile in
    North America   1861

152: Abt.236: Nastaetten: Nr. 1026.   ADS 7:135

SCHEER, Ernst Emil Constantin Theodor (journeyman
baker)

Expulsion from Hamburg as Social Democrat & emi-
    gration to America   1880-1881

39: Polizeibehoerde-Polit.Pol.: S 149/59: Bl. 1-13.
                                               ADS 2:413

SCHEERER, -- (Russian emigrant family)

Affairs of   1878

39: Bestand Senat: Cl.VIII, No.X, Jg.1878 (Regis-
    ter) 6 [Auswanderer-Deputation]   ADS 2:236

SCHEIBECK, Maria (in Buffalo [New York?])

File on [estate?]   1890
227: Rep. MA 1921: A.St.I: Nr. 918.   ADS 9:70

SCHEIBLER, Georg (from Hoesbach)

Release [from Bavarian citizenship]   1807
232: M.P.A.V. 2724.   ADS 9:136

SCHEID, Johann (in New York)

Demand that he return [to Bavaria] & support his
    family   1899
227: Rep. MA 1921: A.St.I: Nr. 1266.   ADS 9:72

SCHEIDEGGER, C. (of Waldmichelbach, Odenwald, HEL)

Emigrant to Texas?   1845
154: CA 4b6: Nr. 2   ADS 7:171

SCHEIDLER, Theodor (in New York)

File on [estate?]   1891-1894
227: Rep. MA 1921: A.St.I: Nr. 932.   ADS 9:70

SCHEINER, Josef (in Bay City, Michigan)

File on [estate?]   1889
227: Rep. MA 1921: A.St.I: Nr. 899.   ADS 9:70

SCHEITZOW, --

Claim made against Union [U.S.] government by J. C.
    L. Scheitzow for son's death benefits   1863
39: Bestand Senat: Cl.VIII, No.X, Jg.1863 (Regis-
    ter) I [Amerika]   ADS 2:183

SCHELINSKI, --

Estate of   1942
148: R XI.   ADS 6:262

SCHELLENBERG, Adolf (from Goddelau, Landkreis Gross-
    Gerau, HEL)

Emigrant to America   1874
173: Ortsbuergerregister.   ADS 7:224

SCHELLENBERG, Moritz (from Goddelau, Landkreis Gross-
    Gerau, HEL)

Emigrant to America   1883
173: Gesinderegister 1853-1894: Deckblatt, Rueck-
    seite.   ADS 7:225

SCHELLHAS, John (died 26 April 1904 in New York)

Estate of   1910
152: Abt.405: Wiesbaden: Rep.405/I,2: Nr. 3517.
                                             ADS 7:146

SCHELLHORN, D. C. (in Little Rock, Arkansas)

Letter from Schellhorn regarding his intention to
    cut off from inheritance his divorced & remarried
    wife Mrs. Joh[anna?] Catharina Maximilliane
    Goerke   1847
39: Bestand Senat: Cl.VIII, No.X, Jg.1847 (Regis-
    ter)   ADS 2:151

SCHENITZKI, John (in Milwaukee, Wisconsin)

Estate of   1895
39: Reichs- u. auswaert.Ang.: E.IV.15 [alt.Sig.]
    [Erbschaften]   ADS 2:275

SCHENK, H. G.

Travel aid paid by Boston consulate [repatriation?]
                                             1902
39: Reichs- u. auswaert.Ang.: U.Ia.1.3 [Unter-
    stuetzungen]   ADS 2:327

SCHENUITT, --

Estate of   1943
148: R XI.   ADS 6:262

SCHEPP, -- (in Cincinnati, Ohio)

Estate of   1910-1911
152: Abt.405: Wiesbaden: Rep.405/I,2: Nr. 3519.
                                             ADS 7:146

SCHERER, Emil (from Ruesselsheim, HEL)

Governmental inquiry as to the financial situation
    of Scherer, an emigrant to North America 1852
182: Abt.XI,4.   ADS 7:235

SCHERN, Rudolph Hermann Friedrich (cigar worker)

Expulsion from Hamburg as Social Democrat & emi-
    gration to America   1881
39: Polizeibehoerde-Polit.Pol.: S 149/343: Bl. 14-
    27.   ADS 2:417

SCHERPEL, Max (an American citizen)

Expulsion from Hamburg   1911
39: Reichs- u. auswaert.Ang.: A.IV.4 [Ausweis-
    ungen]   ADS 2:368

SCHEUCHL, Therese (in America)

Financial aid to 1884

227: Rep. MA 1921: A.V. II: Nr. 3936. ADS 9:81

SCHEUENSTUBL, Konrad (in Brooklyn)

File on [estate?] 1891

227: Re. MA 1921: A.St.I: Nr. 934. ADS 9:70

SCHEUER, -- (ship's carpenter; in New York)

Estate of 1848-1852

52: Han.Des.74 Dannenberg: Abt.II, G, Nr. 15.
ADS 4:65

SCHEUNEMANN, --

Estate of 1943

148: R XI. ADS 6:262

SCHEVEN, Eugen (died 1887 in New York)

File on [estate?] 1888

227: Rep. MA 1921: A.St.I: Nr. 894. ADS 9:70

SCHIEBER, M.

Extradition of Schieber from America to Bavaria
1890

39: Bestand Senat: Cl.VIII, No.X, Jg.1890 (Register) 7. ADS 2:264

SCHIEBLER, --

Emigration to America 1837-1852

500: Schwerin: Mecklenburg-Schwerin: Gorssherzog-
liches Geh. u. Hauptarchiv: Cabinett. Vol. 3.
Not in ADS

SCHIELE, Friedrich (typesetter)

Expulsion from Hamburg as Social Democrat & emigration to America 1880-1881

39: Polizeibehoerde-Polit.Pol.: S 149/61: Bl. 28-
36. ADS 2:413

SCHILD, Wilhelm (from Ruesselsheim, HEL)

Emigration application 1861

182: Abt.XI,4. ADS 7:235

SCHILDGEN, Peter (from Raunheim, Landkreis Gross-
Gerau, HEL)

Emigration to America with wife and five? children 1854

181: Raunheimer Evangelischen Pfarrei Familienregister (Lehrer Johannes Buxbaum) ADS 7:234

SCHILL, Martha Louise Anna. *See* Andersen, Julie,
born Hartmann

SCHILLING, Adam (minor; from Sunna, Thuringia)

Emigration application 1839

500: Weimar: Thueringisches Staatsarchiv: Dermbach
II: Gb.2: Nr. 63-92; 93-114; 144-183; 186;
188. Not in ADS

SCHILLING, Johann (from Ruesselsheim, HEL)

Emigration to North America 1849

182: Abt.XI,4. ADS 7:235

SCHILLING, Johann (carpenter)

File regarding his death aboard the American ship
Wild Rover 1864

48: B.13.b.7: Nr. 32-35; 41-43; 46.
DLC has? ADS 3:26

SCHILLINGER, Karl (from Koenigsbrunn; in America)

Petition for financial assistance for his return
to Germany 1899

227: Rep. MA 1921: A.V. II: Nr. 4027. ADS 9:81

SCHINDLER, John (died 1888 in America)

File on [estate?] 1888-1889

227: Rep. MA 1921: A.St.I: Nr. 890. ADS 9:70

SCHLACHTMEYER, Albert. *See* Lammering, Albert, *alias*
Schlachtmeyer

SCHLEICHER, Anna Maria, born Vogt (from Hillesheim,
Grafschaft Falkenstein, RPL)

Division of the property of Anna Maria Schleicher,
an emigrant, among her relatives remaining in
Germany 1783-1784

201: Archiv.Grafschaft Falkenstein: Akt. Nr. 127.
ADS 8:61

SCHLEIERMACHER, --

New York: Estate of 1913

39: Reichs- u. auswaert.Ang.: E.IV.83 [Erbschaften]
ADS 2:379

SCHLEIFER, Henry (in New York)

File one [estate?] 1894

227: Rep. MA 1921: A.St.I: Nr. 990. ADS 9:70

SCHLENS, L

San Francisco: Estate of 1895

39: Reichs- u. auswaert.Ang.: E.IV.48 [alt.Sig.]
[Erbschaften]   ADS 2:275

SCHLESINGER, Alfred (in New York). *See* von Melle,
Emil

SCHLIEKER, Christian (from Wenzen)

Emigration of delinquent to America   1858
56: L Neu Abt.129A: Gr.11, Nr. 86.   ADS 4:192

SCHLIMME, Heinrich (journeyman blacksmith; from Olx-
heim)

Emigration to America   1860
56: L Neu Abt.129A: Gr.11, Nr. 73.   ADS 4:191

SCHLIMME, L. (from Einbeck?)

Police custory of Schlimme & his transportation to
America   1846
52: Han.Des.80: Hildesheim I: E.DD.37a: Nr. 1295.
ADS 4:89

SCHLITZ, *alias* Mosher, --

Hamburg police search for person wanted in America
1918
39: Reichs- u. auswaert.Ang.: P.III.1.1 [Polizei-
wesen]   ADS 2:395

SCHLOSSER, -- (businessman; from Pfortzheim, BWL)

Permit to emigrate to America   1751
260: Abt.236: Innenmin., Wegzug: Auswand.: Nr.
2651.   ADS 10:3

SCHLOTTER, Georg (in Buffalo [New York?])

File on [estate?]   1898
227: Rep. MA 1921: A.St.I: Nr. 1037.   ADS 9:71

SCHLOTTMANN, Heinrich Ernst Friedrich (bartender)

Expulsion from Hamburg as Social Democrat & emigration to America   1880-1881
39: Polizeibehoerde-Polit.Pol.: S 149/40: Bl. 37-
44.   ADS 2:412

SCHLUETER, Otto

New York: Estate of   1914
39: Reichs- u. auswaert.Ang.: E.IV.52 [Erbschaften]
ADS 2:384

SCHMADEL, *alias* Winter, John. *See* Winter, John
[*correctly* Schmadel]

SCHMEIL, Anna. *See* Lang, Anna, born Schmeil

SCHMELZ, --

Estate of   1941
148: R XI.   ADS 6:262

SCHMELZEISEN, Anna. *See* Schmelzeisen, Anton

SCHMELZEISEN, Anton (of Bauschheim, Landkreis Gross-
Gerau, HEL; in New York)

Register of Removals [Abmelderegister] contains
notation of the emigration of Anton, Anna (wife),
Luise [relationship not stated], Georg (son),
Ottilie (daughter) & Elisabethe Matiza (step-
daughter)   1920
162   ADS 7:212

SCHMELZEISEN, Georg. *See* Schmelzeisen, Anton

SCHMELZEISEN, Josef Anton (of Bauschheim, Landkreis
Gross-Gerau, HEL; in New York)

Land Register [Grundbuch], vol. 5, p. 292, Acker
Flur 1, Parzelle 680, next to "Gemeinen Boslach"
contains notation of ownership
162   ADS 7:212

SCHMELZEISEN, Luise. *See* Schmelzeisen, Anton

SCHMELZEISEN, Ottilie. *See* Schmelzeisen, Anton

SCHMENGER, Johannes (from Erfelden, Landkreis Gross-
Gerau, HEL)

Letter from emigrant Schmenger in North Hampton
[state not give in ADS description] USA to his
father & the mayor regarding conditions encoun-
tered & news of Philipp Rothmann in Kansas
[which see, herein]   1875-1877
169: Abt. XI,4.   ADS 7:219

SCHMID, Peter (of Mittelbuchen, Hessen-Hanau)

Attempt by Christoph Ruppel, of Neuhanau, to re-
cover property [sequestered by the government]
of his sister, the wife of Peter Schmid, emigrants to America   1766-1790
151: Bestand 81: Rep.E: Gemeinde Mittelbuchen:
Nr. 1.   ADS 7:82

SCHMID, Walburga (wife of shoemaker; from Gudelsdorf)

Inheritance in America   1835-1836
226: Rep. MInn 16: Nr. 52649.   ADS 9:19

SCHMIDT, Adolf (from Wengeringhausen, HEL)

Emigration of   n.d.
151: Bestand 180: Arolsen: Acc.1937/76,5: Nr. 118.
ADS 7:112

SCHMIDT, Albert (in Chicago)

Information requested by private person in Germany
regarding   1897
39: Reichs- u. auswaert.Ang.: N.Ib.1.7 [alt.Sig.]
[Nachforschungen]   ADS 2:291

SCHMIDT, Alfred (supervisor; an American citizen)

Forceable expulsion of Alfred & Joseph Schmidt
1910
39: Reichs- u. auswaert.Ang.: A.IV.4 [Ausweisungen]
ADS 2:361

SCHMIDT, Anna Dorothea. *See* Schmidt, Johann Wilhelm

SCHMIDT, Christian

German consular report from New York regarding
Schmidt's complaint of fraud   1869
39: Bestand Senat: C1.VIII, No.X, Jg.1869 (Register) 38 [Nordamerika]   ADS 2:203

SCHMIDT, Conrad (naturalized American citizen; jailed
at Bremen)

Intercession of American consul for Schmidt from
Masseln [Hannover], wanted by Hannoverian
authorities for failure to complete military
service obligation   1853
48: Dd.11.c.2.N.1.a.1: Nr. 47.   ADS 3:61

SCHMIDT, C. A. [*alias* Smith](seaman; went down on
American warship *Maine*)

Estate of C. A. Smith   1898
39: Reichs- u. auswaert.Ang.: E.IV.45 [alt.Sig.]
[Erbschaften]   ADS 2:296

SCHMIDT, Dorothea (widow; in Chicago)

Estate of   1899-1900
149: Rep.49a: Nr. 563/4.   ADS 7:16

SCHMIDT, Emma Luise (from Dornheim, Landkreis Gross-
Gerau, HEL)

Emigration to America   1930
168: Abmelderegister.   ADS 7:218

SCHMIDT, Georg (died in Geneva, Illinois)

File on [estate?]   1885

227: Rep. MA 1921: A.St.I: Nr. 1723.   ADS 9:74

SCHMIDT, Heinrich (in Will County, Illinois)

Estate of   1908
39: Reichs- u. auswaert.Ang.: E.IV.14 [Erbschaften]
ADS 2:351

SCHMIDT, H. A. (businessman)

Extradition from USA to Apenrade, via Hamburg
1906
39: Reichs- u. auswaert.Ang.: A.Ia.9 [Ausliefer-
ungen]   ADS 2:342

SCHMIDT, Henry Clay (a naturalized American citizen)

Expulsion of   1888-1889
39: Bestand Senat: C1.1, Lit.T, No.20, vol.2b,
Fasc. 2. [Ausweisung]   ADS 2:20

Representations of the American [consulate] requesting cancellation of the order expelling
Schmidt from Hamburg   1889
39: Bestand Senat: C1.VIII, No.X, Jg.1889 (Register) 1.   ADS 2:262

SCHMIDT, Jakob (from Bretthausen, HEL)

Applications of Jakob Schmidt & Christ. Reeh?
[Rech?] to emigrate to Texas   1845
152: Abt.233: Marienberg: Nr. 1100.   ADS 7:134

SCHMIDT, Johann Joseph (from Winnweiler, Grafschaft
Falkenstein, RPL; in Franklin County, Pennsylvania)

Division [among relatives remaining in Germany?] of
the property of Schmidt, who emigrated in 1761
1791
201: Archivabt. Grafschaft Falkenstein: Akt Nr.
136.   ADS 8:61

SCHMIDT, Johann Michael (in Pittsburgh [Pennsylvania])

File on [estate?]   1890-1892
227: Rep. MA 1921: A.St.I: Nr. 919.   ADS 9:70

SCHMIDT, Johann Peter. *See* Schmidt, Johann Wilhelm

SCHMIDT, Johann Wilhelm (unknown residence in America)

Guardianship of Joh. Wilh. Schmidt, Anna Dorothea
Schmidt, & Johann Peter Schmidt   1866
152: Abt.236: Nastaetten: Nr. 1005.   ADS 7:135

SCHMIDT, Josef. *See* Ballinger, G.

Schmidt, Joseph. *See* Schmidt, Alfred

SCHMIDT, J. (widow). *See* Schaefer, John E.

SCHMIDT, Karl (in New Orleans)

Letter from Luise Seemann, born Schmidt, of Plauen in Vogtlande, regarding four gold pieces to be claimed from her brother Karl Schmidt [mentions Weissenfels bei Naumburg/Saale]   1841

39: Bestand Senat: Cl.VIII, No.X, Jg.1841 (Register)   ADS 2:142

SCHMIDT, Karl (died May 1914 in New York)

Estate of   1914-1919

152: Abt.405: Wiesbaden: Rep.405/I,2: Nr. 3520.
                                               ADS 7:146

SCHMIDT, Katharina. *See* Schmidt, Nikolas

SCHMIDT, Louise

New York: Estate of   1899

39: Reichs- u. auswaert.Ang.: E.IV.61 [alt.Sig.]
[Erbschaften]   ADS 2:301

SCHMIDT, L. (fron Naunstadt; in America?)

Query by [Wuerttemberg] general consul in Baltimore [Maryland] regarding a certificate of stock ownership in the Wuerttembergische Gesellschaft fuer Ansiedlung in Amerika [Wuerttemberg Society for Settlement in America] made out to Schmidt
                                               1853-1854

263: Rep.E/41-44: Minist. d. AA.II: Verzeichnis 63: Nr. 227.   ADS 10:42

SCHMIDT, Margaretha (died in Baltimore, Maryland)

File on [estate?]   1898

227: Rep. MA 1921: A.St.I: Nr. 1044.   ADS 9:71

SCHMIDT, Mary Eugenie. *See* Runyon, Mary Eugenie, born Schmidt

SCHMIDT, Max

Philadelphia: Estate of   1901

39: Reichs- u. auswaert.Ang.: E.IV.95 [Erbschaften]
                                               ADS 2:316

SCHMIDT, Nikol (in Philadelphia)

Proceedings against Schmidt; herein is requested a statement from his widow Katharina [Schmidt] regarding community property   1862

227: Rep. MA 1898: A.St.: Nr. 6325.   ADS 9:59

SCHMIDT, Philipp (from Darmstadt)

Emigration of   1851

149: Rep. 51 V: Fuerth: Nr. 25 S.   ADS 7:1

SCHMIDT, Wilhelm (pastor; in Columbus, Ohio)

Letter from Schmidt asking for a collection of theological books for the German ministerial & teachers' seminar in Ohio   1832

39: Geistl.Minis.: II,11: Konventsprotokolle: p. 75f.   ADS 2:475

SCHMIDT, -- born Gruebel (in Chicago)

Escaped prisoner   1903

39: Reichs- u. auswaert.Ang.: J. Ic.1 [Justizsachen]   ADS 2:332

SCHMIDT, -- (died in American military service)

Estate of, & pension for his widow in Paderborn
                                               1859-1886

116: Minden: Praes.-regist.: I.C. Nr. 13. ADS 6:53

SCHMIDT, --

Guardianship in USA   1908

39: Reichs- u. auswaert.Ang.: V.III.4 [Vormundschaftswesen]   ADS 2:355

SCHMIDT, --

Estate of   1927

148: R XI.   ADS 6:262

SCHMIDT, -- (pastor; in Pennsylvania). *See* Helmuth, --

SCHMIDTS, Johann Jost (from Suedershausen). *See* Brachthaeuser, Daniel

SCHMIEDEL, --

San Francisco: Estate of   1894

39: Reichs- u. auswaert.Ang.: E.IV.79 [alt.Sig.]
[Erbschaften]   ADS 2:269

SCHMITT, Franz Seraph (in America)

Report on his life & place of residence   1888

227: Rep. MA 1921: A.St.I: Nr. 1735.   ADS 9:74

SCHMITT, Heinrich (from Ruesselsheim, HEL)

Emigration to North America   1847-1848

182: Abt. XI,4.   ADS 7:235

SCHMITT, Ludwig (from Godramstein; died in America)

File on [estate?]   1889
227: Rep. MA 1921: A.St.I: Nr. 900.   ADS 9:70

SCHMOLLER, --

Estate of   1941-1942
148: R XI.   ADS 6:262

SCHMUECKER, J. Georg (president of the General Synod
of the Free North American States [Lutheran?])

Letter to Hamburg Spiritual Ministry [Ministry of
Religion]   1826
39: Geistl.Minis.: III,B Fasc.14: Lose Akten 1826
(2).   ADS 2:479

SCHNAKE, August. See Schnake, Wilhelm Friedrich
August Heinrich

SCHNAKE, Wilhelm Friedrich August Heinrich (waiter;
from Dransfeld)

Releases from citizenship for Wilhelm & August
Schnake (the latter a cook)   1880
71: Mappe 20.G: 051-20.   ADS 4:250

SCHNEIDER, Adam. See Bommarius, Friedrich

SCHNEIDER, Anne Gottliebe (in Philadelphia)

Claim of Gottlieb Friedrich Reichert [in Germany]
to estate of Anne Gottliebe Schneider 1815-1816
500: Berlin-Dahlem: Geh.Staatsarchiv: Rep.XI, 21a,
Conv. 4,1.   Not in ADS

SCHNEIDER, Hermann (in Chicago)

File on [estate?]   1901
227: Rep. MA 1921: A.St.I: Nr. 1080.   ADS 9:71

SCHNEIDER, Johann Nicolaus (from Kaltennorden, Thuringia)

Emigration of   1856
500: Weimar: Thueringisches Staatsarchiv: Dermbach
II: Gb2: Nr. 63-92; 93-114; 144-183; 186;
188.   Not in ADS

SCHNEIDER, Josef (from Regensburg; in America)

Repatriation of   1911
227: Rep. MA 1936: A.St.: MA 98044.   ADS 9:94

SCHNEIDER, Karl (died 3 April 1909 in Newport, Washington)

Estate of   1910

152: Abt.405: Wiesbaden: Rep.405/I,2: Nr. 3521.

SCHNEIDER, Nic. (from Goessenreuth)

Sale of two false $50 bills to Schneider [an emi-
grant] by a certain Levy   1878
39: Bestand Senat: Cl.VIII, No.X, Jg.1878 (Register) 50 [Nordamerika]   ADS 2:238

SCHNEIDER, N. N. (died in Philadelphia, Pennsylvania)

Estate of   1910
152: Abt.405: Wiesbaden: Rep.405/I,2: Nr. 3542.
ADS 7:146

SCHNEIDER, --

Philadelphia: Estate of   1894
39: Reichs- u. auswaert.Ang.: D I h, 35 [Erbschaft-
en]   ADS 2:269

SCHNEIDERS, --

Estate of   1944
148: R XI.   ADS 6:262

SCHNELL, Adam, II (from Ruesselsheim, HEL)

Releases from Hessian citizenship for him and his
family for the purpose of emigrating to America
1889
182: Abt. XI,4.   ADS 7:236

SCHNOHR, Ludwig Carl Heinrich (cigar worker)

Expulsion from Hamburg as Social Democrat & emigration to America   1880-1881
39: Polizeibehoerde-Polit.Pol.: S 149/249: Bl.
171-177.   ADS 2:415

SCHNUTE, Heinrich (cigar maker; from Muenchedorf)

Emigration of delinquent to America   1856-1858
56: L Neu Abt.129A: Gr.11, Nr. 67.   ADS 4:190

SCHOCK, Emil (Lieutenant Colonel in Bavarian First
Infantry Regiment)

Suggestions for the conclusion of his inheritance
matters in America   1911
227: Rep. MA 1936: A.St.: MA 98057.   ADS 9:96

SCHOELCH, Johann Michael (died in Boston, Massachusetts)

Estate of   1910
152: Abt.405: Wiesbaden: Rep.405/I,2: Nr. 3523.
ADS 7:146

SCHOELER, Albert

Galveston [Texas]: Search for Schoeler   1878

<u>39</u>: Bestand Senat: Cl.VIII, No.X, Jg.1878 (Register) 21 [Nordamerika]   ADS 2:237

SCHOENEMANN, G.

Complaint from New York by Schoenemann regarding agent August Behrens, Hamburg   1873

<u>39</u>: Bestand Senat: Cl.VIII, No.X, Jg.1873 (Register) 46 [Nordamerika]   ADS 2:226

New York consular report regarding Schoenemann's complaint against A. Behrens for passing false $20 bill   1874

<u>39</u>: Bestand Senat: Cl.VIII, No.X, Jg.1874 (Register) 9 [Nordamerika]   ADS 2:227

SCHOENEMANN, Wilhelm (from Quickborn; killed in U.S. Civil War)

Estate of   1868

<u>52</u>: Han.Des.74: Dannenberg: Abt.II,G: Nr. 24.   ADS 4:65

SCHOENFELDT, Anna (in Orange Township, Iowa [county not given in ADS description])

Estate of   1906

<u>39</u>: Reichs- u. auswaert.Ang.: E.IV.58 [Erbschaften]   ADS 2:343

SCHOENHERR, -- (from Wolfenbuettel)

Emigration to America   n.d.

<u>56</u>: L Neu Abt.127, Fb.1: Nr. 52.   ADS 4:185

SCHOENING, Hans Hermann Christian (mason)

Expulsion from Hamburg as Social Democrat & emigration to America   1880-1881

<u>39</u>: Polizeibehoerde-Polit.Pol.: S 149/87: Bl. 277-299.   ADS 2:413

SCHOETT, Johann Heinrich (born in Landenhausen)

Declaration of missing person [disappeared in America]   1888

<u>149</u>: Rep.49a: Nr. 566/11.   ADS 7:18

SCHOETTLER, --

Support for Schoettler who proposes to emigrate to America. He was a monk who fled from Paderborn to Hamburg, thereafter studied medicine at Jena, Halle, & Helmstedt, returning to Hamburg after three years, thereafter at University of Dorpat [Latvia] & Petersburg, where he was denied the right to practice medicine. Upon return to Hamburg, he asks for help from Spiritual Ministry

[Ministry of Religion] & other friends for trip to America   1804

<u>39</u>: Geistl.Minis.: II,9: Protokolle 1795-1850: p. 94; *see also* III B Fasc.9: Lose Akten 1804.   ADS 2:474

SCHOLL, H. P. (in Lincoln [Nebraska?])

Gift to the community of Frei-Laubersheim   1921

<u>149</u>: Rep.48/1: Konv.25, Fasz. 3.   ADS 7:15

SCHOLTE-HAMM, -- (Captain). *See* Hamm, Scholte (captain)

SCHOO, -- (mechanic)

Extradition from the USA   1902

<u>39</u>: Reichs- u. auswaert.Ang.: D.Ic,74 [Auslieferungen]

SCHOODT, W. N. F., *alias* John Jansen (a cook)

New York: Estate of   1896

<u>39</u>: Reichs- u. auswaert.Ang.: E.IV.51 [alt.Sig.] [Erbschaften]   ADS 2:282

SCHOPPE, August (vagabond laborer; from Gittelde)

Sentenced to house of correction in Bevern & emigration to America   1860

<u>56</u>: L Neu Abt.129A: Gr.11, Nr. 46.   ADS 4:188

SCHOPPEK (Sopek), -- (an Austrian citizen)

Extradition of Schoppek from the USA to Austria, via Hamburg   1906

<u>39</u>: Reichs- u. auswaert.Ang.: A.Ia.1.1 [Auslieferungen]   ADS 2:342

SCHORK, Georg (in New York)

File on [estate?]   1893

<u>227</u>: Rep. MA 1921: A.St.I: Nr. 961.   ADS 9:70

SCHORR, Friedrich (from Ginsheim-Gustavsburg, Landkreis Gross-Gerau, HEL)

Application for emigration permit   1850

<u>172</u>: Abt. XI,4.   ADS 7:223

SCHOTT, Henry. *See* New, Henny, born Schott

SCHOTT, Moses (from Ruesselsheim, HEL)

Application for emigration to America   1893
Voyage to America   1893

<u>182</u>: Abt. XI,4.   ADS 7:236

SCHRADER, Friedr[ich] Eduard. *See* Dominicus, Friedr[ich] Eduard

SCHRADER, Ludwig (journeyman mason; from Windhausen)

His dissolute life, incarceration in the house of correction at Bevern, & emigration to America 1854-1856

56: L Neu Abt.129A: Gr.11, Nr. 89.   ADS 4:192

SCHRADER, Theodor Louis Franz (mason)

Expulsion from Hamburg as Social Democrat & emigration to America 1881; 1895

39: Polizeibehoerde-Polit.Pol.: S 149/328: Bl. 300-310.   ADS 2:416

SCHRADER, --. *See* Hoffmann, J. N.

SCHRAMM, E. G. M. (in New York)

Information requested by private person [in Germany] regarding 1897

39: Reichs- u. auswaert.Ang.: N.Ib.1.3 [alt.Sig.] [Nachforschungen]   ADS 2:291

SCHRAMM, Kilian (from Bamberg; died in St. Paul [Minnesota])

File on [estate?] 1893-1896

227: Rep. MA 1921: A.St.I: Nr. 971.   ADS 9:70

SCHRAMM, --

Withdrawal of funds from the general welfare fund [in Hamburg] by the German consulate in New York for Schramm 1901

39: Reichs- u. auswaert.Ang.: G.Ia.1.22 [Geschaeftsverkehr]   ADS 2:317

SCHRAY, Georg

San Francisco: Estate of 1906

39: Reichs- u. auswaert.Ang.: E.IV.79 [Erbschaften]   ADS 2:343

SCHRECK, Michael (in Colfax [state not given in ADS description])

Claim against the administration of the County of Whitman [state not given in ADS description] 1899

227: Rep. MA 1921: A.St.I: Nr. 1265.   ADS 9:72

SCHREIBER, August (in Chicago)

Request for information regarding Schreiber made by master baker Johann Koeberlein [in Germany] 1893

227: Rep. MA 1921: A.St.I: Nr. 1764.   ADS 9:74

SCHREIBER, Auguste

New York: Estate of 1914

39: Reichs- u. auswaert.Ang.: E.IV.9 [Erbschaften]   ADS 2:384

SCHREIBER, Christian (from Greene)

Investigation of the brothers Ludwig & Christian Schreiber for theft & the emigration to America of Christian Schreiber & his family 1851-1859

56: L Neu Abt.129A: Gr.11, Nr. 48.   ADS 4:189

SCHREIBER, Josef (died in Utica [New York?])

File on [estate?] 1891

227: Rep. MA 1921: A.St.I: Nr. 937.   ADS 9:70

SCHREIBER, Ludwig (from Greene). *See* Schreiber, Christian

SCHREINER, Elisabeth, widow of Jakob Hartmann, born Lang (died in Port Washington, Ohio)

Estate of 1901

152: Abt.405: Wiesbaden: Rep.405/I,2: Nr. 3524.   ADS 7:146

SCHREYER, -- (member Moravian Brethren; in North Carolina). *See* Danz, --

SCHRIEPER, Johann. *See* Shrieper, Johann

SCHRIEVER, Elisabeth

New York: Estate of 1917

39: Reichs- u. auswaert.Ang.: E.IV.28 [Erbschaften]   ADS 2:392

SCHROEDER, Heinrich (tailor; from Beckstedt, Grafschaft Hoya, NL)

Emigrated to America ca. 1840; returned to Beckstedt in 1883 & opened restaurant "California"; descendants (the Heinrich Roevekamp family) still live in Beckstedt

58 [write directly to Beckstedt]   ADS 4:226

SCHROEDER, Johann Friedrich (tailor)

Expulsion from Hamburg as Social Democrat & emigration to America 1881

39: Polizeibehoerde-Polit.Pol.: S 149/336: Bl. 382-393.   ADS 2:416

SCHROEDER, Wilhelm Georg Carl (nailsmith)

Expulsion from Hamburg as Social Democrat & emigration to America   1880-1885

39: Polizeibehoerde-Polit.Pol.: S 149/240: Bl. 394-406.   ADS 2:415

SCHROEDER, W.

New York: Estate of   1876

39: Bestand Senat: Cl.VIII, No.X, Jg. 1876 (Register) 2 [Nordamerika]   ADS 2:232

SCHROEDER, Wilhelm

Estate of   1877

39: Bestand Senat.: Cl.VIII, No.X, Jg.1877 (Register) 10 [Nordamerika]   ADS 2:234

SCHROEDER, --

Testimony taken in Portland, Oregon, & in St. Thomas in [legal] case   1904

39: Reichs- u. auswaert.Ang.: J.Ib.73 [Justizsachen]   ADS 2:336

SCHROEDL, Friedrich (died in America)

File on [estate?]   1889

227: Rep. MA 1921: A.St.I: Nr. 904.   ADS 9:70

SCHROTTENFELS, Manuel (from Ruesselsheim, HEL)

Emigration to America   1849

182: Abt. XI,4.   ADS 7:235

SCHUBERT, F. C. (died in America)

Estate of   1874

39: Bestand Senat: Cl.VIII, No.X, Jg.1874 (Register) 39 [Nordamerika]   ADS 2:228

SCHUBERTH, Johann (in New York)

File on [estate?]   1892

227: Rep. MA 1921: A.St.I: Nr. 954.   ADS 9:70

SCHUCHARDT, -- (bookbinder)

Extradition from USA to Hamburg; wanted for forgery of certificate   1909

39: Reichs- u. auswaert.Ang.: A.Ib.8 [Auslieferungen]   ADS 2:356

SCHUCHERT, Georg Conrad

New York: Death certificate for   1878

39: Bestand Senat: Cl.VIII, No.X, Jg.1878 (Register) 57 [Nordamerika]   ADS 2:239

SCHUECKING, -- (judge; born in Soegel [Huemling])

Frau Dr. -- Schuecking has a large private collection of papers, books, letters, & newspaper clippings from Delaware, Ohio, & Hudson [?]. The collection deals with a Judge -- Schuecking who emigrated with his four children to North America during the first half of the nineteenth century

138   ADS 6:123

SCHUEDER, Jacob

San Francisco: Estate of   1900

39: Reichs- u. auswaert.Ang.: E.IV.56 [Erbschaften]   ADS 2:307

SCHUENEMANN, Friedrich (day laborer; from Alt-Gandersheim)

Sentencing to house of correction in Bevern & emigration to America   1860-1861

56: L Neu Abt.129: Gr.11, Nr. 21.   ADS 4:186

SCHUETT, Heinrich. See Stuenkel, Heinrich, alias Schuett

SCHUETT, Hermann (in Iowa)

Estate of   1918

39: Reichs- u. auswaert.Ang.: E.IV.127 [Erbschaften]   ADS 2:394

SCHUETT, Ludwig

Surveillance of Schuett, brought here from New York   1897

39: Reichs- u. auswaert.Ang.: P.III.2.6 [alt.Sig.] [Polizeiwesen]   ADS 2:291

SCHUETTERLE, Matthias

Cincinnati [Ohio]: Estate of   1909

39: Reichs- u. auswaert.Ang.: E.IV.6 [Erbschaften]   ADS 2:356

SCHULIUS, Georg (member Moravian Brethren; in South Carolina). See Moravian Brethren [subject index]

SCHULTE, Ernst (in Chicago)

Inquiry by private person [in Germany] regarding   1896

39: Reichs- u. auswaert.Ang.: N.Ib.9 [alt.Sig.] [Nachforschungen]   ADS 2:284

SCHULTZ, August Carl Nicolaus (peddler)

Expulsion from Hamburg as Social Democrat & emigration to America   1880-1881

39: Polizeibehoerde-Polit.Pol.: S 149/105: Bl. 451-470.   ADS 2:414

SCHULTZ, Heinrich (in Hamburg, South Carolina)

Brief from   1836

39: Bestand Senat: Cl.VIII, No.X, Jg.1836 (Register)   ADS 2:133

Request from Heinrich Schultz, founder of town of Hamburg, South Carolina   1838

39: Bestand Senat: Cl.VIII, No.X, Jg.1839 (Register) 4 [Nordamerika]   ADS 2:136

SCHULTZ, Louise

New York: Estate of   1894

39: Reichs- u. auswaert.Ang.: E.IV.91 [alt.Sig.] [Erbschaften]   ADS 2:269

SCHULTZ, Therese ([died?] 1895 in Chicago)

Will of   1898
227: Rep. MA 1921: A.St.I: Nr. 1042.   ADS 9:71

SCHULTZ, --. *See* Kleiner (Kleiber), Mrs. F. H. L., born Schultz

SCHULTZE, Richard F. (married couple; in Brooklyn)

Estate of   1917

39: Reichs- u. auswaert.Ang.: E.IV.20 [Erbschaften]   ADS 2:392

SCHULTZE, -- (in Charlestown, South Carolina)

Letters regarding payment of inheritance portions to Schultze family   1790

500: Berlin-Dahlem: Preus.Geh. Staatsarchiv: Rep. XI, 21a, Conv. 1,7: Nr. h.   Not in ADS

SCHULTZEN, Christoph Emanuel (pastor; in Philadelphia)

File on his pastoral call   1764; 1766

500: Halle: Missionsbibliothek des Waisenhauses: Abteilung H IV (Nordamerika) Fach A, Nr. 9; 19.   Not in ADS

SCHULZ, Anna Katharina (from Goddelau, Landkreis Gross-Gerau, HEL)

Emigration to America   1863

173: Gesinderegister 1853-1894: Deckblatt, Rueckseite.   ADS 7:224

SCHULZ, August (in New Dorp [state not given in ADS description])

Estate of   1910

39: Reichs- u. auswaert.Ang.: E.IV.41 [Erbschaften]   ADS 2:362

SCHULZ, Conrad (from Trier region)

Arrested for encouraging emigration to USA
                              Year XI [of French Republic]

200: Abt.276: Saardepartement, Praefektur Trier: Nr. 1723.   ADS 8:14

SCHULZ, Eduard (in St. Paul, Minnesota)

Estate of   1911

39: Reichs- u. auswaert.Ang.: E.IV.36 [Erbschaften]   ADS 2:368

SCHULZ, Elisabetha (from Geinsheim, Landkreis Gross-Gerau, HEL)

Emigration to America   1836

170: Abt. XI,4.   ADS 7:221

SCHULZ, Emil (from Goddelau, Landkreis Gross-Gerau, HEL)

Emigrant to America   1838

173: Ortsbuergerregister.   ADS 7:234

SCHULZ, Erna (in Illinois)

An orphan receiving a "pension" of 36 marks from the Hamburg Orphanage Administration   1894

39: Reichs- u. auswaert.Ang.: W.I.2 [alt.Sig.] [Waisenhaussachen]   ADS 2:272

SCHULZ, Ernst (in America)

Investigation of   1888

227: Rep. MA 1921: A.St.I: Nr. 1737.   ADS 9:74

SCHULZ, Johann (from Bienenbuettel, Amt Medingen)

Transportation to America   1849-1852

96: B.13.Nr.9: Archiv des Klosters St. Michaelis.   ADS 4:289

SCHULZ, Valentin (from Froeschen, RPL; in America)

Claim of Heinrich Schaar, of Permasens, against Michel Brandstetter as buyer of house & land of Schulz, an emigrant to America   1770

201: Archivabt. Grafschaft Hanau-Lichtenberg: Akt Nr. 2378.   ADS 8:67

SCHULZ, --

Estate of   1941
148: R XI.   ADS 6:262

SCHULZE, Heinrich (sergeant; from Gifhorn)

File on the handling of court matters for Schulze,
    now in America   1861-1862
52: Calenberger Brief-Archiv Des.15: Privatakten
    Band I: S, Nr. 282.   ADS 4:56

SCHULZEN, H? (pastor? in Pennsylvania). *See* Heinzelmann, H.

SCHUMACHER, A. *See* Schaefer, John E.

SCHUMACHER, Dietrich Nicolaus (of San Francisco)

Personal file on   1887
39: Ritzebuettel II: XI,C,Nr. 2 [*now* Bestand Senat:
    Cl.VII, Lit.L^b-, Nr. 17^e; *and* Ritzebuettel I,
    Nr. 386].   ADS 2:466

SCHUMACHER, Franz (from Gernsheim, Landkreis Gross-
Gerau, HEL)

Emigration to America   1922
171: Abmelderegister 1912-1924.   ADS 7:222

SCHUMACHER, Heinr[ich] Louis

San Francisco: Residency of   1876
39: Bestand Senat: Cl.VIII, No.X, Jg.1876 (Regis-
    ter) 23 [Nordamerika]   ADS 2:233

SCHUMACHER, H. L. (in San Francisco)

Inquiry regarding possible Hamburg citizenship
                                                1876
39: Bestand Senat: Cl.1, Lit.T.,No.14, vol. 4,
    Fasc.5b, Invol. 23.   ADS 2:6

SCHUMACHER, -- (Pittsburgh, Pennsylvania)

Estate of   1912
39: Reichs- u. auswaert.Ang.: E.IV.101 [Erbschaften]
                                      ADS 2:373

SCHUMANN, Chr.

Emigration of   n.d.
151: Bestand 180: Arolsen: Acc.1937/76,5: Nr. 461.
                                      ADS 7:113

SCHURZ, Karl

Communist plots. Much personal information [on
    numerous persons, including Schurz]. Activities

for the revolutionary party   1851
226: Rep. MInn 7: 45546.   ADS 9:19

SCHUSTER, Th. C.

Letter from Hamburg consul in America regarding
    Schuster's military & personal situation   1850
39: Bestand Senat: Cl.VIII, No.X, Jg.1850 (Regis-
    ter) I.1 [Nordamerika]   ADS 2:157

SCHWABE, J. H. G. (from Hamburg)

New York: Statements regarding Schwabe's death
    aboard the *Westphalia*   1880
39: Bestand Senat: Cl.VIII, No.X, Jg.1880 (Regis-
    ter) 39 [Nordamerika]   ADS 2:244

SCHWABEL, Walburga (from? Stadtamhof)

Emigration of   1826
253: Fach 24, Fasz. 35.   ADS 9:173

SCHWANER, Julius (died in San Francisco)

Estate of   1895
152: Abt.405: Wiesbaden: Rep.405/I,2: Nr. 3565.
                                      ADS 7:146

SCHWANK, -- (died in Baltimore)

Estate of   1891
227: Rep. MA 1921: A.St.I: Nr. 931.   ADS 9:70

SCHWARZE, Heinr[ich]

New York: Estate of   1906
39: Reichs- u. auswaert.Ang.: E.IV.22 [Erbschaften]
                                      ADS 2:343

SCHWARTZENTRAUBER, -- (a Mennonite)

Report that Schwartzentrauber, a Mennonite, is
    enticing persons to emigrate and is to be ar-
    rested   23 March 1817
200: Abt.655.10: Ehrenbreitstein: Nr. 550.
                                      ADS 8:50

SCHWEITZER, Elisabeth (died in New York)

Estate of   1895
152: Abt.405: Wiesbaden: Rep.405/I,2: Nr. 3566.

SCHWENKE, Johann. *See* Rusch, Johann

SCHWEPENDIECK, Wilhelm Ernst Berthold (cabinet maker)

Expulsion from Hamburg as Social Democrat & emi-
    gration to America   1880-1881
39: Polizeibehoerde-Polit.Pol.: S 149/80: Bl. 1-9.
                                      ADS 2:413

SCHWIEM, H. (in Oakland, California)

Estate of   1908
39: Reichs- u. auswaert.Ang.: E.IV.42 [Erbschaften]
                                            ADS 2:351

SCHWIND, Peter (from Mittelbuchen)

Emigration of Schwind & Johannes Michler to New
    England   1766-1790?
151: Bestand 81: Rep.E: 5.Gem. Mittelbuchen: Nr. 1.
                                            ADS 7:82

SCHWONE, August (from Hohenrode)

Two children left in Germany when Schwone emi-
    grated to America   1864-1865
52: Schaumburg Des.H.2: Rinteln: Neue reg.: V.9.
    Hohenrode.   ADS 4:99

SEAR, Fred (died in Chicago)

Information regarding the heirs of   1915
39: Reichs- u. auswaert.Ang.:E.IV.25 [Erbschaften]
                                            ADS 2:388

von SEEBISCH, -- (major; Third English-Waldeck Regi-
    ment)

Courtmartial for breaches in service   1782
151: Bestand 118: Nr. 985.
     DLC has.   ADS 7:101

SEEHAUSEN, John

Repatriation to Hamburg from San Francisco of
    insane person   1900
39: Reichs- u. auswaert.Ang.: H.IIb.7 [Heimschaff-
    ungsverkehr]   ADS 2:308

SEELAND, Jakob (from Durlach? BWL)

Emigration of   1770-1771
260: 217.   ADS 10:1

SEELIG, Jacob (of Buettelborn, Landkreis Gross-Gerau,
    HEL)

Emigration to America of Jewish person   1884
166: Abmelderegister 1876-1924: Nr. 173. ADS 7:216

SEEMANN, Louise, born Schmidt. See Schmidt, Karl

SEGELCKE, W. F.

San Francisco: Estate of   1907
39: Reichs- u. auswaert.Ang.: E.IV.116 [Erbschaften]
                                            ADS 2:347

SEHER, Louise (from Diez? in Newark, New Jersey)

Large packet of correspondence regarding her gener-
    ous aid to poor residents of Diez after the first
    world war
205   ADS 8:71

SEIBEL, S. (in America). See Bedassem, J.

SEIDEL, Nathanael (member Moravian Brethren; in Amer-
    ica?). See von Marschall, Friedrich

SEIP, --

Estate of   1942
148: R XI.   ADS 6:262

SEIPEL, Ludwig (of Buettelborn, Landkreis Gross-
    Gerau, HEL)

Emigration to America with wife and four daughters
                                             1884
166: Abmelderegister 1876-1924: Nr. 166. ADS 7:216

SEITZINGER, J. L. (pastor, Evangelical Lutheran
    Church, Pepin, Wisconsin)

Letter from   [1919-1922?]
44: Mappe Amerika.   ADS 2:600

SELIGMANN, Joseph (in Chicago). See Guthmann, Eman-
    uel

SEVERIDT, Friedrich (footman; from Ahlshausen)

Punishment & emigration to America   1854-1862
56: L Neu Abt.129A: Gr.11, Nr. 18.   ADS 4:186

SEYFFERT, Herm. (in Hoboken, New Jersey)

Estate of   1912
39: Reichs- u. auswaert.Ang.: E.IV.19 [Erbschaften]
                                            ADS 2:373

SIEBERT, Bernhard (of Hosenfeld "und NN von N"[?])

Complaint of government in Fulda regarding his
    alleged impressment into military service [Frei-
    korps]   1781
151: Bestand 81: Rep.E: 1.Gen.III, Nr. 11.
                                            ADS 7:80

SIEBKEN, Max (in Hilton Head, South Carolina)

Estate of   1907
39: Reichs- u. auswaert.Ang.: E.IV.24 [Erbschaften]
                                            ADS 2:347

SIEBLITZ, Heinr[ich] (journeyman tailor)

Proposed transportation to America   1853
52: Han.Des.80: Hildesheim I: E.DD.36a: Nr. 1240.
                                                   ADS 4:89

SIEBOLD, -- (from Kassel; soldier serving in America)

Petition from wife for exemption from taxation on
  her business (sewing)   1780
151: Bestand 5: Nr. 7002.   ADS 7:39

SIEMEN, Paul (in Philadelphia)

File regarding his death   1830
2: Abt.11: Reg.I: Nr. 695.   ADS 1:9

SIEMERS, E. (married woman; in Waco, Texas)

Transmittal of a claim made by C. Siemers, a businessman in Hamburg, against Mrs. E. Siemers, via
  German consulate in Galveston   1915
39: Reichs- u. auswaert.Ang.: K.II.7 [Kassenwesen]
                                                   ADS 2:389

SEITER, William (died in Sagimay [Saginaw?] Michigan)

File on [estate?]   1891-1892
227: Rep. MA 1921: A.St.I: Nr. 940.   ADS 9:70

SEPPEL, Walburga (in America)

Information regarding   1889
227: Rep. MA 1921: A.St.I: Nr. 1739.   ADS 9:74

SESSDORF, Alfred (died in Chicago)

File on [estate?]   1900
227: Rep. MA 1921: A.St.I: Nr. 760.   ADS 9:69

SEYBOLD, Elise (died in Philadelphia)

File on [estate?]   1885-1886
227: Rep. MA 1921: A.St.I: Nr. 1724.   ADS 9:74

SHRIEPER, Johann (in Pennsylvania)

File on [estate?][surname probably misspelled]1768
227: Bayerische Gesandtschaft London: Nr. 837.
                                                   ADS 9:47

SIEBLER, Auguste (died in New York)

File on [estate?]   1894
227: Rep. MA 1921: A.St.I: Nr. 991.   ADS 9:70

SIEGISMUND, A. (doctor; died beginning of the 1860s
  in Houston, Texas)

Estate of   1895
152: Abt.405: Wiesbaden: Rep.405/I,2: Nr.3567.
                                                   ADS 7:146

SIEMS, L. (in Chicago)

[Hamburg] Orphanage Commission requests information regarding Siems' financial standing   1894
39: Reichs- u. auswaert.Ang.: W.I. [alt.Sig.]
    [Waisenhaussachen]   ADS 2:272

SIEVEKE, Bernhard Christian (died at Kilony? LeSukur?
  Minnesota)

Estate of   1877-1879
116: Minden: Praes.-regist.: I.L: Nr. 248.
                                                   ADS 6:54

SIEVERS, -- (married woman; in Chicago). See Mohr,
  M.

SIGEL, Franz (German revolutionary independence
  leader; Union general in U.S. Civil War; victor at
  the battle of Pea Ridge, Arkansas, 1862)

Memoirs & momentos from the Sigel family of Langen-
  brueck & from American visitors

285   ADS 10:90

Correspondence & manuscript regarding his life
                                                  1888-1901
150: Nachlass Blos.   ADS 7:27

Memoirs of Franz Sigel to be found in the family
  archives of the Sigel family, Bad Langenbruecken,
  BWL.

SILBERMANN, Katherina. See Kumpf, Peter A.

SILKROTH, Caroline, alias Voss (vagabond; unmarried;
  from Hahausen or Hoetensleben)

Emigration to America   1855-1858
56: L Neu Abt.129A: Gr.11, Nr. 49.   ADS 4:189

SILLER, Heinrich

Behavior of American consul -- Bardel in Bamberg
  toward Bavarian officials in the criminal case
  of Siller   1906
227: Rep. MA 1936: A.V.: Nr. 688.   ADS 9:100

SIMMERING, Greetje (from Schwerinsdorf)

Emigration to America   1865-1866
50: Rep.21a: Nr. 8937.   ADS 4:8

SIMON, August

New York: Estate of   1912

  39: Reichs- u. auswaert.Ang.: E.IV.26 [Erbschaften]
                                     ADS 2:373

SIMON, Martin

San Francisco: Death certificate for   1877

  39: Bestand Senat: Cl.VIII, No.X, Jg.1877 (Regis-
    ter) 50 [Nordamerika]   ADS 2:235

SIMONIS, --

New York General Consulate: Loeb Brothers *versus*
  Simonis   1896

  39: Reichs- u. auswaert.Ang.: J.Ib.10.3 [alt.Sig.]
    [Justizsachen]   ADS 2:283

SINGER, Franz (in Omaha, Nebraska, in 1890)

File on [estate?]   1892

  227: Rep. MA 1921: A.St.I: Nr. 953.   ADS 9:70

SINGER, --. *See* Last, --

SINKMAJER, -- (pastor)

Sinkmajer's complaint against emigration agent
                                     1897

  39: Reichs- u. auswaert.Ang.: A.III.8 [alt.Sig.]
    [Auswanderungen]   ADS 2:288

SIROGI, George (married couple)

Hamburg police request information from the American consulate regarding   1911

  39: Reichs- u. auswaert.Ang.: P.III.3 [Polizei-
    wesen]   ADS 2:370

SITTMANN, -- (widow; in Berlin). *See* Neumann, I. F.

SKJOENNEMANN, --, born Hansen (married woman; in Los
  Angeles)

Hamburg Orphanage Commission seeks information regarding   1897

  39: Reichs- u. auswaert.Ang.: W.I.6 [alt.Sig.]
    [Waisenhaus-Ang.]   ADS 2:293

SKOETT, --

Estate of   1943
  148: R XI.   ADS 6:262

SKOW, --

Application of Karl Schloemer, of Hamburg, requesting appointment of an American lawyer to repre-
  sent his interests in the estate of Skow   1914

  39: Reichs- u. auswaert.Ang.: E.IV.46 [Erbschaften]
                                       ADS 2:384

SMITH, Charles

San Francisco: Estate of   1902

  39: Reichs- u. auswaert.Ang.: E.IV.33 [Erbschaften]
                                       ADS 2:323

SMITH, C. A. *See* Schmidt, C. A.

SMITH, Edward J.

Search by Hamburg police for Smith, wanted for
  fraud in USA   1905

  39: Reichs- u. auswaert.Ang.: P.III.1.1 [Polizei-
    wesen]   ADS 2:341

SMITH, Henry J. (an American citizen)

Arrest [in Hamburg?] of   1897

  39: Reichs- u. auswaert.Ang.: P.III.2.9 [alt.Sig.]
    [Polizeiwesen]   ADS 2:292

SMITH, Reuben

Arrest of, in Rendsburg   1807

  2: Abt.65: Nr. 4860, Fasc.1.
    DLC has [see under Kiel]   ADS 1:13

SMITH, V. Hugo (in Seattle, Washington)

Correspondence   1914

  2: Abt.399 (Fam.Archiv Rendtorff): Nr. 212.
                                       ADS 1:34

SNOEING, Fenna. *See* Snoeing, Hermine

SNOEING, Hermine (from Neuenhaus?)

Transportation with sister Fenna to America
                                       1854-1864

  54: Rep.122.II.1: Neuenhaus: Fach 96, Nr. 21.
                                       ADS 4:133

SNYDACKER, Godfrey (in Chicago)

Application for Bremen consulship in Chicago 1862

  48: B.13.6.0: Nr. 18   ADS 3:10
    DLC has?

von SOBBE, George

New York: Estate of   1915

<u>39</u>: Reichs- u. auswaert.Ang.: E.IV.33 [Erbschaften]
ADS 2:388

SOBIETZKI, --

San Francisco: Estate of   1913

<u>39</u>: Reichs- u. auswaert.Ang.: E.IV.101 [Erbschaften]
ADS 2:379

SOELLNER, Friedrich (died in America)

File on [estate?]   1889

<u>227</u>: Rep. MA 1921: A.St.I: Nr. 897.   ADS 9:70

SOERENSEN, -- (a Danish national)

Complaint of Danish ministry in Berlin regarding
treatment [in Hamburg?] of emigrants [to USA?]
Soerensen, Nybor, & Christensen   1913

<u>39</u>: Reichs- u. auswaert.Ang.: C.IIIc,8 [Auswander-
ung]   ADS 2:378

SOLLINGER, Josefine, born Maerz (in America)

Investigation of   1887

<u>227</u>: Rep. MA 1921: A.St.I: Nr. 1734.   ADS 9:74

SOLZE, --

Letter from Lancaster, Pennsylvania, regarding
Solze, who absconded with money   1827

<u>39</u>: Bestand Senat: Cl.VIII, No.X, VI,B (Register)
ADS 2:128

SOMMER, Johann Christian [or Christoph?]. *See* Sommer, Johann Jacob

SOMMER, Johann Jacob (in Philadelphia)

Inheritance from father   1805

<u>39</u>: Bestand Senat: Cl.VII, Lit.C<u>^c</u>, No.15: vol. 4,
Fasc.44.   ADS 2:64

SOMMER, Petrus Nicolaus (from Hamburg)

Acceptance as candidate [pastor] by Spiritual
Ministry [Ministry of Religion, Hamburg]   1739

<u>39</u>: Geistl. Minis.:II,6: Protokoll lit.A, 1720-
1745: p.119 (4).   ADS 2:469

SOMMER, Petrus Nicolaus. *See also* Sommers, Peter
Nicolaus

SOMMER, Wilhelm (a Dragoon in Braunschweig [Bruns-
wick] troops)

Settlement of accounts with Sommer, "von dem Ise-
beck'schen Hofe in Herrhausen" & payment into
the Invalid Fund in Braunschweig   1801-1803

<u>56</u>: L Alt.Abt.8, Bd.3: Seesen, Gr.40, Nr. 136.
ADS 4:167

SOMMER, --

St. Louis: Letter regarding death certificate for
husband of widow -- Sommer   1878

<u>39</u>: Bestand Senat: Cl.VIII, No.X, Jg.1878 (Regis-
ter) 6 [Nordamerika]   ADS 2:236

SOMMERFELD, -- (shoemaker)

Proposed transportation of Sommerfeld & family to
America   1841

<u>52</u>: Han.Des.80: Hildesheim I: E.DD.7i: Nr. 926.
ADS 4:88

SOMMERICH, Dietrich (in St. Louis)

File on [estate?]   1894

<u>227</u>: Rep. MA 1921: A.St.I: Nr. 977.   ADS 9:70

SOMMERS, Johanna (in Illinois)

Estates of Johanna Sommers & Marie Stange   1914

<u>39</u>: Reichs- u. auswaert.Ang.: E.IV.2 [Erbschaften]
ADS 2:384

SOMMERS, Peter Nicolaus

Call as pastor of Scoharie congregation [New York]
1742

<u>39</u>: Geistl.Minis.: II,6: Protokoll lit.A, 1720-
1745: p.151; 152(3); 155; 160; 186; 207(III).
ADS 2:470

SOMMERS, Peter Nicolaus. *See also* Sommer, Petrus
Nicolaus

SONN, Abraham (of Buettelborn, Landkreis Gross-
Gerau, HEL)

Emigration to America of Jewish person with wife
and eight children   1887

<u>166</u>: Abmelderegister 1876-1924: Nr. 229. ADS 7:216

SONNENBERG, Thilde. *See* Lindheimer-Sonnenberg,
Thilde

SOPEK, --. *See* Schoppek, --

SORGENFREI, C. (in Grand Island, Nebraska)

Estate of   1903
39: Reichs- u. auswaert.Ang.: E.IV.14 [Erbschaften]
                                      ADS 2:330

SOTTMANN-KIEWERT, Robert (in Milwaukee, Wisconsin)

Estate of   1903
39: Reichs- u. auswaert.Ang.: E.IV.86 [Erbschaften]
                                      ADS 2:330

SPANGENBERG, Christian (day laborer; from Ahlshausen)

Sentencing to house of correction in Bevern & emigration to America   1858-1859
56: L Neu Abt.129A: Gr.11, Nr. 19.   ADS 4:186

SPANGENBERG, Louise, born Koerner (in Brooklyn)

File on [estate?]   1894-1895
227: Rep. MA 1921: A.St.I: Nr. 980.   ADS 9:70

SPANGENBERG, -- (member Moravian Brethren; in Germany & America)

Letters during four trips to America   1736-1762
Letters & memoranda from Pennsylvania to [Count]
   Zinzendorf & Johannes von Watteville
                          1751-1760; 1753-1755
Miscellanea from Pennsylvania 1742-1756; 1753-1789
500: Herrnhut: Archiv: Rep.14.A.Nr.18; 19; 20; 38;
   39.   Not in ADS

SPANIER, Nikolaus (cigar worker)

Expulsion from Hamburg as Social Democrat & emigration to America   1881-1883
39: Polizeibehoerde-Polit.Pol.: S 149/370: Bl.
   247-273.   ADS 2:417

SPANN, Johannes (watchmaker; born 1828 in Hamburg;
died 1862 in San Francisco)

Estate of   n.d.
39: Bestand Familienarchive: Fam. Spann: Nr. 1.
                                      ADS 2:490

SPECHT, Georg (tanner; from Uslar?)

Transportation to America   1853-1865
52: Han.Des.80: Hildesheim I: E.DD.44a: Nr. 1718.
                                      ADS 4:91

SPECHT, --. See Roesch, --

SPECK, --

Baptismal entry for child of Speck family emigrating to Pennsylvania, via Duisburg-Ruhrort   1744

126: Taufregister der Evang. Gem. Ruhrort.
                                      ADS 6:91

SPEHR, Paul Otto (Lutheran pastor; in Platte Center,
Nebraska)

Letter   [1911-1922?]
44: Mappe Amerika.   ADS 2:599

SPENGLER, Elisabetha (from Geinsheim, Landkreis
Gross-Gerau, HEL)

Report on passport issuance for purpose of emigration to America   1850
170: Abt. XI,4.   ADS 7:221

SPENGLER, Joh. (from Geinsheim, Landkreis Gross-
Gerau, HEL)

Emigration to North America   1833
170: Abt. XI,4.   ADS 7:221

SPENGLER, Johannes (from Goddelau, Landkreis Gross-
Gerau, HEL)

Emigrant to America   1841
173: Ortsbuergerregister.   ADS 7:224

SPERLE, Daniel, II (from Ruesselsheim, HEL)

Emigration to America at cost to the city   1850
182: Abt. XI,4.   ADS 7:235

SPERLE, Jakob, II (from Ruesselsheim, HEL)

Application for emigration to America   1852
182: Abt. XI,4.   ADS 7:235

SPETMANN, W. H. (in Mills County, Iowa)

Estate of   1903
39: Reichs- u. auswaert.Ang.: E.IV.26 [Erbschaften]
                                      ADS 2:330

SPICKER, --

German consular report from New York regarding
   Spicker brothers' accusation of fraud by Hamburg money exchanger Salomon   1870
39: Bestand Senat: Cl.VIII, No.X, Jg.1870 (Register) 43 [Nordamerika]   ADS 2:211

SPIEGELTHAL, C. (in Lexington, Kentucky)

Receipt of statement from C. Spiegelthal regarding
   citizenship of son Caesar Spiegelthal, a music
   teacher sojourning in Hamburg   1890
39: Bestand Senat: Cl.I, Lit.T, No.14, vol.4, Fasc.
   5b, Invol. 43 [Eingabe]   ADS 2:6

SPIEGELTHAL, Caesar. *See* Spiegelthal, C.

SPIER, Nathan (died in January 1898 in Kansas City [Missouri?])

Estate of   1900

152: Abt.405: Wiesbaden: Rep.405/I,2: Nr. 3530.
                                                        ADS 7:146

SPINNER, Maria Magdalena Fidelis, born Brumer (from Brodzelten; in Herkimer, New York)

Heiress of an estate   1818

227: Rep. Bayer. & Pfaelz. Gesandtschaft Paris:
       Nr. 8384.   ADS 9:50

SPITT, Georg

Petition of U.S. consular agent in Cuxhaven for
the pardoning of married couple Spitt for the
crime of avoiding the stamp tax on playing cards
[probably they could not emigrate to the United
States without this pardon]   1908

39: Bestand Senat: Cl.VII, Lit.A$^{\underline{a}}$, No.2: vol.9d,
       Fasc. 39 [Begnadigungsrecht]   ADS 2:63

SPITT, Josephine. *See* Spitt, Georg

SPITT, -- (married couple; American citizens)

Pardon granted to Mr. & Mrs. Spitt, American citizens, for having failed to pay stamp tax on playing cards   1908

39: Reichs- u. auswaert.Ang.: S.VI.4.1 [Steuern]
                                                        ADS 2:355

SPITZNER, --

Estate of   1944

148: R XI.   ADS 6:262

SPRADAU, C. A. (in America)

Estate of   1897

39: Reichs- u. auswaert.Ang.: E.IV.35 [alt.Sig.]
       [Erbschaften]   ADS 2:290

SPREISER, G. A. (soldier in 15th Infantry Regiment Prinz Johann von Sachsen)

Investigation of his emigration to America   1850

231: Rep.270/II: Reg. Mittelfranken: Kammer des
       Innern: Abgabe 1932: Nr. 628.   ADS 9:134

SPRENGELMEYER, -- (from Hagen)

Emigration of Sprengelmeyer family to North America
with help of Hannoverian consul in Bremen   1845

54: Rep.122,V: Iburg: Fach 311, Nr. 7.
       DLC has.   ADS 4:135

STAACK, --

Testimony in case of Staack *versus* Staack; at the
request of German consulate in St. Paul, Minnesota   1912

39: Reichs- u. auswaert.Ang.: J.Ia.4.10 [Justiz-
       sachen]   ADS 2:374

STABEL, Anna, born Brueckner (died in Chicago)

File on [estate?]   [1890s?]

227: Rep. MA 1921: A.St.I: Nr. 955.   ADS 9:70

STACEY, --

Hamburg police search for Stacey, wanted in USA
                                                        1912

39: Reichs- u. auswaert.Ang.: P.III.2.2 [Polizei-
       wesen]   ADS 2:375

STACH VON HOLTZHEIM, Moritz (in Hoboken, New Jersey)

Letter to Ernst William Merck in Hamburg 1934-1935

39: Bestand Familienarchive: Fam. Merck: II.7.
       Konv. 14.   ADS 2:488

STAEHLER, C. (pastor; of Hintermeilingen, Amt Hademar, HEL)

Contract as pastor to be sent out to the Texas German colonies   1846

154: C A 4cl: Nr. 5.
       DLC has.   ADS 7:174

STAEMMLER, F. (in Wanzeka, Wisconsin)

Estate of   1898

39: Reichs- u. auswaert.Ang.: E.IV.14 [alt.Sig.]
       [Erbschaften]   ADS 2:296

STAHL, Justus (from Ginsheim-Gustavsburg, Landkreis Gross-Gerau, HEL)

Application of his widow & her family for permit
to emigrate to North America   1852

172: Abt. XI,4.   ADS 7:223

STAHMANN, J. H. A. (in Bergedorf, HH)

Transmittal of a report from the German consul in
Chicago   1901

39: Reichs- u. auswaert.Ang.: G.Ia.1.2 [Geschaefts-
       verkehr]   ADS 2:316

STAHMER, John (in Argentine, Kansas)

Estate of   1904

39: Reichs- u. auswaert.Ang.: E.IV.45 [Erbschaften]
                                                        ADS 2:335

STERN, M.

Extradition of Stern from the USA to Bavaria, via
Hamburg   1905

39: Reichs- u. auswaert.Ang.: A.I.a.4 [Ausliefer-
ungen]   ADS 2:338

STERNFELS, Arthur (from Erfelden, Landkreis Gross-
Gerau, HEL)

Letter from Hamburg-Amerika Linie [steamship line]
to mayor [of Erfelden] regarding release from
military service obligation upon emigration to
America   1936

169: Abt. XI,4.   ADS 7:219

STERNFELS, Leonhardt (from Erfeldne, Landkreis Gross-
Gerau, HEL)

Application of Jewish person for release [from
citizenship] as emigrant & for issuance of a
passport   1887

169: Abt. XI,4.   ADS 7:219

von STETTEN, Rudolf (from Stetten, BWL; in America
after 1870)

Correspondence & description of his participation
in a cavalry expedition against the Indians
[19th century]

(Apply to the Archiv der Freiherren von Stetten,
Schloss, Stetten, Wuerttemberg)   ADS 10:91

STICHWEH, Chr.? Heinr[ich] (hobo; from Bramsche)

Emigration to America   1862-1863

54: Rep.122: Han.Amt Bersenbrueck: Nr. 147.
ADS 4:132

STICK, Wilhelm (in Chicago)

Investigation of   1897
227: Rep. MA 1921: A.St.I: Nr. 1784.   ADS 9:74

STICKEL, Peter. *See* Buechenbach (parish office)

STOCK, Francis (from San Jose [California]; died in
Wiesbaden)

File on [estate?]   1899-1900
227: Rep. MA 1921: A.St.I: Nr. 1049.   ADS 9:71

STOCKBAUER, -- (died in Montana)

File on [estate?]   1899
227: Rep. MA 1921: A.St.I: Nr. 1050.   ADS 9:71

STOCKFISCH, Hinrich (from Ostendorf)

Pardoning of convict under condition that he emigrate to America immediately   1863

62: Di, Fach 102, vol.17: Regiminalia (Bd.II).
ADS 4:231

STOCKINGER, Hansy (in Kansas)

Inquiry regarding his whereabouts   1914

227: [Rep. not given]: Nr. 251.   ADS 9:99

STOEHER, Max (formerly a shopkeeper)

Expulsion from Hamburg as Social Democrat & emigration to America   1880-1881

39: Polizeibehoerde-Polit.Pol.: S 149/37: Bl. 506-
512.   ADS 2:412

STOLL, F.

San Francisco: Estate of   1887

39: Bestand Senat: Cl.VIII, No.X, Jg.1887 (Register) 2 [Erbschafts-Amt]   ADS 2:258

STOLLWERK, Richard

Emigrant?   1928-1929
115: Rep.D9 III: Abt.E: Nr. 145.   ADS 6:34

STOLZ, --

Estate of   [1940s?]
148: R XI.   ADS 6:262

STOPPEL, Barbara, born Eschenfelder (in Pittsburgh,
Pennsylvania)

Estate of   1901
39: Reichs- u. auswaert.Ang.: E.IV.43 [Erbschaften]
ADS 2:315

STRASSEN, H. (in New Orleans)

Accusation by Strassen against J. F. W. Hermann
of Hamburg   1847

39: Bestand Senat: Cl.VIII, No.X, Jg.1847 (Register) [Nordamerika]   ADS 2:151

STRASSNER, --

Estate of   1941
148: R XI.   ADS 6:262

STRATMANN, Heinrich (from Blendern [Blender?])

Transportation to America   1842
52: Han.De.74: Westen-Thedinghusen: I.56: Nr. 41.
ADS 4:80

STRAUSS, Martha (from Astheim, HEL; in Chicago)

Register of Removals [Abmelderegister] contains
notation of her emigration   1937
161: Abt. XI, Abschnitt 4.   ADS 7:211

STRAUSS, Max (from Moerfelden, Landkreis Gross-Gerau,
HEL)

Emigration to New York   1937; 1941
179: Erhebung ueber Konzentrations- und Zwangsar-
     beitslager, Ordner 13/Nr. 22.   ADS 7:232

Emigration to America   1939
179: Wiedergutmachungsakten, Ordner Nr. 061/00/05.
                                          ADS 7:232

Emigration to America   1938
179: Wiedergutmachungsakten, Ordner Nr. 061/00/05.
                                          ADS 7:232

STRAUSS, Siegfried (from Gernsheim, Landkreis Gross-
Gerau, HEL)

Emigration of three persons, via London, to North
America   1933
171: Judenkartei.   ADS 7:222

STRAUSS, Walter (from Nauheim, Landkreis Gross-Gerau,
HEL)

Emigration of   1936
180: Wiedergutmachungsakten.   ADS 7:233

STRAUSS, --

Chicago: Estate of   1915
39: Reichs- u. auswaert.Ang.: E.IV.34 [Erbschaften]
                                          ADS 2:388

STREICHER, Josef (in America)

Information regarding   1887
227: Rep. MA 1921: A.St.I: Nr. 1733.   ADS 9:74

STRELAND, W. J.

Philadelphia: Request for information regarding
personal data on   1877
39: Bestand Senat: Cl.VIII, No.X, Jg.1877 (Regis-
    ter) 14 [Nordamerika]   ADS 2:234

STROEBEL, Georg (in New York)

File on [estate?]   1894
227: Rep. MA 1921: A.St.I: Nr. 972.   ADS 9:70

STROEHER, Peter (from Starkenberg)

Emigration of   1748
201: Zweibruecken III: Akt Nr. 3358: unterakt "Die
     Emigranten in Ungarn und andere Colonien . .
     . ."   ADS 8:60

STROHN, Ad. Aug. (in Kings County, New York)

Estate of   1904
39: Reichs- u. auswaert.Ang.: E.IV.42 [Erbschaften]
                                          ADS 2:335

STROMSKI, -- (sailor; impressed at Danzig)

Representations of American consul general in
Copenhagen regarding   1817-1818
500: Berlin-Dahlem: Geh.Staatsarchiv: Auswaertige
     Amt III: Rep.10.Justiz.   Not in ADS

STRUPP, Anna (from Lohr; died in New York)

File on [estate?]   1893-1895
227: Rep. MA 1921: A.St.I: Nr. 969.   ADS 9:70

von STRUVE, Gustav (in America)

Portion of a file from the Mannheim censor & the
police; also contains his legal appeal   1846
226: Rep. MInn 7: 25115c.   ADS 9:18

von STRUVE, Gustav (German revolutionary; in America).
See German Political Refugees [subject index]

STUBBE, Gustav (in Sterling Illinois)

Estate of   1913
39: Reichs- u. auswaert.Ang.: E.IV.6 [Erbschaften]
                                          ADS 2:379

STUDEMANN, J. L.

Consular report from New York: Claim of damages of
Studemann's heirs against the Staten Island Railroad Company   1872
39: Bestand Senat: Cl.VIII, No.X, Jg.1872 (Regis-
    ter) 52 [Nordamerika]   ADS 2:222

STUENKEL, Heinrich, *alias* Schuett (released convict)

Transportation to America   1862-1864
100: Gem. Oberndorf: XIII,1: Nr. 3e.   ADS 4:312

STUERCKE, -- (an American citizen)

Inquiry of U.S. consul regarding Stuercke, now in
Hamburg [police] custody   1842
39: Bestand Senat: Cl.VIII, No.X, Jg.1842 (Regis-
    ter) I.a.1 [Nordamerika]   ADS 2:144

STUERMER, Johann (in America)

    Possible extradition from America   1900

    <u>227</u>: Rep. MA 1921: A.St.I: Nr. 1233.   ADS 9:71

STUMPF, Max (in Philadelphia)

    File on [estate?]   1897

    <u>227</u>: Rep. MA 1921: A.St.I: Nr. 1026.   ADS 9:70

STURM, Johann N. (from Goddelau, Landkreis Gross-Gerau, HEL)

    Emigrant to America   1868

    <u>173</u>: Ortsbuergerregister.   ADS 7:224

SUDHEIMER, Adam (from Wolfskehlen, Landkreis Gross-Gerau, HEL)

    Emigration to America   1854

    <u>173</u>: Gesinderegister 1837-1843 [of Goddelau]: Nr. 102.   ADS 7:224

SULTAN, -- (an American citizen)

    Expulsion of   1913

    <u>39</u>: Reichs- u. auswaert.Ang.: A.IV.5 [Ausweisungen]   ADS 2:378

SULZBACH, --

    Emigration of   1854?

    <u>152</u>: Abt.228: Amt Hoechst: Nr. 331.   ADS 7:132

von SUMINSKI, Wladimir (deranged American citizen)

    Repatriation of; refused by USA   1894

    <u>39</u>: Reichs- u. auswaert.Ang.: H.IIa.11 [alt.Sig.] [Heimschaffung]   ADS 2:270

SWARTZENDRUBER. *See* Schwartzentrauber

SZAMOSSY, -- (married couple)

    Extradition [from Hamburg] to Hungary of Szamossy couple excluded from America   1912

    <u>39</u>: Reichs- u. auswaert.Ang.: A.Ia.1.25 [Ausliefer-ungen]   ADS 2:372

SZCZUTOWSKI, Albert (in Cincinnati)

    Hamburg General Poorhouse seeks information re-garding   1896

    <u>39</u>: Reichs- u. auswaert.Ang.: N.Ia.16 [alt.Sig.] [Nachforschungen]   ADS 2:284

SZOELLOESY, John (seaman)

    Request by court in Passaic, New Jersey, for the arrest of   1906

    <u>39</u>: Reichs- u. auswaert.Ang.: P.III.1.4 [Polizei-wesen]   ADS 2:344

TAMM, Edward Grower (born in St. Louis, Missouri)

    Investigation, at the request of the U.S. ambassa-dor, of the expulsion of Tamm by Hamburg police   1910

    <u>39</u>: Bestand Senat: Cl.VII, Lit.L^b.2, vol.110 [Ver-mittlung]   ADS 2:93

    Expulsion of   1910

    <u>39</u>: Reichs- u. auswaert.Ang.: A.IV.3 [Ausweisungen]   ADS 2:361

TAMSEN, --

    New York consular fees in case against Tamsen 1901

    <u>39</u>: Reichs- u. auswaert.Ang.: J.Ib.10.1 [Justiz-wesen]   ADS 2:318

TANENHAUS, --

    Hamburg police search for Tanenhaus, wanted in America   1912

    <u>39</u>: Reichs- u. auswaert.Ang.: P.III.2.8 [Polizei-wesen]   ADS 2:375

TANNHAEUSER, S. *See* Thannhaeuser, S.

TATZL, Karl (an Austrian national)

    Extradition to Switzerland from the USA, via Ham-burg   1911

    <u>39</u>: Reichs- u. auswaert.Ang.: A.Ia.22 [Ausliefer-ungen]   ADS 2:367

TAUBER-HARPER, Otto (in America)

    Inquiry regarding his great-grandfather   1908

    <u>227</u>: Rep. MA 1936: A.St.: MA 98072.   ADS 9:97

TEICHWITZ, Anna, born Franzen (married woman; in Philadelphia)

    Hamburg Orphanage Commission seeks information re-garding financial standing of   1896

    <u>39</u>: Reichs- u. auswaert.Ang.: W.I.6 [alt.Sig.] [Waisenhaus-Ang.]   ADS 2:287

TEMPEL, Herm. (in Walfish Bay [state not given in ADS description)

    Estate of   1911

    <u>39</u>: Reichs- u. auswaert.Ang.: E.IV.93 [Erbschaften]   ADS 2:369

TEUTENBERG, --

Estate of   1941
<u>148</u>: R XI.   ADS 6:262

TEWITZ, -- (from Hunteburg). *See* Boehmer, --

THANNHAEUSER, S. (banker; in San Francisco)

File on [estate?]   1894-1895
<u>227</u>: Rep. MA 1921: A.St.I: Nr. 387.   ADS 9:67

THEIS, Johann (of Siegen)

File of Dillenburg administration regarding release of their subject [Theis] from the Hesse-
   Hanau Freikorps   1782
<u>151</u>: Bestand 81: Rep.E: 1.Gen.III: Nr. 13.
                                           ADS 7:80

THEISSEN, --

Estate of   1941
<u>148</u>: R XI.   ADS 6:262

THIEDE, --

Fees due Chicago consulate in case of Thiede *versus* Thiede   1902
<u>39</u>: Reichs- u. auswaert.Ang.: J.Ib.10.4 [Justiz-
   sachen]   ADS 2:325

THIELE, Edmund Georg Heinrich (father was a naturalized American citizen)

Expulsion of   1888-1889
<u>39</u>: Bestand Senat: Cl.I, Lit.T, No.20, vol.2b,
   Fasc.3 [Ausweisung]   ADS 2:20

Representations of the America [consul] requesting
   cancellation of order expelling Thiele from Hamburg   1889
<u>39</u>: Bestand Senat: Cl.VIII, No.X, Jg.1889 (Regis-
   ter) 1.   ADS 2:262

THIELEBOERGER, Heinr[ich] (from Moringen)

Proposed transportation of Thieleboerger, a prisoner in the workhouse   1854
<u>52</u>: Han.Des.80: Hildesheim I: E.DD.40a: Nr. 1526.
                                           ADS 4:90

THIEN, Jul.

New York: Estate of   1887
<u>39</u>: Bestand Senat: Cl.VIII, No.X, Jg.1887 (Regis-
   ter) 24 [Erbschafts-Amt]   ADS 2:258

THIESSEN, L. (of New Orleans)

Letter from Kirchspielvogtei [parish office] Meldorf regarding the estate of L. Thiessen, written on behalf of his mother, the widow -- Thiessen, born Oeser   1841
<u>39</u>: Bestand Senat: Cl.VIII, No.X, Jg.1841 (Regis-
   ter).   ADS 2:141

THIESSEN, -- (in Mountain Lake [state not given in
ADS description)

Transmittal of 50 [Russian?] rubles to   1897
<u>39</u>: Reichs- u. auswaert.Ang.: A.III.10 [alt.Sig.]
   [Auswanderung]   ADS 2:288

THILO, -- (physician; in [Ebenezer?] Georgia)

File on   1737
<u>500</u>: Halle: Missionsbibliothek des Waisenhauses:
   Abteilung H IV (Nordamerika) Fach G, Nr. 5.
                                          Not in ADS

THODE, Eduard Jacob (born 18 September 1838 in
Otterndorf; in Charleston, South Carolina)

Letters to parents   1857; 1858
<u>100</u>: Magistrat Otterndorf: Fach 140, Nr. 2, vol. I.
                                           ADS 4:302

THOENERT, J. O.

Philadelphia: Estate of   1903
<u>39</u>: Reichs- u. auswaert.Ang.: E.IV.7 [Erbschaften]
                                           ADS 2:330

THOENSEN, Pet[er] (in Lyons, Iowa)

Estate of   1897
<u>39</u>: Reichs- u. auswaert.Ang.: E.IV.9 [alt.Sig.]
   [Erbschaften]   ADS 2:290

THOLEY, Franz (of St. Wendel; an American citizen)

Petition for renaturalization [in Germany]
                                        1887-1905
<u>200</u>: Abt.442: Regierung Trier: Nr. 8665. ADS 8:41

THOMAS, Stephen (from Hoheneiche)

File regarding the transmittal of funds from Bremen, via Baltimore consulate, to emigrant Thomas
   who had lost money in Hessian territory during
   his trip to America   1855
<u>48</u>: B.13.b.2: Nr. 18; *also* P.8.B.8.c.1.a. ADS 3:18

THOMPSON, Ida

San Francisco: Estate of   1906
<u>39</u>: Reichs- u. auswaert.Ang.: E.IV.29 [Erbschaften]
                                           ADS 2:343

THOMSEN, Fritz Heinrich (in Brooklyn)

Inquiry by Hamburg Orphanage Commission regarding
Thomsen's financial condition   1895

39: Reichs- u. auswaert.Ang.: W.I.8 [alt.Sig.]
    [Waisenhausang.]   ADS 2:280

THORDSEN, Markus

Mentioned in memoirs of Peter Christian Hansen
17   ADS 1:52

THORUN, W. A.

New York: Thorun's accusation against the Hamburger
Volksbank for fraudulent exchange of money 1876

39: Bestand Senat: Cl.VIII, No.X, Jg.1876 (Regis-
    ter) 20 [Nordamerika]   ADS 2:232

THRON, Johann Hinrich. *See* Throw, Johann Hinrich

THROW [Thron?] Johann Hinrich (from Basdachel?)

Expulsion to America   1864
62: D i, Fach 102, vol.20: Regiminalia (Bd.II).
                                            ADS 4:231

THURN, -- (physician; died in New Orleans)

File on [estate?]   1891
227: Rep. MA 1921: A.St.I: Nr. 366.   ADS 9:67

THUERRIEGEL, Johann Caspar

File on Thuerriegel's plan to entice citizens [of
Baden & Kurpfalz] to Spain   1767-1770

260: Abt.236: Innenmin., Wegzug: Auswand.: 6740-
     6741.   ADS 10:4

TIEDEMANN, Johann Heinrich Andreas

Exile to North America or Brazil of serval prisoners for theft: J. H. A. Tidemann; J. Fr. Lange;
-- Meintzen; -- Danziger   1832

39: Polzeibehoerde-Krim.: Jg.1832, No. 62; *also*
    Jg.1833, No. 211.   ADS 2:410

TIENSCH, Carl Johannes Conrad (businessman; born 27
December 1852 in Hoya/Weser; died 13 July 1934 in
Maplewood, New York; married first: Louisa Kolle
in Louisville, Kentucky; married second: Amanda,
widow of Siebern, born 9 September 1850; entered
USA at Baltimore; lived 1869-1879 in Louisville,
Kentucky; 1879-1887 in Cincinnati, Ohio; 1887-1927
in Brooklyn, New York; 1927-1934 in Maplewood, New
York)

His letters to parents & brother Georg in Germany
are preserved by Rektor i.R. Richard Tiensch,
Scholienstrasse 50, Otterndorf, Niedersachsen
                                       ADS 4:326-328

TIENSCH, -- (physician; from Morsum)

Application of Dr. -- Tiensch to practice medicine
in Harpstedt & to dispense medicines; also his
journey to America   1831-1834

52: Han.Des.80: Harpstedt: B.1.III.C.3a: Nr. 43.
                                            ADS 4:84

TILLMANNS [Tillmann] Gustav (in Lippe, Indiana)

Correspondence   1901
120: Akte Frank Buchman (D II B)   ADS 6:75

TIMM, Anna Catherina (from Westerende-Otterndorf)

Application of Anna Timm & Catharina Maria Kuerver,
both welfare recipients, for emigration to America   1846

100: Kirchspielsgericht Westerende-Ott.: Loc.17:
     Nr. 5.   ADS 4:308

TIMM, --

Divorce proceedings in Chicago   1918
39: Reichs- u. auswaert.Ang.: J.Ib.69 [Justizang.]
                                            ADS 2:394

TIMMERMANN, H. W.

San Francisco: Estate of   1910
39: Reichs- u. auswaert.Ang.: E.IV.46 [Erbschaften]
                                            ADS 2:362

TINSDAHL, Henry

New York: Estate of   1879
39: Bestand Senat: Cl.VIII, No.X, Jg.1879 (Regis-
    ter) 1 [Nordamerika]   ADS 2:240

TINTELMANN, --

Fees due Chicago consulate in case of Tintelmann
versus Tintelmann   1902

39: Reichs- u. auswaert.Ang.: J.Ib.10.2 [Justiz-
    sachen]   ADS 2:324

TISCHER, Johann (died in America)

Death certificate for   1887
227: Rep. MA 1921: A.St.I: Nr. 341.   ADS 9:67

TITTMANN, -- (widow; in Egg Harbor City, New Jersey)

Estate of   1894
39: Reichs- u. auswaert.Ang.: E.IV.66 [alt.Sig.]
    [Erbschaften]   ADS 2:269

TROPPMANN, Josef (died in New York)

File on [estate?]   1892
227: Rep. MA 1921: A.St.I: Nr. 372.   ADS 9:67

TRUMMER, -- (from Feuchtwangen; a Bavarian national)

Issuance of travel papers by the American notary
public J. A. Nesseler in New York for Trummer
227: Rep. MA 1890: A.V.: Nr. 328.   ADS 9:54

TRUMPFHELLER, Elisabeth (from Goddelau, Landkreis
Gross-Gerau, HEL)

Emigration of Jewish person to America   1923
173: Abmelderegister 1916-1924: Nr. 1025.ADS 7:224

TRUMPFHELLER, Heinrich (from Goddelau, Landkreis
Gross-Gerau, HEL)

Emigration of Jewish person to America   1923
173: Abmelderegister 1916-1924: Nr. 1025.
ADS 7:224

TUERK, Georg (from Thuringia)

Emigration application   1839
500: Weimar: Thueringisches Staatsarchiv: Dermbach
II: Gb.2: Nr. 63-92; 93-114; 144-183; 186;
188.   Not in ADS

TUIBEL, -- (in Philadelphia)

Estate of   1817
227: Rep. Bayer. & Pfaelz. Gesandtschaft Paris:
Nr. 8382.   ADS 9:50

TURRIEGEL, Johann Caspar. See Thurriegel, Johann
Caspar

von TYSKA, Dr. Fr. [Friedrich or Freiherr?](Lieutenant, retired; in Sheperd [?] Texas)

Estate of   1909
39: Reichs- u. auswaert.Ang.: E.IV.39 [Erbschaften]
ADS 2:357

UFER, Ed[uard?] (in New York)

Damage claim [to Hamburg] by Ufer for expulsion
from Paris   1871
39: Bestand Senat: Cl.VIII, No.X, Jg.1871 (Register) 52 [Nordamerika]   ADS 2:217

UHDE, Heinrich Ludwig Christian (from Ildeshausen)

Emigration to America   1860-1861
56: L Neu Abt.129A: Gr.11, Nr. 54.   ADS 4:189

UHL, Leonhard (in Brooklyn)

Inquiry regarding whereabouts   1902
227: [Rep. not given] Lit.U: Nr. 9.   ADS 9:99

ULLMANN, Christof (died in New Orleans)

File on [estate?]   1887
227: Rep. MA 1921: A.St.I: Nr. 1083.   ADS 9:71

ULLRICH, -- (from Dornheim, Landkreis Gross-Gerau,
HEL)

Emigration of family of two persons   1880
168: Abmelderegister.   ADS 7:218

ULRICH, Doris (unmarried; from Peine)

Transportation to America   1845
52: Han.Des.80: Hildesheim I: E.DD.8a: Nr. 966.
ADS 4:88

ULRICH, Georg (from Philadelphia). See Farny, Johann Jakob

ULRICH, Philipp. See Lambert, Balthasar

UNDHEIM, --

Estate of   1941
148: R XI.   ADS 6:262

UNKELBACH, P., born Lappe

New York: Estate of   1901
39: Reichs- u. auswaert.Ang.: E.IV.70 [Erbschaften]
ADS 2:316

UNRUH, N. (from Oberlustadt [Herrschaft])

Petition to be allowed to emigrate to Pennsylvania
1771
201: Archivabt. Johanniterorden: Akt Nr. 36.
ADS 8:61

UNTERSCHMITTNER, Katharina, born Neumaier (died 1906
in Pittsburgh)

Issuance of a death certificate   1907
227: [Rep. not given]: Lit.U: Nr. 10.   ADS 9:99

UTZ, Georg (pastry baker; in Philadelphia)

Claim of -- Wiegel et al in Ansbach to estate of
Utz   1803-1817
500: Berlin-Dahlem: Geh. Staatsarchiv: Rep.XI,
21a, Conv. 3,6.   Not in ADS

UTZ, Georg (pastry baker; in Philadelphia)

Estate of   1811-1812

227: Rep. Bayer. & Pfaelz. Gesandtschaft Paris:
8366.   ADS 9:50

File on [estate?]   1817-1841

227: Bayerische Gesandtschaft London: Nr. 853.
                         ADS 9:47

VAGTS, Alfred (born in Otterndorf)

His writings are in 100   ADS 4:318-319

VAHLBERG, Wilhelmine (vagrant; daughter of day laborer
Christoph Vahlberg, from Volkersheim)

Emigration to America   1854-1856

56: L Neu Abt.129A: Gr.11, Nr. 84.   ADS 4:191

VAHLTEICH, Julius (member of the Reichstag)

Hamburg police surveillance over [mainly newspaper clippings]   1881; 1891-1910

39: Polizeibehoerde-Polit.Pol.: S 275: Bl. 1-11.
                        ADS 2:418

de VALLANT, Isadore E.

Investigation in Hamburg, at request of New York
police authorities, regarding murder of de Vallant; correspondence with American authorities
in court case   1909

39: Reichs- u. auswaert.Ang.: P.III.6 [Polizei-
wesen]   ADS 2:359

VANSELOW, --

Inquiry of Landrat Loewenberg, Silesia [now called
Lwówek Slaski, Poland] as to whether Vanselow
has shipped out to America   1850

39: Bestand Senat: Cl.VIII, No.X, Jg.1850 (Register).   ADS 2:157

VARANI, B. (in America). *See* Bedassem, J.

VARNY, Johann Jakob. *See* Farny, Johann Jakob

VASARY, Eugen (Doctor). *See* Weiss, Eugen (Doctor)

VATER, Carl Leberecht (mason)

Expulsion from Hamburg as Social Democrat & emigration to America   1880-1881

39: Polizeibehoerde-Polit.Pol.: S 149/66: Bl. 243-
265.   ADS 2:413

VAUPEL, -- (sergeant in Hanau military unit; in
America)

Letter from   1782

500: Hanau: Geschichtsverein: Ms Nr. 559g.
                       Not in ADS

von VELTHEIM, Hans (son of *Hofjaegermeister* [master
of the hunt for ducal court]

Death reported by Brunswick consulate in St. Louis
                           1865

56: L Neu Abt.19F: St. Louis: Nr. 2.   ADS 4:183

VERDELMANN, Leonhard (died in Holland, Michigan)

Estate of   1900-1901

152: Abt.405: Wiesbaden: Rep.405/I,2: Nr. 3534.
                        ADS 7:147

VETTER, -- (married woman)

New York: estate of   1905

39: Reichs- u. auswaert.Ang.: E.IV.56 [Erbschaften]
                        ADS 2:339

VICK, E. H. *See* Vick, V.

VICK, V.

Chicago: death certificates for V. Vick & E. H.
Vick   1879

39: Bestand Senat: Cl.VIII, No.X, Jg.1879 (Register) 52 [Nordamerika]   ADS 2:242

VIOGT, G. (in New York)

Request of Notary -- Martin [in Hamburg?] to have
a letter transmitted to Viogt   1916

39: Reichs- u. auswaert.Ang.: G.Ia.1.3 [Geschaefts-
verkehr]   ADS 2:391

VIVELLI, --

Estate of   1941

148: R XI.   ADS 6:262

VOGEL, Charles F.

Complaint regarding postoffice   1805

39: **Bestand Senat:** Cl.I, Lit.Nc: Nr. 58b [Beschwer-
den]
DLC has.   ADS 2:1

VOGEL, Joh. Conr[ad] (in Chicago)

Information regarding Vogel sought by private
person [in Germany]   1894

39: Reichs- u. auswaert.Ang.: N.Ib.17 [alt.Sig.]
[Nachforschungen]   ADS 2:271

VOGEL, John (died in San Francisco)

File on [estate?]   1898-1900
227: Rep. MA 1921: A.St.I: Nr. 519.   ADS 9:67

VOGEL, Wilhelm

Correspondence of Hamburg Senate with   1849-1858
39: Bestand Senat: Cl.VI, No.16p: vol.4b: Fasc.1 d
[Korrespondenz]   ADS 2:47

VOGT -- (married couple; in New York)

Hamburg Orphanage Commission seeks information on
1897
39: Reichs- u. auswaert.Ang.: W.I.2 [alt.Sig.]
[Waisenhausang.]   ADS 2:293

VOGELY, C.

New Orleans: Death certificate for   1879
39: Bestand Senat: Cl.VIII, No.X, Jg.1879 (Register) 37 [Nordamerika]   ADS 2:241

VOGENITZ, Friedrich (cigar worker)

Expulsion from Hamburg as Social Democrat & emigration to America   1880-1881
39: Polizeibehoerde-Polit.Pol.: S 149/46: Bl. 297-
308.   ADS 2:412

VOGL, M. (from Lenggries; died in San Francisco)

File on [estate?]   1895-1896
227: Rep. MA 1921: A.St.I: Nr. 495.   ADS 9:67

VOGT, Anna Maria. See Schleicher, Anna Maria, born
Vogt

VOGT, Otto (in Hamburg)

New York: Death certificate [of person unnamed in
ADS description] sent to Vogt   1878
39: Bestand Senat: Cl.VIII, No.X, Jg.1878 (Register) 1 [Nordamerika]   ADS 2:236

VOIGT, Max (in Emmetsburg, Iowa)

Inquiry by private person in Germany regarding
1899
39: Reichs- u. auswaert.Ang.: N.Ib.1.11 [Nachforsch-
ungen]   ADS 3:303

VOIGT, Oswald (of Bischofsheim, Landkreis Gross-
Gerau, HEL)

Emigration with family to Reading [Pennsylvania?]
1923

165: Abmelderegister.   ADS 7:215

VOIGT, -- (pastor; in Pennsylvania)

Pastoral calls & transportation of -- Voigt & --
Krug as pastors to Pennsylvania, with attached
statement of costs of transportation   1763
500: Halle: Missionsbibliothek des Waisenhauses:
Abteilung H IV (Nordamerika) Fach A, Nr. 7.
Not in ADS

VOLK, Berthold (died in Chicago, Illinois)

Estate of   1900
152: Abt.405: Wiesbaden: Rep.405/I,2: Nr. 3536.
ADS 7:147

VOLKERS, W. See Folkers, W.

VOLKMANN, Johann Caspar (from Dermbach, Thuringia)

Application for emigration with son Johannes
Volkmann   1848
500: Weimar: Thueringisches Staatsarchiv: Dermbach
II: Gb.2: Nr. 63-92; 93-114; 144-183; 186;
188.   Not in ADS

VOLKMANN, Johannes. See Volkmann, Johann Caspar

VOLTZ, Daniel (of Biebesheim, Landkreis Gross-Gerau,
HEL)

Emigrant to America (shown in list of former residents who have lost citizenship)   1855
164: Abt. XI,4.   ADS 7:214

VOLZ, Wilhelm (from Ginsheim-Gustavburg, Landkreis
Gross-Gerau, HEL)

Voyage to USA for four-year period   1851
172: Abt. XI,4.   ADS 7:223

VOLZE, Heinrich (vagabond hand laborer; from Gehren-
rode)

Emigration to America   1857-1859
56: L Neu Abt.129A: Gr.11, Nr. 42.   ADS 4:188

VORNBERGER, Georg (cigar worker)

Expulsion from Hamburg as Social Democrat & emigration to America
39: Polizeibehoerde-Polit.Pol.: S 149/347: Bl.
352-363.   ADS 2:417

VOSS, August (journeyman shoemaker; from Hallensen?)

Emigration to America with Ludwig Voss, Christian
Kaufmann, Heinrich Buenger & family, Christian
Koch, & Christian Ludwig Baye   1852
56: L Neu Abt.129A: Gr.11, Nr. 50.   ADS 4:189

VOSS, Caroline. *See* Silkroth, Caroline, *alias* Voss

VOSS, Ludwig (servant; from Hallensen?). *See* Voss,
August

VOSS, Otto (in New York)

Inquiry by private person in Germany regarding
                                                 1899
39: Reichs- u. auswaert.Ang.: N.Ib.1.9 [Nachforsch-
ungen]   ADS 2:303

VOSS, -- (retired First Lieutenant; from Meppen)

Investigation of police action against Voss   1841
54: Rep.122,VIII: Reg. Meppen: Nr. 1926.   ADS 4:141

VOSS, -- (in America)

Estate of   1912
39: Reichs- u. auswaert.Ang.: E.IV.73 [Erbschaften]
                                            ADS 2:373

VRANKEN, O.

Emigrant?   1891
115: Rep. D9 III: Abt.E: Nr. 146.   ADS 6:34

VRIES, Anna, born Mundi. *See* Fries, Anna, born
Mundi

WAALKES, -- (of Greetsiel)

Transport of family Waalkes to America at the expense of the welfare organization of Greetsiel
                                                 1866
50: Rep.21a: 7528.   ADS 4:7

WABNITZ, Hermann (businessman)

Expulsion from Hamburg as Social Democrat & emigration to America   1880-1881
39: Polizeibehoerde-Polit.Pol.: S 149/68: Bl. 436-
446.   ADS 2:413

WACHTMANN, F. Chr. (in Chicago)

Inquiry by private person in Germany regarding
                                                 1900
39: Reichs- u. auswaert.Ang.: N.Ib.1.2 [Nachforsch-
ungen]   ADS 2:309

WADE, John (member Moravian Brethren; in Pennsylvania?)

Letters & reports from John Wade, [Johannes] Bechtel, -- Post & from a so-called Voyageur à l'Eter-
nité from Oly   1742
500: Herrnhut: Archiv: Rep.14.A: Nr. 23.
                                            Not in ADS

WAGECK, Johannes (in New York)

File on [estate?]   1895
227: Rep. MA 1921: A.St.I: Nr. 1109.   ADS 9:71

WAGNER, Anna Margaretha (from Raunheim, Landkreis
Gross-Gerau, HEL)

Emigration to America with two illegitimate children & three grandchildren   1854
181: Raunheimer Evangelischen Pfarrei Familienregister (Lehrer Johannes Buxbaum) ADS 7:234

WAGNER, Cosima (married woman)

Use of archival materials in a civil trial in
America by Mrs. Cosima Wagner   1903
227: Rep. MA 1921: I.V.: Nr. 2702.   ADS 9:77

WAGNER, J. F. A.

Application made by Wagner for duplicate of a lost
medal awarded him   1872
39: Bestand Senat: Cl.VIII, No.X, Jg.1872 (Register) 48 [Nordamerika]   ADS 2:222

WAGNER, -- (member Moravian Brethren; in North Carolina). *See* Danz, --

WAITZ, Carola (daughter of an artist; from Munich;
in America)

Repatriation of   1913
227: Rep. MA 1936: A.St.: MA 98050.   ADS 9:94

WAIZENEGGER, --

Estate of   1944
148: R XI.   ADS 6:262

WALDEMANN, Heinr[ich] (a convict)

Application for financial assistance to emigrate
to America   1866
52: Han.Des.80: Hildesheim I: E.DD.36a: Nr. 1248.
                                            ADS 4:89

WALDENMEYER, Andreas (from Durlach? BWL)

Emigration of   1770-1771
260: Nr. 217.   ADS 10:1

WALDER, Adalbert

Extradition to Hungary from the USA, via Hamburg, charged with counterfeiting   1908

39: Reichs- u. auswaert.Ang.: A.Ia.22 [Ausliefer-ungen]   ADS 2:350

WALDSCHMIDT, --

Heirs of Waldschmidt; payment of emigration tax
                                                    1794

119: Litt.W, Nr. 7.   ADS 6:74

WALKENHORST, -- (from St. Annen)

Complaint of -- Walkenhorst, linen weaver [in Ger-many] regarding the emigration of his daughter without his permission   n.d.

54: Rep.122,VII,B: Groenenberg-Melle: Regiminalia: Fach 15, Nr. 3.
    DLC may have this.   ADS 4:138

WALLACE, --. *See* Gronewaldt, --

WALLON, Louis (American Methodist missionary; in Friedrichsdorf, Hessen-Homburg)

Petition of Wallon to be allowed to proselytize?
                                                    1854

152: Abt.310: Hessen-Homburg: X.C.10.   ADS 7:137

WALTER, --

General Consulate in New York: Case of Walter *ver-sus* Gausmann   1898

39: Reichs- u. auswaert.Ang.: J.Ib.10 [alt.Sig.] [Justizsachen]   ADS 2:297

WALTHER, Carl Hermann Rudolf (Otto)(cigar dealer)

Expulsion from Hamburg as Social Democrat & emi-gration to America   1880-1881

39: Polizeibehoerde-Polit.Pol.: S 149/222: B1. 444-459.   ADS 2:415

WALTHER, Charles (in New York)

File on [estate?]   1899

227: Rep. MA 1921: A.St.I: Nr. 1139.   ADS 9:71

WALTHER, Ch. (in New Ulm, Minnesota)

Estate of   1909

39: Reichs- u. auswaert.Ang.: E.IV.88 [Erbschaften]
                                             ADS 2:357

WALTHER, Georg (from Hesse)

Emigration of   1852

149: Rep.51 V: Nr. 93 A-Z.   ADS 7:1

WALTHER, -- (pastor; from Hannover region)

Collection in Hamburg to help him emigrate to America   1850

39: Geistl.Minis.: II,11: Konventsprotokolle: p. 181.   ADS 2:476

WANGLER, Arthur (in Genf [Geneva, state not given in ADS description])

Investigation regarding his [Wuerttemberg] mili-tary service obligation   1900-1913

263: Rep.E 46-48: Minist. d. AA III: Verz. 1: 148/17.   ADS 10:43

WANNEMACHER, Jakob (from Walldorf, Landkreis Gross-Gerau, HEL)

Emigration of   19th century

184: Ortsbuergerregister (19. Jhdt.): Nr. 131.
                                             ADS 7:238

WANTZELIUS, --. *See* Kainer, --

WARCHAM, E., born Getzmann (married woman; in Brook-lyn [New York])

Estate of   1903

39: Reichs- u. auswaert.Ang.: E.IV.91 [Erbschaften]
                                             ADS 2:330

WARE, R. (an American citizen)

Pardon application   1918

39: Reichs- u. auswaert.Ang.: J.Ic.26 [Justiz-sachen]   ADS 2:394

WARLING, Caspar (from St. Annen)

Emigration to America of Caspar & Claere Warling

54: Rep.122,VII.B: Groenenberg-Melle: Regiminalia: Fach 15, Nr. 4.   ADS 4:138

WARLING, Claere. *See* Warling, Caspar

WARNECKE, A. F. C. (in New York)

Consular report on the death & estate of   1873

39: Bestand Senat: C1.VIII, No.X, Jg.1873 (Regis-ter) 32 [Nordamerika]   ADS 2:225

WARNECKE, Emilie

Boston: Information regarding   1877
39: Bestand Senat: Cl.VIII, No.X, Jg.1877 (Register) 47 [Nordamerika]   ADS 2:235

WARTENSLEBEN, Josef (from Gernsheim, Landkreis Gross-Gerau, HEL)

Emigration with family of four persons   1933
171: Judenkartei.   ADS 7:222

WASSERMANN, Klara (from Goddelau, Landkreis Gross-Gerau, HEL)

Emigration of Jewish person to America   1925
173: Abmelderegister 1916-1924: Nr. 217. ADS 7:224

WASWO, Carl (in New York)

Inquiry of Hamburg Orphanage Commission regarding financial standing of   1895
39: Reichs- u. auswaert.Ang.: W.I.6 [alt.Sig.] [Waisenhausang.]   ADS 2:279

Inquiry of Hamburg Orphanage Commission regarding   1902
39: Reichs- u. auswaert.Ang.: W.I.1 [Waisenhausang.]   ADS 2:327

von WATTEVILLE, Johannes (member Moravian Brethren). See Spangenberg, --

WEBER, Friederike (servant girl; from Badenhausen)

Sentenced for theft & emigrated to America   1862-1867
56: L Neu Abt.129A: Gr.11, Nr. 26.   ADS 4:187

WEBER, Hattie A. A. (in Laurens, New York)

Inquiry by Mrs Weber for information regarding the estate of her grandfather, John Hanns Crill 1897
149: Rep.49a: Nr. 565/17.   ADS 7:17

WEBER, Johanne (spinster; from Oberhuette)

Punishment for theft, imprisonment in the house of correction at Bevern, & emigration to America   1852-1857
56: L Neu Abt.129A: Gr.11, Nr. 72.   ADS 4:190

WEBER, Martin (from Affennest; in Trenton [New Jersey])

Complaint of   1911
227: Rep. MA 1936: A.St.: MA 98046.   ADS 9:94

WEBER, --

Request of Weber family for permission to emigrate to North America   1883
227: Rep. MA 1921: A.V. IV: Nr. 11707.   ADS 9:85

WEBER, -- (in North America)

Weber brothers [apparently transmittal of information for] Jakob Weber [in Bavaria?]   1778
227: Bayerische Gesandtschaft London: 838.   ADS 9:47

WEBER, -- (a German national)

Prosecution for fraud; refused admittance into USA   1910
39: Reichs- u. auswaert.Ang.: J.Ic.13 [Justizsachen]   ADS 2:364

WEBER, --

Estate of   1941
148: R XI.   ADS 6:262

WEBER, -- (from Kelsterbach, Landkreis Gross-Gerau, HEL). See Frenzel, --

WEBER, --. See Jarmulowski, --

WEBSTER, Jane E. See von Melle, Emil

WEDEKIND, --. See Pelitowski, --, born Wedekind

WEDEMEIER, Karl. See Wedemeier, Karoline

WEDEMEIER, Karoline (unmarried; from Gandersheim)

Emigration with son Karl to America   1868
56: L Neu Abt.129A: Gr.11, Nr. 41.   ADS 4:188

WEDEMEYER, --. See Paulmann, Ludwig

WEGENER, Friedrich Andreas (newspaper peddler)

Expulsion from Hamburg as Social Democrat & emigration to America   1880-1881
39: Polizeibehoerde-Polit.Pol.: S 149/38: Bl. 200-208.   ADS 2:412

WEGNER, Heinrich (of Chicago)

Application for citizenship in Hamburg   1914
39: Reichs- u. auswaert.Ang.: S.III.7 [Staatsangehoerigkeit]   ADS 2:387

WEHDE, Martin

Proposed transportation to America   1849

<u>52</u>: Han.Des.80: Hildesheim I: E.DD.7i: Nr. 940.
ADS 4:88

WEHNKE, Rudolf (in San Francisco)

Consular report regarding   1861

<u>39</u>: Bestand Senat: Cl.VIII, No.X, Jg.1861 (Register) I [Amerika]   ADS 2:179

WEHRBEIN, Wilhelm

Correspondence regarding Wehrbein's incarceration in the house of correction in Bevern & his subsequent transportation to America   1855-1859

<u>74</u>: Z.IV [Schriftwechsel]   ADS 4:254

WEHRENBERG? [*may read* Ehrenberg] Diedrich

San Francisco: Estate of   1900

<u>39</u>: Reichs- u. auswaert.Ang.: E.IV.70 [Erbschaften]
ADS 2:307

WEIDEMANN, Henry

Detroit: Estate of   1896

<u>39</u>: Reichs- u. auswaert.Ang.: E.IV.32 [alt.Sig.] [Erbschaften]   ADS 2:282

WEIDEMANN, Otto

Possible pardon for Weidemann [in Hamburg prison] if opportunity & money [promised by father] for transportation to America is forthcoming   1841

<u>39</u>: Bestand Senat: Cl.VIII, No.X, Jg.1841 (Register) B.2 [Zuchthaus]   ADS 2:142

WEIDINGER, Susanna

Transportation of Susanna Weidinger, an insane person, to America   1869

<u>227</u>: Rep. MA 1921: A.V. IV: Nr. 11688.   ADS 9:84

WEIGAND, Johannes (of Kempfenbrunn)

Confiscation of the property of soldier who deserted in America   1786

<u>151</u>: Bestand 82: Rep.E: 55.Amt Bieber X: Nr. 7.
ADS 7:83

WEIL, Amalie (from Gernsheim, Landkreis Gross-Gerau, HEL)

Emigration to New York   1938

<u>171</u>: Judenkartei.   ADS 7:222

WEIL, Hermann (from Gernsheim, Landkreis Gross-Gerau, HEL)

Emigration of four [Jewish] persons   1935

<u>171</u>: Judenkartei.   ADS 7:222

WEILER, Kurt (of Bischofsheim, Landkreis Gross-Gerau, HEL)

Emigration to Evansville [state not given in ADS description]   1938

<u>165</u>: Judenkartei.   ADS 7:215

WEILERT, Karl Aug. (pipesmith; from Flecken Herzberg)

Proposed transportation to America   1852-1853

<u>52</u>: Han.Des.80: Hildesheim I: E.DD.47a: Nr. 1750.
ADS 4:91

WEINECK, C. R.

New York: Weineck's support by public charity 1878

<u>39</u>: Bestand Senat: Cl.VIII, No.X, Jg.1878 (Register) 23 [Nordamerika]   ADS 2:237

New York: Support by public welfare   1879

<u>39</u>: Bestand Senat: Cl.VIII, No.X, Jg.1879 (Register) 10.

WEIS, Guido (journeyman carpenter; from Dermbach, Thuringia)

Emigration application   1848

<u>500</u>: Weimar: Thueringisches Staatsarchiv: Dermbach II: Gb.2: Nr. 63-92; 93-114; 144-183; 186; 188.   Not in ADS

WEISE, Friedr[ich]

Chicago: Estate of   1912

<u>39</u>: Reichs- u. auswaert.Ang.: E.IV.71 [Erbschaften]
ADS 2:373

WEISER, Conrad (member Moravian Brethren; in Pennsylvania)

Letters to [Count] Zinzendorf   1742-1743; 1745

<u>500</u>: Herrnhut: Archiv: Rep.14A: Nr. 25. Not in ADS

WEISS [*or* Vasary] Eugen (doctor)

Request by criminal court in Budapest for the extradition of Weiss from the USA to Hungary, via Hamburg   1910

<u>39</u>: Reichs- u. auswaert.Ang.: A.Ia.1.16 [Auslieferungen]   ADS 2:360

WEISS, H.

German consulate in Louisville, Kentucky: Sworn statements in case of H. G. C. Neb *versus* H. Weiss   1874

39: Bestand Senat: Cl.VI, No.16p: vol.4b: Fasc. 21.
                                                    ADS 2:60

WEISS, Margarethe (in Brooklyn)

Investigation by Georg Pedall   1921

227: [Rep. not given]: Lit.W: Nr. 101.   ADS 9:99

WEISSENFELS, C. (an American citizen)

Appeal for pardon & remittance of remainder of prison sentence by   1900

39: Reichs- u. auswaert.Ang.: J.Ic.3 [Justizsachen]
                                                    ADS 2:309

WEISSMANN, Friedrich Wilhelm Albert (newspaper editor)

Expulsion from Hamburg as Social Democrat & emigration to America   1880-1884

39: Polizeibehoerde-Polit.Pol.: S 149/67: Bl. 452-484.   ADS 2:413

WEITLING, Wilhelm (journeyman tailor; German revolutionary; in America?)

Weitling & the communist unrest in Zuerich   1843

226: Rep. MInn 7: 25131f.   ADS 9:18

Communist plots: Committee of Liberation [Befreiungsbund] in North Germany, vol. I
                                        January-July 1850

226: Rep. MInn 7: 45543.   ADS 9:19

File on   1850-1851

227: Rep. MA 1957: Laufende Registratur: Nr. 829.
                                                    ADS 9:108

WEITZ, Sebastian (from Schoenthal; in New Orleans)

Petition for pardon   1847-1852

227: Rep. MA 1898: A.St.: Nr. 6326.   ADS 9:59

WELCH, P. A. *See* George, A. Frieda Eva

WELKE, --

Estate of   1942

148: R XI.   ADS 6:262

WELLMANN, Jobst Heinr[ich] (from Bomte)

Transportation of convict Wellmann to North America   1842

54: Rep.122,X.B: Wittlage-Hunteburg: I.Fach 46: Nr. 18.
DLC may have this.   ADS 4:143

WELTZ, Julia P. (in West Hoboken [New Jersey])

Bequest to the Gewerbeschule [handicraft school] in Speyer   1917

227: Rep. MA 1936: A.V.: Nr. 1439.   ADS 9:101

WENDEL, Ella

Estate of   1934-1940

148: R XI.   ADS 6:262

WENDEL, Ferd.

File on application to emigrate to America   1836

54: Rep.122,VIII: Reg.Meppen: Nr. 308.
DLC may have this.   ADS 4:141

WENDEL, Johann Nep.

Estate of   1941

148: R XI.   ADS 6:262

WENDT, --

Estate of, said to be millions [of dollars in value]   1912

39: Reichs- u. auswaert.Ang.: E.IV.65 [Erbschaften]
                                                    ADS 2:373

WENINGER, Josef (died in Detroit)

File on [estate?]   1894-1895

227: Rep. MA 1921: A.St.I: Nr. 1097.   ADS 9:71

WENKE, -- (married couple)

Estate of; abstract requested of retired Lieutenant -- Wolff in Hoboken [New Jersey]   1896

39: Reichs- u. auswaert.Ang.: E.IV.43 [alt.Sig.] [Erbschaften]   ADS 2:282

WENSCH, H. C.

Arrest of, & intended extradition to America 1865

39: Bestand Senat: Cl.VIII, No.X, Jg.1865 (Register) I.   ADS 2:187

WENZ, James (in Minnesota?)

Application to be Bremen vice consul in Minnesota   1855

48: B.13.b.O.: Nr. 14.
DLC has.   ADS 3:10

WENZ J[ames] (in Buffalo, New York)

Seeks Hamburg consulate in Minnesota   1856-1857
39: Bestand Senat: Cl.VI, No.16p: vol.4b: Fasc. 16.
ADS 2:59

WENZEL, Gottlieb (from Flecken Lauterberg?)

Proposed transportation to America   1856
52: Han.Des.80: Hildesheim I: E.DD.48a: Nr. 1779.
ADS 4:91

WERNER, Johann (carpenter & cabinetmaker; from Alf-
stedt)

Emigration to America   1853
62: A, Fach 6, vol. 7.   ADS 4:232

von WERNER, Ludwig (born 5 January 1879 [in Germany];
died in America)

Mentioned in family papers
149: Rep.73e: Abt.Depositum u. Nachlass von Werner:
Nr. 4.   ADS 7:20

WERNER, --

Extradition to Berlin from the USA, via Hamburg
1910
39: Reichs- u. auswaert.Ang.: A.Ia.29 [Ausliefer-
ungen]   ADS 2:361

WERTHMANN, Johann (beer brewer; from Pautzfeld; died
in Pittsburgh)

Information regarding his descendants   1903
227: Rep. MA 1921: A.St.I: Nr. 1845.   ADS 9:75

WESENDONCK, H. (in America). *See* Bedassem, J.

WESSELHOEFT, M. H.

San Francisco: Estate of   1911
39: Reichs- u. auswaert.Ang.: E.IV.97 [Erbschaften]
ADS 2:369

WESTENBERGER, Joh[ann] (died 2 October 1906 in Portland, Oregon)

Estate of   1906-1913
152: Abt.405: Wiesbaden: Rep.405/I,2: Nr. 3538.
ADS 7:147

WESTENBERGER, Peter J. (died July 1905 in Cleveland,
Ohio)

Estate of   1909-1910
152: Abt.405: Wiesbaden: Rep.405/I,2: Nr. 3539.
ADS 7:147

WESTENDORFF, Charles P. L. (in Charleston [South
Carolina])

Letters to brother Dr. Carl Westendorff & sister
Sophie Haufft, wife of Major -- Haufft, in Hamburg   1821-1835
39: Bestand Familienarchive: Fam. Westendorff: 2.
ADS 2:490

WESTERFELD, Karl Friedrich Wilhelm (from Flecken
Syke)

Issuance of emigration permit   1862-1867
52: Han.Des.80: Syke: B.aa.VI.4: Nr. 462. ADS 4:85

WETZEL, Johann Anton (from Hof; in Philadelphia)

File on [estate?]   1900
227: Rep. MA 1921: A.St.I: Nr. 1142.   ADS 9:71

WHITE, Georg Christoph (in Raleigh, Virginia)

Estate of [claims of Kuehn, from Ackerburg, with
documents relating to estate of Maria Dildey &
to her mother's brother Georg Christoph White's
property in Raleigh, Virginia]   1780-1817
500: Berlin-Dahlem: Preus.Geheimes Staatsarchiv:
Rep.XI, 21a, Conv.1,2.   Not in ADS

WICHTE, Luise (from Dornheim, Landkreis Gross-Gerau,
HEL)

Emigration to America   1921
168: Abmelderegister.   ADS 7:218

WICHTE, Maria Luise (from Dornheim Landkreis Gross-
Gerau, HEL)

Emigration to California   1920
168: Abmelderegister.   ADS 7:218

WIDEEN, K. Ch., *alias* Erikson (in New York)

Inquiry by private person in Germany regarding
1899
39: Reichs- u. auswaert.Ang.: N.Ib.1.8 [Nachforsch-
ungen]   ADS 2:303

WIECHMANN, K. E. E. E. (in Ohio)

Cincinnati: Death certificate for   1876
39: Bestand Senat: Cl.VIII, No.X, Jg.1876 (Register) 21 [Nordamerika]   ADS 2:233

WIEGAND, Elisabetha (of Schmalnau, HEL)

Emigration law extended to allow women to be issued emigration permits as a consequence of Elisabetha Wiegand's case   1806
151: Bestand 98: Fulda X: Acc.1875/27: Nr. 47.
ADS 7:91

WIEGAND, --. *See* Issler, --

WIEGNER, Christ. (member Moravian Brethren; in Germantown, Pennsylvania)

Letters to Schwenckfelder [groups]   1734-1737
Letters from -- Wieme, Joh. Bechtel, Georg Neisser
                                              1738-1739

500: Herrnhut: Archiv: Rep.14.A: Nr. 21.
                                          Not in ADS

WIEME, -- (member Moravian Brethren; in Pennsylvania?)
See Wiegner, Christ.

WIESBAUER, --

Estate of   1941
148: R XI.   ADS 6:262

WIESE, Wilhelm (in Brooklyn, New York)

Estate of   1915
39: Reichs- u. auswaert.Ang.: E.IV.57 [Erbschaften]
                                          ADS 2:388

WIESEN, Joh. Georg

Arrest of Wiesen & Anna Katharina Hammer(in) on
   charges of attempted emigration   1749
200: Abt.35: Reichsgrafschaft Wied: Nr. 3348.
                                          ADS 8:12

WIESENECKER, -- (of Bischofsheim, Landkreis Gross-
Gerau, HEL)

Emigration to Pittsburgh [Pennsylvania?]   1926
165: Abmelderegister.   ADS 7:215

WIESENFELD, Walter (from Ginsheim-Gustavsburg, Landkreis Gross-Gerau, HEL)

File on indemnification of Jewish person who emigrated to USA in 1938
172: Wiedergutmachungsakten.   ADS 7:223

WIESINGER, Fr. K. (from Lehrberg; died in Louisville
[Kentucky?])

File on [estate?]   1849
231: Rep.270/II: Reg. Mittelfranken: Kammer des
        Innern: Abgabe 1932: Nr. 603.   ADS 9:134

WIETING, -- (in America)

[Illegal?] migration of soldier [or desertion?]
                                              1797
500: Berlin-Dahlem: Preus. Geh. Staatsarchiv: Rep.
        XI, 21a, Conv. 2,4.   Not in ADS

WIGGER, -- (an American citizen)

Complaint by Wigger regarding fine imposed by Hamburg police   1913
39: Reichs- u. auswaert.Ang.: P.III.7 [Polizei-
   wesen]   ADS 2:381

WILBRECHT, J.

San Francisco: Death & estate of   1877
39: Bestand Senat: Cl.VIII, No.X, Jg.1877 (Register) 9 [Nordamerika]   ADS 2:234

WILD, Heinrich. *See* Buexler, Friedrich

WILD, Vitus

St. Louis: Search for   1879
39: Bestand Senat: Cl.VIII, No.X, Jg.1879 (Register) 41 [Nordamerika]   ADS 2:241

WILDER, Ernst

New Orleans: Death of Wilder on the *Constantia*
                                              1878
39: Bestand Senat: Cl.VIII, No.X, Jg.1878 (Register) 32 [Nordamerika]   ADS 2:237

WILHELM, Friedrich Bernhard (cigar worker)

Expulsion from Hamburg as Social Democrat & emigration to America   1882
39: Polizeibehoerde-Polit.Pol.: S 149/407: Bl.
   351-366.   ADS 2:417

WILKEN, Ben (in Coleridge, Nebraska)

Estate of   1904
39: Reichs- u. auswaert.Ang.: E.IV.9 [Erbschaften]
                                          ADS 2:335

WILKENS, J. O. (born 11 September 1866 in New York)

Personal file on   1904-1906
39: Land. Bergedorf II: XI.C.186.   ADS 2:467

WILLIAMS, Edward (Negro ship's clerk)

Complaint of U.S. consul regarding sentencing of
   Andrew Owen, yeoman of ship *Coriolanus*, Captain
   Wadsworth, commanding, for killing of Williams
                                              1837
39: Bestand Senat: Cl.VIII, No.X, Jg.1837 (Register) 1.a [Nordamerika]   ADS 2:134

WILLIAMS [*or* Stepney](escaped slave from New Orleans)

Consular file regarding Williams who escaped aboard
a Bremen ship   1842

48: Dd.11.c.2.N.1.a.1. Nr. 27; *also* Bestand 4,48:
A.1.c.1.d.1.b.   ADS 3:59

WILLKOMM, Ernst Julius Maximilian (of Geestlande?)

File regarding his release from Hamburg citizenship
upon resettlement to America   1872

39: Geestlande I: XI (I, Nr. 3095).   ADS 2:460

WILLE, Johann Karl (cabinetmaker)

Expulsion from Hamburg as Social Democrat & emigration to America   1880-1881

39: Polizeibehoerde-Polit.Pol.: S 149/221: Bl.
367-376.   ADS 2:415

WILLIG, --

Estate of   1941

148: R XI.   ADS 6:262

WILLMANN, A. C. (in Milwaukee, Wisconsin)

Seeks Hamburg consulate   1857-1858

39: Bestand Senat: Cl.VI, No.16p: vol.4b: Fasc. 18.
                                          ADS 2:59

WILMANN, A. G. (Hannoverian consul in Milwaukee, Wisconsin)

Application to be Bremen consul for Wisconsin,
Illinois, Iowa, Indiana, Michigan, & Minnesota
[*see also* Willmann, A. C.; probably same person]
                                               1857

48: B.13.b.O.: Nr. 16.
DLC has.   ADS 3:10

WILTBROK, L.

Inquiry regarding American laws of inheritance
                                               1906

39: Reichs- u. auswaert.Ang.: E.IV.41 [Erbschaften]
                                          ADS 2:343

WIND, Anna (from Koedrig, Oberpfalz [Bavaria]; died
in America). *See* Haeckl, Georg

WIND, Georg (farmer; from Koedrig, Oberpfalz [Bav-
aria]; died in America). *See* Haeckl, Georg

WINHEIM, Heinrich (in Philadelphia)

File on [estate?]   1893-1895

227: Rep. MA 1921: A.St.I: Nr. 1092.   ADS 9:71

WINIKER, -- (in Fort Oglethorp [state not given in
ADS description])

Estate of   1918

39: Reichs- u. auswaert.Ang.: E.IV.138 [Erbschaften]
                                          ADS 2:394

WINKE, C. M. O. (a minor)

Request of Hamburg Orphanage Administration to
have Winke, en route to America, delivered to
his stepfather by German consulate in New York
                                               1895

39: Reichs- u. auswaert.Ang.: W.I.4 [alt.Sig.]
[Waisenhausang.]   ADS 2:279

WINKELMANN, Wilhelm (in Hoboken, New Jersey)

Estate of   1910

39: Reichs- u. auswaert.Ang.: E.IV.54 [Erbschaften]
                                          ADS 2:362

WINKELMANN, --

New York: Estate of   1913

39: Reichs- u. auswaert.Ang.: E.IV.104 [Erbschaften]
                                          ADS 2:379

WINKLER, --

Estate of   1941

148: R XI.   ADS 6:262

WINTER, Franz Friedrich

Hamburg consulate in Richmond, Virginia: Information regarding Winter   1882

39: Bestand Senat: Cl.VI, No.16p: vol.4b: Fasc. 7c,
Invol. 1 [Ermittlung]   ADS 2:54

WINTER, H. (in Sedalia, Missouri)

Estate of   1918

39: Reichs- u. auswaert.Ang.: E.IV.147 [Erbschaften]
                                          ADS 2:394

WINTER, Johann (released convict)

Transportation to America   1843-1865

100: Gem. Oberndorf: XIII, 1: Nr. 3e.   ADS 4:312

WINTER [*correctly* Schmadel] John (in America)

Investigation of   1888

227: Rep. MA 1921: A.St.I: Nr. 1826.   ADS 9:74

WITTMANN, Johann (American soldier; from Mitwitz [Germany]; died 1862 in Corinth, Mississippi)

Back pay for   1886

227: Rep. MA 1921: A.St.I: Nr. 1272.   ADS 9:72

WITTMANN, Johann Adam (from Schoenbrunn, BYL)

Emigration of peasant's son to Upper Canada; unmarried   1848

222: Rep.132: Neustadt a.d.W.N.: Nr. 1203. ADS 9:3

WOECKENER, August Bernhard. *See* Jacke, August Bernhard, *alias* Woeckener

WOELKY, Carl (shoemaker)

Expulsion from Hamburg as Social Democrat & emigration to America   1879-1881; 1899

39: Polizeibehoerde-Polit.Pol.: S 149/507: Bl. 427-
460.   ADS 2:418

WOERLEN, Johann Gottfried (butcher; in Natchez [Miss-
issippi])

Estate of (& also that of his brother Johann Balthasar Woerlen)   1817-1818

227: Rep. Bayer. & Pfaelz. Gesandtschaft Paris:
Nr. 8379.   ADS 9:50

WOERLEN, Johann Balthasar. *See* Woerlen, Johann
Gottfried

WOERMER, Friedrich. *See* Woermer, -- (widow)

WOERMER, -- (widow; in New York)

Inquiry by Hamburg Orphanage Commission regarding
widow Woermer & her son Friedrich Woermer   1902

39: Reichs- u. auswaert.Ang.: W.I.22 [Waisenhaus-
ang.]   ADS 2:328

WOERRISHOFFER, --

Estate of   1943-1944

148: R XI.   ADS 6:262

WOHLERS, Heinrich

Chicago: Estate of   1880

39: Bestand Senat: Cl.VIII, No.X, Jg.1880 (Register) 38 [Nordamerika]   ADS 2:244

WOHLERS, Sophie

New York: Estate of   1914

39: Reichs- u. auswaert.Ang.: E.IV.13 [Erbschaften]
ADS 2:384

WOLF, Bernhard (in Seattle [Washington])

Estate of   1909

39: Reichs- u. auswaert.Ang.: E.IV.31 [Erbschaften]
ADS 2:357

WOLF, Louis Leo (Bavarian citizen? from New York)

Eccentric portable grinding device   1846

227: Rep. Ma 1895: A.V.: Nr. 719.   ADS 9:56

WOLF, Ph. H. (from Algenroth, HEL)

Emigration to America   1854

152: Abt.231: Langenschwalbach: Nr. 1287.
ADS 7:133

WOLF, William (from Pittsburgh)

Admission to insane assylum in Goettingen   1866

52: Han.Des.9: sog. Geheime Reg.: A: Nr. 49.
ADS 4:104

WOLFF, Felix (in New York)

Inquiry by private person in Germany regarding
1898

39: Reichs- u. auswaert.Ang.: N.Ib.1.7 [Justiz-
sachen]   ADS 2:297

WOLFF, H.(in New York)

Complaint of   1873

39: Bestand Senat: Cl.VIII, No.X, Jg.1873 (Journal) 13 [Auswanderer-Deputation]   ADS 2:222

Complaint against agent Aug. Behrens, Hamburg 1873

39: Bestand Senat: Cl.VIII, No.X, Jg.1873 (Register) 29 [Nordamerika]   ADS 2:225

WOLFF, Johann August (pastor of Raritan congregation)

Call to Raritan congregation   7 May 1734

39: Geistl.Minis.: II,6: Protokoll lit.A, 1720-
1745: p.76(3).   ADS 2:469

Ordination of   11 May 1734

39: Geistl.Minis.: II,6: Protokoll lit.A, 1720-
1745: p.77.   ADS 2:469

Letter, dated 1 August 1739, from congregation
complaining about Pastor Wolff 9 September 1740

39: Geistl.Minis.: II,6: Protokoll lit.A, 1720-
1745: p.135(2).   ADS 2:469

American files, particularly regarding Wolff's
dispute with congregation [256 pages] 1734-1747

39: Geistl.Minis.: III,A2k: Gebundene Akten: Nr.
VIII, pp.44ff.   ADS 2:477

[continued]

Letter from Pastor -- Berckemeyer regarding difficulties between Wolff & his congregation, including many appendices [appendices may contain names
of genealogical interest]   8 June 1742

<u>39</u>: Geistl.Minis.: II,6: Protokoll lit.A, 1720-
1745, p.150f(4); pp.270(XI); 272; 291; 292(III);
297; 298(I); 301(III); 303; 304(IV).

ADS 2:470

Letter from Pastor H. Kraeuter (Trinity Church,
London) regarding deposal of Wolff by an American synod   4 March 1746

<u>39</u>: Geistl.Minis.: II,7: Protokoll lit.B, 1746-
1758: pp. 5; 7(II); 62f(V).   ADS 2:470

WOLKAU, J. E.

German consular report from San Francisco regarding death & estate of   1872

<u>39</u>: Bestand Senat: Cl.VIII, No.X, Jg.1872 (Regis-
ter) 9 [Nordamerika]   ADS 2:220

WOLZE, Wilhelm (farmhand; from Klein Rhueden). *See*
Cleve, Johann Heinrich

WOOD, John

St. Louis: Information regarding   1880

<u>39</u>: Bestand Senat: Cl.VIII, No.X, Jg.1880 (Regis-
ter) 15 [Nordamerika]   ADS 2:243

WOOD, R. F.

Hamburg police search for Wood, wanted in America
1909

<u>39</u>: Reichs- u. auswaert.Ang.: P.III.1.3. ADS 2:358

WRAGE, -- (died in America)

Alleged estate of   1914

<u>39</u>: Reichs- u. auswaert.Ang.: E.IV.30 [Erbschaften]
ADS 2:384

WUELBERN, Anna (in Brooklyn [New York])

Estate of   1898

<u>39</u>: Reichs- u. auswaert.Ang.: E.IV.11 [alt.Sig.]
[Erbschaften]   ADS 2:296

WUELPERN, Hinrich (from Huetten)

Release from citizenship   1849
<u>62</u>: A, Fach 6, Vol. 3.   ADS 4:231

WUERDEMANN, -- (widow)

Correspondence with Bremen consul in St. Louis regarding   1853-1854

<u>48</u>: B.13.b.11: Nr. 6-15.
DLC has?   ADS 3:30

WUERFEL, C. R. (barber)

Consular report from New York regarding death &
estate of Wuerfel, aboard the *S.S. Hammonia* 1874

<u>39</u>: Bestand Senat: Cl.VIII, No.X, Jg.1874 (Regis-
ter) 32 [Nordamerika]   ADS 2:228

WUERTENBAECHER, Georg (died in New Orleans)

File on [estate?]   1830-1832

<u>227</u>: Rep. Bayerische Gesandtschaft London: Nr. 884.
ADS 9:47

WULF, Emilie Karoline, born Denker (in Chicago)

Inquiry by private person [in Germany] regarding
1896

<u>39</u>: Reichs- u. auswaert.Ang.: N.Ib.19 [alt.Sig.]
[Nachforschungen]   ADS 2:284

WULF, Johs. (in Chicago)

Hamburg Orphanage Commission seeks information regarding financial condition of   1896

<u>39</u>: Reichs- u. auswaert.Ang.: W.I.4 [alt.Sig.]
[Waisenhausang.]   ADS 2:287

WULFF, C. E. H. (in New York)

Civil registry matter [marriage?]   1872

<u>39</u>: Bestand Senat: Cl.VIII, No.X, Jg.1872 (Regis-
ter) 34 [Nordamerika]   ADS 2:221

WULLNER, -- (teacher; in Buch i.E.)

Inquiry regarding a [U.S.?] military pension for
himself   1894

<u>227</u>: Rep. MA 1921: A.St.I: Nr. 1827a.   ADS 9:74

WUNSCH, Ernst (assistant switchman)

Extradition from USA to Schneidemuehl, via Hamburg   1909

<u>39</u>: Reichs- u. auswaert.Ang.: A.Ia.11 [Ausliefer-
ungen]   ADS 2:355

WURKLER, Joh[ann] (in USA)

Estate of Wurkler claimed by Barbara Pachmann, in
Boul, Alsace   n.d.

<u>227</u>: Rep. Bayer. & Pfaelz. Gesandtschaft Paris:
Nr. 8550.   ADS 9:50

WURZBACH, Franz Justus

Deposit in Mannheim [of an emigrant?] & his claim
for damages   1846-1853
154: CA4c3: Nr. 5.   ADS 7:179

WUTTKE, Adolf (an American citizen)

Exemption from [German] military service obligation
                                         1915
39: Reichs- u. auswaert.Ang.: M.II.4 [Militaer-
wesen]   ADS 2:389

YORK, John

San Francisco: Estate of   1896
39: Reichs- u. auswaert.Ang.: E.IV.68 [alt.Sig.]
[Erbschaften]   ADS 2:282

YOUNG, --

Estate of   n.d.
148: R XI.   ADS 6:262

ZABEL, Emilie. *See* Berner, Emilie, born Zabel

ZADIK, Zadek Peritz

Expulsion of a former Prussian subject, now an
American citizen   1897-1898
39: Bestand Senat: Cl.VII, Lit.L^b, No. 28.2, vol.
110 [Ausweisung]   ADS 2:93

ZAHLHAS, Georg (in Chicago)

File on [estate?]   1897
227: Rep. MA 1921: A.St.I: Nr. 1214.   ADS 9:71

ZAHN, Christine. *See* Zahn, Friedrich

ZAHN, Friedrich (in New York)

File on Friedrich & Christine Zahn   1893
227: Rep. MA 1921: A.St.I: Nr. 1207.   ADS 9:71

ZANONI, Karl (from Amorbach; died in America)

Death & estate of   1864-1865
227: Rep. MA 1921: A.St.I: Nr. 1845.   ADS 9:75

ZECHOW, Jacob (in State of Washington)

Estate of   1895
39: Reichs- u. auswaert.Ang.: E.IV.25 [alt.Sig.]
[Erbschaften]   ADS 2:275

ZEH, Michael

Philadelphia: Estate of   1880
39: Bestand Senat: Cl.VIII, No.X, Jg.1880 (Register) 14 [Nordamerika]   ADS 2:243

ZEIDLER, Wilhelm Oswald (coal barger worker)

Expulsion from Hamburg as Social Democrat & emigration to America   1882-1883
39: Polizeibehoerde-Polit.Pol.: S 149/412: Bl.
506-519.   ADS 2:418

ZEIER, John

St. Louis: Estate of   1917
39: Reichs- u. auswaert.Ang.: E.IV.31 [Erbschaften]
                                         ADS 2:392

ZEINER, Wendelin

San Francisco: Estate of   1906
39: Reichs- u. auswaert.Ang.: E.IV.32 [Erbschaften]
                                         ADS 2:343

ZEITLER, Anna. *See* Glaessl, Johann Michael

ZEITZ, Albert

Chicago: Estate of   1918
39: Reichs- u. auswaert.Ang.: E.IV.152 [Erbschaften]
                                         ADS 2:394

ZELLER, --

Estate of   1944
148: R XI.   ADS 6:262

ZELZNER, Anna (in New York)

Inquiry regarding her   1900
227: Rep. Ma 1921: A.St.I: Nr. 1848.   ADS 9:75

ZENGER, Katharina, born Haept (in New Orleans)

Request for information regarding the life & death
of her parents   1869-1870
227: Rep. MA 1921: A.St.I: Nr. 1847.   ADS 9:75

ZENK, Marie (in Buffalo [New York])

File on [estate?]   1900-1901
227: Rep. MA 1921: A.St.I: Nr. 1220.   ADS 9:71

ZENTNER, H. (businessman)

Extradition to Nuernberg from the USA, via Hamburg   1910

<u>39</u>: Reichs- u. auswaert.Ang.: A.Ia.18 [Auslieferungen]   ADS 2:361

ZERPIES, Paul (in Brooklyn [New York])

Estate of   1901

<u>39</u>: Reichs- u. auswaert.Ang.: E.IV.36 [Erbschaften]   ADS 2:315

ZERRENNER, Johannes (carpenter; from Volkersheim)

Emigration to America   1851

<u>56</u>: L Neu Abt.129A: Gr.11, Nr. 82.   ADS 4:191

ZEYN, J. H. (in Beemer, Nebraska)

Estate of   1910

<u>39</u>: Reichs- u. auswaert.Ang.: E.IV.8 [Erbschaften]   ADS 2:362

ZIEG, Georg (died 1896 in Illinois)

File on [estate?]   1898-1899

<u>227</u>: Rep. MA 1921: Nr. 1217.   ADS 9:71

ZIEGLER, Andreas (died in New York)

Estate of   1910

<u>152</u>: Abt.405: Wiesbaden: Rep.405/I,2: Nr. 3541.
                                                ADS 7:147

ZIEGLER, Christian (in New York)

File on [estate?]   1890

<u>227</u>: Rep. MA 1921: A.St.I: Nr. 1201.   ADS 9:71

ZIEGLER, Martin (from Sondheim, Thuringia)

Emigration of   1840

<u>500</u>: Weimar: Thueringisches Staatsarchiv: Dermbach
     II: Gb.2: Nr. 63-92; 93-114; 144-183; 186;
     188.   Not in ADS

ZIMMERMANN, Georg (servant; in Philadelphia)

Estate of   1837

<u>227</u>: Rep. Bayer. & Pfaelz. Gesandtschaft Paris:
     Nr. 8510.   ADS 9:50

ZIMMERMANN, Paul Eduard Hermann (cigar dealer)

Expulsion from Hamburg as Social Democrat & emigration to America [includes list of 42 Social
     Democrats expelled from Altona]   1880-1881

<u>39</u>: Polizeibehoerde-Polit.Pol.: S 149/227: Bl.
     568-585.   ADS 2:415

ZIMPEL, Theodor (poet; from Land Hadeln)

An article entitled "Theodor Zimpel, ein Sohn
     unserer Heimat, als Dichter in den Vereinigten
     Staaten" [Theodor Zimpel, a son of our land, as
     poet in the USA] *Mitteilungen des Stader Geschichts- und Heimats-Vereins*, 33 Jahrgang (1958)
     Heft 2-3, pp. 43-45.   ADS 4:335

ZINCK, Johann Heinrich Friedrich (machinist)

Expulsion from Hamburg as Social Democrat & emigration to America   1880-1881

<u>39</u>: Polizeibehoerde-Polit.Pol.: S 149/32: Bl. 586-
     596.   ADS 2:412

ZINK, -- (engineer; in Texas)

Service contract with Texas Verein [German colonization venture]   1844

<u>154</u>: CA4c1: Nr. 23.
     DLC has.   ADS 7:176

ZINZENDORF, Nikolaus Ludwig *Graf* [Count](born Dresden 26 May 1700; died Herrnhut 9 May 1760; founder
     of the Moravian Brethren)

Correspondence regarding Georgia   1734-1735; 1737
Trip to Pennsylvania   1741-1747
Trips to Pennsylvania   1742-1760
Sojourn in Bethlehem [Pennsylvania]   1742
Letters from North America   1742-1743; 1745-1754

<u>500</u>: Herrnhut: Archiv: Rep.14.A: Nr. 3; 13; 14; 15;
     16; 17.   Not in ADS

Mentioned in letter from Roxbury   1724

<u>119</u>: Litt.K: Nr. 36.   ADS 6:74

ZINZENDORF, Nikolaus Ludwig *Graf* [Count]. *See* Wied,
Duchy of [Geographic Index]

ZIPPLING, Julius (handworker; from Gandersheim)

Convicted because of his *Lebenswandel* [manner of
     living, not otherwise described]; confined to
     house of correction in Bevern; emigration to
     America   1858-1859

<u>56</u>: L Neu Abt.129A: Gr.11, Nr. 34.   ADS 4:187

ZIPPRICH, Karl. *See* Pfannenschlag, Therese

ZOELLER, August (from Frankenthal; died in New York)

Request for a death certificate for   1866

<u>227</u>: Rep. MA 1921: A.St.I: Nr. 1846.   ADS 9:75

ZOELLIKOFFER, Elfriede, born Faber. *See* von Melle, Emil

ZOELLIKOFFER, Oskar (in New York). *See* von Melle, Emil

ZOLLES, Johann (in Baltimore)

File on [estate?]   1886
<u>227</u>: Rep. MA 1921: A.St.I: Nr. 1194.   ADS 9:71

ZORN, Georg Heinrich (died in Cincinnati, Ohio)

Estate of   1901

<u>152</u>: Abt.405: Wiesbaden: Rep.405/I,2: Nr. 3543.
ADS 7:147

ZSCHOCKE, Emilie (died in Chicago, Illinois)

Estate of   1895
<u>152</u>: Abt.405: Wiesbaden: Rep.405/I,2: Nr. 3568.
ADS 7:147

ZWAHL, Maria (butcher's wife; in New York)

Petition for diplomatic assistance to make possible
her return to Germany   1893
<u>227</u>: Rep. MA 1921: A.V. II: Nr. 3989.   ADS 9:81

Geographic Index

ACHERN (village)   BWL

  Emigration files   1818-1882; 1901-1931
  288: Kommunalarchiv Achern.   ADS 10:99

ACHKARREN (village; Kreis Freiburg)   BWL

  Emigration   1840-1848; 1857; 1872-1891
  288: Kommunalarchiv Achkarren.   ADS 10:99

ACKERHOLZ (Amt)   HEL

  Residency permits   1833-
  151: Bestand 82: Acc.1882/17: Nr.51.   ADS 7:85

ADELSREUTE (village; Landkreis Ueberlingen)   BWL

  Emigration files   1841-1938
  288: Kommunalarchiv Adelsreute.   ADS 10:99

ADENAU (Kreis)   RPL

  Naturalizations   1821-1872; 1902-1913
  200: Abt.441: Regierung Koblenz: Nr. 23560; 23669.
                                            ADS 8:27

AFFOLTERBACH (village)   HEL

  Emigration   1864-1902
  149: Rep.40,2: Ablief.2: Bd.291.   ADS 7:22

AHAUS (Kreis)   RWL

  List of certificates of citizenship issued
                                         1912-1922
  116: Muenster Reg.: Nr. 3405.   ADS 6:56

  Emigration   1892-1929
  116: Muenster Reg.: Nr. 1780-1781.   ADS 6:56

  Nr. 1059: Inheritances by foreigners   1880-1912
  Nr. 344: Death certificates for persons dying
            abroad   1817-1885
  Nr. 2099: Passports   1816-1820
  Nr. 2100: Passports   1816-1837
  Nr.   10: Emigration files   1817-1857
  Nr.   91: Certificates of original domicile &
            citizenship   1838-1912
  Nr.   90: Ex post facto issuance of certificates
            of original domicile   1840-1878
  116: Muenster Reg.: Ahaus: 1.A: files as above.
       DLC has Nr. 10 only.   ADS 6:64

  Nr.   60: Certificates of original domicile &
            citizenship   1873-1911
  Nr. 317: Passports   1888-1910
  Nr.   32: Foundling children   1910-1912
  116: Muenster Reg.: Ahaus: files as above.
                                          ADS 6:64

  Nr. 182: Foreign residents   1896-1912
  Nr. 1263: Foreigners subject to military service
                                         1914-1938

  Nr. 663: Expulsions, vol. 1   1898-1911
            Expulsions, vol. 2   1911-1912
  116: Muenster Reg.: Ahaus: 1.A: files as above.
                                          ADS 6:65

AHAUSEN (village; Amtsgericht Ueberlingen)   BWL

  Estate files   1832-1942; 1888-1934
  288: Kommunalarchiv Ahausen   ADS 10:99

AHRENSBOEK, see Ploen & Ahrensboek   SHL

AHRWEILER (Kreis)   RPL

  Naturalizations   1820-1913
  200: Abt.441: Regierung Koblenz: Nr. 21973; 21981;
       21974-21978.   ADS 8:27

AICHEN (village)   BWL

  Emigration   1862-
  288: Kommunalarchiv Aichen.   ADS 10:99

ALBBRUCK (village)   BWL

  Emigration   1870-1941
  288: Kommunalarchiv Albbruck.   ADS 10:99

ALBERSDORF   SHL

  Nr. 460: Emigration files   1889-1897
  Nr. 461: Petitions for, & losses of, citizenship
                                         1889-1905
  3: VIII: file numbers as above.
     DLC has?   ADS 1:42

ALBISHEIM (village; Landkreis Kirchheimbolanden; Kurpfalz)   RPL

  Names of emigrants   18th century
  201: Archivabt. Ausfautheiakten   ADS 8:63

ALKEN (village)   RPL

  Naturalizations & releases from citizenship
                                         1819-1903
  200: Abt.655,7: Brodenbach: Nr. 202.   ADS 8:50

ALLENSBACH (village; Landkreis Konstanz)   BWL

  Emigration files   1865-1868; 1883-1912; 1900-1931
  288: Kommunalarchiv Allensbach.   ADS 10:99

ALLMANSWEILER (village; Kreis Lahr)   BWL

  Emigration; naturalization   1829-1933
  288: Kommunalarchiv Allmansweiler.   ADS 10:100

ALTENA (Landratsamt)   RWL

   Nr. 467: Gains & losses of Prussian citizenship
                                      1843-1882
   Nr. 1133: Gains & losses of Prussian citizenship
                                      1905-1913

   116: Arnsberg: files as above.   ADS 6:57

   Passports, naturalizations, certificates of orig-
     inal domicile   1863-1913
   116: Arnsberg: Nr. 1140; 1139; 1143; 468-469; 1138;
     1145; 1135.   ADS 6:57

   Certificates of citizenship   1905-1913
   116: Arnsberg: Nr. 1144.   ADS 6:57

   Emigration & certificates of original domicile;
     applications for releases from citizenship
                                       1850-1875
   117: Nr. 616.   ADS 6:72

   Emigration   1850-1905
   116: Arnsberg: Nr. 1141; 1136; 1148.   ADS 6:57

ALTENDAMME (Oldenburg) See Oldenburg   NL

ALTENDIEZ (Gemeinde; Kreis Limburg)   HEL

   Naturalizations & releases from citizenship
                                        1847-1852
   152: Abt.232: Limburg: Nr. 2010.   ADS 7:134

ALTENGRONAU   HEL

   Correspondence re emigration, mainly illegal; gives
     grounds for emigration [may contain names of some
     emigrants]   1792
   151: [Bestand 80?] Rubrik XXXIII Manumissiones:
     Lit.D.   ADS 7:79

ALTENHASSLAU (Amt)   HEL

   Emigrants   1741
   151: Bestand 86: Nr. 2085.   ADS 7:89

   Emigration [reverse pages contain names] 1751
   151: Bestand 81: Rep.Ba: Gef.9, Nr. 8.   ADS 7:80

   Exemptions from emigration tax. See Hesse-Kassel.

ALTENKIRCHEN (Kreis)   RPL

   Naturalizations   1857-1913
   200: Abt.441: Regierung Koblenz: Nr. 21982-21986.
                                          ADS 8:27

ALTENKIRCHEN (village; Landkreis Waldmohr, Kurpfalz)
                                                       RPL
   Names of emigrants   18th century
   201: Archivabt. Ausfautheiakten.   ADS 8:63

ALTENKIRCHEN. See Sayn-Altenkirchen

ALTENMUHR bei Gunzenhausen. See Rehn, Jacob   BYL

ALTHEIM (village; Landkreis Ueberlingen)   BWL

   Emigration   1860-1876; 1888-1939
   288: Kommunalarchiv Altheim.   ADS 10:100

ALTOETTING (Landratsamt)   BYL

   List of passports and pass cards issued 1887-1896
   229: Rep. Landratsaemter: Altoetting: 46.979.
                                          ADS 9:117

ALTON (Madison County, Illinois)   USA

   Inheritance case [name not given in ADS descrip-
     tion]   1858
   39: Bestand Senat: C1.VIII. No.X, Jg. 1858 (Regis-
     ter) II.1.   ADS 2:174

ALTONA   HH

   Ship muster rolls; Seamen's mustering out of ser-
     vice, 17 vols.   1764-1850
   40: Bestand 2.VII: Prot. Wasserschouts: b.
                                          ADS 2:501

   Ship muster rolls: seamen mustering aboard, 14
     vols   1787-1850
   40: Bestand 2.VII: Prot. Wasserschouts: a.
                                          ADS 2:501

   List of Altona men subject to sea duty [records
     news from America regarding deaths, desertions,
     estates of seamen]   1815-1842
   40: Bestand 102-105: Verzeichnis.   ADS 2:502

   Alphabetical name list of seamen and estates regis-
     tered in the Altona office of the Hamburg Wasser-
     schout   1802-1847
   39: Bestand Wasserschout: III.J.   ADS 2:454

   List of 42 Social Democrats expelled from Altona
     [included in file of Paul Eduard Hermann Zimmer-
     mann]   [1880-1881]
   39: Polizeibehoerde-Polit.Pol.: S 149/227:
     Bl. 568-585.   ADS 2:415

ALTSCHWEIER (village; Kreis Buehl)   BWL

   Emigration   1824-1873; 1854-1949
   288: Kommunalarchiv Altschweier.   ADS 10:100

ALTSIMONSWALD (village; Kreis Emmendingen)   BWL

   Emigration   1853-1893; 1904-1943
   288: Kommunalarchiv Altsimonswald.   ADS 10:100

ALZENAU (Amtskellerei; Kurmainz)   RPL

Reports of the confiscation of the property of
citizens who have emigrated secretly [illicitly];
list of debts incurred by emigrants   1789-1791

232: Rep. 77: Aschaffenburger Archivreste: Ger.
Alzenau: Fasz.15/1573   ADS 9:137

AMBDORFF. *See* Dillenburg

AMBERG (Land Oberpfalz)   BYL

Nr. 1428: Emigration & immigration 1802; 1809-1810
Nr. 1442: Emigration & immigration   1807-1808
Nr. 1454: Emigration & immigration   1802
Nr. 1491: Emigration & immigration   1803-1805
Nr. 1508: Emigration & immigration   1808-1809
Nr. 1558-1561: Emigration & immigration 1802-1806
Nr. 8832: Emigration   1803

222: Rep. Oberpfaelzische Administrativakten:
files as above.   ADS 9:1

AMBERG (Landgerichtsamt)   BYL

Citizens [*Untertanssoehne*=sons of subjects] who
have escaped [i.e., emigrated illicitly to avoid
military service obligation] in the last 24
years, 1-15   1780-1805

222: Rep.123: Amberg: Nr. 264.   ADS 9:2

Activities of foreign emissaries [religious mis-
sionaries], Nr? 1, 7, 8   1855; 1859; 1864

AMBERG (city)   BYL

Taxation of emigration & immigration   1873-1888

222: Regierung. Kammer des Innern: Rep.116: Nr.
11517.   ADS 9:1

AMOLTERN (village; Kreis Emmendingen)   BWL

Emigration   1883-1916

288: Kommunalarchiv Amoltern.   ADS 10:100

AMORBACH   BWL

Material of an undescribed nature regarding emigration to the USA is available (one fascicle
has to do with the Texas-Gesellschaft [Texas
Company], in which a number of noblemen were
stockholders)   [19th century?]

233   ADS 9:139

ANDERNACH (city)   RPL

There is a card catalog of citizens who have emigrated   20th century

202: Zettelsammlung StA. Weidenbach.   ADS 8:68

ANGERBURG (Kreis)   East Prussia (Gumbinnen)

Applications & permits to emigrate   1832-1874

51: Rep.12: Abt.I, Titel 3, Nr. 8, Bd. 1. ADS 4:32

ANNWEILER (city)   RPL

Two letters regarding emigrants   1764; 1817

203: Alte Akten: Nr. 54.   ADS 8:69

Note: The Neuere Archivabteilung (1850-1930) also
contains some information on emigrants.

ANSBACH (Oberamt)   BYL

38: Files of the Landgericht regarding Jewish persons   1807-1873
87: Formalities to be observed for the renunciation
of the Jewish faith [*Abschwoerung des Juden-
eides in der Synagoge*]   1747

231: Rep. 212: Bezirksamt Ansbach: Rep.212/1:
files as above.   ADS 9:128

Emigration & immigration   1862

231: Rep. 270/II: Reg. Mittelfranken: Kammer des
Innern: Abgabe 1932: 686.   ADS 9:134

ANSBACH (Emigration collection point [*Lager*])   BYL

Emigration & immigration, vols I, II, III (with
appended papers for latter two volumes)
1846-1854

231: Rep. 270/II: Reg. Mittelfranken: Kammer des
Innern: Abgabe 1932: 23.   ADS 9:134

APENRADE (Landratsamt)   SHL

Bemerk. 112: Emigration files   1874-1890
Bemerk. 77: Emigration files   1891-1905

2: Abt.320 (Apenrade): files as above.
DLC has?   ADS 1:25

APPENHOFEN (Kreis Billigheim; Kurpfalz)   RPL

List of emigrants   18th century

201: Archivabt. Ausfautheiakten.   ADS 8:62

ARFELD (Amt; Kreis Wittgenstein)   RWL

Nr. 113: Emigration & immigration   1845-1855
Nr. 114: Emigration & immigration   1855-1866

116: Arnsberg Reg.: Wittgenstein: files as above.
ADS 6:62

ARFELD. *See also* Wittgenstein

ARLEN (village). *See* Rielasingen   BWL

ARNSBERG (Regierung)   RWL

Estates of persons dying abroad   1914-1919

116: Regierung Arnsberg: I 8 Nr. 1.   ADS 6:48

ARNSBERG (Kreis)   RWL

   Nr. 2: Gains & losses of citizenship   1914-1930
   Nr. 3: Certificates of original domicile
          & citizenship   1914-1930
   Nr. 4: Certificates of original domicile
          & citizenship   1914-1930
   116: Arnsberg Reg.: B Fach 31: files as above.
                                           ADS 6:58

   Nr. 308: Certificates of original domicile
            & citizenship   1907-1912
   Nr. 309: Gains & losses of citizenship 1903-1913
   116: Arnsberg Reg.: A: files as above. ADS 6:58

AROLSEN (Landratsamt)   HEL

   Nr. 235: List of emigrants to America   n.d.
   Nr. 248: Emigration   n.d.
   Nr. 334a: Emigration   n.d.
   Nr. 78: Emigration   n.d.
   Nr. 79: Naturalizations; certificates of
            original domicile   n.d.
   151: Bestand 180: Arolsen: Acc.1937/76,5: files as
        above.   ADS 7:113

   Releases from citizenship   n.d.
   151: Bestand 180: Arolsen: Acc.1937/76,5: Nr. 88;
        95; 116-117; 125; 135; 167; 190; 255; 308;
        316; 412; 162; 177-179; 182-185; 565.
                                           ADS 7:113

   Nr. 37: Emigration (11 packages of documents) n.d.
   Nr. 38: Notarizing of papers for citizens
           abroad   n.d.
   Nr. 39: Losses of citizenship   n.d.
   151: Bestand 180: Arolsen: Acc.1937/76,1: Fach 2:
        files as above.   ADS 7:112

ASCHAFFENBURG (Vicedomsamt [cathedral office]) BYL

   Fasz. 34/XXI: Emigration to Hungary and America;
                 applications for permits to emi-
                 grate; reports of secret [illicit]
                 emigration   1801-1803
   Fasz. 205/II: Passports   1815-1816
   232: Rep. 77: Aschaffenburger Archivreste: files
        as above.   ADS 9:136

   Emigration & the exportation of personal property
                                           1814-1817
   232: Rep. 77: Aschaffenburger Archivreste: Adm.
        Fasz. 747/17332.   ADS 9:136

   Applications for emigration permits and decisions
        thereon   1814-1816
   232: Rep. 77: Aschaffenburger Archivreste: Adm.
        Fasz. 694/15663.   ADS 9:136

   Emigration of persons subject to military service
                                           1815
   232: Rep. 77: Aschaffenburger Archivreste: Adm.
        Fasz. 747/17335.   ADS 9:136

ASCHBACH   HEL

   Emigration   1864-1902
   149: Rep. 40,2: Ablief.2: Bd. 291.   ADS 7:22

ASCHENDORF (Kreis)   NL

   Releases from citizenship   1887
   54: Rep.116,I: Nr. 13418.   ADS 4:123

ASCHENDORF (Amt)   NL

   Emigration permits [for permits before 1871,
        see Osnabrueck]   1871-1879
   54: Rep.116,I: Nr. 6067.   ADS 4:120

   Releases from citizenship   1880-1886
   54: Rep.116,I: II.4.a.1e: Nr. 75.   ADS 4:121

ASSENHEIM (village; Grafschaft? Leiningen-Hardenburg)
                                                    RPL

   Inventories; partitions; estates   1760?-1796
   201: Archivabt. Leiningen-Hardenburg: Akt Nr. 133-
        135.   ADS 8:62

ASTHEIM (Landkreis Gross-Gerau)   HEL

   Register of emigrants to North America 1837-1887
   Applications for release certificates for purpose
        of emigration to America   1839-1915
   161: Abt.XI: Abschnitt 4.   ADS 7:211

AU (village; Kreis Freiburg)   BWL

   Emigration   1853-1942
   288: Kommunalarchiv Au.   ADS 10:100

AUENHEIM (village; Kreis Kehl)   BWL

   Emigration   1880-1933; 1882-1937; 1914-1939
   288: Kommunalarchiv Auenheim.   ADS 10:100

AUF DEN CAMP (congregation? New York State)   USA

   See Rhynbeck (congregation; New York State)

AUERBACH.   See Heppenheim (Kreis)   HEL

AUGGEN (village)   BWL

   Emigration   1850-1935; 1862-1878; 1890-1931
   288: Kommunalarchiv Auggen.   ADS 10:100

AUGSBURG (city)   BYL

   Emigration files   19th century

235: S.D.A.A., Sachregister, Abg.1930: Nr. 40-41;
44-49; 70: A 65 G II 10/11.   ADS 9:142

AUGSPURG. *See* Karlsruhe (amt)

AURICH (Ostfriesland)   NL

Files on the granting & loss of citizenship, as
follows:

File	Letter	Year	File	Letter	Year
105	A	1842	117	N	1866
106	B	1833	118	O	1838
107	C	1858	119	P	1867
108	D	1850	120	R	1828
109	E	1838	121	S	1850
110	F	1866	122	T	1865
111	G	1828-1873	124	U	1869
112	H	1829	125	V	1868
113	I-J	1840	126	W	1841
114	K	1836	127	Z	1853
115	L	1868			
116	M	1828			

50: Rep.26a: Abt.II.G: Files as above.   ADS 4:11

Naturalization & renaturalization, 30 vols.
                                       1895-1931
50: Rep.21a: file numbers, as listed in ADS.
                                       ADS 4:6

Releases from Hannoverian citizenship & liberation
from military service obligations   1824-

50: Rep.6: Dimissoriales: 3.   ADS 4:2

Nr. 5570: Loss of citizenship, vol. 1   1870-1896
Nr. 5594: Loss of citizenship, vol. 2   1914-1928
Nr. 5240: Loss of Reich citizenship, vol. 1
                                       1901-1912

50: Rep.21a: files as above.   ADS 4:6

Emigration files   1784-

50: Rep.6: Dimissoriales:(Best. Nr. 4391). ADS 4:2

Emigration permits   1824-

50: Rep.6: Dimissoriales.   ADS 4:2

Reg.Nr. 21-04; Lauf.Nr. 157: Emigration permits
                                       1867-1878
Reg.Nr. 21-05; Lauf.Nr. 154: Citizenship & resi-
dency certificates   1878-1926
Reg.Nr. 21-10; Lauf.Nr. 4: Emigration permits
                                       1879-1926

50: Depositum XXXIV: I: files as above.   ADS 4:16

Emigration permits, as follows:

File	Letter	Vol.	
9028	A	I	1867-1887
8829	A	II	1888-1906
9029	B	I & II	1865-1886
9030	B	III	1887-1892
9031	B	IV	1893-1912
9032	C	I	1867-1887
9033	C	II	1888-1913
9034	D	I	1868-1887
9035	E	I	1842-1867
9036	E	II	1888-1919
8831	F	I	1867-1880

File	Letter	Vol.	
9037	F	II	1881-1887
9038	F	III	1888-1907
9039	G	I	1868-1880
9040	G	II	1881-1887
9041	G	III	1888-1913
9042	H	I	1842-1880
9043	H	II	1888-1913
8827	J	II	1888-1914
9044	J	III	1895-1913
8828	K	II	1881-1887
9045	K	III	1888-1912
9046	L	II	1888-1914
9047	M	I	1867-1884
9048	O	I	1868-1887
9049	P	I	1868-1887
9050	P	II	1888-1913
9051	R	I	1842-1857
9052	R	III	1888-1913
9053	S	III	1888-1894
9054	S	IV	1895-1914
9055	U	I	1886-1901
9056	V	-	1867-1914
9057	W	II	1881-1887

50: Rep.21a: files as above.   ADS 4:10

Emigration, vols. II & III   1866-1936

50: Rep.21a: Nr. 9027; 5856.   ADS 4:9

Transit of emigrants   1817

50: Rep.11: VII.2.530 [Durchzug]   ADS 4:5

Deaths of Aurich citizens abroad   1820-

50: Rep.26b: Abt.II,G,1: 34 [Ableben]   ADS 4:12

Death certificates for persons dying abroad 1825-

50: Rep.6: Varia: Gen.: 65 (Best.Nr. 14498-14499).
                                       ADS 4:5

Files on the estates of missing persons & on those
dying abroad   1814-1858

50: Rep.6: Erbschafts-Sachen: (Best.Nr. 4466).
                                       ADS 4:2

Nr. 9493-9536: Estates of Germans dying abroad,
               vol. I   1887-1928
Nr. 9492: Estates of Germans dying abroad, vol. II
                                       1905

50: Rep.21a: files as above.   ADS 4:6

Overseas settlement of asocial persons   1865

50: Rep.26a: Abt.VIII,G,5: 1093.   ADS 4:12

AUSTRIA. *See* Galicia

S.S. *AUSTRIA*

List of emigrants saved & lost at sea from ship
                                       1859

39: Bestand Senat: Cl.VIII, No.X, Jg.1859: Register 1 [Auswanderer-Verhaeltnisse]   ADS 2:175

BACHARACH (city)   RPL

Data on a number of women, children, and movable

property taken to the "Island of Pennsylvania"
by emigrants   1709

200: Abt.613: Stadt Bacharach: Nr. 52, Blatt 43 u.
     44.   ADS 8:45

BACKNANG (Oberamt)   BWL

59a: Lists of emigrants   1817-1884
60-82: Individual emigration cases, A-Z
                                      19th century
59c: Applications for public assistance for emi-
      gration to North America   1849-1850

261: F 152: Backnang: files as above.   ADS 10:16

BAD EMS (baths)   HEL

Treatment & guest lists   1867-1896

152: Abt.405: Wiesbaden: Rep.405/I,4: Nr. 1217.
                                        ADS 7:148

BADEN (Kingdom)   BWL

Emigration rights & emigration affairs [contains
   general regulations 1536-1798]   1536-1813

260: Rep.Abt.74: Baden Gen.: Fasz. 20-232.
                                       ADS 10:10

Subjects who have emigrated without permits & the
   confiscation of their properties   1739

260: Abt.236: Innenmin., Wegzug: Auswand.: 6529.
                                        ADS 10:4

Emigration to America   1727-1809

260: Rep.Abt.74: Baden Gen.: Fasz. 9847-9856.
                                       ADS 10:10

Emigration files   1762-1808

260: Abt.236: Innenmin., Wegzug: Auswand.: 6538-
     6541.   ADS 10:4

Emigration files   1804-

260: Rep.Abt. 332: Polizeidirektion Baden.
                                       ADS 10:15

933.IV.1: Emigration permits issued   1809-1826
935:      Emigration to the various parts of
          America   1816-1832
1029:   Emigration to Russia and other countries
                                       1824-1827

260: Abt.236: Innenmin., Wegzug: Auswand.: files
     as above.   ADS 10:3

Emigration to various parts of America, parts 2-4
                                       1818-1847

260: Abt.236: Innenmin., Wegzug: Auswand.: 639-641
                                        ADS 10:1

Emigration to the American states   1817-1864
260: Zugang 1891: Nr. 24. V. Gen.   ADS 10:1

Helpless emigrants abroad; medical & food costs,
   I. part   1842-1850

260: Abt.236: Innenmin., Wegzug: Auswand.: 643
                                        ADS 10:1

Legal action [in Rheinprovinz] against emigrants
   from southern Germany to America   1827-1851

200: Abt.441: Regierung Koblenz: Nr. 5106-5109.
                                        ADS 8:32

Emigration to America via Belgium   1847-1869

260: Rep.Abt.233: Minist. des Innern: Fasz. 14970.
                                       ADS 10:13

Emigration to America via Bremen   1832-1866

260: Rep.Abt.233: Minist. des Innern: Fasz. 14971.
                                       ADS 10:13

Emigration to America via England   1855-1878

260: Rep.Abt.233: Minist. des Innern: Fasz. 5389.
                                       ADS 10:13

Emigration to America via France   1831-1867

260: Rep.Abt.233: Minist. des Innern: Fasz. 14972.
                                       ADS 10:13

Germany emigrants in LeHavre   1847-1849

260: Rep.Abt.233: Minist. des Innern: Fasz. 5387.
                                       ADS 10:12

Emigration to North America via The Netherlands
                                       1828-1866

260: Rep.Abt.233: Minist. des Innern: Fasz. 5388.
                                       ADS 10:12

Emigration to Canada   1905-1925

260: Rep.Abt.233: Minist. des Innern: Fasz. 11405.
                                       ADS 10:13

Emigration to Texas (German Colonial Society in
   Biebrich)   1848-1855

260: Rep.Abt.233: Minist. des Innern: Fasz. 2619.
                                       ADS 10:13

Miscellaneous matters--baptisms, education of children, marriages   1766

260: Abt.236: Innenmin., Wegzug: Auswand.: 4229.
                                       ADS 10:3

Notifications of inheritances [in Germany] of persons who have emigrated to America 1834-1850

260: Zugang 1899: Nr. 53. II. Gen. 47. ADS 10:1

BADENWEILER (town)   BWL

Emigration files   1752-1902; 1902-1934; after 1945

288: Kommunalarchiv Badenweiler   ADS 10:101

BAD GRIESBACH (village)   BWL

Immigration; certificates of original domicile
                                       1819-1863

288: Kommunalarchiv Bad Griesbach   ADS 10:100

BAD HOMBURG (baths)   HEL

Treatment & guest lists   1867-1896

152: Abt.405: Wiesbaden: Rep.405/I,4: Nr. 1217.
                                        ADS 7:148

BAD KROZINGEN (village)   BWL

   Emigration   1849-1947; 1880-1942
   Naturalizations   1898-1934
   Emigration   1905-1917
   Naturalizations; citizenship; emigration   1948

   288: Kommunalarchiv Bad Krozingen.   ADS 10:100

BAD LANGENSALZA (town)   Saxony

   Emigration file on the Langen Salzer Company   1706

   500: Dresden: Saechsisches Hauptstaatsarchiv: Loc.
      9905 [film]. Learned, 299.   Not in ADS

BAD PETERSTAL (town)   BWL

   Emigration files   1840; 1847-1855; 1877-1896
   Naturalizations   1918-1933
   Releases from [German] citizenship   1918-1943
   Financial aid to Germans abroad; list of citizens
      residing abroad   1920-1939

   288: Kommunalarchiv Bad Peterstal.   ADS 10:101

BAD PYRMONT (town)   NL

   A.VIII.18: Emigration to America   1847-1864
   A.VIII.19: Emigration to California   1885
   OA.V.9:   Emigration to America   1846-1847

   101: files as above.   ADS 4:329

BAD RIPPOLDSAU (town)   BWL

   Emigration files   1900?-1950

   288: Kommunalarchiv Bad Rippoldsau.   ADS 10:101

BAD SALZUFLEN (town)   NL

   Applications for emigration permits (with state-
      ments as to age, military service obligation,
      assets, reputation)   1859-1869; 1884-1887
   104   ADS 4:331

BAD SCHWALBACH. *See* Langenschwalbach

BAHLINGEN (town)   BWL

   Emigration files   1864-1890
   Alphabetical register of emigrants overseas
                                               19th century

   288: Kommunalarchiv Bahlingen.   ADS 10:101

BAITENHAUSEN (town)   BWL

   Emigration to the USA   1866-1905; 1907-1942

   288: Kommunalarchiv Baitenhausen.   ADS 10:101

BALINGEN (Oberamt)   BWL

   129: Lists [of emigrants?]   1817-1851
   133: Public assitance to emigrants   1848-1882

   130: Proceedings [for the disposal of] property
                                               1843-1872

   261: F 153: Balingen: files as above.   ADS 10:16

   Emigration   1857-1941

   262: Wue 65/4: Balingen: 6135.   ADS 10:37

BALLRECHTEN (village)   BWL

   Emigration files   1858-1864; 1867-1910; 1880-1937

   288: Kommunalarchiv Ballrechten.   ADS 10:101

BALTIMORE (Maryland)   USA

   Letter from emigrant in   1759

   149: F 8 B.   ADS 7:1

   List of matters dealt with by the Royal [Wuerttem-
      berg] Consulate in Baltimore   1847-

   263: Minist. d. Ae.: 1910 eingekommene Akten: Verz.
      58, 33.   ADS 10:41

   List of matters dealt with by the Royal [Wuerttem-
      berg] Consulate in Baltimore (Nr. 98)
                                               [19th century]

   263: Minist. d. Ae.: 1910 eingekommene Akten: Verz.
      58, 42.   ADS 10:41

   Hamburg Consul in Baltimore: Estate matters; pen-
      sion matters   1867

   39: Bestand Senat: Cl.VIII, No.X, Jg.1867 (Regis-
      ter) C.4; C.12. [Nordamerika]   ADS 2:193

   Hamburg Consul in Baltimore: estate matters   1868

   39: Bestand Senat: Cl.VIII, No.X, Jg.1868 (Regis-
      ter) C.5 [Nordamerika]   ADS 2:198

BAMBERG (city)   BYL

   Emigration material [not otherwise described in
      ADS]

   236   ADS 9:143

BAMBERG (Bezirksamt I)   BYL

   Emigration to North America and other countries
                                               1865-1912

   223: Rep. K.5.ª/Verz. XII: Nr. 103-118.   ADS 9:8

BAMBERG (Bezirksamt II)   BYL

   Nr. 787-798: Changes of name; exportation of prop-
      erty [as an incident of emigration]; emigration
      to North America   1836-1859
   petitions for emigration   1859
   releases from citizenship   1881
   Nr. 800: Emigration & immigration   1891

   223: Rep. K.5.b/Verz.XII: files as above. ADS 9:7

BAMBERGEN (village)   BWL

   Emigration to the USA   1850-1854; 1857-1932
   <u>288</u>: Kommunalarchiv Bambergen.   ADS 10:101

BAMLACH (village)   BWL

   Emigration files   1839-1952
   <u>288</u>: Kommunalarchiv Bamlach.   ADS 10:101

BANKHOLZEN (village)   BWL

   Emigration   1901-1919
   <u>288</u>: Kommunalarchiv Bankholzen.   ADS 10:101

BARBELROTH (congregation)   RPL

   Marginal entries on emigration to be found in the
   Lutheran or Reformed parish registers
                                   18th century
   <u>217</u>   ADS 8:86

BARMEN (city)   RWL

   Emigration   1863-1883; 1889-1893
   <u>115</u>: Rep.D8: Neuere Akt.: Abt.1, C3: Nr. 12039-
      12041; 12043.   ADS 6:26

BARNTRUP. *See* Sternberg-Barntrup

BARTERODE (parish)   NL

   Emigrant list   1840-
   <u>57</u>: I.1.11.A 110.   ADS 4:226

BATTENBERG (Amtsgericht)   HEL

   Emigration files arranged by village [not listed
   in ADS description]   1823-
   <u>151</u>: Bestand 275: Battenberg: Acc.1912/19: Nr. 12.
                                       ADS 7:124

BAUMHOLDER (Landkreis)   RPL

   Naturalizations of foreigners   1924-1925
   <u>200</u>: Abt.442: Regierung Trier: Nr. 10286.
                                        ADS 8:40

BAVARIA (Kingdom)   BYL

   Emigration & immigration; exportation of personal
   property   1767-1805
   <u>227</u>: Rep. Bayerische Gesandtschaft Wien: Nr. 622/
      1-25.   ADS 9:51

   Various grants of mercy & special dispensation to
   mercenary troops sent to America   1779-1784
   <u>500</u>: Hist. Verein fuer Mittelfranken, Ms. hist.

---

     487 [film][Verschiedene Gnadenerteilg.]
                              Not in ADS

   Emigration   1781-1843; 1799
   <u>227</u>: Rep. Kasten Schwarz: 651/1-835, 836-928.
                                       ADS 9:42

   Emigration & immigration; exportation of personal
   property   [1781-1843?]
   <u>227</u>: Rep. Kasten Schwarz: 653/1-1544.   ADS 9:42

   Salary advances to various interned officers [in
   America?]   1794-1795
   <u>500</u>: Hist. Verein fuer Mittelfranken, Ms. hist.
      486 [film][Gehaltsvorschuesse]   Not in ADS

   Emigration & immigration; exportation of personal
   property   n.d.
   <u>226</u>: Rep.MInn 5: 15354.   ADS 9:18

   South German emigrants & their movement through
   Austria   1818
   <u>227</u>: Gesandtschaft Stuttgart Nr. 618/248. ADS 9:41

   Emigration due to mystical compulsion   1820
   <u>227</u>: Gesandtschaft Stuttgart Nr. 618/251. ADS 9:41

   Legal action [in Rheinprovinz] against emigrants
   from southern Germany to America   1827-1851
   <u>200</u>: Abt.441: Regierung Koblenz: Nr. 5106-5109.
                                       ADS 8:32

   Financial aid to Bavarian emigrants on the ship-
   wrecked Dutch ship *Helena Maria* near Falmouth
                                       1828
   <u>227</u>: Rep. Bayerische Gesandtschaft London: 473.
                                       ADS 9:46

   Emigration of [Bavarian?] citizens to North America   1831
   <u>227</u>: Gesandtschaft Stuttgart Nr. 618/259.
                                       ADS 9:41

   Emigration of unemployed Bavarians   1841-1849
   <u>227</u>: Rep. MA 1890: A.St.: Nr. 361.   ADS 9:54

   Emigration to North & South America, Conv. 7-14
                                       1843-1851
   <u>227</u>: Rep. MA 1921: A.V. IV: Nr. 11646-11653.
                                       ADS 9:83

   Emigration   first half 19th century
   <u>226</u>: Rep. MInn 7: 24384-24759.   ADS 9:18

   Nr. 1643: Sinking of the emigrant ship *Powhatan*
                                       1855-1856
   Nr. 1647: Sinking of the emigrant ship *Luna* [Amer-
        ican flag] and the Bavarian citizens
        thereon   1860
   [included herein because some emigrants may have
   had relatives already in America]
   <u>227</u>: Rep. Bayer. & Pfalz. Gesandtschaft Paris:
      files as above.   ADS 9:48

   Return of two emigrants from North America   1869
   <u>227</u>: Rep. MA 1921: A.V. IV: Nr. 11689.   ADS 9:84

Inquiries from, and information to, would-be emigrants, Conv. I   1882-1895

<u>227</u>: Rep. MA 1921: A.V. IV: Nr. 11705.   ADS 9:85

Nr. 845: Emigration to the USA   1892-1917
Nr. 847: Emigration to Florida   1904-1906
Nr. 848: Emigration to Texas   1905-1907

<u>227</u>: Rep. MA 1936: A.V.: files as above. ADS 9:101

Emigration to the United States & to Central America, Vol. II   1926-1930

<u>227</u>: Rep. MA 1943: A.V. (1918-1933): Nr. 486.
ADS 9:104

Extraterritorial personages   1921-1932

<u>227</u>: Rep. MA 1943: A.V. (1918-1933): Nr. 8.
ADS 9:103

Bavarian citizens abroad   1800-1820

<u>227</u>: Rep. Kasten Schwarz: 647/314 - 648/134a.
ADS 9:41

Representations made by the American government on behalf of naturalized citizens born in Bavaria who have been conscripted for [Bavarian] military service   1854-1857

<u>227</u>: Rep. MA 1898: A.St.: Nr. 9.   ADS 9:57

Travelers to the American states, 2 vols.
1920-1932

<u>227</u>: Rep. MA 1943: A.V. (1918-1933): Nr. 541-542.
ADS 9:104

Passports & travel to America   1933

<u>227</u>: Rep. MA 1957: Laufende Registratur: Nr. 169,12.
ADS 9:106

Register of commercial passports issued   1771

<u>229</u>: Rep. H.R. Hofamts-Registratur: Fasz. 273.6.
ADS 9:110

2447.4568-4723: Releases from Bavarian citizenship
1883-1887
2448.4724-4853: Releases from Bavarian citizenship
1888-1891
2449.4854-5021: Releases from Bavarian citizenship
1892-1895
2450.5022-5215: Releases from Bavarian citizenship
1896-1899

<u>229</u>: Rep. Reg. Akten von Oberbayern: Kammer des Innern: files as above   ADS 9:117

Certificates of original domicile & citizenship, files A-Z   1919-1926

<u>227</u>: Rep. MA 1943: A.V. (1918-1933): Nr. 374.
ADS 9:104

Nr. 71640: Expulsions   1854-1918
Nr. 71641: Expulsions, vol. II   1913-1925
Nr. 71642: Expulsions, vol. III   1926-1937
Nr. 71643: Expulsion of foreigners, vol. II
1906-1914
Nr. 71644: Expulsion of dangerous foreigners, vol. I   1927-1938

<u>226</u>: Rep. MInn 20: files as above.   ADS 9:20

Expulsions to America; miscellaneous cases
1891-1905

<u>227</u>: Rep. MA 1921: A.V. IV: Nr. 10250.   ADS 9:82

Estates of American citizens in Bavaria 1878-1900

<u>227</u>: Rep. MA 1914, Bd. II: Fasz. 173-174. ADS 9:61

Estates of American citizens in Bavaria 1919-1931

<u>227</u>: Rep. MA 1943: A.V. (1918-1933).   ADS 9:105

Marriages of American citizens in Bavaria, Conv.II
[19th century]

<u>227</u>: Rep. MA 1921: A.V. II: Nr. 9827.   ADS 9:82

Marriages of Bavarian citizens in the United States
1858-1904

<u>227</u>: Rep. MA 1921: A.V. II: Nr. 9490-9533.
ADS 9:82

Official requests for assistance, conv. I-IV
1860-1900

<u>227</u>: Rep. MA 1914, Bd. II: Fasz. 151-153. ADS 9:61

The Jewish question & the Jewish boycott 1935-1938

<u>227</u>: Rep. MA 1957: Laufende Registratur: Nr. 178,9.
ADS 9:106

Jewish bank accounts   1938-1940

<u>226</u>: Rep. MWi I: 412.   ADS 9:30

Petitions for permission to use foreign academic titles [in Bavaria]   1901-1904

<u>226</u>: Rep. MK 3: 11143-11145.   ADS 9:36

Acceptance of foreign students in Bavarian universities & technical schools, 9 vols.   1904-1926

<u>226</u>: Rep. MK 3: 11095-11101.   ADS 9:36

BAVARIA. *See also* Mittelfranken, Oberbayern, Rheinpfalz, Unterfranken

BAYREUTH (Bezirksamt)   BYL

  Verz.VI: Nr. 1655-1711: Emigration to North America   1861-1865
  Verz.IX: Nr. 25-26: Emigration to North America
1866-1912

  <u>223</u>   ADS 9:8

BEAVERTOWN (Pennsylvania?)   USA

Letter from a young man in Beavertown who wishes to inherit from Professor -- Dalencon (woman)
1834

<u>39</u>: Bestand Senat: Cl.VIII, No.X, Jg.1834 (Register)   ADS 2:132

BECHTOLDSKIRCH. *See* Mengen   BWL

BECKUM (Kreis)   RWL

    Nr. 31-36: Emigration   1847-1851; 1856-1914
    Nr. 38:   Foreigners   1910-1920
    Nr. 631-633: Foreigners   1930-1935

    <u>116</u>: Muenster Reg.: Beckum: 1.: files as above.
                                   ADS 6:65

    Emigration   1866-1880; 1889-1929

    <u>116</u>: Muenster Reg.: Nr. 1661; 2660; 1658; 1678-
        1679.   ADS 6:56

    List of certificates of citizenship issued
                             1899-1923

    <u>116</u>: Muenster Reg.: Nr. 3393.   ADS 6:56

BEDERKESA. *See* Bremen

BEILNGRIES (Bezirksamt)   BYL

    Summary of emigration to North America   1843-1861
    <u>222</u>: Rep. 124: Beilngries: Nr. 7.   ADS 9:2

BELGIUM

    Emigration & immigration to Belgium; exportation
       of personal property   1833
    <u>229</u>: Rep. Reg. Akten von Oberbayern: Kammer des
       Innern: 1794.10   ADS 9:116

BELLHEIM (village; Landkreis Germersheim, Kurpfalz)

    Names of emigrants   18th century
    <u>201</u>: Archivabt. Ausfautheiakten.   ADS 8:63

BELLINGEN (town)   BWL

    Emigration   1890-1936; 1897-1934
    <u>288</u>: Kommunalarchiv Bellingen.   ADS 10:101

BELLINGS (Amt)   HEL

    Exemptions from emigration tax. *See* Hesse-Kassel

BENTHEIM (Amt)   NL

    Releases from citizenship, vol. I   1864-1873
    <u>54</u>: Rep. 122 II: Bentheim: Fach 30, Nr. 3.
                                    ADS 4:132

    Emigration permits [for permits before 1871, *see*
       Osnabrueck]   1871-1879
    <u>54</u>: Rep. 116 I: Nr. 6068.   ADS 4:120

    Releases from citizenship   1880
    <u>54</u>: Rep. 116 I: Nr. 13419.   ADS 4:121

BERCHTESGADEN. *See* Salzburger emigrants   BYL

BERG (Grandduchy)   RWL

    Nr. 11001: Passports issued   1809
    Nr. 11002: List of passports & visas issued
                            1810-1813

    <u>115</u>: Rep. D3.d: I.Div.b: files as above. ADS 6:20

    Nr. 10206: Emigration & immigration, 4 vols.
                            1812-1813
    Nr. 10207: Population changes, 4 vols.   1810-1813
    <u>115</u>: Rep. D3.d: I.Div.a: Verwaltung C: files as
       above.   ADS 6:20

BERG (Generalgouvernment)   RWL

    List of foreigners arriving in Duesseldorf
                     January-May 1816

    <u>115</u>: Rep.D5: XV: Nr. 119.   ADS 6:21

BERG. *See also* Juelich-Berg

BERGEDORF (town)   HH

    File on tithes taxes (10% & 5%) on inheritances
       sent abroad   1614-1809
    <u>39</u>: Bergedorf I: Pars II, Sectio I, vol. VIII.7.
       Decimation: a.   ADS 2:466

    Passports, visas, fate of citizens   1814-1872
    <u>39</u>: Bergedorf II: Nr. 294 [Passbuecher] ADS 2:467

    Fasc. 2: Releases from Bergedorf residency [other
           than to Hamburg]   1818-1871
    Fasc. 4: Citizenship of persons subject to mili-
           tary service upon release of their par-
           ents from citizenship   1844-1856
    Fasc. 3: Reports on persons released from the off-
           ice & village of Bergedorf   1852-1871
    <u>39</u>: Bergedorf II: Nr. 36: files as above.
                                 ADS 2:467

    Losses of citizenship   1877-1893
    <u>39</u>: Stadt Bergedorf-Magistrat: C.II.c.1.
                                  ADS 2:468

BERGEN (Amt)   HEL

    Emigrants   1741
    <u>151</u>: Bestand 86: Nr. 2085.   ADS 7:89

    File on the property of a soldier from Bergen who
       deserted in America [not named in ADS descrip-
       tion] [Hesse-Hanau military unit?]   1788
    <u>151</u>: Bestand 81: Rep.E: 21.Amt Bergen VIII: Nr. 3.
       [Vermoegen]   ADS 7:82

BERGHAUPTEN (town)   BWL

    Emigration files   1832-1890; 1880-1937
    Immigration files   1913-1940
    <u>288</u>: Kommunalarchiv Berghaupten.   ADS 10:102

BERGHEIM (Kreis)   RWL

Emigration   1857-1868
115: Rep.D9: Abt.I.e: Nr. 221.   ADS 6:31

BERGISCH GLADBACH   RWL

Emigration; applications for emigration permits;
   releases from Prussian citizenship; summaries
                                                  1846-1854
118: C 289.   ADS 6:73

Family records: Fues, Koch, Pfeifer-Reynolds
                                                  1863-1933
118: C 1142.   ADS 6:73

BERGOESCHINGEN (town)   BWL

Emigration files   1837-1940
288: Kommunalarchiv Bergoeschingen.   ADS 10:102

BERLEBURG (Grafschaft)   RWL

Litt.A, Nr. 2: Emigration tax   [18th century]
Litt.A, Nr. 20: List of emigrants to America 1773
Litt.F, Nr. 91: List of emigrants to America
                     [estates]   1726-1730
Litt.F, Nr. 92: List of emigrants to America
                     [estates] 2 vols.   1745
119: files as above.   ADS 6:74

BERLIN (Hamburg Resident in)

Passport & visa issuance; notarizing of signatures
                                                  1820-1852
39: Hanseat. u. Hamb.dipl.Vertretungen: A.11
    [Pass-Protokoll]   ADS 2:396

Passports submitted   1822-1841
39: Hanseat. u. Hamb.dipl.Vertretungen: E.2.
                                          ADS 2:396

Small file of papers on North America   1839-1849
39: Hanseat. u. Hamb.dipl.Vertretungen: B.15.
                                          ADS 2:396

BERLIN (Hanseatische Gesandtschaft in Berlin)

Passports & residency matters   1859-1875
   [files for 1861-1867 are in Staatsarchiv Luebeck
   & for 1859-1870 in Staatsarchiv Bremen]
39: Hanseat. u. Hamb.dipl.Vertretung: S.1
    [Aeltere Reg.]   ADS 2:398

BERLIN. See also Hamburg (Hanseatic) for much mate-
           rial on emigration, citizenship, residency
           certificates

BERMATINGEN (town)   BWL

Emigration files   1873-1939; 1896-1923; 1907-1945
288: Kommunalarchiv Bermatingen.   ADS 10:102

BERMERSBACH (village; Kreis Offenburg)   BWL

Emigration files   1819-1857; 1840-1843; 1812-1849
288: Kommunalarchiv Bermersbach.   ADS 10:102

BERMERSBACH (village; Kreis Rastatt)   BWL

Emigration files (7 fascicles)   1851-1937
Germans abroad   1928-1939
288: Kommunalarchiv Bermersbach.   ADS 10:102

BERNKASTEL (Landkreis)   RPL

Emigration   1854-1857; 1864-1893
200: Abt.442: Regierung Trier: Nr. 191-193; 6071-
       6076.   ADS 8:41

Manuscript on emigration from Kreis Bernkastel by
   Josef Mergen   1955
200: Abt.701: Nachlaesse, etc.: Nr. 964.
                                          ADS 8:53

BERSENBRUECK (Amt)   NL

Emigration   1832-1889
54: Rep.122: LRAmt Bersenbrueck: Nr. 20.
                                          ADS 4:132

Nr. 307: Emigration certificates   1868-1871
Nr. 308: Emigration certificates   1868-1881
54: Rep.122: Han.Amt Bersenbrueck: files as above.
                                          ADS 4:132

Emigration permits [for permits before 1871, see
   Osnabrueck]   1871-1879
54: Rep.116.I: Nr. 6069.   ADS 4:120

Nr. 139-160: Emigration permits, vols. I-IX
                                                  1878-1885
Nr. 146a: Emigration permits   1880-1884
54: Rep.122: Han.Amt.Bersenbrueck: files as above.
                                          ADS 4:131

Emigration permits   1885-1909
54: Rep.122: LRAmt Bersenbrueck: Nr.22-34
                                          ADS 4:132

Applications for releases from citizenship
                                                  1880-1881
Applications for releases from citizenship
                                                  1881-1883
54: Rep.116,I: Nr. 11083; 11084.   ADS 4:121

Nr. 88: Releases from citizenship   1881
Nr 90: Releases from citizenship   1881-1883
Nr. 96: Releases from citizenship   1883-1884
Nr. 104: Releases from citizenship   1885-1886
Nr. 108: Releases from citizenship   1887
Nr. 112: Releases from citizenship   1888
Nr. 115: Releases from citizenship   1889
Nr. 118: Releases from citizenship   1890-1891
Nr. 123: Releases from citizenship   1892-1893
Nr. 126: Releases from citizenship   1894-1899
54: Rep.116,I: II.4.A.1e: files as above.
                                          ADS 4:121-123

Releases from citizenship   1900

 54: Rep.116,I: Nr. 13420   ADS 4:124

Nr. 36: Persons dying abroad & their estates,
        vol. I   1844-1900
Nr. 37: Persons dying abroad & their estates,
        vol. II   1900-1914

 54: Rep.122: LRAmt Bersenbrueck: files as above.
                                       ADS 4:132

BESIGHEIM (Oberamt)   BWL

Nr. 165: Applications [for emigration] which have
         been denied or to which protest has been
         made; documents regarding Wuerttemberg
         citizenship; cancelled releases from citi-
         zenship; naturalization of Wuerttemberg
         citizens in other countries; entrance of
         [Wuerttemberg citizens] in foreign mili-
         tary or governmental service 1846-1900
Nr. 166-240: Emigration files (55 large packets)
         arranged chronologically   1810-1902
Nr. 167: Lists [of emigrants]   1815-1880
Nr. 168: Individual cases [of emigration]
                                       1830-1831

261: F 154: Besigheim: files as above. ADS 10:17

BEUREN (town)   BWL

Emigration files   1873-1940
Immigration files   1889-1928

288: Kommunalarchiv Beuren.   ADS 10:102

BIBERACH (Oberamt)   BWL

Nr. 443-469a: Emigration files (12 large packets);
         naturalizations in Wuerttemberg; exemp-
         tions from emigration tax   1807-1923
Nr. 470: Releases from [Wuerttemberg] citizenship;
         losses of citizenship; residency abroad
         with retention of [Wuerttemberg] citizen-
         ship   1843-1923

261: F 155: Biberach: files as above.   ADS 10:17

BIBERACH (town)   BWL

Emigration book   1848-
266   ADS 10:66

BIBERACH (village; Landkreis Wolfach)   BWL

Emigration files   1827-1928
Immigration files   1886-1938

288: Kommunalarchiv Biberach   ADS 10:102

BICKEN. *See* Dillenburg   HEL

BIEBER (Amt)   HEL

Emigration [reverse pages contain names]   1751
151: Bestand 81: Rep.Ba: Gef. 9, Nr. 8. ADS 7:80

Permits to emigrate to Pennsylvania for some citizens but not for those having property worth
   more than 500 florins   1751

151: Bestand 80: Lit.A, Nr. 7 [Verschiedene]
                                       ADS 7:78

Emigration matters   1764-1767

151: Bestand 81: Rep.E: 55.Amt Bieber VII: Nr. 1.
                                       ADS 7:83

Prohibition against emigration of persons worth
   more than 100 gulden. List of families who have
   emigrated   1765-1767

151: Bestand 80: Lit.A, Nr. 14.   ADS 7:79

Sequestration of properties of emigrants to America   1784

151: Bestand 81: Rep.E: 55.Amt Bieber I: Nr. 6.
                                       ADS 7:83

Files on costs occasioned by emigration   1789

151: Bestand 81: Rep.Ba: Gef. 18-21, Nr. 50.
                                       ADS 7:80

BIEBESHEIM (Landkreis Gross Gerau)   HEL

List of 25 persons emigrating to America (married
   or otherwise; with or without permit) 1854-1857

164: Abt.XI,4.   ADS 7:214

BIEDENKOPF (Landratsamt)   HEL

Nr. 64: Affairs of individual citizens with for-
        eign governmental offices   1810-1855
Nr. 245: Emigration to Hungary & North America
                                       n.d.
Nr. 233: Emigration in general   1785-1864
Nr. 234: Emigration in general   1740-1866
Nr. 235: Emigration from various *Gemeinde* [commun-
        ities]   n.d.
Nr. 243 Emigration from various *Gemeinde* [commun-
        ities]   n.d.

151: Bestand 180: Biedenkopf: files as above.
                                       ADS 7:113

Small number of entries recording removals of emigrants   1810-1815

153   ADS 7:160

BIEDERBACH (village)   BWL

Emigration files   1848-1887; 1885-1944
Immigration files   1892-1923
Fifty years of emigration   1932-1934
Germans abroad   1916-1945

288: Kommunalarchiv Biederbach.   ADS 10:102

BIELEFELD (city)   RWL

Nr. 5: Emigration & immigration, general
                                       1818-1881
Nr. 6: Expulsions & emigration   1763-1805
Nr. 6: Emigration & immigration   1817-1837
Nr. 7: Emigration & immigration   1808-1867

Nr. 8: Emigration & immigration   1838-1847
Nr. 9: Emigration & immigration   1847-1858
Nr. 10: Emigration & immigration   1848-1856
Nr. 11: Emigration & immigration   1856-1874

121: Rep.IB: files as above.   ADS 6:76

BIELEFELD (Kreis)   RWL

Issuance of certificates of original domicile
                                         1863-1924
Emigration   1874-1924

116: Minden: Praes.Regist.: I.A. Nr. 10-16; 115-
     117.   ADS 6:50-52

BIELEFELD (Landratsamt)   RWL

Nr. 227: Emigration & immigration   1876-1919
Nr. 233: Summaries of emigration & immigration
                                         1872-1886

116: Minden Reg.: Bielefeld: files as above.
                                         ADS 6:62

BIELEFELD (city)   RWL

Nr. 118-120: Emigration   1874-1924
Nr. 17-27:   Issuance of certificates of original
             domicile   1887-1924
Nr. 64:   Certificates of citizenship 1899-1905
Nr. 65-72:   Certificates of citizenship 1908-1924

116: Minden: Praes.-regist.: I.A.: files as above.
                                         ADS 6:50

BIENGEN (village)   BWL

Emigration files   1834-1927

288: Kommunalarchiv Biengen.   ADS 10:102

BIERSDORF (village; Reichsgrafschaft Sayn-Altenkirchen). See Daaden   RPL

BIESENDORF (village)   BWL

Emigration files   1854-1913

288: Kommunalarchiv Biesendorf.   ADS 10:103

BIETINGEN (village)   BWL

Emigration files   1854-1938
Immigration files   1904-1944

288: Kommunalarchiv Bietingen.   ADS 10:103

BILLAFINGEN (village)   BWL

Emigration to the USA   1820-1854; 1881-1943

288: Kommunalarchiv Billafingen.   ADS 10:103

BILLIGHEIM (Landkreis)   RPL

Names of emigrants   18th century

201: Archivabt. Ausfautheiakten.   ADS 8:62

BILLIGHEIM (Landkreis). See also Appenhofen; Impflingen; Klingen; Muehlhofen; Rohrbach bei Lindau;
Steinweiler; Wollmesheim; Heuchelheim (Kreis Bergzabern); Ilbesheim (Kreis Landau)

BILLIGHEIM (parish)   RPL

The Lutheran church book contains marginal nota-
     tions regarding emigrants to America 1711-1793

201: Archivabt. Kirchenbuecher.   ADS 8:63

BINNINGEN (village)   BWL

Emigration to the USA   1851-1852; 1890-1915

288: Kommunalarchiv Binningen.   ADS 10:103

BINZEN (village)   BWL

Emigration files   1852-1947

288: Kommunalarchiv Binzen.   ADS 10:103

BIRKENAU   HEL

Emigration   1860-1901

149: Rep.40,2: Ablief.2: Bd.292.   ADS 7:22

BIRKENDORF (village)   BWL

Emigration files   1819-1943

288: Kommunalarchiv Berkendorf.   ADS 10:103

BIRKENFELD   NL

Releases from citizenship   1821-1855
     [for America in particular, 1834-1837]

53: Bestand 32: Birkenfeld: 32-10-3-2.
    DLC has.   ADS 4:115

BIRKENFELD (Regierung)   RPL

Nr. 855: Passports   1819-1854
Nr. 856: Passports   1854-1880
Nr. 858: Foreigners residing in the Duchy of Bir-
         kenfeld   1851-1858
Nr. 862: Immigration & immigration of women
                                         1822-1846
Nr. 866: Collection of inheritances abroad
                                         1842-1884
Nr. 872: Emigration overseas (Algiers, Brazil,
         America)   1845-1881
Nr. 873: Emigration overseas   1854-1881

200: Abt.393: Regierung zu Birkenfeld: files as
     above.   ADS 8:15

Nr. 877-974: Emigration from 90 communities within
             governmental area   1818-1882
Nr. 975-1123: Naturalizations in 90 communities
             within governmental area   1818-1882

Nr. 1152: List of certificates of original domicile issued   1869-1900
Nr. 1154-1172: Certificates of original domicile
                                       1853-1872

200: Abt.393: Regierung zu Birkenfeld: files as
above.   ADS 8:16

Nr. 2359: Claims by former French soldiers for
wages & rights of inheritance for soldiers killed in American military service   1819-1865
Nr. 866: Claims of inheritance abroad 1842-1884
Nr. 2272: Marriages with foreigners   1839-1858

200: Abt.393: Regierung zu Birkenfeld: files as
above.   ADS 8:16

BIRKENFELD (Amt)   RPL

Naturalizations & releases from citizenship; residency of foreigners   1818-1856

200: Abt.395,1: Amt Birkenfeld: Nr. 15.   ADS 8:16

BIRKINGEN (village)   BWL

Emigration files   1857-1942

288: Kommunalarchiv Birkingen.   ADS 10:103

BIRNDORF (village)   BWL

Emigration files   1921-1939
Immigration files   1890-1928

288: Kommunalarchiv Birndorf.   ADS 10:103

BIRSTEIN (Kreis Gelnhausen)   HEL

Files of Joseph Hess, Junior, emigration agent
                                       1884-1889

151: Bestand 180: Gelnhausen: Acc.1958/36 B: Nr. 3.
                                       ADS 7:117

BIRSTEIN (Amt)   HEL

Residency permits   1832-

151: Bestand 82: Acc.1882/17: Nr. 50.   ADS 7:85

BISCHHEIM (village; Landkreis Kirchheimbolanden,
Kurpfalz)   RPL

Names of emigrants   18th century

201: Archivabt. Ausfautheiakten.   ADS 8:63

BISCHOFFINGEN (village)   BWL

Emigration files   1849-1937

288: Kommunalarchiv Bischoffingen.   ADS 10:103

BISCHWEIER (village)   BWL

Immigration files   1918-1936

Emigration files   1934-1937

288: Kommunalarchiv Bischweier.   ADS 10:103

BITBURG (Landratsamt)   RPL

Emigration   1861-1925
Immigration   1890-1894; 1897-1907

200: Abt.460: Bitburg: Nr. 64; 66-68; 70-71.
                                       ADS 8:43

Diplomatic negotiations regarding inheritances
                                       1891-1921

200: Abt.460: Bitburg: Nr. 60.   ADS 8:44

BITBURG (Landkreis)   RPL

Emigration   1855-1893

200: Abt.442: Regierung Trier: Nr. 196-199; 6117-
6123; 8655-8656.   ADS 8:39-40

Manuscript on emigration from Kreis Bitburg   1954

200: Abt.701: Nachlaesse, etc.: Nr. 964. ADS 8:53

BLANKENLOCH (Amt)   BWL

Emigration; manumissions   1770-1771

260: 217.   ADS 10:1

BLANKENLOCH. See also Durlach

BLANSINGEN (village)   BWL

Emigration files   1850-1939

288: Kommunalarchiv Blansingen.   ADS 10:104

BLAUBEUREN (Oberamt)   BWL

Emigration files   1818-1869

261: F 156: Blaubeuren: 78-88.   ADS 10:17

BLEIBACH (village)   BWL

Emigration to the USA   1850-1933
Germans abroad; certificates of citizenship
                                       1925-1938
Germans abroad   1937-1951

288: Kommunalarchiv Bleibach.   ADS 10:104

BLOMBERG   RWL

Nr. 1: Passports   1765-1832
Nr. 2: Passport applications & receipts 1832-1838

122: Altes Archiv: D.I.g: files as above. ADS 6:77

Nr. 1: Passports   1830-1850
Nr. 3: Foreign passports   1890-1915
Nr. 6: Controversial domicile matters (& certifi-
cates of original domicile)   1857-1868
Nr. 7: Matters of domicile & citizenship
                                       1877-1916

Nr. 8: Deposited certificates of original citizenship   1867-1878

<u>122</u>: Neues Archiv: D.I.g: files as above. ADS 6:77

Nr. 9: Emigration & immigration   1813-1912
Nr. 11: Emigration & immigration   1844-1869

<u>122</u>: Neues Archiv: D.I.g: files as above. ADS 6:77

BLOMBERGE   RWL

List of citizens & their children who have emigrated [of 132 persons, apparently only one came
 to America; his name was Johann Christoph Wit-
 ten]   1704

<u>114</u>: Lipp.Reg.: Aelt.Regis.: Fach 206, Nr.2.
               ADS 6:2

BLUMEGG (village)   BWL

Emigration files   1827-1950

<u>288</u>: Kommunalarchiv Blumegg.   ADS 10:104

BLUMENAU-BOKELOH (Amt)   NL

Nr. 417: Applications for release from citizenship
  vol. I   1829-1840
Nr. 418: Applications for release from citizenship
  vol. II   1841-1855

<u>52</u>: Han.Des.80: Hannover: Blumenau-Bokeloh: B.a.
 IV.4: files as above.   ADS 4:83

BLUMENBACH (Kreis Gelnhausen)   HEL

Files of Karl Franz Isaak, emigration agent
             1886-1888

<u>151</u>: Bestand 180: Gelnhausen: Acc.1958/36 B: Nr.1.
             ADS 7:117

BLUMENTHAL (Amt)   NL

Emigration applications & permits   1835-1860

<u>55</u>: Han.Des.74: Blumenthal: IX.J.1: Fach 30, No.
 10.   ADS 4:153

Transportation of asocial persons to America
             1844-1851

<u>52</u>: Han.Des.74: Fach 292, Nr.9.   ADS 4:64

List of persons of military age who have been
 issued emigration permits [contains 28 names,
 dates, etc.]   1852-1858

<u>55</u>: Han.Des.74: Blumenthal: I.Fach 13, Nr.6.
             ADS 4:150

Nr. 18f: Visa register [contains 818+1212+1291 entries; gives name, birthplace, destination, etc.]   1853; 1854; 1855
Nr. 18g: Visa register [mostly internal Germany]
             1858

<u>55</u>: Han.Des.74: Blumenthal: IX.H.: Fach 297: files
 as above.   ADS 4:153

Nr. 18a: Register of temporary residents 1860-1862
Nr. 18b: Register of temporary residents 1863-1864
Nr. 18c: Register of temporary residents 1865-1867
Nr. 18d: Register of temporary residents 1878-1885

<u>55</u>: Han.Des.74: Blumenthal: IX.H.: Fach 297: files
 as above.   ADS 4:153

Releases from citizenship   1871-1874

<u>55</u>: Han.Des.74: Blumenthal: I: Fach 13, Nr. 21.
             ADS 4:151

BLUMENTHAL. *See* Bremen

BOCKENHEIM (Kreis?)   HEL

Residency permits   1833-1861

<u>151</u>: Bestand 82: Acc.1882/17: Nr. 47.   ADS 7:85

BODERSWEIER (village)   BWL

Emigration files   1797-1937; 1881-1937; 1937-1939
Jewish emigration   1937-1939

<u>288</u>: Kommunalarchiv Bodersweier.   ADS 10:104

BODMAN (village)   BWL

Emigration files   1839-1925
Immigration files   1884-1937
List of former residents who have emigrated
           prepared 1932

<u>288</u>: Kommunalarchiv Bodman.   ADS 10:104

BOEBLINGEN (Oberamt)   BWL

70-95a: Emigration files arranged by village
            1816-1909
96: Lists [of emigrants?]   1849-1872
98: Prevention of escape [emigration] to America
 via France   1853-1859
99: Lists of immigrants   1850-1872

<u>261</u>: F 157: Boeblingen: files as above. ADS 10:17

BOECKWEILER (village; Landkreis Hornbach, Kurpfalz)

Names of emigrants   18th century

<u>201</u>: Archivabt. Ausfautheiakten.   ADS 8:63

BOEHRINGEN (village)   BWL

Emigration files   1840-1877; 1875-1941
Immigration files   1875-1951; 1913-1947

<u>288</u>: Kommunalarchiv Boehringen.   ADS 10:104

BOETZINGEN (town; includes Oberschaffhausen)   BWL

Emigration files   1787-1941
Immigration files   1901-1933

<u>288</u>: Kommunalarchiv Boetzingen.   ADS 10:105

BOGEN (Bezirksamt)   BYL

    Nr. 1-136: Emigration to America   1846-1862
    Nr. 141-147: Applications for emigration permits
           to North America   1846-1854

    225: Rep.164: Verz.1: files as above.   ADS 9:13

    Emigration   [19th century]

    225: Rep.164: Verz.1.   ADS 9:15

BOGEN (Landgericht)   BYL

    Applications for emigration permits   1815-1846

    225: Rep.168: Verz.1.   ADS 9:16

BOHLINGEN (village)   BWL

    Emigration files   1822-1871; 1888-1919
    Immigration files   1892-1942

    288: Kommunalarchiv Bohlingen.   ADS 10:104

BOHLSBACH (village)   BWL

    Emigration files   1863-1930
    Germans abroad   1933-1939

    288: Kommunalarchiv Bohlsbach.   ADS 10:104

BOKELOH. *See* Blumenau-Bokeloh   NL

BOLDIXUM (Wyk auf Foehr)   SHL

    Nr. 4406: Registration of departing persons
                    1905-1908
    Nr. 5060: Emigration   1902-1920

    38: Verz.Nr. 3: Flecken Boldixum: files as above.
                          ADS 1:72

BOLLENBACH (village)   BWL

    Emigration files   1819-1867; 1846-1883

    288: Kommunalarchiv Bollenbach.   ADS 10:104

BOLLSCHWEIL (village)   BWL

    Emigration files   1831-1879
    Emigration & immigration   1905-1943

    288: Kommunalarchiv Bollschweil.   ADS 10:105

BONN (Landratsamt)   RWL

    Nr. 60-61: Emigration   1855-1882
    Nr. 62:   Register of emigration 1852-1892
    Nr. 64-65: Renaturalization of former emigrants to
           America   1902-1913
    Nr. 1587: Extradition & transmittal of criminals
                      1873-1930
    Nr. 1611: Emigration: special cases   1919-1930
    115: Rep.G29/6: Landratsamt Bonn: files as above.
                          ADS 6:37

BONN (Kreis)   RWL

    Emigration   1857-1866

    115: Rep.D9: Abt.I.e: Nr. 222-225.   ADS 6:31

BONN (city)   RWL

    P 1163: Emigration permits   1853-1860
    P 1164: Emigration permits   1814-1852
    P 1329: Emigration permits   1891-1901
    P 1214: Emigration permits   1867-1880
    P 1166: Emigration permits   1881-1890
    P 20/632: Emigration (list)   n.d.
    P 20/633: Emigration (list)   1924-1926
    P 20/558: Emigration (list)   1923-1931
    P 20/564: Emigration (list)   1928-1931

    123: files as above.   ADS 6:79

BONSWEILER   HEL

    Emigration   1860-1901

    149: Rep.40,2: Ablief.2: Bd. 292.   ADS 7:22

BORDESHOLM (Landratsamt)   SHL

    L 64: Emigration list   1847-1891
    L 134-135: Emigration & immigration   1880-1915

    2: Abt.320 (Bordesholm): files as above.
        DLC has?   ADS 1:25

BORKEN (Kreis)   RWL

    Emigration   1890-1929

    116: Muenster Reg.: Nr. 1676-1677.   ADS 6:56

    Emigration & immigration   1862-1889

    116: Muenster Reg.: Borken: Nr. 13.   ADS 6:65

BORKEN. *See also* Liedern

BORNHEIMER BERG?   HEL

    Emigrations to Nova Scotia & Pennsylvania   1752

    151: Bestand 86: Nr. 51974.   ADS 7:88

BORSTEL. *See* Jork

BOSTON (Massachusetts)   USA

    List of Hamburg vessels arriving in port of Boston
                              1850
    List of Hamburg vessels arriving in port of Boston
                              1860

    42: Bestand H516: Hamburg Consulate in Boston: (1).
                         ADS 2:521-541

    Hamburg Consulate in Boston: Estate matters; legal
        affairs   [1881?]

    39: Bestand Senat: Cl.VI, No.16p: vol.4b: Fasc.6h,
        invol. 1-6, 8-15.   ADS 2:54

BOTTENAU (village)   BWL

   Emigration files   1890-1949

   <u>288</u>: Kommunalarchiv Bottenau.   ADS 10:105

BOTTROP (Kreis)   RWL

   Emigration   1923-1928

   <u>116</u>: Muenster Reg.: Nr. 1672.   ADS 6:57

BRACKENHEIM (Oberamt)   BWL

   151-176: Emigration files from villages in Oberamt
                                   1807-1888
   176a: Lists of emigrants; emigration files; re-
      leases from Wuerttemberg citizenship
                                   1824-1881

   <u>261</u>: F 158: Brackenheim: files as above. ADS 10:17

BRACKENHEIM (Amtsgericht)   BWL

   The following *Bueschel* [bundles] contain emigra-
      tion matters: 101-103; 252; 257-258; 260-261;
      271; 273; 276-277; 284; 288; 299; 304; 307-310;
      315; 317; 321-328; 333-334; 372; 375-377; 381-
      382; 384; 386-389; 391-393; 395-396; 400-402;
      404-405; 411; 414-415   n.d.

   <u>261</u>: F 258: Brackenheim: files as above. ADS 10:29

BRAKE (Amt)   RWL

   Fach 17, Nr. 3-14: Emigration & immigration
                                1847-1876
   Fach 22, Nr. 3:   Passports   1844-1864
   Fach 7, Nr. 4:   Reports of emigration 1881-1887
   Fach 7, Nr. 5:   Emigration   1880-1887

   <u>114</u>: Brake: files as above.   ADS 6:15

BRAMSCHE   NL

   Emigration to America   1833-1849

   <u>54</u>: Depositum 59: XIa. Nr. 4.   ADS 4:148

BRAUBACH (Amt)   HEL

   Nr. 3503: Marriages, naturalizations, & releases
         from citizenship   1852-1853
   Nr. 3582: Releases from citizenship   1872-1883

   <u>152</u>: Abt. 220: Braubach: files as above. ADS 7:130

BRAUNFELS   HEL

   A.85.6.Nr. 1: Emigrations to Poland & America from
         Greifenstein & Braunfels
                 1784-1785; 1803-1805
   A.85.6.Nr. 2: Emigration to America   1773; 1805

   <u>154</u>: A. Haupt-Archiv: files as above.   ADS 7:161

   List of names of the Friedrichsburger [colonists
     in Texas?]   1847

   <u>154</u>: CA4b1: Nr. 1, Blatt 170.
      DLC has.   ADS 7:165

Emigrant [to Texas colony] correspondence [alpha-
    betically arranged]   n.d.

   <u>154</u>: CA4a3: Nr. 1-17
      CA4a4: Nr. 1-12
      CA4a5: Nr. 1-17
      DLC has   ADS 7:163-165

BRAUNSCHWEIG   NL

   Passports, vol. 1   1813-1837
   Passports, vol. 2   1838-1850
   Passports, vol. 3   1851-1861

   <u>56</u>: L Neu Abt. 12A, Fb.2: E. Varia: Nr. 13.
                                   ADS 4:171

   Emigration permits, 17 vols.   1851-1871

   <u>56</u>: L Neu Abt. 12A, Fb.2: E. Varia: Nr. 15.
                                   ADS 4:171

   Issuance of certificates of residency
                      1820; 1837-1849

   <u>56</u>: L Neu Abt. 12A, Fb.2: E. Varia: Nr. 12.
                                   ADS 4:171

   Death certificates, 4 vols.   1832-1867

   <u>56</u>: L Neu Abt. 12A, Fb.2: E. Varia: Nr. 9.
                                   ADS 4:171

   Special files on the administration of enemy prop-
     erty [some belonging to Americans]   1914-

   <u>56</u>: Hs Abt.VI: Gruppe 9, Nr. 88.2 (vol. X)
                                   ADS 4:164

BRAUNSCHWEIG (Kreis)   NL

   Nr. 52: Emigration permits   1830-1851
   Nr. 56: List of emigrants   1855-1868

   <u>56</u>: L Neu Abt.126, Fb1: Gr. II: files as above.
      [This collection to be rearranged; catalog
      numbers will be changed.]   ADS 4:184

BRAUNSCHWEIG (city)   NL

   About 1000 emigration permits (chronologically
     filed; alphabetical name list)   1837-1924
   Beginning in middle of 19th century emigrants re-
     quired to publish their permits to emigrate in
     official gazette *Braunschweigischen Anzeiger*
   Emigration lists from 1848 period were burned in
     1945
   Manuscript entitled "Braunschweiger in Nordamerika"
     [Brunswick Citizens in North America] by Dr. --
     Moderhack   1948

   <u>60</u>   ADS 4:228

BRAZIL   South America

   Emigration & immigration to Brazil; exportation of
     personal property   1824

   <u>229</u>: Rep. Reg. Akten von Oberbayern: Kammer des
      Inneren: 1794.11.   ADS 9:116

   Emigration   1868

   <u>263</u>: Rep. E 70: Gesandtschaftsakten: Verz. 52:
      Bern: 8.   ADS 10:54

Emigration from Bezirksamt Feuchtwangen   1884

<u>231</u>: Rep. 212/6: Feuchtwangen: Abgabe 1937, Verz.
I: 2/27/4.   ADS 9:132

Emigration from Uffenheim   1889

<u>231</u>: Rep. 212/18: Uffenheim: 18. Abgabe, 1930: 87.
                                             ADS 9:133

BREDSTEDT (Amt). *See* Helgoland   SHL

BREITFURT (village; Landkreis Hornbach, Kurpfalz)
                                                     RPL

Names of emigrants   18th century

<u>201</u>: Archivabt. Ausfautheiakten.   ADS 8:63

BREMBERG (village)   HEL

Manumissions   1783-1808

<u>152</u>: Abt.351: Nassau-Vierherrisch: Gen.VIIIe 44.
                                             ADS 7:139

BREMEN   HB

Emigration records [includes emigrants from German
states passing through Bremerhaven port]
   a.   before 1840
   b.   1840-1850
   c.   after 1871

<u>48</u>: P.8.B.8: a; b; c.
     DLC has *a* and *b* only   ADS 3:51

Issuance of emigration permits:
   a. General & miscellaneous (for emigrants not
      citizens of Bremen, see P.8.B.10)   1847-1867
   b. Individual cases:
      1. A-L
      2. M-Z
      3. In accordance with general German emigra-
         tion law of 25 June 1849

<u>48</u>: P.8.A.12: a; b.
     DLC has *a* only.   ADS 3:50

Emigration consents (from Vegesack)
   1. To 1860
   2. From 1861

<u>48</u>: P.13.b.29.b: files as above.   ADS 3:54

Emigrants from rural areas [of Bremen land] n.d.

<u>48</u>: Q.1.qq.7.   ADS 3:54

Emigration to the USA [not adequately described
in ADS]

<u>48</u>: B.13.d *and* P.8.B.8 *and* H.4.Q.
     DLC may have parts of these series.   ADS 3:43

Police permits, etc. [not otherwise described in
ADS]   [19th century]

<u>48</u>: D.20.b.13.a; b; c; *also see* P.8.
     DLC may have part of this series.   ADS 3:46

Applications for emigration permits (after 1867
law prohibiting employment of substitutes for

military service obligation) contains:
Register of persons released from citizenship
                                              n.d.
Register of certificates of release from citizen-
   ship   1871-1875

<u>48</u>: P.8.12.A.d.   ADS 3:50

Passenger lists   n.d.

<u>49</u>: A.4.(i) [Auswanderungswesen]   ADS 3:110

Emigrant passenger lists   1907-1911
   [originals destroyed during World War II]

<u>48</u>: P.8.B.8.c.1.
     DLC has   ADS 3:51

Inquiries for heirs, death certificates, reports
on emigrants [ADS does not describe these materials but states that researchers should ask for
individual files]   [19th century]

<u>48</u>: D.20.c.8.   ADS 3:46

Inquiries regarding specific emigrants   n.d.

<u>49</u>: A.4. (k) [Auswanderungswesen]   ADS 3:110

Expulsion from America of German nationals   1918-

<u>49</u>: A.4. (r) [Auswanderungswesen]   ADS 3:110

Issuance of emigration permits to men who have
completed their one-year military service obligation   n.d.

<u>48</u>: P.8.A.12.e.   ADS 3:51

Status of persons subject to Bremen military obligation who have emigrated [contains a list of
men, born 1826-1831, who have not fulfilled
their miligary service obligation]   n.d.

<u>48</u>: P.8.A.12.c.   ADS 3:50

Transport [exile] of criminals to the USA   1845

<u>48</u>: B.13a: Nr. 23; *and* P.8.B.8b.   ADS 3:2

1. Register of persons granted Bremen citizenship
                                         1883-1912
2. Issuance of release [emigration] permits
                                         1883-1912
3. Individual cases of releases (Vegesack)
                                         1881-1904
6. Register of persons in Vegesack . . . losing
   citizenship   1882-1904
7. Release certificates (individual cases)
                                         1873-1876

<u>48</u>: P.8.F.1: files as above.   ADS 3:53

Deserters   n.d.

<u>48</u>: D.19.e.   ADS 3:54

Desertions from German ships in the port of New
York   1904

<u>48</u>: A.3.N.3: Nr. 214.   ADS 3:90

Hanseatic Ministry in Washington: Reparation
claims of Bremen citizens   1863-1867

<u>39</u>: Hanseat. u. Hamb.dipl.Vertretungen: C.6
     [Reklamations-Vorderungen]   ADS 2:406

Files regarding complaints from emigrants about
ship accommodations, exchange of money, agents
[666 files, almost all having to do with American immigrants]   1875-1941

49: A.4 (a); (b); (c); (d) [Auswanderungswesen]
                                          ADS 3:109

BREMEN (*including* Vegesack, Blumenthal, Bederkesa)

Loss of Bremen citizenship:
   C.4: through residence abroad   1841-1874
   C.5: through trans-Atlantic removal 1841-1870
   C.6: miscellaneous releases from citizenship
                                        1867-
   C.7: emigration of persons subject to military
        service   n.d.

48: P.8.A.10: files as above.   ADS 3:49

BREMEN. *See also* U. S. Civil War; Hamburg (Hanseatic) Ministry of the Four Cities to Bundestag in
Frankfurt/Main (geographic entry)

BREMERHAVEN   HB

Emigration of persons subject to military service
                                          1848-

61: 25 (Lehe): Nr. 5 [Wehrmacht-Erfassungswesen]
                                          ADS 4:229

BREMERVOERDE   NL

Files on emigration of persons subject to military
   service   1838-1844

62: H, Fach 14, vol. 6.   ADS 4:231

Vol. 5: Passports cancelled   1838
Vol. 8: Visa register   1845-1847
Vol. 10: Passport matters   1852
Vol. 11: Register of passports issued   1847-1865
Vol. 12: Register of visas   1854-1860
Vol. 13: Register of visas   1860-1862
Vol. 14: Register of visas   1863-1865

62: D i, Fach 106: volumes as above [Regiminalia,
   Bd. II]   ADS 4:231

Files on releases from citizenship or from military service obligation   1847-1857

62: A, Fach 6: Vol. 8.   ADS 4:232

File of notices from foreign governments as to the
   naturalization [by them] of former Bremervoerde
   citizens [file missing?]   1858

62: A, Fach 6: Vol. 16.   ADS 4:233

List of emigrants   1859

62: A, Fach 6: Vol. 18.   ADS 4:233

Files on permits for emigrants & releases from
   citizenship & from military service obligation
                                         1855-1861

62: A, Fach 6: Vol. 17.   ADS 4:233

Emigration permits & releases from citizenship:
   Vols. 20-51   1866-1896
   Vol. 53   1897

62: A, Fach 6: volumes as above.   ADS 4:234

BREMERVOERDE (Amt)   NL

Administration of Germans returning from America
   & their expulsion:
   Nr. 49: Vols. I-IV   1887-1906
   Nr. 50:   1896

52: Han.Des.74: A, Fach 6: files as above.
                                          ADS 4:64

BREMERVOERDE (city)   NL

Emigrants from city [name, age, parents, year,
   where emigrated]   1855-1874

63   ADS 4:240

Emigrant lists of Bremervoerde & the parish of
   Oerel   n.d.

In possession of Oberstudienrat in R. Dr. Diedrich
   Mueller, Kehdinger Muehren 30, Stade, Niedersachsen.   ADS 4:336

BREMGARTEN (village)   BWL

Emigration files   1851-1940; 1854-1940; 1880-1936

288: Kommunalarchiv Bremgarten.   ADS 10:105

BRENSCHELBACH (village; Landkreis Hornbach, Kurpfalz)
                                                RPL

Names of emigrants   18th century

201: Archivabt. Ausfautheiakten.   ADS 8:63

BRETWIG bei DRESDEN. *See* C. F. Gaebler

BREUNIGWEILER (village; Landkreis Kirchheimbolanden,
   Kurpfalz)   RPL

Names of emigrants   18th century

201: Archivabt. Ausfautheiakten.   ADS 8:63

BRILON (Kreis)   RWL

Passport register   1840-1860

116: Arnsberg Reg.: Brilon: L.A. Fach 172, Nr. 3.
                                          ADS 6:58

BRITZINGEN (village)   BWL

Emigration files   1891-1939
Immigration files   1918-1919

288: Kommunalarchiv Britzingen.   ADS 10:105

BRODENBACH (village)   RPL

Nr. 228: Certificates of original domicile
                                         1827-1928
Nr. 197: Emigration & immigration   1828-1928

Nr. 237: Naturalizations & releases from citizenship   1830-1894

<u>200</u>: Abt. 655.7: Brodenbach: files as above.
                                      ADS 8:50

BROMBACH (village)   BWL

Emigration files   1804-1948
Immigration files   1900-1948

<u>288</u>: Kommunalarchiv Brombach.   ADS 10:105

BUCHENBACH (village)   BWL

Emigration files   1860-1932; 1929-1947
Immigration & residency   1933-1936

<u>288</u>: Kommunalarchiv Buchenbach.   ADS 10:105

BUCHHEIM (village; Kreis Freiburg)   BWL

Emigration files   1817-1858; 1867-1932
Immigration files   1894-1937

<u>288</u>: Kommunalarchiv Buchheim.   ADS 10:105

BUCHHEIM (village; Landkreis Stockach)   BWL

Emigration files   1880-1925
Immigration files   1870-1914
Emigration to the USA, etc.   1873-1894

<u>288</u>: Kommunalarchiv Buchheim.   ADS 10:105

BUDBERG. *See* Orsoy   RWL

BUECHENBACH (parish office)   BYL

Emigration to North America; incitement to emigrate by Peter Stickel   1902

<u>227</u>: Rep. MA 1921: A.V. IV: Nr. 11736.   ADS 9:85

BUECHENBEUREN (village; Grafschaft Sponheim). *See*
Sohren (village)

BUECHERTHAL (Amt)   HEL

Nr. 1796: Emigration tax   1688-1730
Nr. 51974: Emigrants to Nova Scotia & Pennsylvania
                                  1752

<u>151</u>: Bestand 86: files as above.   ADS 7:88

BUECHIG (Amt)   BWL

Emigration; manumissions   1770-1771
<u>260</u>: 217.   ADS 10:1

BUECHIG. *See also* Durlach   BWL

BUEDINGEN (ducal estate)   HEL

Emigration lists   19th century

155   ADS 7:190

BUEHL (town)   BWL

Emigration files   1831-1886; 1917-1937
<u>288</u>: Kommunalarchiv Buehl.   ADS 10:106

BUEHLERTAL (village)   BWL

General emigration files   1832-1938
Emigration files   1843-1934
Emigration to the USA   1851-1855
General files   1851-1940
Emigration files   1860-1939
Immigration files   1898-1944
Estate files also include emigration data
                                  1800-1852

<u>288</u>: Kommunalarchiv Buehlertal.   ADS 10:106

BUEHLKAU (parish)   NL

Various applications for releases from citizenship
                                  1868-1897
Lists of emigrants & immigrants (name, age, profession, destinations or places of residence)
                                      n.d.

<u>100</u>: Gem. Buelkau: V., Nr. 3.   ADS 4:311

BUER (Amtsvogtei)   NL

Emigrations to America   1842-1849

<u>54</u>: Rep. 122, VII.1: Landratsamt Melle: A: Nr. 66.
DLC has.   ADS 4:140

BUER. *See* Gelsenkirchen-Buer

BUEREN (Kreis)   RWL

Nr. 73-74: Certificates of citizenship 1863-1924
Nr. 28-31: Issuance of certificates of original
domicile   1863-1924
Nr. 121-122: Emigration   1873-1924

<u>116</u>: Minden: Praes.-regist.: I.A.: files as above.
                                  ADS 6:50; 52

BUERGEN (village)   RPL

Naturalizations & releases from citizenship
                                  1822-1902

<u>200</u>: Abt. 655.7: Brodenbach: Nr. 221.   ADS 8:50

BUESUM   SHL

Nr. 202: Citizenship & emigration 1811-1910
Nr. 207: Passports   1841-1867
<u>4</u>: files as above.   ADS 1:42

BUETTELBORN (Landkreis Gross-Gerau)   HEL

Applications for emigration to America 1848-1853

<u>166</u>: Abt. XI, 4.   ADS 7:216

BUGGENSEGEL (village)   BWL

Emigration files   1856-1933
<u>288</u>: Kommunalarchiv Buggensegel.   ADS 10:106

BUGGINGEN (village; Kreis Freiburg)   BWL

Emigration files   1833-1882; 1894-1941
<u>288</u>: Kommunalarchiv Buggingen.   ADS 10:106

BUGGINGEN (village; Kreis Muellheim)   BWL

Emigration files   1935-1943; 1943-1949
<u>288</u>: Kommunalarchiv Buggingen (Kr. Muellheim).
                                             ADS 10:106

BUPHEVER   SHL

Sale of property of a person in America (a mid-
    wife)   1875
<u>28</u>: Band X, Archiv-Verzeichnis No. 10, p. 272.
                                             ADS 1:60

BURG (village; Kreis Freiburg)   BWL

Emigration files   1839-1877
General files   1876-1941
<u>288</u>: Kommunalarchiv Burg.   ADS 10:106

BURG AUF FEHMARN   SHL

Rubr. XVII, Nr. 24: Passports   1794-1860
Rubr. XIX, Nr. 4a: Emigration files 1860-1875
<u>5</u>: files as above.   ADS 1:43

BURGBERG (village; Kreis Villingen)   BWL

Emigration files   1852-1942
<u>288</u>: Kommunalarchiv Burgberg (Kr. Villingen).
                                             ADS 10:106

BURGGRAEFENRODE (Amt)   HEL

For examptions from emigration tax, *see* Hesse-
    Kassel (Kurhessen)

BURGHAUN (Amt)   HEL

Applications for emigration permits 1806-1810
<u>151</u>: Bestand 98: Fulda X: Acc.1875/27: Nr. 62-63.
                                             ADS 7:91

BURGLENGENFELD (Bezirksamt)   BYL

Citizens who have emigrated with permission, also
    citizens of the Kingdom of Italy from whom

inheritances accrue in the Kingdom of Bavaria
                                         1812
<u>222</u>: Rep.125: Burglengenfeld: Nr. 125.   ADS 9:2

BURGWEILER (village)   BWL

Emigration files   1831-1931
Immigration files   1894-1920
<u>288</u>: Kommunalarchiv Burgweiler.   ADS 10:106

BURTSCHEID (canton)   RWL

Population movements   1801-1807
<u>115</u>: Rep.D.2.2: II.Div., 1.Bureau, 2.Bevoelkerung:
    Nr. 93.   ADS 6:19

BUTSCHBACH (village)   BWL

Emigration files   1911-1941
<u>288</u>: Kommunalarchiv Butschbach.   ADS 10:106

BUTTERSTAEDTER-HOEFE [?]   HEL

For exemptions from emigration tax, *see* Hesse-
    Kassel (Kurhessen)

BUTZBACH (town)   HEL

Emigrant list   1870-1895
<u>156</u>   ADS 7:191

BUXTEHUDE   NL

Financial aid for the transportation of persons to
    America (RR 1067)   1852
<u>52</u>: Han.Des.80: Stade: P, Titel 31, Nr. 7.
                                             ADS 4:93

CADOLZBURG. *See* Kadolzburg

CALENBURG (Amt)   NL

Releases from citizenship   1822-1868
<u>52</u>: Han.Des.80: Calenberg: B.c.VI.4: Nr. 378.
                                             ADS 4:84

Emigration of individuals from region   1845-
<u>52</u>: Han.Des.74: Calenburg: E.XIII.1, Nr. 3.
                                             ADS 4:65

CALW (Oberamt)   BWL

AZ 6135: Emigration   1852-1944
AZ 1040.32: Releases from citizenship 1881-1900
<u>262</u>: Wue 65/7: Calw: files as above.   ADS 10:37

<u>145</u>: Emigration files   1888-1923
<u>146</u>: Lists of emigrants & immigrants 1847-1878
<u>147</u>: Applications for release from Wuerttemberg
        citizenship   1846-1920

261: F 159: Calw: files as above.   ADS 10:18

CANADA

Emigration & immigration to & from Canada; exportation of personal property   1855

229: Rep. Reg. Akten von Oberbayern: Kammer des
Innern: 1794.12.   ADS 9:116

Emigration from Bezirksamt Feuchtwangen   1884

231: Rep. 212/6: Feuchtwangen: Abgabe 1937, Verz.
I: 2/27/4.   ADS 9:132

CANNSTATT (*now called* Bad Cannstatt)   BWL

Monthly lists of foreigners in Cannstatt and other
Residenzstaedte [cities in which the king re-
sided][*See also* Wuerttemberg]   1818

263: Rep. E 2: Polizeiminist.: 22,5; 23,4.
ADS 10:52

Nr. 97a: Lists of passports issued   1844-1923
Nr. 99-: Supporting documents for passports issued
1901-1923
Nr. 118: Losses of [Wuerttemberg] citizenship
1887-1906
Nr. 118a: Losses of [Wuerttemberg] citizenship
1815-1865
Nr. 118b: Losses of [Wuerttemberg] citizenship;
lists of passports [*Passkarten*]; lists
of visas   1866-1923
Nr. 457: Lists of certificates of original domicile & of citizenship (with supporting
documentation)   1889-1904
Nr. 465-475: Lists of certificates of original
domicile   1887-1912

261: F 160: Cannstatt: files as above.   ADS 10:18

CARLSRUHE-AUGSPURG. *See* Karlsruhe (Amt)

CAYENNE. *See* Guiana   South America

CELLE (Amt)   NL

Releases from citizenship   1821-1899

66: 1.c.I.Nr.6-10: Fach 25-26 [Entlassungs-Urkunde]
ADS 4:242

Releases from citizenship   1827-1869

66: 1.c.I.Nr.1: Fach 25 [Entlassungs-Urkunde]
ADS 4:242

Emigration matters   1847-1869

66: 1.c.II.Nr.1: Fach 26 [Auswanderungen]
ADS 4:242

Persons who have emigrated without receiving releases from citizenship   1883-

66: 1.c.I.Nr.11: Fach 26 [Entlassungs-Urkunde]
ADS 4:242

CHAM (Bezirksamt)   BYL

Tabular summary of emigration & immigration
1825-1854

222: Rep. 126: Cham: Nr. 675.   ADS 9:2

CHARLESTON (South Carolina)   USA

List of Bremen ships in   1855

48: B.13.b.6: Nr. 6.
DLC has?   ADS 3:24

List of Hamburg ships arriving at   1860

42: Bestand H516: Hamburg, Consulate in Charleston:
(1)(Anlage).   ADS 2:541

Hamburg Consulate in Charleston: Estate & inheritance matters   1870-1875; 1892

39: Bestand Senat: Cl.VI, No.16p: vol.4b: Fasc. 5e:
Invol. 2 [Nachlass]   ADS 2:53

CHICAGO (Illinois)   USA

German Society of Chicago: Membership list & list
of Ladies' Auxiliary   1878-1879

114: Lipp.Reg.: I.Abt.: Auswaert.Ang.: 7a, Nr. 29.
ADS 6:12

Schwabenverein Chicago [Swabian Club]: Membership
roll   1923-1924

264: Abt.63: Deutsch-Amerikaner: 4.   ADS 10:62

CINCINNATI (Ohio)   USA

German Consulate for Ohio, Indiana, Illinois, &
Wisconsin: Legal affairs, estates, etc.
1871-1893

39: Bestand Senat: Cl.VI, No.16p: Vol.4b: Fasc.
15d: Invol. 1-10.   ADS 2:59

CLAUSTHAL (mining district)   NL

Emigration of mining & smelting personnel to the
English colonies in America   1750-1756

68: VI.2.b: Fach 154, Nr. 2 [Auswanderung]
ADS 4:244

Emigration of individual Harz [Mountain] inhabit-
ants; subsidy for transportation:
Vol. I: Principal negotiations   1848-1849
Vol. II: Special   1848-1849
Vol. III: Commission fees   1848

68: VI.2.b: Fach 154, Nr. 5; Fach 155, Nr. 6.
ADS 4:245

Emigration of workers from estates of noblemen:
Nr. 8: [to America]   1849
Nr. 10: to southern Australia & North America
1850-1851
Nr. 11-13: to southern Australia & North America
1850-1852

68: VI.2.b: Fach 155 [for Nr. 8]; Fach 156 [for Nr.
10-13.   ADS 4:245

Emigration of laborers from the estates of noblemen:

Nr. 14-31: to southern Australia & North America
                                        1851-1855

68: VI.2.b: Fach 157-161.   ADS 4:246

Promotion of the emigration of inhabitants of the
  Harz [Mountain region]   1848-1853

68: VI.2.b: Fach 154, Nr. 4 [Befoerderung]
                                        ADS 4:244

Rejected applications for financial assistance to
  emigrate to America   1852-1867

68: VI.2.b: Fach 162, Nr. 35 [Abgeschlagene Ge-
  suche]   ADS 4:246

CLAUSTHAL-ZELLERFELD   NL

Zellerfeld: Rights to move [emigrate]
                        17th & 18th centuries

67: Zellerfeld: A.I.d [Fach 93][Abzugsrecht]
                                        ADS 4:243

Rights to move [emigrate]   1720-1801

67: Clausthal: A.I.e [Fach 131][Abzugsrecht]
                                        ADS 4:243

Emigration matters   n.d.

67: Zellerfeld: C.III: [Fach 117]   ADS 4:243

Recruiting of workers from the estates of noblemen
  for British service in America   1756-1758

68: VII: Fach 318, Nr. 3.   ADS 4:247

Emigration matters:
Nr. 1: Witnesses for emigrants   1845-1852
Nr. 3: List of emigrants   1845-1846
Nr. 6: Arrangements for families of persons who
       plan to emigrate   1870-1896

67: Clausthal: C.III.a: Files as above [Fach 153]
                                        ADS 4:243

Emigration matters:
Nr. 2: To America   1830-1845
Nr. 13: Emigrants   1853

67: Clausthal: C.III.b: files as above [Fach 153]
                                        ADS 4:243

Emigration matters:
Nr. 1: Citizens who have been expelled by the
       government   1854
Nr. 3: Releases from citizenship   1854-1905
Nr. 4: Emigration & immigration   1854
Nr. 5: Releases from citizenship   1890-1930

67: Clausthal: C.III.c: files as above [Fach 153]
                                        ADS 4:244

Emigration of workers from estates of noblemen
                                        1856-1866

68: VI.2.b: Fach 162, Nr. 42.   ADS 4:247

Applications for releases from citizenship
                                        1854-1872

67: Clausthal: C.II.b.1: (Fach 152)[Gesuche]
                                        ADS 4:243

CLEVE (Arrondissement)   RWL

Population movements   1801-1802; 1811-1812

115: Rep.D2.2: II.Division, 1.Bureau, 2.Bevoelker-
     ung: Nr. 103.   ADS 6:19

COBURG (Principality)   BYL

Nr. 127: Applications for emigration permits
                                        1817; 1819
Nr. 134/1: Emigration plans of a number of persons
           (including applications for emigration
           permits & public assistance to emigrate)
                                             1833
Nr. 134/2: Emigration plans, vol. II.   1852-1856
Nr. 134/3: Emigration plans, vols. III & IV
                                        1856-1880

224: Ministerial-Archiv: Loc.G.Tit.IV.2: files as
     above.   ADS 9:11

Nr. 48: Passports, vol. XVIII   1847
Nr. 63: Passports, vol. XXX   1857

224: Landesregierungs-Archiv: Tit.III.c.8: files
     as above.   ADS 9:10

COBURG (city)   BYL

Applications for emigration permits   1846

224: Landesregierungs-Archiv: Tit.III.4.f.44.
                                        ADS 9:10

COBURG.   See also Saxony-Coburg-Gotha

COESFELD (Kreis)   RWL

Emigration   1885-1925

116: Muenster Reg.: Nr. 1674-1675.   ADS 6:57

COESFELD (Landratsamt)   RWL

Nr. 825: Foreigners   1898-1912
Nr. 795: Foreigners   1910-1912
Nr. 857: Foreigners   1914-1922
Nr. 122: Certificates of original domicile & citi-
         zenship   1910-1914
Nr. 694: Certificates of original domicile & citi-
         zenship   1914-1927

116: Muenster Reg.: Coesfeld: files as above.
                                        ADS 6:66

COLMBERG (Oberamt; later Kameralamt of Bezirksamt
  Ansbach)   BYL

Nr. 19: Files on reception of Jews; letters of
        protection; protection money   1582-1738
Nr. 57: Files on the Jewish community   1804-1805

231: Rep. 212: Bezirksamt Ansbach: Rep. 212/1:
     files as above.   ADS 9:128

COPPENBRUEGGE (Amt)   NL

   Releases from citizenship   1824-1858

   52: Han.Des.80: Coppenbruegge: B.d.VI.4: Nr. 166.
                               ADS 4:84

CRAILSHEIM (Oberamt)   BWL

   Nr. 387: Lists of emigrants   1824-1870
   Nr. 386-396 [*Bueschel*]: Emigration files (arranged
       chronologically?)   1811-1856

   261: F 161: Crailsheim: files as above.   ADS 10:18

CREFELD (Arrondissement)   RWL

   Population movements   1801-1802; 1811; 1812

   115: Rep. D2.2: II.Division, 1.Bureau, 2.Bevoelker-
      ung: Nr. 117.   ADS 6:19

CUXHAVEN. *See* Ritzebuettel (Amt)

DAADEN (village; Reichsgrafschaft Sayn-Altenkirchen)
                                     RPL

   Emigration of poor inhabitants of the villages of
     Daaden & Biersdorf to America; permissions to
     sell their homes   1753

   200: Abt. 30: Reichsgrafschaft Sayn-Altenkirchen
     (u. Hachenburg): Nr. 39.   ADS 8:11

DAHN (Kreis). *See* Rumbach (village)

DANDENFELS (village; Landkreis Kirchheimbolanden,
Kurpfalz)   RPL

   Names of emigrants   18th century

   201: Archivabt. Ausfautheiakten.   ADS 8:63

DANNENBERG (Amt)   NL

   Citizens dying abroad   1827-

   52: Han.Des.74: Dannenberg: Abt.II,G, Nr. 23.
                               ADS 4:65

DANNSTADT (village). *See* Kurpfalz   RPL

DARKEHMEN (Kreis)   East Prussia (Gumbinnen)

   Emigration [file missing?]   1832-1874

   51: Rep.12: Abt.I, Titel 3, Nr. 8, Bd. 2. ADS 4:32

DARMSTADT (city)   HEL

   Applications for emigration, 9 vols. [alphabeti-
     cally arranged]   1823-1879

   157: Rep. Stadtarchiv Darmstadt: XI, 4. ADS 7:192

DARMSTADT. *See also* Hesse-Darmstadt

DARSBERG   HEL

   Emigration   1864-1903

   149: Rep. 40,2: Ablief.2: Bd. 293.   ADS 7:22

DATTINGEN (village)   BWL

   Emigration files   1886-1910; 1907-1917
   Immigration files   1887-1899

   288: Kommunalarchiv Dattingen.   ADS 10:106

DAUN (Landkreis)   RPL

   Emigration to America   1855; 1868-1910

   200: Abt.442: Regierung Trier: Nr. 201; 6135-6139.
                             ADS 8:41

   Nr. 974: Emigrants   19th century
   Nr. 964: Manuscript on emigration from Kreis Daun,
       by Josef Mergen   1958

   200: Abt.701: Nachlaesse, etc.: files as above.
                             ADS 8:53

DAVERDEN (Kirchenkreis Vechta)   NL

   Emigrants to America   1849
   Emigration & immigration   1849-1914

   69: I.A.1.10.Fasc.1, Nr. 105.   ADS 4:249

DEGERNAU (village)   BWL

   Emigration files   1907-1941

   288: Kommunalarchiv Degernau.   ADS 10:107

DEGGENDORF (Landgericht)   BYL

   Applications for emigration permits   1815-1846

   225: Rep.168: Verz.1.   ADS 9:16

DEGGENDORF (Bezirksamt)   BYL

   Emigration   [19th century]

   225: Rep. 168: Verz.1.   ADS 9:15

   Nr. 83: Emigration   1870-1913
   Fasz. 43, Nr. 43: Emigration   1870-1914
   Fasz. 45, Nr. 70: Emigration   1870-1914
   Fasz. 60, Nr. 411: Emigration   1870-1914
   Fasz. 61, Nr. 520: Emigration   1870-1914

   225: Rep. 164: Verz.2: files as above.   ADS 9:13

DEGGENDORF (city)   BYL

   Emigration   [19th century]
   225: Rep. 168: Verz. 1.   ADS 9:15

DEGGENHAUSEN (village)   BWL

   Emigration to the USA   1851-1866; 1857-1866
   Emigration files   1900-1943
   Immigration files   1901-1942

   288: Kommunalarchiv Deggenhausen.   ADS 10:107

DEHRN. *See* Limburg   HEL

DEISENDORF (village)   BWL

   Emigration to the USA   1852
   Emigration to various [countries]   1852-1904

   288: Kommunalarchiv Deisendorf.   ADS 10:107

DELVE (Kreis Norder-Dithmarschen)   SHL

   Nr. 25,3: File on emigrants   1872-1928
   Nr. 25,4: Granting of passports   1867-1928

   6: files as above   ADS 1:44

DEMARARA   South America

   Emigration & immigration to & from Demarara; exportation of personal property   1839

   229: Rep. Reg. Akten von Oberbayern: Kammer des
      Innern: 1794.13.   ADS 9:116

DENKINGEN (village)   BWL

   Emigration to overseas [countries]
             1836-1868; 1888-1937

   288: Kommunalarchiv Denkingen.   ADS 10:107

DENZEN (village; Grafschaft Sponheim)   RPL

   From the register of the communal official [*Schul-
   theis*] order manumitting certain inhabitants for
   the purpose of emigration & threat of expulsion
   in the event that they did not leave within three
   weeks   1722-1771

   200: Abt.33: Vordere Grafschaft Sponheim u. beide
      Grafschaften gemeinschaftlich: Nr. 2753.
                      ADS 8:11

DENZLINGEN (village)   BWL

   Emigration files   1833-1932
   Immigration files   1910-1931

   288: Kommunalarchiv Denzlingen.   ADS 10:107

DERMBACH (Bezirk)   Thuringia

   Individual emigration applications   1836-1840

   500: Weimar: Thueringisches Staatsarchiv: Dermbach
      II: Gb.2.: Nr. 63-92; 93-114; 144-183; 186;
      188.   Not in ADS

DESSIGHOFEN (village)   HEL

   Manumissions   1806

   152: Abt.351: Nassau-Vierherrisch: Gen.VIII.e.55.
                    ADS 7:139

DETMOLD (Landratsamt)   RWL

   Nr. 1: Emigration & immigration   1870-1899
   Nr. 2: Acceptances & losses of citizenship
                     1872-1907
   Nr. 5: Passports & certificates of domicile, vol.1
                     1886-1905

   114: Detmold: Fach 4: files as above.   ADS 6:15

DETMOLD (city)   RWL

   Nr. 1: Emigration & immigration   1864-1884
   Nr. 2: Emigration to America   1873
   Nr. 3: Emigration & immigration   1853-1858
   Nr. 4: Emigration & immigration   1859-1867
   Nr. 5: Emigration & immigration   1847-1852
   Nr. 6: Emigration & immigration   1885
   Nr. 7: List of emigrants & immigrants   1847
   Nr. 8: Emigration & immigration   1910

   124: D.Akten: Fach 55: files as above.   ADS 6:84

   Nr. 2: Expulsions of foreigners   1902
   Nr. 3: Expulsions of foreigners   1902
   Nr. 4: Residencies & expulsions of Germans abroad
                     1905
   Nr. 5: Citizenship (in Lippe)   1871
   Nr. 6-7: Citizenship (in Lippe)   1912-1913
   Nr. 8-10: Certificates of original domicile
                 1868; 1900-1915

   124: D.Akten: Fach 56: files as above.   ADS 6:85

   Passports   1914
   124: D.Akten: Fach 24, Nr. 6.   ADS 6:84

DETTIGHOFEN (village)   BWL

   Emigration files   1846-1943

   288: Kommunalarchiv Dettighofen.   ADS 10:107

DETZELN (village)   BWL

   Emigration files   1887-1935

   288: Kommunalarchiv Detzeln.   ADS 10:107

DIEPENAU (Amt)   NL

   Releases from citizenship:
   Nr. 195: Vol. I   1823-1836
   Nr. 196: Vol. II   1837-1859

   52: Han.Des.80: Diepenau: B.e.VI.4: files as above.
                    ADS 4:84

DIEPHOLZ (Amt)   NL

    Nr. 2: Emigration permits   1823-1830
    Nr. 4: Emigration permits   1831-1840
    Nr. 6: Emigration permits   1841-1849

    <u>52</u>: Han.Des.74: Diepholz: Abt.III,F: files as
    above.   ADS 4:66

    Applications for releases from citizenship:
       Nr. 622: Vol. II,1   1850-1863
       Nr. 623: Vol. II,2   1864-1870
    <u>52</u>: Han.Des.80: Diepholz: B.f.VI.4: files as above.
                      ADS 4:84

DIERSBURG (village)   BWL

    Emigration files   1833-1897
    Immigration files   1878-1920
    Jewish emigration   1937

    <u>288</u>: Kommunalarchiv Diersburg.   ADS 10:107

DIESSEN (village)   BWL

    Emigration   1786-1836

    <u>262</u>: Ho 201: Glatt: I 3944.   ADS 10:37

DIETLINGEN. *See* Weilheim I   BWL

DIEZ (Landratsamt)   HEL

    Nr. 224: Support of disabled veterans of the U.S.
          Civil War   1865-1867
    Nr. 226: Emigration   1888-1893
    Nr. 228: Emigration   1891-1897
    Nr. 227: Emigration   1897-1913
    Nr. 463: Emigration   1876-1886

    <u>152</u>: Abt.417: Diez: files as above. ADS 7:157

DIEZ (town)   RPL

    Emigration   1852-1885
    <u>205</u>   ADS 8:71

DILLENBURG   HEL

    Emigration of some subjects [to] Pennsylvania 1709
    [According to Learned, "This is a most valuable
    collection of original papers dating from the
    period of the great exodus of the Palatines.
    The package contains original requests . . . of
    those wishing to emigrate, some 60 papers in
    all. The last sheet contains a valuable list of
    names with the following title: *Verzeichnis*
    deren Unterthanen, so umb den Abzug angehalten
    und darzu die Bewilligung erhalten, als die
    Aemter Herborn, Offenbach, Bicken, Medenbach,
    Gontersdorf, und Marckenbach, Ambdorff, Statt
    Wiedorf, Goffenhayn, Mademuehlen, Rabenschiedt.
    Summa 35 Mann, 37 Weiber, 107 Kinder, noch 2
    ledig von Mademuehlen.

    <u>152</u>: Abt.171: Altes Dillenburger Archiv: P493.
        DLC has.   ADS 7:128

    Protocols regarding emigration & immigration of
      subjects   n.d.
    <u>152</u>: Abt.171: Altes Dillenburger Archiv: L 104
      [Protocollum]   ADS 7:128

DILLENBURG (Landesregierung)   HEL

    Files on emigration to foreign colonies, 10 vols.
                       1750-1806

    <u>152</u>: Abt.172: Nr. 2664.
        DLC has.   ADS 7:128

DILLENBURG (Amt)   HEL

    Losses & granting of citizenship   1816-1867

    <u>152</u>: Abt.222: Amt Dillenburg: Nr. 1694-1698.
                      ADS 7:130

DILLINGEN (Bezirksamt)   BYL

    Emigration   [19th century]

    <u>230</u>: Rep. Akten der Bezirksaemter: Dillingen: 2028.
                      ADS 9:122

DILLKREIS (Landratsamt)   HEL

    Naturalizations & releases from citizenship, vols.
      VI-IX [Vols. I-V in Abt.222: Amt Dillenburg]

    <u>152</u>: Abt.410: Dillkreis: Nr. 179.   ADS 7:154

DINGOLFING (Bezirksamt)   BYL

    Emigration   1802-1899

    <u>225</u>: Rep.164: Verz.Nr.3: Nr. 123-142. ADS 9:13

    Emigration   [19th century]

    <u>225</u>: Rep.168: Verz.1.   ADS 9:15

DINGOLFING (Landgericht)   BYL

    Applications for emigration permits   1845-1846

    <u>225</u>: Rep.168: Verz.1.   ADS 9:16

DINGELSDORF (village)   BWL

    Emigration files   1889
    Immigration files   1900-1934

    <u>288</u>: Kommunalarchiv Dingelsdorf.   ADS 10:107

DINKELSBUEHL (Bezirksamt)   BYL

    Emigration & immigration, 3 parts   1824-1890

    <u>231</u>: Rep.212/3: Dinkelsbuehl: 1067.   ADS 9:128

    Emigration & immigration; applications for emigra-
      tion permits   1871-1900

    <u>231</u>: Rep.212/213: Dinkelsbuehl: Abgabe 1943: 1699-
      1703.   ADS 9:133

DINKELSBUEHL (Emigration collection point [*Lager*])
                                                        BYL

Emigration & immigration   1853

231: Rep. 270/II: Reg. Mittelfranken: Kammer des
     Innern: Abgabe 1932: 24.   ADS 9:134

DINKELSBUEHL (town)   BYL

Nr. 17: Emigration files (individual)   1839-1908
Nr. 16-18; 24; 30; 36; 51; 54; 59; 62-64: Emigra-
     tion files   n.d.
Nr. 67; 69; 75: Emigration files   n.d.
Nr. 63: Applications for emigration permits   1853
Nr. 66: Notices of emigration   1853

237: Tit. II: files as above.   ADS 9:146

DISSEN (parish)   NL

Emigration to America   1838

70: II.1.11.A 114.   ADS 4:249

DISSEN-HILTER (Vogtei)   NL

Nr. 6: Emigration files   1829-1852
Nr. 7: Emigration to America   1834-1858
Nr. 8: Emigration files   1853-1874
Nr. 9: Transportation of persons of evil reputa-
       tion at public cost   1845-1863

54: Rep.122, V.2: Dissen-Hilter: files as above.
    DLC has 6, 7, & 9.   ADS 4:136-136

DITHMARSCHEN, NORDER   SHL

Deserters   1708-1774

2: Abt.101: Nr. 363 [Werber u. Deserteure]
   DLC has?   ADS 1:17

DITHMARSCHEN, NORDER. *See also* Delve, Hennestedt,
Norderwoehrder, Weddingstedt, Wesselburg

DITHMARSCHEN, SUEDER. *See* Suederhasted

DOBRINGK (Niederlausitz). *See* Fischer, G.

DOEGGINGEN (village)   BWL

Emigration files   1817-1937
Immigration files   1905-1935

288: Kommunalarchiv Doeggingen.   ADS 10:108

DOERSDORF (village; Amt Nastaetten)   HEL

Naturalizations & releases from citizenship
                                        1851-1860

152: Abt. 236: Nastaetten: Nr. 173.   ADS 7:135

DOGERN (village)   BWL

Emigration files   1928-1937

288: Kommunalarchiv Dogern.   ADS 10:107

DONAUESCHINGEN (town). *See* Fuerstenberg   BWL

DONAUWOERTH (Bezirksamt)   BYL

Nr. 1058: Emigration & immigration; exportation of
          personal property   1811-1819
Nr. 2025: Register of emigration & immigration
                                        1853-1860
Nr. 36; 39-43; 52; 55: Individual applications for
          emigration to North America 1870-1880

230: Rep. Akten der Bezirksaemter: Donauwoerth:
     files as above   ADS 9:123

DORHEIM (Amt)   HEL

For exemptions from emigration tax, *see* Hesse-
Kassel [Kurhesse]

DORMAGEN (canton)   RWL

Population movements   1801-1802; 1811-1812

115: Rep. D2.2: II.Division, 1.Bureau, 2.Bevoelker-
     ung: Nr. 113.   ADS 6:19

DORNHOLZHAUSEN (village)   HEL

Manumissions   1803

152: Abt.351: Nassau-Vierherrisch: Gen.VIIIe 53.
                                       ADS 7:139

DORTMUND (Kreis)   RWL

A. Nr. 100: Emigration   1828-1896
B. Nr. 313: Emigration   1900-1912
A. Nr.   6: Emigration & immigration   1856-1866

116: Arnsberg Reg.: Dortmund: 1.A.: files as above.
                                       ADS 6:58

DOSSENBACH (village)   BWL

Emigration files   1879-1933

288: Kommunalarchiv Dossenbach.   ADS 10:108

DOTTINGEN (village)   BWL

Emigration files (four fascicles)   1818-1950

288: Kommunalarchiv Dottingen.   ADS 10:108

DREISAM (Kreis). *See* Freiburg (city)   BWL

DREISEN (village; Landkreis Kirchheimbolanden, Kur-
     pfalz)   RPL

Names of emigrations   18th century

201: Archivabt. Ausfautheiakten.   ADS 8:63

DUCHTLINGEN (village)   BWL

  Emigration files (three fascicles)   1817-1942
  288: Kommunalarchiv Duchtlingen.   ADS 10:108

DUDERSTADT (town)   NL

  List of 27 Prussian nationals who died in U.S.
  service without known heirs (reported by German
  consul in New York)   1862
  See Duderstaedter Wochenblatt, Nr. 57, 23 July
  1862.   ADS 4:250

  Emigration & immigration   1876-1887
  52: Han.Des.74: Gieboldehausen: VIII.A.III: Nr. 14.
                                        ADS 4:69

DUENEBURG (estate)   NL

  Passports issued   n.d.
  54: Depositum 33: Archiv Gutes Duenenburg: I: Nr.
  225.   ADS 4:147

DUEREN (Landratsamt)   RWL

  Emigration   1838-1856
  115: Rep. G 29/8: Nr. 7.   ADS 6:37

DUEREN (canton)   RWL

  Population movements   1801-1807
  115: Rep.D2.2: II.Division, 1.Bureau, 2.Bevoelker-
  ung: Nr. 94.   ADS 6:19

DUERKHEIM (Bezirksamt)   RPL

  Emigration files   1862-1930
  201: Archivabt. Bezirksaemter: Duerkheim: Akt Nr.
  39-41.   ADS 8:64

DUESSELDORF (Regierung)   RWL

  Nr. 465-499: Immigration applications, 35 vols.
                                        1823-1851
  Nr. 500-528: Emigration applications, 29 vols.
                                        1816-1846
  115: Rep.D8: I.Abt.e: files as above. ADS 6:24

  Emigration applications   1834-1835
  115: Regierung Duesseldorf: I c a: Nr. 30616.
                                        ADS 6:29

  Nr. 9153: Releases from citizenship & emigration
            applications [therein are special cases
            of families with minors later to be sub-
            ject to military service] Vol. I
                                        1820-1842
  Nr. 9154: Releases from citizenship & emigration
            applications [therein complaints regard-
            ing the emigration of sons of wealthy
            persons to avoid military service] Vol.
            II   1843-1872

  Nr. 9155: Investigations of illicit emigration of
            persons subject to military service
                                        1884-1885
  115: Regierung Duesseldorf: I C 3: Files as above.
                                        ADS 6:29

  Non-German immigrations, vol. 1   1904-1913
  115: Regierung Duesseldorf: I c a: Nr. 30536.
                                        ADS 6:28

  Emigration & immigration [contains information on
  certain emigrant families to America] 1905-1924
  115: Regierung Duesseldorf: C B I: Fach 28, Nr. 1.
                                        ADS 6:29

DUESSELDORF (Landratsamt)   RWL

  Emigrations; repatriation   1908-1918
  115: Rep. G 29/9: Nr. 146.   ADS 6:37

DUESSELDORF (Kreis)   RWL

  Emigration   1860-1898
  115: Rep. D8: Neuere Akt.: Abt. 1 C 3: Nr. 11926-
  11939.   ADS 6:26

DUISBURG (Kreis)   RWL

  Emigration   1860-1865
  115: Rep. D8: Neuere Akt.: Abt. 1 C 3: Nr. 11913.
                                        ADS 6:26

DUISBURG (city)   RWL

  Nr. 92: Emigration   1824-1841
  Nr. 93-98: Emigration   1844-1902
  115: Rep. G 24/4: A.: files as above. ADS 6:31

DUNDENHEIM (village)   BWL

  Emigration files   1880-1936
  288: Kommunalarchiv Dundenheim.   ADS 10:108

DURBACH (village)   BWL

  Emigration files   1890-1948
  Immigration files   1898-1931
  List of persons emigrating to USA 1840; 1848-1889
  288: Kommunalarchiv Durbach.   ADS 10:108

DURLACH (Amt)   BWL

  Information regarding the English colonies of
  Pennsylvania & Carolina, etc., collected at the
  request of German subjects & migrants from Switzerland [advantages & difficulties for would-be
  emigrants; may contain some information on very
  early emigrants]   1727-1737
  260: Abt.236: Innenmin., Wegzug: Auswand.: 9847.
                                        ADS 10:5

Emigration   1770-1771
260: 217.   ADS 10:1

Emigration of numerous citizens   1771
260: 216 A.   ADS 10:1

Petitions for emigration permits   1794-1810
260: Abt.236: Innenmin., Wegzug: Auswand.: 1729.
                                              ADS 10:3

DURLACH (city & Amt)   BWL

Emigration from Durlach, Blankenloch, Buechig,
    Groetzingen, to Pennsylvania & Prussia   1770-
260: Rep.Abt.136: Durlach Stadt u. Amt: [file num-
    bers not given]   ADS 10:11

DURLACH. *See also* Oberdurlach

EBENEZER (Effingham County, Georgia)   USA

Diary of [Lutheran congregation] 1736; 1738; 1741;
                    1742; 1744; 1745; 1763
Letters to & from   1761-1769; 1749

500: Halle: Missionsbibliothek des Waisenhauses:
    Abteilung H IV (Nordamerika), Fach J, Nr. 2;
    4a; 5; 6; 8; 10; 11; 12; 15.   Not in ADS

Files on congregation   1780-1799
Letters   1753 (1737)-1796; 1799-1807

500: Halle: Missionsbibliothek des Waisenhauses:
    Abteilung H IV (Nordamerika): unverzeichnet
    [not catalogued]   Not in ADS

EBENEZER (Effingham County, Georgia). *See* Moravian
Brethren, Strasburger Emigrants, Strobel, P. A.

EBERFINGEN (village)   BWL

Emigration files   1907-
Emigration files   1895-
Immigration files   1866-
General files   1890-

288: Kommunalarchiv Eberfingen.   ADS 10:108

EBERHAUSEN (chaplain's archive). *See* Barterode   NL

EBERMANNSTADT (Bezirksamt)   BYL

Individual applications for emigration   1807-1869
223: Rep. K.8.$\frac{1}{-}$: Verz. 1: Nr. 58-1669.   ADS 9:7

EBERSBERG (Landratsamt)   BYL

Nr. 19.181: Emigration & immigration: collection
                of decisions   1908
Nr. 21.253: Naturalizations & losses of Bavarian
                citizenship   1897-1927
Nr. 22.280: Emigration to other countries   1900
Nr. 84.1772: List of passports issued   1915-1930
Nr. 205. 4856: Emigration & immigration; whereabouts

of Bavarian citizens abroad & for-
    eigners in Bavaria   1912-1932
Nr. 205.4857: Emigration & immigration   1862-1912
Nr. 205.4861: Expulsion of foreigners   1921-1932

229: Rep. Landratsaemter: Ebersberg: files as
    above.   ADS 9:118

EBERSTEINBURG (village)   BWL

Emigration files   1845-1873
288: Kommunalarchiv Ebersteinburg.   ADS 10:108

EBERSWEIER (village)   BWL

Emigration files   1832-1864
288: Kommunalarchiv Ebersweier.   ADS 10:108

EBNET (village)   BWL

Emigration to the USA
            1853-1855; 1854-1939; 1910-1938
288: Kommunalarchiv Ebnet.   ADS 10:108

EBRINGEN (village)   BWL

Emigration files   1847-1932
Emigration to the USA   n.d.
Emigration files   1876-1940
Immigration files   1918-1941
Jewish emigration   1936-1941

288: Kommunalarchiv Ebringen.   ADS 10:109

ECKERNFOERDE   SHL

Nr. 4: Old passports   1810-1811
Nr. 5: Passports   1862-1900
Nr. 6: Passports & identification papers   1902-1922

7: Abt.II, G: files as above.   ADS 1:44

EDENKOBEN. *See* Neustadt (Oberamt)   RPL

EDENKOBEN. *See* Kurpfalz   RPL

EFRINGEN-KIRCHEN (villages)   BWL

Emigration files   1872-1944
Emigration files, Kirchen   1808-1941
Immigration files, Efringen   1911-1928

288: Kommunalarchiv Efringen-Kirchen.   ADS 10:109

EGGENFELDEN (Landgericht)   BYL

Applications for emigration permits   1811-1846
225: Rep.168: Verz.1.   ADS 9:16

EGGENFELDEN (Bezirksamt)   BYL

Individual emigration cases to America   1851-1867

225: Rep.164: Verz.4: Nr. 101-.   ADS 9:13

Emigration   [19th century]
225: Rep.168: Verz.1.   ADS 9:15

EGRINGEN (village)   BWL

Emigration file   1839-1945
Germans abroad   1938-1939
288: Kommunalarchiv Egringen.   ADS 10:109

EHINGEN (town)   BWL

Emigration files   1850-1920
General files   1880-1930
Immigration files   1901-1923
288: Kommunalarchiv Ehingen.   ADS 10:109

EHINGEN (Oberamt)   BWL

AZ 1040.3: Emigration from various villages
                                             1892-1938
AZ 6135:   Emigration   1810-1938
262: Wue 65/9: Ehingen, Abt.1: files as above.
                                              ADS 10:38

Passports   1889-1903
261: F 162: Ehingen: 66.   ADS 10:18

EHRENBREITSTEIN (Nassau Regierung)   RPL

Nr. 261: Files on marriages, naturalizations, &
          emigration permits   1809-1816
Nr. 265: Files on emigration & releases from
          citizenship   1811-1812
200: Abt.332: Nassauische Regierung (Ehrenbreit-
          stein): files as above.   ADS 8:14

EHRENBREITSTEIN (mayor's office)   RPL

Petitions for emigration permits
                                      [early 19th century]
200: Abt. 655.10: Ehrenbreitstein: Nr. 550.
                                              ADS 8:50

EHRENBURG (Amt)   NL

Planned emigration of several persons   1829
52: Han.Des.80: Ehrenburg: B.g.III.A.3: Nr. 34.
                                              ADS 4:84

Emigration certificates & releases from citizen-
     ship   1845-1853
52: Han.Des.74: Freudenberg: III,F: Nr. 4.
                                              ADS 4:68

EHRENSTETTEN (village)   BWL

Emigration files   1853; 1854-1934
Emigration to the USA   1854
Emigration & immigration   1878-1934

288: Kommunalarchiv Ehrenstetten.   ADS 10:109

EHRSBERG (village)   BWL

Emigration files; National Union of German Emigra-
     tion & Settlement files   1834-1949
Transportation of dissolute females with their
     children to North America   1852-
288: Kommunalarchiv Ehresberg.   ADS 10:109

EICHSTAETT (Bezirksamt)   BYL

Emigration & immigration, 4 Konvolute   1801-1850
231: Rep.212/4: Eichstaett: Nr. 936.   ADS 9:128

EICHSTAETT (Hochstift, Kloester, Domkapitel [Clois-
     ter & cathedral office])   BYL

Emigration & immigration   1803-1804
231: Rep.190/II: Hochstift . . . Eichstaett: 463.
                                              ADS 9:127

EICHSTAETT (Hochstift Kastenamt Spalt)   BYL

Emigration [to North America?]   1646
231: Rep.225: Hochstift Eichstaett'sche Kastenamt
          Spalt: 1760.   ADS 9:127

EIDERSTEDT (rural area). See Helgoland, Toenning,
     Westerhaver (church)   SHL

EIMELDINGEN (village)   BWL

Emigration files (four fascicles)   1801-1950
288: Kommunalarchiv Eimeldingen.   ADS 10:109

EINBACH (village)   BWL

Emigration files   1855-1925; 1932-1934
288: Kommunalarchiv Einbach.   ADS 10:109

EISENACH. See Saxony-Weimar-Eisenach

EISENTAL (village)   BWL

Emigration files (3 fascicles)   1828-1950
288: Kommunalarchiv Eisental.   ADS 10:110

ELBERFELD (Kreis)   RWL

Emigration   1860-1895
115: Rep.D8: Neuere Akt.: Abt.1, C3: Nr. 11969-
          11977.   ADS 6:26

ELBERFELD (city)   RPL

List of foreigners in Elberfeld   1832
200: Abt.403: Oberpraes.Koblenz (Rheinprov.): 2088.
                                              ADS 8:19

ESCHERSHAUSEN (congregation)   NL

  Removal [including emigration] of members of the
    congregation   1853-1854

  <u>75</u>: I.1.A.105.   ADS 4:254

ESCHERSHEIM (Amt)   HEL

  For exemptions from emigration tax, *see* Hesse-
  Kassel [Kurhessen]

ESCHHOFEN. *See* Limburg   HEL

ESCHWEGE (Kreis)   HEL

  Naturalization, 7 vols.   1822-1867

  <u>151</u>: Bestand 17(b): Gef. 91, Nr. 2.   ADS 7:69

ESCHWEGE (Regierungsbezirk)   HEL

  Naturalizations   1849

  <u>151</u>: Bestand 16/II: Klasse 15, Nr. 18.   ADS 7:66

ESCHWEILER (canton)   RWL

  Population movements   1801-1807

  <u>115</u>: Rep. D2.2: II.Division, 1.Bureau, 2.Bevoelker-
    ung: Nr. 95.   ADS 6:19

ESENS   NL
  Deaths abroad of citizens; death certificates; &
    transmittals of estates to heirs living in Esens
    district   1851-1870

  <u>50</u>: Rep.29: Abt.II.G: Nr. 193.   ADS 4:13

ESSEN (Landratsamt)   RWL

  Nr. 58: Emigration   1869-1882
  Nr. 62: Emigration   1868-1882
  Nr. 139: Expulsions & repatriations   1910-1929

  <u>115</u>: Rep. G29/14: Essen: files as above. ADS 6:38

ESSEN-OBERHAUSEN (Kreis)   RWL

  Emigration   1873-1885

  <u>115</u>: Rep. D8: Neuere Akt.: Abt. 1, C3: Nr. 12002-
    12007.   ADS 6:26

ESSLINGEN (Oberamt)   BWL

  Nr. 73: Emigration lists; indexed   1816-1873
  Nr. 74-81: Emigration files   1851-1890

  <u>261</u>: F 164: Esslingen: files as above. ADS 10:19

ESSLINGEN (town)   BWL

  Material on emigration to be vound in the *Gemeinde-
  rats-Protokollen* [proceedings of the town

council] and in the local newspaper collection
                    [19th century]

<u>269</u>   ADS 10:69

ETTENHEIM (village)   BWL

  Emigration files (2 fascicles)   1851-1895
  Emigration to the USA; emigration to North & South
    America (2 fascicles)   1852-1854
  General files   1853-1947
  Immigration files; Germans abroad   1897-1944
  Emigration documentation   1919-1920; 1941-1949
  Emigration accounts [public financial aid?]
                          1857-1858

  <u>288</u>: Kommunalarchiv Ettenheim.   ADS 10:111

ETTENHEIMMUENSTER (village)   BWL

  Emigration files (3 fascicles)   1852-1932
  Emigration with permit   1852-1949

  <u>288</u>: Kommunalarchiv Ettenheimmuenster. ADS 10:111

EUSKIRCHEN (Landratsamt)   RWL

  Emigration   1865-1892

  <u>115</u>: Rep. G29/15: Euskirchen: Nr. 105. ADS 6:39

EUTIN   SHL

  Nr. 9: Applications & permits to be released from
    citizenship:
    Vol. I   1809 (1819)-1853
    Vol. II   1854-1861
    Vol. III   1862-1864
    Vol. IV   1865-1869
    Vol. V   1870

  <u>2</u>: Abt.Reg. Eutin: 25.1.A.: Nr. 9.
  DLC has?   ADS 1:10

  Recognition & loss of citizenship   1872

  <u>2</u>: Reg. Eutin: 25.1.A.: Nr. 66.   ADS 1:11

  Recognition & loss of citizenship; releases; nat-
    uralizations:
    Conv. I: 1881
    Conv. II: 1882-1888
    Conv. III: 1889-1897
    Conv. IV: 1898-1902
    Conv. V: 1903-1907
    Conv. VI: 1908

  <u>2</u>: Reg. Eutin: 25.1.A.: Nr. 63.   ADS 1:11

EVANSVILLE (Indiana)   USA

  Financial support to Catholic residents of Evans-
    ville for the building of a church   1841

  <u>227</u>: Rep. MA 1895: A.V.: Nr. 1325. ADS 9:57

FALKENSTEIG (village)   BWL

  Emigration & immigration   1875-1942

  <u>288</u>: Kommunalarchiv Falkensteig.   ADS 10:112

FALKENSTEIN (Grafschaft)   RPL

  Emigration & immigration
          1725; 1746-1750; 1757-1761; 1763

  <u>201</u>: Archivabt. Grafschaft Falkenstein: Akt Nr.
    126.   ADS 8:61

FAUTENBACH (village)   BWL

  Emigration files   1863-1948; 1908-1937; 1938-1950
  General files   1911-1949
  <u>288</u>: Kommunalarchiv Fautenbach.   ADS 10:112

FEDERAL (village? Oberpfalz)   BYL

  Emigration tax   1826-1832

  <u>222</u>: Regierung: Kammer der Finanzen: Rep. 118: II.
    Teil: Nr. 9720.   ADS 9:2

FEHMARN. *See* Burg auf Fehmarn

FELDBERG (village; Kreis Muellheim)   BWL

  Emigration files   1896-1950

  <u>288</u>: Kommunalarchiv Feldberg.   ADS 10:112

FESSENBACH (village)   BWL

  Emigration files   1850-1949
  General files   1851-1937

  <u>288</u>: Kommunalarchiv Fessenbach. ADS 10:112

FEUCHTWANGEN (town)   BYL

  Emigration & immigration, vol. VI   1853

  <u>231</u>: Rep. 270/II: Reg. Mittelfranken: Kammer des
    Innern: Abgabe 1932: Nr. 29.   ADS 9:134

FEUCHTWANGEN (Bezirksamt)   BYL

  Nr. 789: Bavarian citizens killed in the [American]
        war with Mexico   1854
  Nr. 1116: Death notices of Bavarian citizens killed
        at the battle of Chickamauga   1863-1864

  <u>231</u>: Rep. 212/6: Feuchtwangen: files as above.
                      ADS 9:128

  Applications for emigration permits to America
                    1867-1868

  <u>231</u>: Rep. 212/6: Fechtwangen: Nr. 1641-1642; 1646;
    1652; 1663; 1673-1674.   ADS 9:128

  2/24/7: Reports on Americans living in Bavaria &
        the induction of returning emigrants for
        military service   1907
  2/52/2: Emigration & immigration   1875
  2/27/4: Emigration to Brazil & Canada   1884
  3/27/1-3: Applications for emigration permits
                    1876-1915

  <u>231</u>: Rep. 212/6: Feuchtwangen: Abgabe 1937, Verz.
    I: files as above.   ADS 9:132

FINTEL   NL

  Emigration certificates   1853-1884

  <u>52</u>: Han.Des.74: Rotenburg: I.A.1.i.: Fach 22, Nr.
    28.   ADS 4:76

FISCHBACH (village)   BWL

  Emigration files   1899-1949
  Correspondence with emigrants abroad   1931-1942

  <u>288</u>: Kommunalarchiv Fischbach.   ADS 10:112

FISCHBACH (mayor's office; Amt Oberstein)   RPL

  Naturalizations & releases from citizenship
                    1853-1856

  <u>200</u>: Abt. 395,2: Amt Oberstein: Nr. 67.   ADS 8:17

FISCHERBACH (village)   BWL

  Emigration file   1832-1883
  Emigration without permits; emigration files
                    1851-1937
  Emigration to the USA   1865
  Emigration files   1865-1937; 1892-1918

  <u>288</u>: Kommunalarchiv Fischerbach. ADS 10:112

FISCHHAUSEN (Landratsamt)   East Prussia?

  Emigration; release certificates   1887-1929
    [contains name, profession, place of residency,
    family members, military service & other infor-
    mation]

  <u>51</u>: Rep. 18: Abt.VIII: Nr. 4.   ADS 4:46

FLENSBURG (Landratsamt)   SHL

  Nr. 752: Emigration permits   1872-1875
  Nr. 751: Emigration permits   1875-1881
  Nr. 747: Emigration permits   1881-1885
  Nr. 734: Applications for certificates of domicile
                    1881-1886
  Nr. 735: Applications for certificates of domicile
                    1886-1891
  Nr. 733: Applications for certificates of domicile
                    1892-1903
  Nr. 732: Applications for certificates of domicile
                    1903-1907

  <u>2</u>: Abt.320: Flensburg: files as above.
    DLC has?   ADS 1:26

FLENSBURG   SHL

  Nr. 30: Releases from Danish & Prussian citizen-
        ship   1776-1886
  Nr. 863: Reports on Flensburger citizens abroad on
        various missions   1763-1866

  <u>8</u>: Alte Abt.: files as above.   ADS 1:45-46

  Emigration; citizenship   1866-1901

  <u>8</u>: Neue Abt.: I C: Nr. 33-48.   ADS 1:46

FLENSBURG. *See also* Butenschoen, C.

FLOERSBACH   HEL

   Applications for emigration to Pennsylvania   1748
   151: Bestand 80: Lit.A, Nr. 6 [Verschiedene]
                                              ADS 7:78

FLORIDA. *See* Louisiana   USA

FOEHR. *See* Boldixum, Suederende, Wyk

FOEHRENTAL (village)   BWL

   Emigration files   1847-1868
   288: Kommunalarchiv Foehrental.   ADS 10:112

FOERSTE (estate)   NL

   Lease of estate   1847
   [*See also* von Oldershausen, F. J. W. O.]
   99: Nr. 317.   ADS 4:298

FOUNTAIN CITY (Wisconsin)   USA

   "History of Fountain City, Wisconsin," [in German]
   by Dr. Peter Schoenhofen, editor of the *Buffalo
   County Republikaner*   1911
   [apparently a typewritten copy, 2 pages]
   200: Abt.701: Nachlaesse, etc.: Nr. 968. ADS 8:53

FRANKENBERG (Landratsamt)   HEL

   Nr. 3: Certificates of original domicile & citizen-
          ship   1828-1907
   Nr. 42: Citizens dying abroad   1870-1898
   Nr. 39: Naturalizations   1870-1885
   Nr. 40: Naturalizations; special files 1874-1908
   Nr. 287: Passports   1906-1923
   Nr. 288: Emigration   1874-1914
   Nr. 289: List of ships' contracts approved
                                         1854-1887
   Nr. 290: List of emigrants   1854-1887
   151: Bestand 180: Frankenberg: Acc.1938/30: files
        as above.   ADS 7:114

   Identification of illicit emigration (avoidance of
     military service obligation)   1869-1878
   151: Bestand 180: Frankenberg: Acc.1938/30.
                                         ADS 7:114

   Releases from citizenship   1886-1912
   151: Bestand 180: Frankenberg: Acc.1938/30: Nr.
        291-312.   ADS 7:114

FRANKFURT/MAIN (city)   HEL

   Emigration matters, 3 vols.   1817-1867
   158: Acta Senatus: B 120: Nr. 16.
        DLC has [originals in 158 destroyed]
                                         ADS 7:119

   List of foreigners in Frankfurt   1920

   152: Abt. 405: Wiesbaden: Rep.405/I,6: Nr. 6770.
                                         ADS 7:152

   List of foreigners in Frankfurt   1934
   152: Abt. 405: Wiesbaden: Rep.405/I,2: Nr. 2927.
                                         ADS 7:152

FRANKFURT/MAIN (Prussian Polizei Praesidium)   HEL

   List of foreigners in Frankfurt, Nr. 149   1863
   152: Abt. 407: Frankfurt: Nr. 721.   ADS 7:154

FRECKENFELD (village; Landkreis Kandel, Kurpfalz)RPL

   Names of emigrants   18th century
   201: Archivabt. Ausfautheiakten.   ADS 8:63

FRECKENFELD (parish)   RPL

   The Lutheran church book contains marginal notations regarding emigrants to America 1722-1788
   201: Archivabt. Kirchenbuecher.   ADS 8:63

FREIBURG AN DER ELBE   NL

   Releases from citizenship:
     Nr. 1: Vol. I   1833-1854
     Nr. 2: Vol. II   1855-1881
     Nr. 6: Vol. III   1882-1883
     Nr. 7: Vol. IV   1884-1886
     Nr. 8: Vol. V   1887-1889
     Nr. 9: Vol. VI   1890-1894
     Nr. 10: Vol. VII   1895-
   52: Han.Des.74: Freiburg/Elbe: I.G6, Fach 87:
       files as above.   ADS 4:67-68

   List of emigration certificates granted   1863
   52: Han.Des.74: Freiburg/Elbe: I.G6, Fach 87:
       Nr. 5.   ADS 4:67

   Information regarding citizens abroad   [n.d.]
   52: Han.Des.74: Freiburg/Elbe: I.G6, Fach 87:
       Nr. 3.   ADS 4:67

   Expulsion of various families:
     Vol. I   1792-1855
     Vol. II   1856-1926
   52: Han.Des.74: Freiburg/Elbe: I.G5, Fach 86:
       Nr. 1; 2.   ADS 4:67

   Emigration & immigration   1859
   52: Han.Des.74: Freiburg/Elbe: I.E, Fach 55: Nr.
       12.   ADS 4:67

FREIBURG (city)   BWL

   Nr. 1053: Applications for emigration permits;
             taxation   1790-1807
   Nr. 1954: Applications for emigration permits
                                        1800-1807
   260: Abt. 236: Innenmin., Wegzug: Auswand.: files
        as above.   ADS 10:3

Information on emigration to the USA   1817-1818

270: *Freiburger Wochenblatt* [weekly newspaper]:
Nr. 13; 26; 29; 84.   ADS 10:70

FREINSHEIM (village). *See* Kurpfalz   RPL

FREIOLSHEIM (village)   BWL

Emigration to the USA   1850-1892

288: Kommunalarchiv Freiolsheim.   ADS 10:113

FREISBACH-GOMMERSHEIM (congregation)   RPL

Marginal entries on emigration to be found in
either the Lutheran or Reformed parish registers   18th century

217   ADS 8:86

FREUDENSTADT (Oberamt)   BWL

Nr. 182-182a: Disposal of property [as a prerequisite of emigration]; renaturaliza-
tions   1801-1841
Nr. 183:   [Emigration files from] Pfalzgrafen-
weiler-Woernersberg   1806-1841
Nr. 183-198: Emigration files   1806-1866

261: F 165: Freudenstadt: files as above.
                                       ADS 10:19

FRIEDBERG (Bezirksamt)   BYL

Individual files (totaling 17) on emigration to
North America   [19th century]

230: Rep. Akten der Bezirksaemter: Friedberg:
Nr. 4718-.   ADS 9:123

FRIEDRICHSBURG (*or* Fredericksburg, Texas)   USA

Nr. 1, Blatt 170: List of names [of colonists]
                                           1847
Nr. 2, Blatt 280: Members of Lutheran congregation
                                           1850

154: CA 4b 1: files as above.
DLC has.   ADS 7:165

FREREN (Amt)   NL

Nr. 87: Emigration   1866-1882
Nr. 88: Emigration   1883-1885
Nr. 115: Emigration without having obtained re-
lease certificates   1883-1885

54: Rep.122, VI.1: Freren: A.3: files as above.
                                       ADS 4:138

Nr. 6078: Emigration permits [for permits issued
before 1871, *see* Osnabrueck]   1871-1879
Nr. 9274: Releases from citizenship   1880

54: Rep. 116,I: files as above.   ADS 4:120-121

FREUDENBERG (Amt)   NL

Emigration certificates & releases from citizen-
ship:
Nr. 1: Vol. I   1819-1839
Vol. II   1840-1859
Nr. 5:   1859-1865
Nr. 6:   1866

52: Han.Des.74: Freudenberg: III,F: files as above.
                                       ADS 4:68

Emigration permits; emigration of persons subject
to military service obligation:
Nr. 7:   1863-
Nr. 8: Vol. I   1867-[1870]
Vol. II   1871-1873
Vol. III   1874-1880
Vol. IV   1881-1883
Vol. V   1884-1885

52: Han.Des.74: Freudenberg: III,F: files as above.
                                       ADS 4:68

FREUDENSTADT (Kreis)   BWL

Some letters from emigrations   [19th century?]

271   ADS 10:71

FRICKINGEN (village)   BWL

Emigration   1847-1933

288: Kommunalarchiv Frickingen.   ADS 10:113

FRIESENHEIM (village)   BWL

Immigration files (6 fascicles)   1805-1872
Emigration files (9 fascicles)   1819-1940
Illicit emigration (2 fascicles)   1844-1859

288: Kommunalarchiv Friesenheim.   ADS 10:113

FRISIA, EAST (rural areas). *See* Ostfriesland

FRITZLAR (Regierungsbezirk)   HEL

Naturalizations   1849

151: Bestand 16.II: Klasse 15: Nr. 19.   ADS 7:66

Files of the higher administrative levels regarding releases from citizenship, vol. II   1851

151: Bestand 17(b): Gef.98: Nr. 51.   ADS 7:72

Inheritances; aid to citizens abroad   1828-1898

151: Bestand 180: Fritzlar: Acc.1931/33: Nr. 37.
                                       ADS 7:114

Inheritances; aid to foreigners in district
                                       1835-1844

151: Bestand 180: Fritzlar: Acc.1931/22.
                                       ADS 7:114

Releases from citizenship (from various villages)
                                        n.d.

151: Bestand 180: Fritzlar: Acc.1958/33: Nr. 18-
55.   ADS 7:114

FRITZLAR (Landratsamt)   HEL

Releases from citizenship   1833-1836

151: Bestand 180: Fritzlar: Acc.1958/33: Nr. 105.
                                               ADS 7:115

Certificates of original domicile & citizenship
                                 1847-1913; 1912-1923

151: Bestand 180: Fritzlar: Acc.1958/33: Nr. 472-
474; 476-477; 478.   ADS 7:116

List of emigrants to America, prepared by emigration agents   1853-1896

151: Bestand 180: Fritzlar: Acc.1958/33: Nr. 68.
                                               ADS 7:115

Nr. 63: Renaturalizations of released citizens
                                             1869-1880
Nr. 65: Renaturalizations of released citizens
                                             1833-1869
Nr. 64: Emigration & immigration   1869-1914
Nr. 67: Emigration & immigration   1914

151: Bestand 180: Fritzlar: Acc.1958/33: files as
above.   ADS 7:115

Nr. 71: Foreigners [in Kreis]   1888-1902
Nr. 75: Foreigners   1903-1904
Nr. 76: Foreigners control   1904-1905
Nr. 78: Foreigners   1907-1916
Nr. 79: Foreigners   1907-1916
Nr. 84: Foreigners   1906
Nr. 80: Foreigners, control of   1908-1912
Nr. 82: List of foreigners in Kreis   n.d.
Nr. 88: Foreigners in Germany at outbreak of war
                                             1914-1915
Nr. 89: Foreigners in Germany at outbreak of war
                                             1917-1922

151: Bestand 180: Fritzlar: Acc.1958/33: files as
above.   ADS 7:115

Releases from Prussian citizenship & issuance of
passports   1893-1913

151: Bestand 180: Fritzlar: Acc.1958/33: Nr. 60.
                                               ADS 7:115

FRITZLAR (Kreis)   HEL

Naturalizations, 5 vols.   1822-1867

151: Bestand 17(b): Gef.91: Nr. 3.   ADS 7:69

FRITZLAR (Kreis). *See also* Homberg   HEL

FRITZLAR-HOMBERG (Landratsamt)   HEL

Nr. 56: Releases from citizenship from villages in
the Kreis   1876-1877
Nr. 58: Releases from citizenship from villages in
the Kreis   1848-
Nr. 57: Petitions for releases from Kurhesse citizenship   1839-1842

151: Bestand 180: Fritzlar: Acc.1958/33: files as
above.   ADS 7:114

Register of emigrants' ship transportation contracts approved [by government]   1858-1897

151: Bestand 180: Fritzlar: Acc.1958?33: Nr. 69.
                                               ADS 7:115

FROITZHEIM (canton)   RWL

Population movements   1801-1807

115: Rep.D2.2: II.Division, 1.Bureau, 2.Bevoelker-
ung: Nr. 96.   ADS 6:19

FUERSTENAU (Amt)   NL

Emigration permits, vols. I-IV   1824-1865
[for 1873-1885, *see* Bersenbrueck, Nr. 139-160]

54: Rep.122,III.1.B: Fuerstenau: Fach 37, Nr. 20.
                                               ADS 4:133

Nr. 6071: Emigration permits [for permits before
1871, *see* Osnabrueck]   1871-1879
Nr. 9275: Releases from citizenship   1880

54: Rep.116,I: files as above.   ADS 4:121

FUERSTENBERG (Principality)   BWL

Manumission material for emigration
                              [18th? & 19th century]

267   ADS 10:67

FUERTH   BYL

Emigration files, alphabetically arranged by name
                                  [19th century?]

239   ADS 9:150

Emigration   1859-1927

149: Rep.40,2: Ablief.2: Bd. 294.   ADS 7:22

FUERTH (city)   BYL

Naturalizations & losses of citizenship 1871-1889

231: Rep. 270/II: Reg. Mittelfranken: Kammer des
Innern: Abgabe 1932: Nr. 728.   ADS 9:134

FUERTH (Bezirksamt). *See* Kadolzburg

FUESSEN (Bezirksamt)   BYL

Emigration from villages of the Amtsbezirk to
North America   1839-1885

230: Rep. Akten der Bezirksaemter: Fuessen: Nr.
437/I.   ADS 9:123

FULDA (Province)   HEL

Nr. 1: Settlement of foreigners in Province Fulda
                                 1831; 1859-1860
Nr. 2: Matters pertaining to property [estates?]
of former Kurhesse citizens resident abroad
                                             1822
Nr. 3: Naturalizations   1839-1860

Nr. 6: Naturalizations   1849-
Nr. 5: Releases from citizenship   1823-

151: Bestand 16.II: Klasse 17: files as above.
                                      ADS 7:66

FULDA (Regierung)   HEL

Nr. 4618: Table of emigration taxes (1789)[may
          contain names]   1790?
Nr. 5467: Travel permits for persons subject to
          military service   1815-1867

151: Bestand 100: files as above.   ADS 7:93

Emigration   1816-1820

151: Bestand 100:a: Nr. 107; 109; 316; 334-336.
                                      ADS 7:95

Nr. 3408: Emigration to America (General)
                                      1822-1853
Nr. 3409: Emigration (General)   1853-1867
Nr. 3399: Emigration (Special)   1823-1843
Nr. 3400: Emigration (Special)   1843-1852
Nr. 3410: Emigration (Special)   1852-1864
Nr. 3411: Emigration (Special)   1865-1867

151: Bestand 100: Acc.1903/9: files as above.
                                      ADS 7:95

Persons who have emigrated to America & have returned [to Germany]   1850-1851

151: Bestand 100:m: Nr. 87.   ADS 7:96

Certificates of original domicile
                    1836-1867; 1852-1865

151: Bestand 100: Nr. 3564; 3565.   ADS 7:92

Releases from citizenship, 2 vols.   1840-1866

151: Bestand 100: Nr. 4624-4625.   ADS 7:93

Death certificates of citizens dying abroad:
    Nr. 4122: Vol. I   1833-1843
    Nr. 4123: Vol. II   1844-1866

151: Bestand 100: files as above.   ADS 7:92

Nr. 6922: Emigration, 18 vols.   1816-1849
Nr. 4626-4632: Naturalizations, vols. III-IX
                                      1822-1867
Nr. 5470: Travel permits   1824-1866

151: Bestand 100: files as above.   ADS 7:93

Naturalizations, 2 vols.   1817-1821

151: Bestand 100:a: Nr. 273-274.   ADS 7:95

FULDA (Landratsamt)   HEL

Emigration to America   1818-1845

151: Bestand 180: Fulda: Acc.1896/19 B: Nr. 15.
                                      ADS 7:116

Emigration to the USA   1833

151: Bestand 180: Fulda: Acc.1896/18: Nr. 203.
                                      ADS 7:116

Emigration tax   1836-1852

151: Bestand 180: Fulda: Nr. 2.   ADS 7:116

Releases from citizenship   1867

151: Bestand 180: Fulda: Acc.1918/21: Nr. 66.
                                      ADS 7:116

Nr. 126: Certificates of citizenship, 4 vol.
                                      1912-1932
Nr. 126?: Certificates of original domicile, 5
          vols.   1903-1932
No number: Register of certificates of original
          domicile issued abroad   1866-1876
No number: List of foreigners   1914-1916

151: Bestand 180: Fulda: Acc.1940/29: files as
     above.   ADS 7:116

No number: Identity papers issued   1921-1935
No number: List of passports   1920-1923
A III 8297: Emigration & immigration 1922-1930

151: Bestand 180: Fulda: Acc.1940/19: files as
     above.   ADS 7:117

FULDA (city)   HEL

Travel permits   1825-1866

151: Bestand 100: Nr. 5469.   ADS 7:93

Emigration to America   1836-1845

159: VI D.   ADS 7:209

Applications for emigration permits:
    Nr. 48-51:   1803-1810
    Nr. 53-61:   1808-1810
    Nr. 77-82:   n.d.

151: Bestand 98: Fulda X: Acc.1875/27: files as
     above.   ADS 7:92

FULDA (Bishopric)   HEL

General Repositur: A22: File on emigrants & provi-
     sion of pastors for them   1868-
In the Pfarrepositturen [parish registers] of
     individual parishes there is much material on
     emigrants & their fate in America; see Personen-
     standbuecher therein   18th & 19th centuries

160   ADS 7:210

FURSCHENBACH (village)   BWL

Emigration files   n.d.

288: Kommunalarchiv Furschenbach.   ADS 10:113

FUERSTENBERG (town)   BWL

Emigration (2 fascicles)   1847-1934
Emigration to the USA   1847-1866
Emigration & immigration   1880-1921
Immigration files   1921-1934

288: Kommunalarchiv Fuerstenberg.   ADS 10:113

GAIENHOFEN (village)   BWL

Emigration to the USA   1858-1924
Immigration   1911-1919

288: Kommunalarchiv Gaienhofen.   ADS 10:113

GAILDORF (Oberamt)   BWL

Nr. 159-163: Emigration files, by village
[19th century?]

261: F 166: Gaildorf: files as above.   ADS 10:19

GAISBURG (*also* Gaisberg). *See* Stuttgart (Polizeipraesidium); Stuttgart (Oberamt)

GALICIA   Austro-Hungarian Empire

Emigration of a number of Galician political refugees to America with the aid of the Austrian
government   1849

39: Bestand Senat: Cl.VIII, No.X, Jg.1849 (Register). 2 [Auswanderer-Verhaeltnisse] ADS 2:154

GALLENWEILER (village)   BWL

Emigration files (2 fascicles)   1835-1924

288: Kommunalarchiv Gallenweiler.   ADS 10:113

GALVESTON (Texas)   USA

Lists of European ships arriving at Galveston 1867

42: Bestand H516: Konsulats-Berichte 1868, Band 1,
A-K (Galveston)(1)(b)(Anlage).   ADS 2:562

Hamburg Consulate in Galveston: estates; searches
for missing persons; legal affairs   1871-1893

39: Bestand Senat: Cl.VI, No.16p: Vol. 4b: Fasc.
9c: Invol. 1-9.   ADS 2:55

GAMMERTINGEN (Oberamt)   BWL

Emigration from various villages (16 fascicles)
1801-1859

262: Ho 193: Gammertingen: I: Nr. 11312-11313;
11315-11316; 11319; 11329; 11413; 11508;
11680; 11904-11906; 11973; 10673; 10946; &
II: 8481.   ADS 10:35

Nr. 13269: Expulsion of a shoemaker to North America because of his dissolute ways
1851
Nr. 9037: [Wuerttemberg citizens] born or dying
abroad   1883-1917
Nr. 9066-9067: Reports on emigration & immigration   1856-1898
Nr. 9072: Emigration & immigration   1851-1922

262: Ho 193: Gammertingen: I: files as above.
ADS 10:35

Nr. 40-58: Emigration from various villages
1853-1936
Nr. 98:   Reservists & persons subject to military service obligation & their emigration without permission   1806-1905
Nr. 465:   Wuerttemberg citizens dying abroad &
their property situation   1869-1925

262: Ho 193: Gammertingen: Acc.3/1956: files as
above.   ADS 10:36

Residence & property situation of persons dying
abroad   1869-1925

262: Ho 193: Gammertingen: IX, 18, 1.   ADS 10:36

GANDERSHEIM (Kreis)   NL

Releases from Braunschweig citizenship:
Surnames A-K, Vol. 1:   1852-1862
Surnames L-W, Vol. 2:   1853-1867

56: L Neu Abt. 129A: Gr.11: Nr. 17.   ADS 4:186

GANDERSHEIM (Amt)   NL

Emigration to America:
Vol. 2:   1847
Vol. 3:   1850-1853
Vol. 4:   1858-1865
Vol. 5:   1866-1870

56: L Neu Abt. 129A: Gr.11: Nr. 9.   ADS 4:185

GANDERSHEIM (town)   NL

Emigration to America:
Vol. 1:   1833-1854
Vol. 2:   1855-1864

56: L Neu Abt. 129A: Gr.11: Nr. 15.   ADS 4:186

GARMISCH (Landratsamt)   BYL

List of emigrants (127 names), many to America
19th century

229: Rep. Landratsaemter: Garmisch: Nr. 98.
ADS 9:118

GAUSBACH (village)   BWL

Emigration files (3 fascicles)   1833-1934
Immigration files   1899-1932

288: Kommunalarchiv Gausbach.   ADS 10:113

GEESTLANDE   HH

Citizenship & residency; files on certificates of
residency, citizenship, & releases from citizenship   1823-1853

39: Bestand Geestlande I: XI (I, Nr. 27) [Proto-
koll]   ADS 2:460

GEILENKIRCHEN (canton)   RWL

Population movements   1801-1807

115: Rep. D2.2: II.Division, 1.Bureau, 2.Bevoelker-
ung: Nr. 97.   ADS 6:19

GEILENKIRCHEN (Landratsamt)   RWL

Nr. 258: Emigration & immigration   1850-1859

Nr. 162: Emigration & immigration 1857-1862
Nr. 300: Emigration & immigration 1855-1858
Nr. 83: Emigration 1817-1833
Nr. 183: Emigration 1835-1850
Nr. 204: Emigration 1877-1892
Nr. 205: Emigration 1892-1903

115: Rep. G 29/16: Geilenkirchen: files as above.
ADS 6:39

GEINSHEIM (Landkreis Gross-Gerau) HEL

Notations on 24 emigration cases 1840-1869
170: Ortsbuergerregister 1796-1881. ADS 7:221

GEISIG HEL

Manumissions 1804-1810
152: Abt.351: Nassau-Vierherrisch: Gen.VIII.e.54.
ADS 7:139

GEISINGEN (village) BWL

Emigration files 1862-1942
Emigration files 1879-1934
General files 1890-1941
Immigration files 1895-1933
288: Kommunalarchiv Geisingen. ADS 10:113

GEISLINGEN (Oberamt) BWL

Emigration; agents; applications for emigration
permits; lists 1811-1902
261: F 167: Geislingen: Nr. 189a-189e. ADS 10:19

GEISSLINGEN (village) BWL

Emigration files 1882-1952
288: Kommunalarchiv Geisslingen. ADS 10:114

GELDERN (Duchy) RWL

Exit taxation, vols. I, IV, & V 1716-1794
115: Rep. C.21: Geldern: Nr. 220. ADS 6:18

GELDERN (Kreis) RWL

Emigration applications 1852-1861
115: Rep. D8: I.Abt.e: Nr. 560-561. ADS 6:25

Emigration 1882-1892
115: Rep. D8: Neuere Akt.: Abt.1, C3: Nr. 11992.
ADS 6:26

GELNHAUSEN (Kreis) HEL

Applications for releases from Kurhessian citizen-
ship:
Nr. 10140: Fasc. I 1822-1825
Nr. 10199: Fasc. II 1823
Nr. 10118: Fasc. III 1824-1825

Nr. 10141: Fasc. IV 1826
Nr. 10092: Fasc. V 1827-1828
Nr. 10091: Fasc. VII 1829
Nr. 10090: Fasc. VIII 1830-1831
Nr. 10089: Fasc. IX 1832
Nr. 10088: Fasc. X 1833
Nr. 10087: Fasc. XI 1834-1835
Nr. 10086: Fasc. XIII 1836-1837
Nr. 10085: Fasc. XIV 1837
Nr. 10143: Fasc. XVIII 1838-1840
Nr. 10142: Fasc. XV 1837
Nr. 10198: Fasc. XIX 1840-1842
Nr. 10084: Fasc. XX 1843-1846
Nr. 10117: Fasc. XVI 1837
151: Bestand 86: files as above. ADS 7:86-87

Naturalization petitions 1867
151: Bestand 82: Acc.1904/13: Nr. 117. ADS 7:84

GELNHAUSEN (Kreis, excluding several Aemter) HEL

Residency permits 1846
151: Bestand 82: Acc.1882/17: Nr. 49. ADS 7:85

GELNHAUSEN (Landratsamt) HEL

Nr. 17: Files of Jean Berk, emigration agent
1870-1881
Nr. 78: Files of foreign emigration agents
1869-1886
Nr. 36: Passports 1924-1934
151: Bestand 180: Gelnhausen: Acc.1958/36 B: files
as above. ADS 7:118

GELNHAUSEN (Amt) HEL

For exemptions from emigration tax, see Hesse-
Kassel [Kurhesse]

GELSENKIRCHEN-BUER (city) RWL

Emigration 1925-1929
116: Muenster Reg.: Nr. 1671. ADS 6:57

GEMMINGEN BWL

Emigration files 1578-1860
260: Rep. Abt. 229: Spezialakten: Gemmingen: Fasz.
31646-31679. ADS 10:11

GEMUEND (canton) RWL

Population movements 1801-1807
115: Rep. D2.2: II.Division, 1.Bureau, 2.Bevoelker-
ung: Nr. 98. ADS 6:19

GEMUENDEN (village; Amt Rennerod) HEL

Nr. 2208: Naturalizations & releases from citizenship, Vol. I 1834-1867
Nr. 2209: Naturalizations & releases from citizenship, particularly Jews, Vol II
1834-1867

Nr. 2210: Naturalizations & releases from citizenship, Vol. III   1834-1867
Nr. 2211: Naturalizations & releases from citizenship, Vol. IV   1834-1867

<u>152</u>: Abt. 237: Rennerod: files as above. ADS 7:135

GEORGIA. *See* Ebenezer (Effingham County), Moravian
Brethren, Salzburger Emigrants   USA

GERABRONN (Oberamt)   BWL

Nr. 409: Emigration lists   1818-1871
Nr. 410-435: Emigration files, chronologically
arranged   1841-1870
Nr. 436: Incomplete emigration   1846-1892

<u>261</u>: F 168: Gerabronn: files as above. ADS 10:19

GERMANY (National Congress)

Nr. 58: Petition from German Union in LeHavre,
France, asking for aid to hundreds of emi-
grants (about 60 signers)   1848-1849
Nr. 63: Emigration petitions   1848-1849

<u>150</u>: Bestand Nationalversammlung: files as above.
ADS 7:25

Nr. 79: Private affairs of Germans in America
1848-1849
Nr. 143: Private affairs of Americans in Germany
1848-1849

<u>150</u>: Bestand Reichsmin. des Aeussern: files as
above.   ADS 7:26

GERMERSHEIM (Landkreis). *See* Bellheim, Ottersheim,
Weingarten, Zeiskam

GERNSHEIM (Landkreis Gross-Gerau)   HEL

The *Familien- und Gemeindebuch der Realschule
Gernsheim zur Einweihung des neuen Schulgebaeude,
1911* [Commemorative Publication to Mark Opening
of New School Building, 1911] lists 27 cases of
emigration to North America   1911?

<u>171</u>   ADS 7:222

GERSFELD (Bezirksamt)   HEL

Passports   1862-1865

<u>151</u>: Bestand 112: Acc.1903/9 LIV: Nr. 12.
ADS 7:97

GERSFELD (Landratsamt)   HEL

Nr. 607: Naturalizations & renunciation of citizen-
ship   1867-1882
Nr. 613: Emigration & immigration 1868-1887
Nr. 609: Illicit emigration   1868-1878
Nr. 605: Passports   1867-1884
Nr. 606: Passports   1880
Nr. 610: Passports   1878-1883
Nr. 611: Passports   1871
Nr. 615: Passports (Passkarten)   1869-1874

Nr. 614: Citizenship certificates   1890-1899
Nr. 612: Certificates of original domicile
1875-1887
Nr. 616: Releases from citizenship   1890-1898
Nr. 691: Releases from citizenship   1895-1901
Nr. 623: Releases from citizenship   1886
Nr. 620: Citizenship   1896-1904

<u>151</u>: Bestand 180: Gersfeld: Acc.1927/26: files as
above.   ADS 7:117

GEVELSBERG (town)   RWL

Emigration & immigration   1884-1907

<u>127</u>: I-2-4.   ADS 6:93

GIEBOLDEHAUSEN (Amt)   NL

Emigration & releases from citizenship:
Nr. 1:   1831-1863
Nr. 2:   1833-1866
Nr. 3: [also for Katlenberg-Lindau & Duderstadt]
1839-1866
Emigration & immigration:
Nr. 10:   1867-1870
Nr. 11:   1871-1873
Nr. 12:   1874-1877
Nr. 13: List of travel documents issued 1874-1911

<u>52</u>: Han.Des.74: Gieboldehausen: VIII.A.III: files
as above.   ADS 4:69

GIESING (town)   BYL

Emigration to North America   1846

<u>229</u>: Rep. Reg. Akten von Oberbayern: Kammer des
Innern: 1245.94.   ADS 9:115

GIMBSHEIM   HEL

Emigration of several families   1848-1849

<u>149</u>: Rep.39/1: Konv.135, Fasz.8.   ADS 7:10

GINSHEIM-GUSTAVSBURG (Landkreis Gross-Gerau)   HEL

Voyages to America of a number of unmarried persons, apparently not for purposes of emigration
1851

<u>172</u>: Abt. XI,4.   ADS 7:223

GLADBACH (Kreis)   RWL

Emigration applications   1852-1860

<u>115</u>: Rep. D8: I.Abt.e: Nr. 564.   ADS 6:25

GLADBACH-RHEYDT (Kreis)   RWL

Emigration   1864-1896

<u>115</u>: Rep. D8: Neuere Akt.: Abt. 1, C3: Nr. 12046-
Nr. 12052.   ADS 6:27

GLADBACH. *See also* Bergisch-Gladbach

GLADBECK (city)   RWL

Emigration   1925-1929
116: Muenster Reg.: Nr. 1670.   ADS 6:57

GLANDORF (Vogtei)   NL

Emigration & immigration   1816-1838
54: Rep. 122.V.3: Glandorf: Nr. 7.   ADS 4:136

GMUEND (Oberamt)   BWL

Emigration files, alphabetically arranged
                                            1819-1872
261: F 169: Gmuend: Nr. 66-70.   ADS 10:19

GOCHSHEIM (town)   BWL

Emigration files   1472-1855
260: Rep. Abt. 229: Spezialakten: Gochsheim Stadt:
        Fasz. 32240-32456.   ADS 10:11

GOEPPINGEN (Oberamt)   BWL

Nr. 279-288: Emigration lists   1817-1872
Nr. 376-383: Emigration files; passport files;
             visas   1897-1915
261: F 170: Goeppingen: files as above.   ADS 10:19

GOEPPINGEN (town)   BWL

Four packages of emigration files
                            18th-20th centuries
272   ADS 10:72

GOERWIHL (village)   BWL

General emigration files   1890-1931
Emigration files   1869-1942
The last five decade of emigration [manuscript?]
                                           1930-1932
288: Kommunalarchiv Goerwihl.   ADS 10:114

GOETTINGEN (Georg-August-Universitaet)   NL

Record book [Stammbuch] of the American colony in
   Goettingen [lists of American students]
                                           1766-1888
77: 8º Ms. Hist. Lit. 108.   ADS 4:259

Information regarding the university careers of
   German-American students can be obtained by
   writing to
      Archiv der Georg-August-Universitaet,
      Goettingen,
      Wilhelmsplatz 1,
      Niedersachsen

GOETTINGEN (town & district)   NL

The Stadtarchiv (Theaterplatz 5, Goettingen) plans

to prepare a file on the emigrants to America tak-
information from tax lists, newspaper clippings,
etc. [reported as of 1958]
                                           ADS 4:265

GOFFENHAYN. See Dillenburg   HEL

GOLDAP (Kreis)   East Prussia (Gumbinnen)

Emigration [file missing?]   1853-1875
51: Rep. 12: Abt.I, Titel 3, Nr. 8, Bd. 3.
                                           ADS 4:32

GOMMERSHEIM. See Freisbach-Gommersheim

GONDELSHEIM BEI BRETTEN (Herrschaft)   BWL

Emigration files   [18th & 19th centuries]
273   ADS 10:76

GONTERSDORF (Dillkreis?). See Dillenburg   HEL

GOTHA. See Saxony-Coburg-Gotha

GOTTENHEIM (village)   BWL

Emigration files   1856-1899
288: Kommunalarchiv Gottenheim.   ADS 10:114

GRAFENAU (Landgericht)   BYL

Applications for emigration permits   1815-1846
225: Rep.168: Verz.1.   ADS 9:16

GRAFENAU (Bezirksamt)   BYL

Emigration   1869-1896
225: Rep.164: Verz.5: Nr. 20-53.   ADS 9:13

Emigration   [19th century]
225: Rep.168: Verz.1.   ADS 9:15

GRASBEUREN (village)   BWL

Emigration files   1882-1918
288: Kommunalarchiv Grasbeuren.   ADS 10:114

GRASDORF AN DER LEINE   NL

Emigration matters   1880-1949
78: I.3.34.A340 [Allgemeines]   ADS 4:266

GREDING   BYL
Emigration files   [19th century?]
241   ADS 9:152

GREENE (Amt)   NL

Emigration to America, 5 vols.   1836-1877
56: L Neu Abt. 129A: Gr.11, Nr. 13.   ADS 4:186

GREFFERN (village)   BWL

Emigration files (2 fascicles)   1836-1937
Immigration   1900-1922
288: Kommunalarchiv Greffern.   ADS 10:114

GREGENSTEIN (Amt)   HEL

Emigration to South Carolina   1753
151: Bestand 17(b): Gef.104a: Nr. 43. ADS 7:73

GREIFENSTEIN (Grafschaft). See Braunfels

GREIN   HEL

Emigration   1859-1927
149: Rep.40,2: Ablief.2: Nr. 294.   ADS 7:22

GRENZACH (village)   BWL

Emigration files   1801-1944
Immigration files   1812-1945
Emigration to the USA (2 fascicles)   1817-1932
288: Iommunalarchiv Grenzach.   ADS 10:114

GREVENBROICH (Kreis)   RWL

Emigration applications   1852-1860
115: Rep.D8: I.Abt.e: Nr. 566.   ADS 6:25

GREVENBROICH (Landratsamt)   RWL

Nr. 163: Emigration   1857-1868
Nr. 127: Emigration   1868-1891
115: Rep.G29/19: Grevenbroich: files as above.
                                      ADS 6:41

GREYTOWN

Negotiations regarding damage claims of Hanseatic
citizens who suffered losses by American bom-
bardment of Greytown on 18 July 1854   1855?
48: B.13.a: Nr. 52.   ADS 3:5

GRIESBACH (Bezirksamt)   BYL

Emigration   [19th century]
225: Rep.168: Verz.1.   ADS 9:15

GRIESBACH (Landgericht)   BYL

Applications for emigration permits   1819-1846

225: Rep.168: Verz.1.   ADS 9:16

GRIESBACH. See also Bad Griesbach

GRIESHEIM (village)   BWL

Emigration files   1854-1945
288: Kommunalarchiv Griesheim.   ADS 10:114

GRIESSEN (village)   BWL

Emigration files   1858-1935
288: Kommunalarchiv Griessen.   ADS 10:114

GRIMMELSHOFEN (village)   BWL

Emigration files   1879-1941
288: Kommunalarchiv Grimmelshofen.   ADS 10:114

GRISSHEIM (village)   BWL

Emigration files (4 fascicles)   1829-1944
Immigration files   1919-1938
288: Kommunalarchiv Grissheim.   ADS 10:114

GROENENBERG (Amt)   NL

Estates of persons dying abroad   1800
54: Rep.122, VII.A.: Groenenberg: Fach 18, Nr. 1.
                                      ADS 4:138

Nr. 77: Releases from citizenship   1880-1881
Nr. 94: Releases from citizenship   1882-1883
Nr. 98: Releases from citizenship   1883-1884
Nr. 103: Releases from citizenship   1885-1886
54: Rep.116,I: II.4.A.1 e: files as above.
                                      ADS 4:121-122

GROENENBERG-MELLE   NL

Nr. 2-32: Emigration & immigration permits [may be
           missing]   1817; 1823-1860
Nr. 10:   Emigration to North America 1846-1888
Nr. 33: Issuance of emigration permits to persons
        subject to military service, 2 vols. 1868
Nr. 34: Issuance of emigration permits to persons
        subject to military service, 2 vols. 1869
Nr. 35: Issuance of emigration permits to persons
        subject to military service, 2 vols. 1870
Nr. 36-52: Releases from citizenship   1871-1892
54: Rep.122,VII.B: Groenenberg-Melle: Regiminalia:
    Fach 159: files as above.   ADS 4:139-140

Emigration permits [for permits issued before 1871,
  see Osnabrueck]   1871-1879
54: Rep.116,I: Nr. 6072.   ADS 4:120

GROETZINGEN. See Durlach   BWL

GROSSAUHEIM (village)   HEL

Petitions of residents for releases from serfdom
& emigration permits   1803-1815

151: Bestand 81: Rep.E: 7.Gemeinde Grossauheim:
Nr. 8. [Gesuche]   ADS 7:82

GROSSAUHEIM (Amt)   HEL

For exemptions from emigration tax, see Hesse-
Kassel [Kurhesse]

GROSSBUNDENBACH (village; Landkreis Homburg, Kurpfalz)   RPL

Names of emigrants   18th century

201: Archivabt. Ausfautheiakten.   ADS 8:63

GROSSCHOENACH (village)   BWL

Emigration files   1913-1929

288: Kommunalarchiv Grosschoenach.   ADS 10:115

GROSSENHAUSEN (Amt)   HEL

For exemptions from emigration tax, see Hesse-
Kassel [Kurhesse]

GROSSENLUEDER (Amt)   HEL

Applications for emigration permits   1810-

151: Bestand 98: Fulda X: Acc.1875/27: Nr. 64-67.
                                        ADS 7:91

GROSS-GERAU (Kreis)   HEL

Applications for emigration permits   1815-1852
Applications for dispensations from military ser-
vice, marriage, & emigration permits   1818-1819
Applications for emigration to North America
                                        1840-1852
Applications for release from Hessian citizenship
for the purpose of emigrating to North America
                                        1840-1857
Lists of debts incurred by emigrants to North America   1845-1846
Funds to assist a number of emigrants to North America   1846
Applications of a number of residents for assistance in emigrating to North America, with appendices (cost estimates)   1846
Notifications to mayors regarding issuance of release [from citizenship] certificates for purpose of emigration to America   1849-1873
List of emigrants with details on travel costs
                                        about 1850
List, by village, of the releases from citizenship
for emigrants   1851-1864
List, by village, of emigrants who have left without obtaining releases from Hessian citizenship
                                             1855

174: Abt.XI,4.   ADS 7:226-227

GROSS-GERAU (Landkreis)   HEL

Names of 36 Jewish persons who have received indemnification for forced emigration to USA
                                        after 1945

174: Wiedergutmachungakten.   ADS 7:227

GROSSKROTZENBURG (village)   HEL

Petitions of residents for releases from serfdom
& emigration permits   1805-1811

151: Bestand 81: Rep.E: 8.Gemeinde Grosskrotzenburg: Nr. 9 [Gesuche]   ADS 7:82

For exemptions from emigration tax, see Hesse-
Kassel [Kurhesse]

GROSSTADELHOFEN (village)   BWL

Emigration to the USA with community guarantees
                                        1847-1851
Emigration files   1875-1933

288: Kommunalarchiv Grosstadelhofen.   ADS 10:115

GROSS-UMSTADT   HEL

Emigrations to Pennsylvania   1793
149: Rep.51: StZ: Z29: A-Z.   ADS 7:1

Emigrants to St. Louis   1862
149: Rep.51: StZ: Z7: U-Z.   ADS 7:1

Emigrants to New Orleans   1874
149: Rep.51: StZ: Z6: S-T.   ADS 7:1

GROSS-ZIMMERN   HEL

Applications of a great number of residents for
assistance in emigrating   1846-1847

149: Rep.39/1: Konv.135, Fasz.5.   ADS 7:10

GROZINGEN (Amt)   BWL

Emigration   1770-1771
260: 217.   ADS 10:1

GRUENINGEN (village)   BWL

Emigration files   1834-1862
General files   1905-1938

288: Kommunalarchiv Grueningen.   ADS 10:115

GRUMBACH (mayor's office)   RPL

Summary of property owned by would-be emigrants;
emigration tax; inheritance claims   1834

200: Abt.655.89: Grumbach: Nr. 37.   ADS 8:52

GRUNERN (village)   BWL

  Emigration files (4 fascicles)   1934-1923
  Immigration files   1919-1942
  288: Kommunalarchiv Grunern.   ADS 10:115

GUDOW   SHL

  Emigration & immigration, Vol. 1   1731-1816
  29: Archiv zu Gudow: Nr. 3.   ADS 1:61

GUENTERSEN (parish archives). See Barterode   NL

GUENZBURG (Bezirksamt)   BYL

  Emigration & immigration   1853-1861
  230: Rep. Akten der Bezirksaemter: Guenzburg: 1605.
                                                   ADS 9:123

GUENZGAU. See Stetten (with Guenzgau)   BWL

GUIANA   South America

  Emigration to Guiana & Island of Cayenne   1764
  151: Bestand 98: Fulda: Acc.1875/27: Nr. 2.
                                                  ADS 7:91

GUMBINNEN (Regierung)   East Prussia

  Applications for emigration to America [498 pages;
    often gives reasons for wishing to emigrate]
                                1845-1852
  51: Rep.12: Abt.I, Titel 3, Nr. 10, Bd. 1. See
    also Bd. 2 & Bd. 3.   ADS 4:33

  Various applications for releases from citizenship
                                1889-1890
  51: Rep.12: Abt.I, Titel 3, Nr. 10, Bd. 9.
                                      ADS 4:39

GUMBINNEN (Kreis)   East Prussia

  Emigration [file missing?]   1853-1875
  51: Rep.12: Abt.I, Titel 3, Nr. 8, Bd. 4. ADS 4:32

GUMMERSBACH (Landratsamt)   RWL

  Nr. 1014: Emigration permits   1816-1850
  Nr. 603: Emigration   1850-1868
  Nr. 307: Emigration   1887-1926
  Nr. 308: Emigration   1926-1931
  Nr. 304: Citizenship matters [including list of
      residents, including some Americans]
                                1914-1929
  115: Rep. G 29/20: Gummersbach: files as above.
     DLC may have Nr. 1014.   ADS 6:41

GUENDLINGEN (village)   BWL

  Emigration files   1908-1944

288: Kommunalarchiv Guendlingen.   ADS 10:115

GUETTINGEN (village)   BWL

  Emigration to the USA & other countries 1881-1912
  288: Kommunalarchiv Guettingen.   ADS 10:116

GUNDELFINGEN (town)   BWL

  Immigration files   1817-1869
  General emigration files   1841-1869
  Emigration files   1870-1940
  288: Kommunalarchiv Gundelfingen.   ADS 10:115

GUNDHOLZEN (village)   BWL

  Emigration to the USA   1840-1906
  288: Kommunalarchiv Gundholzen.   ADS 10:115

GUNZENHAUSEN (Bezirksamt)   BYL

  Nr. 1244: Naturalizations & losses of citizenship
                                  1874-1890
  Nr. 1245-1247: Issuances of certificates of orig-
           inal domicile   1898-1911
  231: Rep. 212/8: Gunzenhausen: files as above.
                                      ADS 9:129

GUNZENHAUSEN   BYL

  Emigration & immigration   1852
  231: Rep. 270/II: Reg. Mittelfranken: Kammer des
      Innern: Abgabe 1932: Nr. 27.   ADS 9:134

GUNZENHAUSEN (town)   BYL

  Nr. 16: General files on Jews   1935-
  Nr. 17: List of 21 Jews who emigrated, with de-
      tails   1934-1935
  242: files as above.   ADS 9:153

GUNZENHAUSEN (Mittelfranken). See Bek, Christoph

GURTWEIL (village)   BWL

  Immigration giles   1919-1924
  288: Kommunalarchiv Gurtweil.   ADS 10:115

GUSTAVSBURG. See Ginsheim-Gustavsburg   HEL

GUTACH (village: Kreis Emmendingen)   BWL

  Immigration files   1815-1875
  Emigration files   1817-1867
  Foreigners   1839-1850
  288: Kommunalarchiv Gutach.   ADS 10:115

GUTACH (village; Kreis Wolfach)   BWL

Emigration to the USA   1853
General emigration files (3 fascicles)   1790-1942
Immigration files   1887-1938
Germans abroad & list of Germans resident abroad
                                           1931-1939
Emigration during the last five decades [manu-
    script?]   1932
Returning emigrants   1932-1942

288: Kommunalarchiv Gutach.   ADS 10:116

GUTENSTEIN (village)   BWL

Immigration files   1878-1930
Emigration files   1879-1942

288: Kommunalarchiv Gutenstein.   ADS 10:116

HAAGEN (village)   BWL

Emigration files   1834-1951
Summary of emigration in the last five decades
    [manuscript?]   1932
Emigration to the USA   1923-1951

288: Kommunalarchiv Haagen.   ADS 10:116

HACHENBURG (Amt)   HEL

Nr. 3974: Emigration   1817
Nr. 3978: List of persons naturalized or released
          from citizenship   1860-1866

152: Abt. 224: Amt Hachenburg: files as above.
                                       ADS 7:131

HACHENBURG. See also Sayn-Altenkirchen (und Hachenburg)

HADAMAR (Amt)   HEL

Nr. 74: Naturalizations & releases from citizen-
        ship   1840-1846
Nr. 54: Emigration   1846-1884
Nr. 351: Report on persons naturalized or released
         from citizenship   1853-1854

152: Abt.225: Amt Hadamar: files as above.
                                       ADS 7:131

HADELN (Land). See Otterndorf   NL

HADERSLEBEN (Landratsamt)   SHL

Nr. 25: Files regarding releases from citizenship
        & naturalization applications   1868-1915
Nr. 26: Files regarding releases from citizenship
        & naturalization applications   1917-1919
Nr. 69: Files on granting & loss of citizenship
                                         1896-1913

2: Abt.320 (Hadamar): files as above.
   DLC has?   ADS 1:26

HAGEN (Kreis)   RWL

Nr. 42: Emigration   1831-1869
Nr. 43: Emigration   1852-1860
Nr. 44: Immigration of foreigners   1859-1874
Nr. 44a: Immigration of foreigners   1817-1863
Nr. 45: Immigration of foreigners   1819-1852

116: Arnsberg Reg.: Hagen: 1.A: files as above.
                                       ADS 6:58-59

HAGEN (Amt)   NL

Applications for releases from citizenship & certi-
    ficates granted   1854-1856

52: Han.Des.74: Hagen: Regiminalia, Fach 41: Nr. 4.
                                       ADS 4:69

HAGEN (Vogtei)   NL

Emigration permits   1836-1845

54: Rep.122.V.4: Iburg: Nr. 11.
    DLC has.   ADS 4:136

HAIGERLOCH (Oberamt)   BWL

Emigration   1640-1854

262: Ho 202: Haigerloch: II: Nr. 1158-1159.
                                       ADS 10:37

Emigration from various villages of district
                                          1779-1924

262: Ho 202: Haigerloch: Reg. Nr. II 8: Zettel-
     reportorium.   ADS 10:37

Nr. 3918: Emigration   1831-1854
Nr. 4102: Reports on emigrants   1864-1870
Nr. 1459-4500: 2627 death certificates for emi-
               grants   1853-1912

262: Ho 202: Haigerloch: I: files as above.
                                       ADS 10:37

HAILER (Amt)   HEL

Emigration applications   1817-1821

151: Bestand 86: Nr. 9963.   ADS 7:86

HALL (Oberamt)   BWL

Nr. 115: Emigration & immigration lists 1818-1890
Nr. 123-141: Emigration files, aphabetically ar-
             ranged   19th century
No Nr.: Naturalization files   1803-1844

261: F 171: Hall: files as above.   ADS 10:20

HALLE (Kreis)   RWL

Emigration & immigration   1853-1876

116: Arnsberg Reg.: Halle: Nr. 358.   ADS 6:63

Nr. 32-34: Issuance of certificates of original
           domicile   1863-1924
Nr. 75: Certificates of citizenship   1899-1924
Nr. 123-126: Emigration   1873-1924

116: Minden: Praes.-regist.: I.A.: files as above.
ADS 6:52

HALLENBERG (town)   RWL

Emigration   1825-1866
128: II.11a, fasc.16: III.1b, fasc. 1.   ADS 6:95

HAMBACH   HEL

Emigration   1865-1931
149: Rep.40,2: Ablief.2: Nr. 295.   ADS 7:22

HAMBORN. *See* Duisburg   RWL

HAMBURG [emigration matters]   HH

Emigration lists, direct [via port of Hamburg] 399
  vols.   1850-1914
Emigration lists, indirect [via other north German
  ports] 118 vols.   1854-1910
  [These lists are the most important sources of
  information on emigration from Germany during
  the period covered.]

39: Bestand Auswandereramt.
    DLC has volumes 1850-1873.   ADS 2:458

Hamburg Consulate in New York: Emigration matters;
  various requests by private persons 1846-1868

39: Hanseat. u. Hamb.dipl.Vertretungen: Nr. 16
    [formerly Cl.VI, No.16p, vol.4c, Fasc.2]
    DLC may have part of this file. ADS 2:407

Hanseatic Ministry in Gerlin: Emigration matters
                                    1867-1877

39: Hanseat. u. Hamb.dipl.Vertretungen: HG VI,
    v. 1 [Auswanderungssachen]   ADS 2:399

Hanseatic Ministry in Berlin: Emigration, vol. I
  [partly in Staatsarchiv Luebeck]   1868-1894
  Vol. II   1896-1897

39: Hanseat. u. Hamb.dipl.Vertretungen: V.b.2.
                                     ADS 2:398

HAMBURG [prisoners transported]   HH

Transport [exile] of Hamburg prisoners to America
                                    1752
  [in *Hamburgische Geschichts- und Heimatsblaet-
  ter*, 16.Jg. (1957) Nr. 3, pages 49-54] ADS 2:607

Applications of prisoners from the workhouse & the
  jail to be transported to America 1773

39: Bestand Senat: Cl.VIII, No.X, Jg.1773 (Regis-
    ter)[Gesuche]   ADS 2:98

Transportation to America of some prisoners from
  the workhouse   1787?

39: Bestand Senat: Cl.VIII, No.X, Jg.1787 (Regis-
    ter)   ADS 2:103

Application of a shipper for a young male prisoner
  to be sent to America under a six-year contract

[indenture]   1792
39: Bestand Senat: Cl.VIII, No.X, Jg.1792 (Regis-
    ter) [Erbieten]   ADS 2:104

Attempts by certain prisoners to be transported
  overseas   1832

39: Bestand Senat: Cl.VIII, No.X, Jg.1832 (Regis-
    ter): A.I.4 [Versuch]   ADS 2:131

Regarding criminals sent from Hamburg to America
                                    1833

39: Bestand Senat: Cl.VIII, No.X, Jg.1833 (Regis-
    ter): I.1. [Nordamerika]   ADS 2:132

HAMBURG [inquiries regarding private persons] HH

Hamburg Consulate in San Francisco: Correspondence
  with various Hamburg offices & with private persons   1850-1862

39: Hanseat. u. Hamb.dipl.Vertretungen: 6 [former-
    ly Cl.VI, No.16p, Vol.4c, Fasc.1]
    DLC may have part of this file. ADS 2:408

Hanseatic Ministry in Washington: Various inquiries regarding private persons   1863-1867

39: Hanseat. u. Hamb.dipl.Vertretungen: C.11.
                                     ADS 2:406

Hanseatic Ministry in Washington: Inquiries regarding persons from the Hanseatic cities &
  Oldenburg   1863-1867

39: Hanseat. u. Hamb.dipl.Vertretungen: C.8.
                                     ADS 2:406

Hamburg Consulate in Baltimore, Maryland: Searches
  for information regarding persons   n.d.

39: Bestand Senat: Cl.VI, No.16p: Vol.4b: Fasc.4c:
    Invol. 5. [Auskunft]   ADS 2:52

HAMBURG [passports, visas, citizenship matters] HH

Ministry of the Four Cities to the Bundestag in
  Frankfurt/Main:
  Nr. a.3: Passport & citizenship matters
                                    1852-1858
  Nr. a.5: Passport matters   1864

39: Hanseat. u.Hamb.dipl.Vertretungen: Fasc. III:
    files as above.   ADS 2:403

Hamburg Consulate in New York: Passports & residency matters; individual cases   1861-1863

39: Hanseat. u.Hamb.dipl.Vertretungen: 18 [for-
    merly Cl.VI, No.16p, Vol.4c, Fasc. 2]
    DLC may have part of this file. ADS 2:408

Hamburg Consulate in New York: List of passports &
  visas issued   1864

42: Bestand H516: Konsulats-Berichte 1865, Band 2,
    M-Z (New York)(H)(Anlage).   ADS 2:554

Hamburg Consulate in New York: List of certificates of citizenship issued to Hamburg citizens
                                    1864

42: Bestand H516: Konsulats-Berichte 1865, Band 2,
    M-Z (New York)(G)(Anlage).   ADS 2:553

Hanseatic Ministry in Berlin: Citizenship matters
[file transferred to Staatsarchiv Luebeck] 1870-

39: Hanseat. u. Hamb.dipl.Vertretungen: [Neu-
ere Reg.]   ADS 2:398

Hanseatic Ministry in Berlin: Hamburg citizenship
matters, 4 vols.   1879-1920

39: Hanseat. u. Hamb.dipl.Vertretungen: GHVII.n.1
[Neuere Reg.]   ADS 2:399

Hanseatic Ministry in Berlin: Certificates of residency [file transferred to Staatsarchiv Luebeck]
1909

39: Hanseat. u. Hamb.dipl.Vertretungen: N.9 [Neu-
ere Reg.]   ADS 2:398

Domicile & citizenship of Hamburg citizens abroad
n.d.

39: Bestand Senat: Cl.I, Lit.T, No.14, Vol.4,
Fasc. 5b [Heimat]   ADS 2:5

HAMBURG [marriages]   HH

Marriages of emigrants   1860

39: Bestand Senat: Cl.VIII, No.X, Jg.1860 (Register) 4 [Auswanderer Verhaeltnisse]   ADS 2:177

Note: The original records of the American Consulate in Hamburg, now in the National Archives,
Washington, D.C., also contain many marriages of
emigrants performed by consular officials.

HAMBURG [death & estate matters abroad]   HH

Hanseatic Ministry in Washington: Obtaining &
legalizing [notarizing] death certificates for
citizens of the Hanseatic cities   1863-1867

39: Hanseat. u. Hamb.dipl.Vertretungen: C.9.
ADS 2:406

Estate matters   1872

39: Bestand Senat: Cl.VIII, No.X, Jg.1872 (Register) 55 [Nordamerika]   ADS 2:222

Deaths abroad   1895

39: Reichs- u. auswaert.Ang.: G.I.a.1 [alt.Sig.]
[Geschaeftsverkehr]   ADS 2:275

HAMBURG [births & deaths aboard ship]   HH

Abstracts from ships' logs regarding births &
deaths on Hamburg ships [reports to the civil
Standesamt]   1868-1875

39: Bestand Wasserschout: VIII.   ADS 2:455

Births & deaths on Hamburg ships   1873

39: Bestand Senat: Cl.VIII, No.X, Jg.1873 (Register) 7 [Nordamerika]   ADS 2:224

Hamburg Consulate in New York: Report on births &
deaths aboard Hamburg ships   1874

39: Bestand Senat: Cl.VIII, No.X, Jg.1874 (Register) 11 [Nordamerika]   ADS 2:227

HAMBURG [ship arrivals & departures]   HH

Lists of all ships calling at the port of Hamburg
with [description of] cargoes [may include early
ship lists]   1778-1811

39: Bestand Senat: Cl.VIII, No.I: Nr. 108-137.
DLC has.   ADS 2:456

Register of all ships entering & leaving the port
of Hamburg [There are many of these registers
arranged in several ways; for further information, see ADS citation.]   1801-1899

39: Bestand Zolljacht.   ADS 2:456-458

Ship arrivals & departures [many volumes]
1814-1900

39: Handelsstatistisches Amt: Ablief. v. 1919: G:
Seeschiffahrt.   ADS 2:447-450

Records of the Shipping & Port Commission, 17 vols.
1814-1864

39: Bestand Senat: Cl.VIII, No.XXXVIII [Protokolle
der Schiffahrt- und Hafen Deputation]
ADS 2:265-266

Ships' papers   1798-1830

39: Bestand Senat: Cl.VII, Lit.K^e, No. 7b.
ADS 2:266

Records of the Commission for Ships' Papers, several volumes   1837-1863

39: Bestand Senat: Cl.VIII, No.XLII [Protokolle
der Kommission fuer die Schiffspapiere]
ADS 2:266

List of Hamburg vessels arriving at New York during 1845   1846

42: Bestand H516: Hamb.Konsulat in New York: Kon-
sulats-Bericht 1846 (1) Anlagen.   ADS 2:508

List of emigrant ships from the port of Hamburg
1850-1914

39: Bestand Auswandereramt.   ADS 2:459

HAMBURG [crew lists, desertions, estates of sailors]

a-u: Lists   1760-1845
v: Alphabetical lists, according to captains'
names, ships' names, & destinations 1814-1820
w: Register, arranged alphabetically according
to captains' names   1760-1813
x: Register, arranged alphabetically according
to ships' names   1760-1813
y: Register, arranged alphabetically according
to destinations   1760-1813

39: Bestand Wasserschout: I.A.1: files as above.
ADS 2:451

Inserierungsbuecher: Alphabetical registers of
seamen who were mustered aboard ships in Hamburg
1796-1857

[These registers were not continued after 1857,
but material in the Seemannspenionskasse (Seamen's Pension Fund), also in the Hamburg Staatsarchiv, carry on these lists from 1857 to 1874.
After 1874, seamen were listed by the Seemanns-

Amt (Seamen's Office). *See also* the *Inserier-ungsbuch* for seamen who mustered aboard foreign ships in the port of Hamburg, 1857-1873 (I.C.3)]

<u>39</u>: Bestand Wasserschout: I.A.2.a-q.   ADS 2:452

Nr. 1.a-e: Hamburg sailing vessels: muster rolls
1846-1857
Nr. 2.a-d: Hamburg steamships: muster rolls, lists
1841-1865
Nr. 3.a: Muster rolls, 22 vols.   1857-1873
[*See also* the on-and-off muster rolls, 1857-1874, in the Archiv der Seemanns-Pensionskasse (call number VI 10 L) which contains a name register.]
Nr. 3.b: Completed muster rolls, 55 vols.
1831-1873

<u>39</u>: Bestand Wasserschout: I.B: files as above.
ADS 2:452

Nr. 1.a-b: Muster rolls (mustering aboard)
1846-1873
Nr. 2.a-d: Rough copies of muster rolls 1850-1873
Nr. 3:   *Inserierungsbuch* (alphabetical register of seamen mustering aboard foreign ships in Hamburg)   1857-1873

<u>39</u>: Bestand Wasserschout: I.C: files as above.
ADS 2:453

Collection of seamen's books (from consular files) arranged alphabetically   n.d.

<u>39</u>: Bestand Wasserschout: III.K.   ADS 2:455

Nr. A.1-10: Muster rolls for Hamburg & foreign sailing vessels   1792-1845
Nr. A.11:   Register of the off-muster lists [ar-ranged alphabetically according to captains' names, ships' names, and places of origin]   1814-1831
Nr. B.1-4: Off-muster lists for Hamburg sailing vessels [pages 216-229 contain lists for Hamburg steamships for February-May 1857]   1846-1857
Nr. C.:   Off-muster lists for Hamburg steam-ships   1841-1856
Nr. D.:   Off-muster lists for all Hamburg ships
1857-1873

<u>39</u>: Bestand Wasserschout: II: files as above.
ADS 2:453

Nr. A.1: Lists, entitled *Kopiebuch von allerhand Vorfaellen und Totenscheinen* [copy book for all kinds of cases & death certifi-cates]   1768-1831
[Of particular importance to American genealogists are the cost accounts for searches for deserters, 1778-1830. In addition, the book contains death certi-ficates & receipts for the disbursement of monies from seamen's estates, 1783-1894.]
Nr. B.1: Cost accounting for the arrest of seamen [deserters?]   1801-1830
Nr. B.2: Further receipts for disbursement of mon-ies from seamen's estates   1801-1839

<u>39</u>: Bestand Wasserschout: III: files as above.
ADS 2:454

Nr. C.: Statements regarding the deaths of seamen & the administration of their estates, 5

vols. (cases 1-3386)   1832-1873
[also contains extracts from III.E]
Nr. D.: Receipts for money disbursed from seamen's estates, 3 vols.   1832-1876
Nr. E.: Statements regarding the deaths of seamen, 4 vols. (cases 365-3386; 1-491) 1842-1872
Nr. F.: Extracts from Nr. C above   1842-1857
Nr. G.: Rough drafts of Nr. C above   1866-1873

<u>39</u>: Bestand Wasserschout: III: files as above.
ADS 2:454

Correspondence with Hamburg Consul in New York: Deserters from Hamburg ships   1860

<u>39</u>: Bestand Senat: Cl.VIII, No.X, Jg.1860 (Regis-ter) II.1 [Amerika]   ADS 2:177

Hamburg Consul in New York: Report on desertion of [Hamburg] seamen   1861

<u>39</u>: Bestand Senat: Cl.VIII, No.X, Jg.1861 (Regis-ter) I [Consulats-Angelegenheiten] ADS 2:180

For further information on Hamburg seamen's deaths and estates, *see also* Altona

HAMBURG [reparation claims]   HH

Hanseatic Ministry in Washington: Reparation claims   1862; 1866

<u>39</u>: Hanseat. u. Hamb.dipl.Vertretungen: C.4.
ADS 2:406

Hanseatic Ministry in Washington: Reparation claims of Hamburg citizens   1862-1866

<u>39</u>: Hanseat. u. Hamb.dipl.Vertretungen: C.5; B.12. [Reklamations-Vorderungen]   ADS 2:406

HAMBURG [extradition cases]   HH

Nr. 1-10: Hamburg police searches for persons wanted in America   1910
Nr. 1-5: Hamburg police searches for persons wanted in America   1911
Nr. 1-5: Hamburg police searches for persons wanted in America   1912
Nr. 1-9: Hamburg police searches for persons wanted in America   1913
Nr. 1-3: Hamburg police searches for persons wanted in America   1914
Nr. 1-12: Hamburg police searches for persons wanted in America   1915

<u>39</u>: Reichs- u. auswaert.Ang.: P.III.1: files as above [Polizeiwesen; Polizeisachen]
ADS 2:364; 370; 375; 381; 386; 389

HAMBURG [miscellaneous]   HH

Repatriation of five insane German nationals from New York   1918

<u>39</u>: Reichs- u. auswaert.Ang.: H.IIb.5 [Heimschaff-ungsverkehr]   ADS 2:394

List of subscribers in Hamburg to socialist news-papers *Glueckauf*, *Freiheit*, *Mahnruf* 1879-1882

<u>39</u>: Polizeibehoerde: Polit.Pol.: S. 9. ADS 2:411

Marriages of American citizens before German officials   1899

39: Reichs- u. auswaert.Ang.: S.IV.7 [Standesamtsachen]   ADS 2:305

Hanseatic Ministry in Berlin: Hearings & testimony taken abroad, 5 vols.   1895-1913

39: Hanseat. u. Hamb.dipl.Vertretungen: HGVII.o.3.
                                                   ADS 2:399

Hanseatic Ministry in Washington: Index to general reports [may contain some material of genealogical use]   1862-1867

39: Hanseat. u. Hamb.dipl.Vertretungen: B.2.
                                                   ADS 2:404

Hamburg Consulate in New York: Alphabetic index to letter book [letters sent to various private persons]   1852-1854

39: Hanseat. u.Hamb.dipl.Vertretungen: 3 [formerly Cl.VI, No.16p, Vol. 4c, Fasc.2]
     DLC may have parts of this file   ADS 2:407

HAMELN (Amt)   NL

Transportation of persons of bad repute to America
                                                        1836-

52: Han.Des.74: Hameln: VIII.G.5: Nr. 10. ADS 4:71

Nr. 90: Emigrant lists   1859
Nr. 101: Annual report of emigration   1868

52: Han.Des.74: Hameln: I.C.: files as above.
                                                   ADS 4:70

Nr. 6: Emigration certificates   1861
Nr. 7: Releases from citizenship [missing?]   1874
Nr. 8: Certificates of residency [missing?]   1871

52: Han.Des.74: Hameln: III.F: files as above.
                                                   ADS 4:71

HAMELN.   See also Fraenckel, Simon   NL

HAMM (Kreis).   See Unna   RWL

HAMMELBACH   HEL

Emigration   1865-1931

149: Rep.40,2: Ablief.2: Nr. 295.   ADS 7:22

HANAU (Grafschaft)   HEL

File on emigrants to Pennsylvania   1748-1749

151: Bestand 86: Nr. 210.   ADS 7:86

HANAU   HEL

Emigration to America & other foreign countries
                                                        1785

151: Bestand 98: Fulda X: Acc.1875/27: Nr. 9.
                                                   ADS 7:91

HANAU (Regierung)   HEL

Nr. 92: Reasons for desiring releases from citizenship & certificates required   1818-1854
Nr. 93: Emigration to Brazil & North America
                                    1825-1834; 1835-1848

151: Bestand 82: Reg. Hanau: files as above.
                                                   ADS 7:84

HANAU (Province)   HEL

Nr. 3: Releases from citizenship   1829-
Nr. 3a: Releases from citizenship   1825
Nr. 5: Matters pertaining to the property [estates?] of Kurhessian citizens resident abroad   1821
Nr. 7: Naturalizations, 13 vols.   1832-

151: Bestand 16.II: Klasse 18: files as above.
                                                   ADS 7:67

Releases from citizenship   1829

151: Bestand 16.II: Klasse 15: Nr. 11a.   ADS 7:66

Gathering of information regarding Hanau-born persons living abroad & persons living in Hanau born abroad, Vols. II-IV   1842-1867

151: Bestand 82: Acc.1904/13: Nr. 164.   ADS 7:85

Collection by Kurhessian citizens of the proceeds from estates in America   1860-1863

151: Bestand 82: Acc.1904/13: Nr. 165.   ADS 7:85

HANAU (Kreis)   HEL

Petitions for release from Kurhessian citizenship
                                        1837; 1838-1864

151: Bestand 82: Reg.Hanau: Nr. 95.   ADS 7:84

HANAU (city)   HEL

Residency permits   1832-1862

151: Bestand 82: Acc.1882/17: Nr. 48.   ADS 7:85

HANAU (Landratsamt)   HEL

In 1960, when the ADS inventory was being taken, the Archives were being recatalogued and, therefore, material could not be included. It is likely that the Archives contain much data of interest to researchers.

151   ADS 7:118

HANAU-LICHTENBERG (Grafschaft).   See Niederbronn, Froeschen, Scheix

HANAU-MUENZENBERG (Grafschaft)   HEL

Emigration & immigration   1750

151: Bestand 86: Nr. 51964.   ADS 7:88

HANAU. *See also* Hesse-Hanau   HEL

HANNOVER   NL

Emigration of subjects to foreign countries   1713

<u>52</u>: Han.Des.93: Innere Landesverw.: 38,A, Nr. 1d.
          ADS 4:95

Archives contains a large quantity of unarranged
  releases from citizenship & military service
  obligations   1823-1865

<u>52</u>: Han.Des.45: II.   ADS 4:61

Emigration to America   1830-1869

<u>52</u>: Han.Des.80: Hannover: A.II.B.I.3.c: Nr. 421.
  DLC has.   ADS 4:82

Releases from citizenship; emigration:
  Vol. I: 1868-1941
  Vol. II: 1942-1945

<u>52</u>: Han.Des.122a: Oberpraes.Hannover: I: Nr. 98b.
          ADS 4:95

Nr. 43.IV.01: Releases from Prussian citizenship
          1877-1892
Nr. 54.4. 01: Files on emigration   1867-1888

<u>79</u>: Rep.2: files as above.   ADS 4:267

Transportation of useless persons to America
          1834-1868

<u>52</u>: Han.Des.80: Hannover: A.II.B.III.1.c: Nr. 626.
          ADS 4:83

Transportation of criminals & asocial persons to
  America   1845

<u>52</u>: Han.Des.74: Di, Fach 102: Nr. 11.   ADS 4:64

Former Hannoverian citizens & refugees subject to
  military duty now living abroad:
  No. 60: Vol. I   1867-1868
  No. 61: Vol. II   1868
  No. 62: Vol. III   1868-1869
  No. 63: Vol. IV   1870
  No. 64: Vol. V   1870-1878
  No. 65: Vol. VI   1879-1888
  No. 66: -   1870-1916

<u>52</u>: Han.Des.122a: Oberpraes.Han.: XXXIV: files as
  above.   ADS 4:97

Nr. 58: Emigration of Germans subject to military
          service   1866-1917
Nr. 59: Germans subject to military service living
          abroad   1916-1919

<u>52</u>: Han.Des.122a: Oberpraes.Han.: XXXIV: files as
  above.   ADS 4:97

Nr. 2729: Estates of Hannoverian citizens who died
          in the American attack on Mexico   1852
Nr. 2735: Pensions due widows & orphans of de-
          ceased soldiers serving the USA
          1868-1869

<u>52</u>: Han.Des.80: Hannover: A.II.B.III.11.Aa: files
  as above.   ADS 4:83

Information sought by private persons on family
  matters via foreign governments   1868-1901

<u>52</u>: Han.Des.122a: Oberpraes.Han.: II, Nr. 5.
          ADS 4:96

Nr. 93: Passports   1868-1921
Nr. 93a: Passports   1921-1923
Nr. 94: Issuance of passports   1868-1896
Nr. 114: Emergency passports [Zwangspaesse]
          1869-1870

<u>52</u>: Han.Des.122a: Oberpraes.Han.: XI: files as
  above.   ADS 4:96

Issuance of residency certificates   1868-1944

<u>52</u>: Han.Des.122a: Oberpraes.Hannover: I: Nr. 98d.
          ADS 4:96

Expulsions   1934-1942

<u>52</u>: Han.Des.80: Lueneburg II: Praesid.Abt., Regis.
  1890: Hauptabt.2: Nr. 17h, Bd. III [Reichs-
  verweisungen]   ADS 4:93

Extradition of foreign criminals & expelled per-
  sons:
  Nr. 115: Vol. I   1872-1898
  Nr. 115a: Vol. II   1899-1938

<u>52</u>: Han.Des.122a: Oberpraes.Han.: XI: files as
  above.   ADS 4:96

HANNOVER (Amt)   NL

Releases from citizenship   19th century

<u>52</u>: Han.Des.74: Hannover: III.F: Nr. 2-4. ADS 4:71

Emigration   19th century

<u>52</u>: Han.Des.74: Hannover: VIII.H.7: Nr. 1-3.
          ADS 4:71

HANNOVER-LANGENHAGEN   NL

Emigration   19th century

<u>52</u>: Han.Des.74: Han.-Langenhagen: VIII.B.7: Nr.
  1-3.   ADS 4:71

Releases from citizenship   19th century

<u>52</u>: Han.Des.74: Han.-Langenhagen: III.F: Nr. 1-2.
          ADS 4:71

HANSEN (village? Oberpfalz)   RPL

Emigration tax   1826-1832

<u>222</u>: Regierung: Kammer der Finanzen: Rep.118: II.
  Teil: Nr. 9720.   ADS 9:2

HARBURG   HH

Releases of residents from Hannoverian & later
  from Prussian military service obligations
  (Mostly emigration permits):
  Vol. I (alphabetical)   1816-1848
  Vol. II (chronological)   1847-1866

<u>41</u>: Bestand Hauptregistratur: III.B.1.8; *also in*
<u>80</u>: Hauptreg. Stadt Harburg (Bd.1): III.B.1.8.
          ADS 2:503; 4:279

Emigration to America   1837-1872

<u>41</u>: Bestand Hauptregistratur: III.B.1.11; *also in*
<u>80</u>: Hauptreg. Stadt Harburg (I): III.B.1.11.
                                        ADS 2:503; 4:280

Emigration permits & certificates of release:
  Vol. I: 1867-1890
  Vol. II: 1891-1916

<u>41</u>: Bestand Hauptregistratur: III.B.1.23; *also in*
<u>80</u>: Hauptreg. Stadt Harburg (Bd.I): III.B.1.23.
                                        ADS 2:503; 4:280

HARBURG (Amt)   HH

The nine emigrating families in Harburg   1733

<u>52</u>: Han.Des.74: Harburg: Regiminalia V 11: Nr. 15
                                              ADS 4:72

Nr. 174: Passports   1733
Nr. 175: Passports   1834

<u>52</u>: Han.Des.71: Harburg: Regiminalia: V 11: files
     as above.   ADS 4:72

Files on searches for families & individuals who
     have emigrated to America   1847

<u>52</u>: Han.Des.74: Harburg: Regiminalia Ia1: Nr. 95.
                                              ADS 4:71

HARPERSDORF. *See* Schwenckfelder Separatists

HARPSTEDT (Flecken)   NL

1A: Certificates of residency   1806-1872
1B: Persons who have moved away   1855-1867

<u>81</u>: B.III: files as above.   ADS 4:281

HARSEFELD (Amt)   NL

Releases from citizenship   n.d.

<u>52</u>: Han.Des.74: Harsefeld: III. Fach 42. ADS 4:72

HARTHAUSEN (village; Oberamt Trochtelfingen)   BWL

Emigration   1831-1847

<u>262</u>: Ho 197: Trochtelfingen: I 11197. ADS 10:36

HARTHEIM (village; Kreis Freiburg)   BWL

Emigration file   1839-1942
Emigration to the USA   1839-1928
Immigration file   1872-1920
List of emigrants for the last five decades [Manu-
     script?]   1932

<u>288</u>: Kommunalarchiv Hartheim.   ADS 10:116

HARTHEIM (village; Kreis Stockach)   BWL

Emigration files   1881-1937
General files   1882-1939

<u>288</u>: Kommunalarchiv Hartheim.   ADS 10:116

HARZ (mountain region). *See* Clausthal (mining dis-
    trict)

HASELUENNE (Amt)   NL

Emigration permits [for permits before 1871, *see*
     Osnabrueck]   1871-1879

Releases from citizenship [missing?]   1880

<u>54</u>: Rep.116,I: Nr. 6073; 9272.   ADS 4:120-121

HASLACH (village)   BWL

Emigration files (4 fascicles)   1831-1949
General files   1890-1949
Immigration files   1831-1906

<u>288</u>: Kommunalarchiv Haslach.   ADS 10:116

HASLACHSIMONSWALD (village)   BWL

Emigration file   1852-1941

<u>288</u>: Kommunalarchiv Haslachsimonswald. ADS 10:117

HASSLOCH (town; Landkreis Hassloch, Kurpfalz)   RPL

Names of emigrants   18th century

<u>201</u>: Archivabt. Ausfautheiakten.   ADS 8:63

HATTENWEILER (village)   BWL

Emigration files   1821-1864
Immigration files   1912-1926

<u>288</u>: Kommunalarchiv Hattenweiler.   ADS 10:117

HATTINGEN (Kreis)   RWL

Relations with foreign countries:
  Vol. I: 1899-1914
  Vol. II: 1926-1929

<u>116</u>: Arnsberg Reg.: Hattingen: 1: I: Nr. 2.
                                              ADS 6:60

Passports   1921-1926

<u>116</u>: Arnsberg Reg.: Hattingen: 1.B: Nr. 83.
                                              ADS 6:60

HATTINGEN (village; Landkreis Donaueschingen) BWL

General emigration file   1880-1941
Emigration lists; Jewish emigration   1881-1937
Immigration files   1889-1942

<u>288</u>: Kommunalarchiv Hattingen.   ADS 10:117

HATTORF (parish)   NL

Emigration to America   1846

<u>82</u>: I.1.A.105.   ADS 4:281

HATZENWEIER (village). *See* Ottersweier   BWL

HAUENEBENSTEIN (village)   BWL

  Emigration files   1845-1934
  Germans abroad   1933-1936

  288: Kommunalarchiv Hauenebenstein.   ADS 10:117

HAUSEN IM TAL (village)   BWL

  Emigration file (four fascicles)   1881-1938

  288: Kommunalarchiv Hausen im Tal.   ADS 10:117

HECHINGEN   BWL

  Emigration from various villages   1856-1938

  262: Ho 13: Hechingen (Acc.4/1954): II 4, 1-32.
                                          ADS 10:35

HECHINGEN (Landratsamt)   BWL

  Inquiries by foreigners regarding the reputation
    & personal data on Germans   1932

  262: Ho 13: Hechingen: (Acc.4/1954): IX 8, 4.
                                          ADS 10:35

HECKLINGEN (village)   BWL

  Emigration files (2 fascicles)   1866-1931
  Immigration file   1919-1934

  288: Kommunalarchiv Hecklingen.   ADS 10:117

HEDWIGENKOOG   SHL

  Citizenship, birth rights, certificates of resi-
    dency   1857-1895

  9: Nr. 45.   ADS 1:47

HEGNE (village)   BWL

  Emigration file   1850-1865

  288: Kommunalarchiv Hegne.   ADS 10:117

HEIDE   SHL

  Nr. 6: Passports   1781-1867
  Nr. 44: Visaed passports   1841-1861

  11: 4: files as above.   ADS 1:48

  Nr. 80: Citizenship   1834-1905
  Nr. 82: Certificates of residency   1879-1889
  Nr. 83: Certificates of residency   1883-1891
  Nr. 84: Certificates of residency   1883-1891
  Nr. 87: Passports   1868-1902

  10: Abt. VI: files as above.   ADS 1:47

HEIDELBERG   BWL

  Contract between [Rhine River] boatman Horst in

Heidelberg & a number of emigrants for their
  transportation to Rotterdam   1754

  260: Abt.236: Innenmin., Wegzug: Auswand.: 5651.
                                          ADS 10:4

HEIDENHEIM (Oberamt)   BWL

  Nr. 126-149a: Emigration files, chronologically
                  arranged   1806-1885
  Nr. 128:   Lists of emigrants   1816-1853
  Nr. 134:   Residents abroad with retention of
                  Wuerttemberg citizenship 1839-1865

  261: F 172: Heidenheim: files as above. ADS 10:20

HEILBRONN (Oberamt)   BWL

  Emigration files, chronologically arranged
                                      1854-1889

  261: F 173: Heilbronn: Nr. 139-153.   ADS 10:20

HEILIGENBERG (village)   BWL

  List of emigrants of the last five decades [Manu-
    script?]   1880-1935
  Emigration files   1883-1937

  288: Kommunalarchiv Heiligenberg.   ADS 10:117

HEILIGENHAFEN   SHL

  Emigration   1870-1875

  12: Abt.VII: Nr. 3.   ADS 1:48

HEILIGENZELL (village)   BWL

  Emigration files (2 fascicles)   1805-1931
  Illicit emigration   1855-1923

  288: Kommunalarchiv Heiligenzell.   ADS 10:117

HEILSBRONN (Bezirksamt)   BYL

  Naturalizations & losses of citizenship 1871-1889

  231: Rep.270/II: Reg. Mittelfranken: Kammer des
         Innern: Abgabe 1932: Nr. 726.   ADS 9:134

HEINSBERG (canton)   RWL

  Population movements   1801-1807

  115: Rep.D2.2: II.Division, 1.Bureau, 2.Bevoelker-
         ung: Nr. 99.   ADS 6:19

HEINSBERG (Landratsamt)   RWL

  Nr. 287: Emigration   1817-1825
  Nr. 288: Emigration   1830-1851
  Nr. 289: Emigration   1852-1861
  Nr. 290: Emigration   1862-1872
  Nr. 291: Emigration   1872-1880
  Nr. 292: Emigration   1880-1892
  Nr. 574: Emigration   1892-1926

  115: Rep.G29/21: Heinsberg: files as above.
                                          ADS 6:41

HEITERSHEIM (village)   BWL

   Emigration files (5 fascicles)   1825-1937
   Immigration & returning emigrants   1894-1929
   Emigration to the USA   1911-1920

   288: Kommunalarchiv Heitersheim.   ADS 10:118

HELDENBERGEN (Amt)   HEL

   For exemptions from emigration tax, *see* Hesse-
   Kassel [Kurhesse]

HELGOLAND   SHL

   Emigration plans of Helgoland residents   1723

   2: Abt.163: Nr. 1519 [Geplannte Ausw.]   ADS 1:20

HELMLINGEN (village)   BWL

   Emigration file   1854-1931

   288: Kommunalarchiv Helmlingen.   ADS 10:118

HELMSTEDT   NL

   Register of persons; lists of souls, citizens,
   residents; certificates of domicile; emigration,
   29 vols.   1548-1931

   83: B.VII.8-10.   ADS 4:281

   Nr. 11: Emigration permits   1853; 1858-1859
   Nr. 12: Emigration permits   1862-1863
   Nr. 13: Emigration permits   1870-1875 [partial]

   56: L Neu Abt.128A: Gr. 8: files as above.
                                                  ADS 4:185

HEMME   SHL

   Nr. 513: Passports   1811-1867
   Nr. 518: Emigration of German nationals without
        loss of citizenship   1882

   13: files as above.   ADS 1:48

HEMME (Kirchspiel Vogtei)   SHL

   Passport applications   1847-1892
   22: Nr. 418.   ADS 1:56

HEMMENHOFEN (village)   BWL

   Emigration file   1845-1913

   288: Kommunalarchiv Hemmenhofen.   ADS 10:118

HENGERSBERG (Landgericht)   BYL

   Applications for emigration permits   1839-1846

   225: Rep.168: Verz.1.   ADS 9:16

HENNESTEDT (Kreis Norderdithmarschen)   SHL

   Nr. 170: Investigation of persons subject to mili-
        tary service who have emigrated without
        permission   1874-1885
   Nr. 351: Citizenship   1822-1888
   Nr. 352: Citizenship   1889-1919
   Nr. 353: Certificates of citizenship 1889-1921
   Nr. 358: Passports   1820-1853
   Nr. 359: Passports   1826-1829
   Nr. 360: Passports   1830-1831

   14: files as above.   ADS 1:49

HEPPENHEIM   HEL

   Nr. 296: Emigration, A-E   1859-1897
   Nr. 297: Emigration, F-J   1859-1911
   Nr. 298: Emigration, K-N   1864-1893
   Nr. 299: Emigration, O-S   1859-1880
   Nr. 300: Emigration, T-Z   1863-1912

   149: Rep.40,2: Ablief.2: files as above. ADS 7:22

HEPPENHEIM (Kreis)   HEL

   Emigration to North America & Brazil   1825

   149: Rep.40,1: Ablief.1: Abt.XI, Absch.3, Bd.106.
                                    ADS 7:21

   Report of release [from citizenship] certificates
     issued   1872-1892

   149: Rep.40,2: Ablief.2: Bd.284,2.   ADS 7:22

   Affairs of individual citizens with other coun-
     tries [estate matters?]   1885-1938

   149: Rep.40,2: Ablief.2: Bd.48,1 [Angelegenheiten]
                                    ADS 7:21

   Return of Germans from abroad   1894-1900

   149: Rep.40,2: Ablief.2: Bd.284,3.   ADS 7:22

   Nr. 4: Passports for emigrants   1935-1938
   Nr. 5: Passports for Jews & migrants   1937-1941
   Nr. 6: Emigration of specialists   1929-1938

   149: Rep.40,2: Ablief.2: Bd.670.   ADS 7:23

   Bd. 319: Emigration A-E [Auerbach-Elmshausen]
                               1939-1943
   Bd. 320: Emigration H-N [Heppenheim-Neckarsteinach]
                               1938-1943
   Bd. 321: Emigration R-Z [Reichenbach-Zwingenberg]
                               1938-1943

   149: Rep.40,2: Ablief.2: volumes as above.
                                    ADS 7:23

HERBOLZHEIM (village)   BWL

   Emigration file   1852-1904
   Summary of emigration & immigration   1867-1889

   288: Kommunalarchiv Herbolzheim.   ADS 10:118

HERBORN (Dillkreis). *See* Dillenburg   HEL

HERDWANGEN (village)   BWL

  Immigration files   1879-1930
  Emigration files   1841-1930

  <u>288</u>: Kommunalarchiv Herdwangen.   ADS 10:118

HERFORD (Kreis)   RWL

  Nr. 127-135: Emigration   1880-1924
  Nr. 76-80:   Certificates of citizenship 1899-1923
  Nr. 35-38: Issuance of certificates of original
      domicile   1905-1922

  <u>116</u>: Minden: Praes.-regist.: I.A.: files as above.
                                 ADS 6:50; 52

  Passports   1851-1930

  <u>116</u>: Minden: Praes.-regist.: I.P.: Nr. 315.
                                 ADS 6:54

HERINGEN. *See* Limburg   HEL

HERRENBERG (Oberamt)   BWL

  Nr. 82: Lists of emigrants & immigrants 1825-1870
  Nr. 84-87: Emigration files, chronologically ar-
      ranged   1833-1861
  Nr. 88: Financial aid to emigrants   1847-1867
  Nr. 89: Fees for notices [in official gazettes?]
                                 1850-1869

  <u>261</u>: F 174: Herrenberg: files as above. ADS 10:20

HERRIEDEN   BYL

  List of 8 emigrants to the USA (no supporting data
    available)   [19th century?]

  <u>243</u>   ADS 9:154

HERRIEDEN (Emigration collection point [*Lager*]) BYL

  Emigration & immigration, Vol. IV   1854

  <u>231</u>: Rep.270/II: Reg. Mittelfranken: Kammer des
    Innern: Abgabe 1932: 22.   ADS 9:134

HERRSTEIN (mayor's office; Amt Oberstein)   RPL

  Naturalizations & releases from citizenship
                                 1834-1856

  <u>200</u>: Abt.395,2: Amt Oberstein: Nr. 58-61. ADS 8:17

HERSFELD   HEL

  Economic emergency after the [American] Revolution
    particularly in Hersfeld (emigration to America)
                                 1817-1818

  <u>151</u>: Bestand 5: Nr. 312.   ADS 7:38

HERSFELD (Regierungsbezirk)   HEL

  Naturalizations   1849

  <u>151</u>: Bestand 16.II: Klasse 17: Nr. 8.   ADS 7:67

HERSFELD (Verwaltungsbezirk)   HEL

  Issuance of travel permits   1849-1851

  <u>151</u>: Bestand 100: Nr. 134.   ADS 7:96

  Nr. 86: Releases from citizenship   1849-1851
  Nr. 100: Death certificates for citizens dying
      abroad   1849-1851
  Nr. 128: Certificates of original domicile
                                  1849-1851
  Nr. 129: Certificates of original domicile
                                  1850-1851

  <u>151</u>: Bestand 100: m: files as above.   ADS 7:96

HERSFELD (Kreis)   HEL

  Nr. 4693-4695: Naturalizations, 3 vols.   1822-1841
  Nr. 5476:   Travel permits   1831-1866
  Nr. 6923: Emigration & citizenship, 2 vols.
                                  1821-1825
  Nr. 6926: Emigration & citizenship, 2 vols.
                                  1826-1835
  Nr. 6924: Applications for emigration permits, 3
      vols.   1836-1848

  <u>151</u>: Bestand 100: files as above.   ADS 7:94

HERSFELD (Landratsamt)   HEL

  Nr. 39: Emigration   1803
  Nr. 36: List of special files on emigration &
      immigration   1846
  Nr. 37: General file on immigration & emigration
                                  1851-1861
  Nr. 38: General file on immigration & emigration
                                  1814
  Nr. 41: Secret [illicit] emigration to America
                                  1851-
  Nr. 42: Emigration to America   1851
  Nr. 46: Prohibited emigration to North America
                                  1857
  Nr. 48: Notations on sequestered properties of
      persons emigrating secretly [illicity]
                                  1860

  <u>151</u>: Bestand 112: Acc.1886/9: files as above.
                                  ADS 7:96-97

  In 1960, when the ADS inventory was taken, the
  Archives were being recatalogued. It is likely
  that the Archives contain much data of interest
  to German-American genealogical researchers not
  included above.

  <u>151</u>   ADS 7:118

HERSFELD (town)   HEL

  Residency, 4 vols.   1822-1861

  <u>151</u>: Bestand 100: Nr. 7502 [Ortsbuergertum]
                                  ADS 7:94

HERPSTEDT (Amt)   NL

  Emigration certificates & releases from citizen-
    ship: Vol. I   1824-1836
         Vol. II   1838-1839

  <u>52</u>: Han.Des.74: Freudenberg: III,F: Nr. 2.
                                  ADS 4:68

HERTEN (village; Kreis Loerrach)   BWL

Emigration files (2 fascicles)   1852-1953
Immigration files   1909-1920
List of emigrants in the last five decade [Manu-
    script?]   1932

288: Kommunalarchiv Herten.   ADS 10:118

HERTINGEN (village)   BWL

Emigration files (2 fascicles)   1855-1940
Immigration files   1868-1931
Immigration files   1924-1944
General files   1892-1950

288: Kommunalarchiv Hertingen.   ADS 10:118

HESSE-DARMSTADT   HEL

Emigrants to America   1789
149: F 113 L.   ADS 7:1

Emigrants to Pennsylvania   1793
149: F 274 M.   ADS 7:1

Nr. 587/2-3: Extradition to & from America, 2 vols.
                                        1859-1925
Nr. 560/1:   Transmittal of inheritances from
              abroad   1872-1927
Nr. 560/2:   Regulation of payments from estates
              of foreigners in Germany 1872-1927
Nr. 1300/8-9: Purchase of real estate [in Darm-
              stadt] by Americans, 2 vols.
                                        1901-1936
Nr. 564; 565/1-2: Estates of Germans dying in
                  America   1909-1936
Nr. 561/3: Foreign estate matters   1914-1936
Nr. 563/1: Estates of Germans dying in America
                                        1865-1909
Nr. 565/5: General files on estates of Germans dy-
           ing in America   1918-1931

149: Rep.49a: files as above.   ADS 7:15-16; 19

HESSE-HANAU   HEL

Package containing a large number of emigration
    papers; destinations Pennsylvania, Carolina,
    New England   1740?

151: Bestand 80: Lit.B: Nr. 1 [Unterthanen]
                                       ADS 7:79

Secret [illicit] emigration "even to America" 1785

158: Frankfurter Ratsakten: Ugb.A.9: Nr. 7.
    DLC has (original document in 158 has been
    destroyed).   ADS 7:198

Manumissions from various localities [Aemter; may
    contain a large number of such manumissions]
                                    [18th century?]

151: Bestand 81: Rep.B: Nr.46-1/2 to Nr. 64; also
    see Nr. 37-46.   ADS 7:79

Two bundles of papers on emigration (unmarked)
                                     18th century

151: Bestand 80: Lit. C; Lit. D.   ADS 7:79

Temporary residency permits for foreigners & per-
    sons with no connections to Hanau citizens 1813-
    [many cases]

151: Bestand 82: Rep.III: Gef.259-269.   ADS 7:86

Passports issued   1816-1829

151: Bestand 82: Rep.III: Gef.248, No. 1. ADS 7:85

Secret [illicit] emigration of soldiers & citizens
    to America   1818-

151: Bestand 81: Rep. Ba: Gef.9: Nr. 3.   ADS 7:79

Release from citizenship of persons subject to
    military service   1845-1867

151: Bestand 82: Acc.1904/13: Nr. 120.   ADS 7:85

HESSE-HANAU (Province)   HEL

Petitions for release from Kurhessian citizenship,
    6 vols.   1840-1867

151: Bestand 82: Acc.1904/13: Nr. 119.   ADS 7:85

HESSE-HANAU. See also Hanau   HEL

HESSE-HOMBURG (Landgrafschaft)   HEL

III.b.26: List of emigrants   1847
XIV.c. 9: Releases from citizenship   1818-1850
XIV.c.10: Passports   1820-1856
XIV.c.16: Applications for emigration; passports
          visaed, 13 vols.   1834-1863

152: Abt.310: Hessen-Homburg: files as above.
                                       ADS 7:138

HESSE (Kurhessen; Electorate)   HEL

Emigration; releases from citizenship (includes a
    list of emigrants to North America, 1832-1841),
    5 vols.   1821-1855

151: Bestand 16.II: Klasse 14, Nr. 4.   ADS 7:64

Emigration permits; releases from citizenship;
    acceptances as citizens, 2 vols.   1832-1838

151: Bestand 16.II: Nr. 10.   ADS 7:65

Granting of Hessian citizenship:

Nr. 6:	1852	Nr. 18:	1859
Nr. 11:	1853-1855	Nr. 19:	1860
Nr. 13:	1855	Nr. 20:	1861
Nr. 16:	1857	Nr. 26:	1865-1866
Nr. 17:	1858	Nr. 28:	1866-1867

52: Schaumburg Des.H.2: Rinteln: Neue reg.: II,
    Klasse 1: files as above.   ADS 4:99

Legitimation of emigrants from Kurhessen at Aurich
                                             1853

50: Rep.21a: Nr. 8884.   ADS 4:9

Releases from citizenship, 3 vols.   1859-1863

52: Schaumburg Des.H.2: Rinteln: Neue reg.: II,
    Klasse 1: Nr. 10.   ADS 4:98

HESSE-KASSEL (Kurhessen; Electorate)   HEL

Nr. 1: Emigration   1723-1766
Nr. 1a: Emigration   1748-1749
Nr. 2: Emigration   1753
Nr. 4: Emigration   1774-1781
Nr. 5: Emigration   1784-1785
Nr. 6: Emigration   1785-1786
Nr. 7a: Emigration   1787
Nr. 8: Emigration   1791

<u>151</u>: Bestand 17(b): Gef.97: files as above.
                                      ADS 7:70

Applications for exemptions from emigration tax:
  Nr. 1628: Lit. R*   1658-1821
  Nr. 1627: Lit. K*   1641-1817
  Nr. 1336: Lit. G*   1754-1820
  Nr. 1340: Lit. L-M*   1692-1813
  Nr. 1337: Lit. H*   1749-1813
  Nr. 817: Lit. N-O*   1733-1820
  Nr. 290: Lit. B*   1652-1815
  Nr. 820: Lit. S*   1755-1769
  Nr. 1556: Lit. L-M*   1746-1807
  Nr. 1554: Lit. S & W*   1736-1814
  Nr. 2205: Lit. H*   1747-1804
  Nr. 2008: Lit. A-B*   1740-1804
  Nr. 2004: Lit. D-E*   1732-1773
  Nr.   2: Lit. S & W*   1711-1792
       *All Aemter of this (these) letters.

<u>151</u>: Bestand 86: files as above.   ADS 7:87-88

Nr. 30: Certificates of exemption from emigration
        tax   1820
Nr. 31: Some remaining emigration levies 1723-1726
Nr. 32: Investigation of inheritances taken out of
        the country without payment of emigration
        tax   1723

<u>151</u>: Bestand 17(b): Gef.97a: files as above.
                                      ADS 7:71

Alphabetical list of emigration taxes levied [by
  individual states & countries]   18th century

<u>151</u>: Bestand 17(b): Gef.97: Nr. 27.   ADS 7:70

Emigration of Hessian citizens to London & the
  pursuit of their property for the purpose of
  levying emigration tax:
  Vol. I   1765-1822
  Vol. II   1823-1848

<u>151</u>: Bestand 17(b): Gef.101: Nr. 1.   ADS 7:72

Emigration to Russia; also various estate matters
                                     1766; 1789

<u>151</u>: Bestand 98: Fulda X: Acc.1875/27: Nr. 6.
                                      ADS 7:91

Emigration tax:
  Nr. 3:   1770-1771   Nr. 8:   1791
  Nr. 4:   1772-1778   Nr. 9:   1786
  Nr. 5: 1779-1780; 1782   Nr. 11:   n.d.
  Nr. 6:   1801-1805

<u>151</u>: Bestand 19(b): files as above.   ADS 7:74

Applications of citizens to leave the country
  without stating destination   1804

<u>151</u>: Bestand 17(b): Gef.97: Nr. 11.   ADS 7:70

Nr. 12: Emigration   1817-1835
Nr. 18: Emigration, Vol. I   1825-1841
Nr. 29: Emigration, Vol. II-VI   1842-1867
Nr. 30: Emigration, Vol. XII   1842-1843
Nr. 33: Emigration, Vol. XIII   1843-1844
Nr. 37: Emigration, Vol. XV-XXI   1845-1849

<u>151</u>: Bestand 17(b): Gef.98: files as above.
       DLC has Nr. 29 above.   ADS 7:71

Nr. 135: Releases from citizenship, 3 vols.
                                      1822-1852
Nr. 136: Emigration permits, 4 vols. 1818-1853
Nr. 137: Emigration tax   1814
Nr. 139: Retention of Hessian citizenship by emi-
         grants   1839
Nr. 239: Passport regulations, 2vols.   1817-

<u>151</u>: Bestand 20: Verz.2, Acc.1876/13: files as
         above.   ADS 7:75

Nr. 10065: Applications for releases from citizen-
           ship   1828-1829
Nr. 10052: Applications for releases from citizen-
           ship   1835-1836

<u>151</u>: Bestand 86: files as above.   ADS 7:86

Emigration of Kurhessian subjects to America
                                      1832-1833

<u>151</u>: Bestand 5: Nr. 773.
       DLC has.   ADS 7:40

Military service obligations of persons who have
  emigrated, or have been released from citizenship   1836

<u>151</u>: Bestand 5: Nr. 1481.   ADS 7:42

Persons who have emigrated to the East Indies in
  the last 100 years   1842-1843

<u>151</u>: Bestand 86: Nr. 10114.   ADS 7:86

Files of the higher administrative levels regarding emigration of citizens to America, 2 vols.
                                      1849-1851

<u>151</u>: Bestand 17(b): Gef.98: Nr. 48.   ADS 7:72

Persons who have fled the country & who have participated in the political unrest of 1848
                                      1850-1851

<u>151</u>: Bestand 17(g): Gef.27: Nr. 60.   ADS 7:73

Nr. 45: Emigration, Vol. I   1851-1853
Nr. 45a: Emigration, Vol. II   1854-1857
Nr. 45b: Emigration, Vol. III   1858-1863
Nr. 45c: Emigration, Vol. IV   1864-1867

<u>151</u>: Bestand 17(b): Gef.98: files as above.
                                      ADS 7:72

Releases from citizenship, Vol. IV   1853-1857

<u>151</u>: Bestand 20: Verz.6, Acc.1903/9.19: Nr. 99.
                                      ADS 7:75

Nr. 74a: Emigration, Vols. V & VI   1854-1863
Nr. 75a: Releases from citizenship, Vols. V & VI
                                      1858-1867
Nr. 90a: Releases from citizenship, Vol. IV
                                      1837-1842

<u>151</u>: Bestand 20: Verz.3, Acc.1878/20C: files as
         above.   ADS 7:75

Issuance of certificates of original domicile
                                        1822-1866

<u>151</u>: Bestand 16.VII: Klasse 10: Nr. 15.   ADS 7:68

Certificates of original domicile:
    Nr. 7: Vol. I   1828-1838
    Nr. 9: Vol. II   1839-1846
    Nr. 10: Vol. III   1847-

<u>151</u>: Bestand 17(b): Gef.89: files as above.
                                        ADS 7:69

Issuance of passports & visas:
    Nr. 2: Vol. I   1815-1833
    Nr. 4: Vol. II   1834-1840
    Nr. 5: Vol. III   1841-1847
    Nr. 6: Vol. IV   1848-
    Nr. 7: Attestation of visas & visas given by
           consulates, etc.   1852

<u>151</u>: Bestand 17(g): Gef.42: files as above.
                                        ADS 7:73

Nr. 2: Passports issued, 5 vols.   1816-1867
Nr. 3: Passport applications, 16 vols. 1836-1852

<u>151</u>: Bestand 16.VII: Klasse 9: files as above.
                                        ADS 7:67

Statements of residency [18 in number; not other-
    wise described in ADS]   1826-1867

<u>151</u>: Bestand 16.VII: Klasse 11.   ADS 7:68

Nr. 78-84: Naturalizations   1819-1867
Nr. 92:   Naturalizations   1825-1867

<u>151</u>: Bestand 20: Verz.6, Acc.1903/9.19: files as
    above.   ADS 7:75

Naturalizations abroad with retention of Hessian
    citizenship, 2 vols.   1824-1865

<u>151</u>: Bestand 17(b): Gef.89: Nr. 8.   ADS 7:69

Nr. 1468: Investigations, searches, & issuance of
          certificates for living & dead persons,
          Vols. I-IX   1815-
Nr. 1470: Baptismal certificates requested &
          issued, Vols. I-IV   1840-
Nr. 1471: Marriage certificates requested &
          issued   n.d.
Nr. 1472-1476: Death certificates requested &
          issued, 18 vols.   1814-

<u>151</u>: Bestand 5: files as above.   ADS 7:42

Death certificates, 2 vols.   1823-1865

<u>151</u>: Bestand 5: Gesandtschaften Berlin: Nr. 214.
                                        ADS 7:45

Baptismal & death certificates of citizens abroad
                                        1864-1866

<u>151</u>: Bestand 20: Verz.3, Acc.1878/20C: Nr. 94a.
                                        ADS 7:75

Nr. 5: Investigation of inheritance matters
                                        1784-1785
Nr. 7: Collection of reports regarding confiscated
       properties of emigrating citizens 1785-1786
Nr. 9: Investigation of inheritance matters
                                        1795-1796
Nr. 10: Investigation of inheritance matters
                                        1797-1798

Nr. 12: Investigation of inheritance matters
                                        1798-1799
Nr. 24: Administration of confiscated properties
        of emigrating citizens, 4 vols.
                                        1782-1818

<u>151</u>: Bestand 17(b): Gef.97: files as above.
                                        ADS 7:70

Estates of various Kurhessian subjects dying
    abroad & of foreigners dying in Kurhesse
                                        1823-1866

<u>151</u>: Bestand 5: Gesandtschaften Berlin: Nr. 209.
                                        ADS 7:45

Nr. 66a: Requests of citizens for intercession to
         collect inheritances abroad, 2 vols.
                                        1822-1865
Nr. 71a: Conditional expulsion of former Kurhessian
         citizens   1825-1843
Nr. 72a: Issuance of certificates of original
         domicile, 2 vols.   1825-1860

<u>151</u>: Bestand 20: Verz.3, Acc.1878/20C: files as
    above.   ADS 7:75

Estates & inheritances [alphabetically arranged;
    22 Konvolute]   1860-

<u>151</u>: Bestand 5: 9a.: Nr. 7.   ADS 7:43

Nr. 3: Preparation of lists on foreigners in prov-
       ince, Vol. I   1825-1828
Nr. 7: Temporary residency permits for foreigners,
       Vol. I   1827-1828
Nr. 10: Temporary residency permits for foreigners,
       Vol. II   1829

<u>151</u>: Bestand 17(g): Gef.41: files as above.
                                        ADS 7:73

Officer retirements   1725-1791

<u>151</u>: Bestand 12b: Blaetter Nr. 8507; 8509; 8510.
                                        ADS 7:51

Nr. 8526: Exceptions, desertions, emigration, Vol.
          I   1755-1787
Nr. 8527: Exceptions, desertions, emigration, Vol.
          II   1725-1798

<u>151</u>: Bestand 12b: files as above.   ADS 7:51

Nr. 11960: Applications for releases from citizen-
           ship   1814-1817
Nr. 8684-8685: Emigration permits   1815-1816
Nr. 8692: Emigration permits   1817; 1821
Nr. 3145: Applications for exemptions from payment
          of the exit tax   1814-1821
Nr. 766: Applications of citizens abroad regarding
         their private affairs, Vol. I   1819-
                                 Vol. II   1840-

<u>151</u>: Bestand 5: Nr. 766.   ADS 7:40

Nr. 965: Residency permits in Kurhesse   n.d.
Nr. 967-977: Residency of Kurhessian citizens
             abroad   1821-1846
Nr. 978-990: Passports   1817-1867

<u>151</u>: Bestand 5: files as above.   ADS 7:41

Releases from citizenship of persons subject to
    military service, 3 vols.   1832-

<u>151</u>: Bestand 5: Nr. 774.   ADS 7:40

Domiciles of various persons   n.d.

<u>151</u>: Bestand 5: Nr. 795.   ADS 7:41

Nr. 2: Measures taken against persons without pro-
       fession or place of abode, 2 vols.   1824-
Nr. 11: Expulsions   1824
Nr. 19: Measures taken against suspicious or other-
        wise strange & unknown individuals, 2 vols.
                                           1828-1867
Nr. 24: Extraditions   1831
Nr. 34: Extraditions   1849-

<u>151</u>: Bestand 16.VII: Klasse 8: files as above.
                                             ADS 7:67

Nr. 9: Naturalizations in the Province of Nieder-
       hessen with Schaumburg, 8 vols.   1831-1849
Nr. 13: Releases from citizenship, 3 vols.
                                           1827-1847
Nr. 22: Releases from citizenship   1822

<u>151</u>: Bestand 16.II: Klasse 15: files as above.
                                             ADS 7:66

Death certificates for foreigners in Prussian
  territory & for Prussian citizens abroad
                                           1867-1869

<u>151</u>: Bestand 165: Nr. 7596.   ADS 7:110

Illicit emigration by Germans subject to military
  service   1884-1914

<u>151</u>: Bestand 165: I: Nr. 625.   ADS 7:111

There are numerous estates on file wherein place
  of death is not stated in the catalog [Reportor-
  ium]. Researchers will want to check these at
  Marburg.

<u>151</u>: Bestand 165: Nr. 7084-.   ADS 7:111

HESSE-KASSEL (Regierungsbezirk)   HEL

Naturalizations   1849

<u>151</u>: Bestand 16.II: Klasse 15: Nr. 16.   ADS 7:66

HESSE-KASSEL (Prussian Regierung)   HEL

Nr. 60: Releases from Prussian citizenship, 91
         vols.   1867-1931
Nr. 6421: Releases from Prussian citizenship
         (with associated papers) 2 vols.
                                           1920-1931
Nr. 1033: Inheritances: K-L   1917-1931
Nr. 1051: Inheritances: M-N   1920-1931
Nr. 1052: Inheritances: O-P   1916-1927
Nr. 1050: Inheritances: Q-R   1919-1931
Nr. 1053: Inheritances: S-T   1915-1931
Nr. 1049: Inheritances: U-V   1917-1931
Nr. 1048: Inheritances: W-X   1917-1931
Nr. 1047: Inheritances: Y-Z   1918-1929
Nr. 3140: Expulsions from German territory, Vols.
         1-2   1873-1889
Nr. 4114: Expulsions from German territory, Vol.
         5   1902-1903
Nr. 4042: Expulsions from German territory, Vols.
         6-9   1903-1913
Nr. 733: Expulsions; extraditions; repatriations
         of welfare cases   1902-1929
Nr. 59: Preparation of certificates & informa-
        tion on citizens abroad, 3 vols.
                                           1867-1876

Nr. 262: Denaturalizations [?] vols. 3-6 1886-1908

<u>151</u>: Bestand 165: I: files as above.
                                    ADS 7:108-109; 111

Lists of the Landratsaemter on persons released
  from Prussian citizenship   1888-1916

<u>151</u>: Bestand 165: No. 1005.   ADS 7:109

Nr. 6422: Determinations of whether certificates
          of releases from citizenship remain in
          force (supplementary papers)   1916-1918
Nr. 1006: Certificates of original domicile, 44
          vols.   1888-1916
Nr. 1014: Citizenship certificates; certificates
          of original domicile, 100 vols.
                                              1902-1929
Nr. 7596?: Birth certificates, etc., for foreigners
          in Prussia & for Prussian citizens
          abroad   1867-1879
Nr. 266: Issuance of certificates & information
          on citizens abroad, Vols. 4-19 1877-1920
Nr. 1054: Miscellaneous matters concerning Germans
          abroad; collection of consular fees
                                              1868-1930
Nr. 267: Certificates pertaining to status of
          citizens abroad, 2 vols.   1901-1908
Nr. 1028: Status of Germans abroad   1919-1930
Nr. 1037: Births, marriages, & deaths of Germans
          abroad & on German ships   1930-1931
Nr. 3141: Germans expelled from foreign countries
          & from other states of the German Federa-
          tion, 4 vols.   1874-1903
Nr. 1040: Naturalizations, 2 vols.   1882-1883
Nr. 1001: Renaturalizations of returning emigrants
                                              1870-1899
Nr. 1002: Naturalization papers, 54 vols.
                                              1900-1938
Nr. 1003: Naturalization papers, 2 vols. 1928-1931
Nr. 1004: Inquiries by foreign states regarding
          prospective naturalizations of Prussian
          residents, Vol. 2   1908-1919
Nr. 7069: List of bothersome foreigners expelled
          from German territory   1920-1922
Nr. 1029: Status of foreigners   1910-1931
Nr. 718: Marriages of foreigners, Vol. 1
                                              1906-1913
Nr. 6730: Data on inheritances, Vol. 2   1870-1873
Nr. 70: Data on inheritances, Vol. 3   1874-1876
Nr. 6730: Data on inheritances, Vol. 4   1876-1877
Nr. 6730: Data on inheritances, Vol. 13 1892-1893
Nr. 6730: Data on inheritances, Vol. 16   1896
Nr. 6730: Data on inheritances, Vol. 17 1896-1898
Nr. 2860: Data on inheritances, Vol. 22 1908-1910
Nr. 1056: Data on inheritances, Vol. 23 1910-1916

<u>151</u>: Bestand 165: files as above.   ADS 7:108-110

HESSE-KASSEL. *See also* Kassel   HEL

HESSE and KURHESSE   HEL

Preparation, receipts, & issuance of birth certificates for foreigners in Hesse & for Hessian
  citizens abroad, 3 vols.   1823-1867

<u>151</u>: Bestand 16II: Klasse 8, Nr. 7.   ADS 7:63

Information regarding inheritance matters, 4 vols.
                                           1823-1856

<u>151</u>: Bestand 16.II: Klasse 8, Nr. 8.   ADS 7:64

Preparation, investigation, delivery of certificates & information on foreigners in Kurhesse
& on Hessian subjects abroad, 6 vols. 1822-1858

<u>151</u>: Bestand 16.II: Klasse 8, Nr. 9.   ADS 7:64

Preparation, investigation, delivery of death certificates for foreigners in Kurhesse & for Hessian citizens abroad, 5 vols.   1822-1865

<u>151</u>: Bestand 16.II: Klasse 8, Nr. 10.   ADS 7:64

List of persons accepted as citizens & those released from citizenship   1820-1824

<u>151</u>: Bestand 16.II: Klass 14, Nr. 2.   ADS 7:64

HESSE-NASSAU (Province)   HEL

Naturalizations & releases from citizenship:
   Nr. 482: Vol. 1   1867-1886
   Nr. 483: Vol. 2   1886-1900
   Nr. 1607:   1919-1930

<u>151</u>: Bestand 150: files as above.   ADS 7:106

HESSE-NIEDERHESSE. *See* Niederhessen   HEL

HESSE-RUMPENHEIM (Philipsruhe)   HEL

C6, Nr. 2: Naturalizations   1833-1864
C22, Nr. 26: Issuances of passports 1857-1865

<u>151</u>: Bestand 300: Verz.2, Abt.11: files as above.
                                       ADS 7:124

HESSENSTEIN (Herrschaft; *also called* Panker)   SHL

Nr. 367: Passports   1817-1863
Nr. 374: Emigration [*also in* Landesarchiv Schles-
         wig, Abt.322, Nr. 16]   1847-1872
Nr. 375: Releases from citizenship [*also in*
         Landesarchiv Schleswig, Abt.322, Nr. 15]
                                   1869-1882
Nr. 1049: Investigations regarding the whereabouts
          of persons subject to military service
          obligation who have illegally emigrated
                                   1863-1887

<u>27</u>: files as above.   ADS 1:60

HESSHEIM (village). *See* Kurpfalz   RPL

HEUCHELHEIM (Kreis Bergzabern, Kurpfalz)   RPL

Names of emigrants   18th century

<u>201</u>: Archivabt. Ausfautheiakten.   ADS 8:62

HEUCHELHEIM BEI FRANKENTHAL (village)   RPL

File on manumissions; gives data on the emigration
of residents   1771-1788

<u>201</u>: Zweibruecken III: Akt Nr. 2018.   ADS 8:58

HEUWEILER (village)   BWL

Emigration to the USA   1846-1881

<u>288</u>: Kommunalarchiv Heuweiler.   ADS 10:118

HEYDEKRUG (Kreis)   East Prussia (Gumbinnen)

Emigration [file missing?]   1837-1874

<u>51</u>: Rep.12: Abt.I, Titel 3, Nr. 8, Bd. 5. ADS 4:32

HILDERS (Amt)   HEL

Passports   1842-1862

<u>151</u>: Bestand 112: Acc.1903/9 LIV: Nr. 13.
                                        ADS 7:98

HILDERS (Landgerichtsbezirk)   HEL

Emigration   1854

<u>151</u>: Bestand 112: Acc.1886/9: Nr. 40.   ADS 7:97

HILDESHEIM   NL

Emigrants' files, name lists, certificates of residence   19th century
<u>84</u>   ADS 4:285

Inquiries regarding emigrants serving in American
armed forces   1862

<u>52</u>: Han.Des.80: Hildesheim I: E.L.I.: Nr. 224.
                                        ADS 4:86

Emigration to North America & Brazil, 2 vols.
                                   1866-1903

<u>52</u>: Han.Des.80: Hildesheim II: I.Praesid.Abt.II:
         Nr. 479.   ADS 4:92

Estates of persons dying abroad, 9 vols. 1866-1901

<u>52</u>: Han.Des.80: Hildesheim II: I.Praesid.Abt.I:
         Nr. 445.   ADS 4:91

HILTER. *See* Dissen-Hilter

HIMMELPFORTEN (Amt)   NL

Releases from citizenship & emigration:
   Nr. 9: 1866   Nr. 20   1877
   Nr. 10: 1869   Nr. 21   1878
   Nr. 12: 1870   Nr. 22   1879
   Nr. 13: 1871   Nr. 23   1880
   Nr. 14: 1872   Nr. 24   1881
   Nr. 15   1873   Nr. 25   1882
   Nr. 16   1874   Nr. 26   1883
   Nr. 17   1875   Nr. 27   1884
   Nr. 18   1876

<u>52</u>: Han.Des.74: Himmelpforten: Regiminalia: B.8:
         files as above.   ADS 4:72

HINTERMEILINGEN. *See* Limburg   HEL

HIRSCHHORN   HEL

Emigration   1854-1901

149: Rep.40,2: Ablief.2: Bd. 301.   ADS 7:22

HOCHDORF (village)   BWL

Emigration file   1893-1938
Immigration file   1918-1923

288: Kommunalarchiv Hochdorf.   ADS 10:118

HOCHHEIM (Amt; Nassau-Usingen)   HEL

Emigration to America   1805

152: Abt.131: Gen XIVc: Nr. 18.   ADS 7:127
     DLC has.

HOCHSTADT (Amt)   HEL

For exemptions from emigration tax, *see* Hesse-
Kassel [Kurhesse]

HOECHENSCHWAND (village)   BWL

Emigration files   1893-1944

288: Kommunalarchiv Hoechenschwand.   ADS 10:119

HOECHST (Amt)   HEL

Nr. 1009: Releases from Nassau citizenship & per-
          mits to emigrate   1850
Nr. 1010: Releases from Nassau citizenship & per-
          mits to emigrate   1851-1855
Nr. 1011-1016: Releases from Prussian citizenship
                                1870-1886

152: Abt.228: Amt Hoechst: files as above.
                                ADS 7:131

HOEDINGEN (village)   BWL

Emigration to the USA   1865-1883
Immigration & emigration   1918-1942

288: Kommunalarchiv Hoedingen. ADS 10:119

HOELLERICH (Amt)   HEL

For exemptions from emigration tax, *see* Hesse-
Kassel [Kurhesse]

HOEMBERG (village; Kreis Nassau)   HEL

Application of *Gemeinde* [village] Hoemberg for
permission to sell *Gemeinde* property for the
purpose of emigrating to America   1852

152: Abt.235: Nassau: Nr. 111.   ADS 7:134

HOERDEN (village)   BWL

Emigration file   1832-1933
Emigration to the USA   1869
Foreign assistance   1929-1939

288: Kommunalarchiv Hoerden.   ADS 10:119

HOEXTER (Kreis)   RWL

Emigration   1887-1897

116: Arnsberg Reg.: Hoexter: A: Nr. 154. ADS 6:63

Certificates of original domicile:
    Vol. I   1888-1897
    Vol. III-IV   1909-1925

116: Minden: Praes.-regist.: I.A.: Nr. 39-41.
                                ADS 6:50

Certificates of citizenship   1909-1928

116: Minden: Praes.-regist.: I.A.: Nr. 81-82.
                                ADS 6:50

Emigration   1879-1924

116: Minden: Praes.-regist.: I.A.: Nr. 136-141.
                                ADS 6:52

HOFGEISMAR (Landratsamt)   HEL

Releases from citizenship from numerous villages
    of the Kreis   1822-1860

151: Bestand 180: Hofgeismar: Nr. 33-35; 37-67.
                                ADS 7:118

Nr. 71: Return of emigrants to their homes
                                1838-1867
Nr. 74: Emigration from Kurhesse, with or without
        permits, & return of emigrants 1825-1851

151: Bestand 180: Hofgeismar: files as above.
                                ADS 7:118

HOFGEISMAR (Kreis)   HEL

Naturalizations, 9 vols.   1818-1866

151: Bestand 17(b): Gef.92, Nr.2.   ADS 7:69

HOFSGRUND (village)   BWL

Emigration file   1892-1944

288: Kommunalarchiv Hofsgrund. ADS 10:119

HOFSTETTEN (village)   BWL

Emigration files   1868-1941

288: Kommunalarchiv Hofstetten. ADS 10:119

HOFWEIER (village)   BWL

Emigration file   1852-1949

Illicit emigration 1853-1932
Immigration file 1866-1928

288: Kommunalarchiv Hofweier. ADS 10:119

HOFWEIER (village). *See also* Offenburg BWL

HOHENBODMAN (village) BWL

Emigration file 1835-1941

288: Kommunalarchiv Hohenbodman. ADS 10:119

HOHENHAUSEN (Amt) RWL

Nr. 2-5: Emigration 1851-1879
Nr. 7: Domicile 1836-1876
Nr. 13: Passports 1858-1879

114: Hohenhausen: Fach 5: files as above. ADS 6:16

Emigration, 3 vols. 1879-1894

114: Brake: Fach 7: Nr. 7. ADS 6:15

HOHENLOHE BWL

Files on emigrants from Hohenlohe to America via
  port of Emden 1804

50: Rep.6: Stadt Emden: Miscel.: Nr. 35. ADS 4:5

HOHENZOLLERN-SIGMARINGEN (Principality) BWL

Emigration to neighboring countries & to America
  & the levying of property tax 1785-1834

262: Fuerst. Hohenzollern-Sig.: C I 2 d 7.
            ADS 10:34

Nr. 141: Loss of citizenship; entry into the ser-
      vice of foreign states 1847-1927
Nr. 142-143: Releases from citizenship (arranged
      by Oberaemter) 1850-
Nr. 146-152: Releases from citizenship (arranged
      by Oberaemter) to 1907
Nr. 153: Support of emigrants from public funds
            1853-

262: Ho 235 B: Preussische Regierung: Abt. I-I:
    file as above. ADS 10:34

List of Germans in Canada; emigration lists
            1936-1938

262: Ho 235 B: Preussische Regierung: St A 53 5.
            ADS 10:34

HOHENZOLLERN-SIGMARINGEN. *See also* Sigmaringen

HOHSEN (parish) NL

Emigrants 1847

85: I.3.34.A340. ADS 4:285

HOLSTEIN *and* LAUENBURG (duchies; Danish) SHL

Attempts to encourage emigration 1851-1853

2: Abt.80: II.Dept.: 2.7.: Nr. 1035.
  DLC may have. ADS 1:17

HOLZEN (village) BWL

Emigration file 1847-1936

288: Kommunalarchiv Holzen. ADS 10:119

HOLZHAUSEN (Kreis? Emmendingen) BWL

Emigration files 1807-1831

260: Rep.Abt.229: Spezialakten: Holzhausen: Fasz.
    45673-45759. ADS 10:11

HOLZHAUSEN (village) HEL

Releases from citizenship 1852-1896

151: Bestand 151: Fritzlar: Acc.1958/33: Nr. 107.
            ADS 7:115

HOLZHAUSEN (Amt) HEL

For exemptions from emigration tax, *see* Hesse-
  Kassel [Kurhesse]

HOLZHAUSEN (village) BWL

Emigration file 1844-1915

288: Kommunalarchiv Holzhausen. ADS 10:119

HOMBERG (village; Landkreis Ueberlingen) BWL

Emigration file 1892-1942

288: Kommunalarchiv Homberg. ADS 10:119

HOMBURG (Amt) HEL

Nr. 194: Petitions for releases from Prussian cit-
      izenship 1860-1872
Nr. 157: Petitions for dispensation from civil &
      private law requirements upon marriage
      [of emigrating persons?] 1867-1869

152: Abt.317: Amt Homburg: files as above.
            ADS 7:138

HOMBURG (Kreis) HEL

Emigration & immigration 1868-1887

151: Bestand 180: Fritzlar: Acc.1958/33: Nr. 61.
            ADS 7:115

Naturalizations, 3 vols. 1822-1867

151: Bestand 17(b): Gef.92: Nr. 1. ADS 7:69

HOMBURG (Kreis). *See also* Fritzlar HEL

HOMBURG (Landkreis). *See* Grossbundenbach, Lambsborn,
  Limbach, Moersbach

HOMBURG (Landgraefl. Hessischer Geheimer Rat)   HEL

  Nr. 330: Naturalizations   1828-1857
  Nr. 331: Releases from citizenship   1822-

  <u>200</u>: Abt. 386: Landgraefl. Hessischer Geheimer Rat
    zu Homburg: files as above.   ADS 8:15

HOMBURG (Landgrafschaft Hessen-Homburg)   HEL

  Applications for releases from citizenship & nat-
    uralizations   1843-1858

  <u>200</u>: Abt. 388: Landgrafschaft Hessen-Homburg: Nr.
    31-46.   ADS 8:15

HOMBURG VOR DER HOEHE (Landratsamt)   HEL

  Naturalizations & releases from citizenship
                                1867-1891

  <u>152</u>: Abt. 413: Homburg v.d.H.: Sect.I, Tit.III:
    Specialia: Nr. 274.   ADS 7:155

HOMBURG. *See also* Bad Homburg, Hesse-Homburg

HORB (Oberamt)   BWL

  Nr. 38: Emigration, in general   1815-1893
  Nr. 42-69: Lists of emigrants, by village, alpha-
    betically arranged   19th century

  <u>261</u>: F 175: Horb: files as above.   ADS 10:20

  Eight emigration lists (1815-1893); emigration
    files   1849-1867

  <u>262</u>: Wue 65/13: Horb: 244.   ADS 10:38

HORBEN (village)   BWL

  Emigration file   1875-1949

  <u>288</u>: Kommunalarchiv Horben.   ADS 10:119

HORHEIM (village)   BWL

  Emigration file   1864-1944

  <u>288</u>: Kommunalarchiv Horheim.   ADS 10:120

HORN (Amt; near Detmold)   RWL

  Certificates of domicile   1883-1905

  <u>114</u>: Detmold: Fach 4: Nr. 5.   ADS 6:15

HORN (village; Kreis Konstanz)   BWL

  Emigration to the USA   1868-1910

  <u>288</u>: Kommunalarchiv Horn.   ADS 10:120

HORNBACH (Landkreis Hornbach, Kurpfalz)   RPL

  Names of emigrants   18th century

  <u>201</u>: Archivabt. Ausfautheiakten.   ADS 8:63

HORNBACH (Landkreis). *See also* Boeckweiler, Breitfurt, Brenschelbach, Walshausen, Walsheim

HORNBURG   NL

  Two emigrants to America   1764

  <u>87</u>: 38 C.   ADS 4:286

HOYA (Amt)   NL

  Nr. 1: Emigration certificates   1844-1851
  Nr. 2: Emigration certificates   1861
  Nr. 3: Emigration certificates   1868-1869
  Nr. 5: Emigration certificates   Jan-Apr 1870

  <u>52</u>: Han.Des.74: Hoya: II.G: files as above.
                                ADS 4:73

  Emigration matters, 2 vols.   1868-1888

  <u>52</u>: Han.Des.74: Hoya: I.C.: Nr. 76.   ADS 4:73

  Releases from citizenship   1860

  <u>52</u>: Han.Des.74: Hoya: III.21: Nr. 1.   ADS 4:73

HUEGELHEIM (village)   BWL

  Emigration file   1838-1924
  Immigration file   1920-1922

  <u>288</u>: Kommunalarchiv Huegelheim.   ADS 10:120

HUEMMLING (Amt)   NL

  Nr. 1: Emigration permits, Vols. I-IV   1828-1870
  Nr. 3: Emigration certificates   1855-1871

  <u>54</u>: Rep.122,IV: Huemmling: Fach 40: files as above.
                                ADS 4:133

  Emigration permits   1871-1879
    [for permits before 1871, *see* Osnabrueck]

  <u>54</u>: Rep.116,I: Nr. 6074.   ADS 4:120

  Releases from citizenship   1880-1886

  <u>54</u>: Rep.116,I: II.4.A.1e: Nr. 86.   ADS 4:121

  Emigration passports to America   1869-1877

  <u>54</u>: Rep.122,IV: Huemmling: Fach 165, Nr. 13.
                                ADS 4:134

HUEMMLING (Kreis)   NL

  Releases from citizenship   1887

  <u>54</u>: Rep.116,I: Nr. 13416.   ADS 4:123

HUENFELD (Kreis)   HEL

  Nr. 4633-4638: Naturalizations, Vol. II-III
                                1820-1867
  Nr. 5477: Passports, 2 vols.   1820-1865
  Nr. 7493: Residency   1819-1841

  <u>151</u>: Bestand 100: files as above.   ADS 7:94

Naturalizations   1816-1819
151: Bestand 100: a: Nr. 275.   ADS 7:95

HUERRLINGEN (village)   BWL

Emigration files   1908-1914
288: Kommunalarchiv Huerrlingen.   ADS 10:120

HUETTEN (village)   BWL

Emigration file   1851-1932
General file   1932-1942
288: Kommunalarchiv Huetten.   ADS 10:120

HUGSTETTEN (village)   BWL

Emigration file   1825-1920
288: Kommunalarchiv Hugstetten.   ADS 10:120

HUGSWEILER (village)   BWL

Emigration file   1838-1916
Immigration & Germans abroad   1912-1943
288: Kommunalarchiv Hugsweiler.   ADS 10:120

HUNDESKOPF BEI TEMPELBERG. See Rohde, -- [surname
index]   Pomerania

HUNDSHAUSEN (village)   HEL

Releases from citizenship   1852-1894
151: Bestand 180: Fritzlar: Acc.1958/33: Nr. 106.
                                         ADS 7:115

HUNGEN (village)   HEL

Emigration to America   1831-1907
187: Abt. XI,4.   ADS 7:242

HUNSRUECK. See Simmern, Sponheim   RPL

HUNTEBURG. See Wittlage-Hunteburg

HUSUM   SHL

Nr. 1239: Passports   1811-1868
Nr. 1240-1249: Visaed passports   1799-1865
Nr. 1272: Citizenship [among other matters]
                                         1819-1872
Nr. 1287: Emigration & immigration   1860-1870
15: files as above.   ADS 1:50

HUSUM (Amt). See Helgoland   SHL

HUTTINGEN (village)   BWL

Emigration file   1832-1951

288: Kommunalarchiv Huttingen.   ADS 10:120

IBACH (village)   BWL

Emigration files (3 fascicles)   1832-1950
288: Kommunalarchiv Ibach.   ADS 10:120

IBER (parish)   NL

Church register (1833-1875) includes emigrants to
America   1846-1849
88: A.II.B.KB 26.   ADS 4:286

IBURG   NL

Emigration permits   1871-1879
[for permits before 1871, see Osnabrueck]
54: Rep.116,I: Nr. 6075.   ADS 4:120

IBURG (Amt)   NL

Chelsea-Out-Pensioners, 4 vols.   1815-1859
[Contains pension correspondence, etc., for the
King's German Legion. The Legion was formed in
1803 from soldiers from the former Hannoverian
army and fought under British flag until 1815.
For genealogists, the most important items in
the file are lists of pensioners living in Amt
Iburg (gives residence, years in service,
amount of pension, disability).
54: Rep.122,V: Iburg: Fach 61, Nr. 1.   ADS 4:134

Nr. 1: Emigration permits   1819-1828
Nr. 2: Releases from citizenship, Vol. I,
                                         1823-1829
     Also included: Emigration permits for per-
     sons subject to military service   1829-1834
Nr. 14: Emigration permits   1866
54: Rep.122,V: Iburg: Fach 62: files as above.
DLC has Nr. 2.   ADS 4:134

Oppositions to, & complaints regarding, projected
emigration [of various individuals]   1829-1846
54: Rep.122,V: Iburg: Fach 311, Nr. 2.
DLC has.   ADS 4:134

Transportation of asocial persons to America:
Nr. 2: General   1835
Nr. 3: Vol. I   1838-
       Vol. II   1868
54: Rep.122,V: Iburg: Fach 304: files as above.
                                         ADS 4:134

Lists of emigrants & immigrants   1859-1882
54: Rep.122,V: Iburg: Fach 38, Nr. 1.   ADS 4:134

Nr. 69: Releases from citizenship   1880-1884
Nr. 100: Releases from citizenship   1884-1888
54: Rep.116,I: II.4.A.1e: files as above.
                                         ADS 4:120; 122

Nr. 13417: Releases from citizenship   1889
54: Rep.116,I: files as above.   ADS 4:123

IBURG (flecken)   NL

  Emigration file   1830-1870
  <u>54</u>: Depositum 13: Archiv Fl. Iburg: III.G.: Nr. 1.
                                  ADS 4:146

IBURG (Vogtei)   NL

  Nr. 8: Emigration, immigration, residency of for-
        eigners   1829-1866
  Nr. 9: Emigration   1829-1855
  Nr. 12: Emigration permits   1838-1845
  <u>54</u>: Rep.122,V.4: Iburg: files as above. ADS 4:136
  DLC has Nr. 8.

ICHENHEIM (village)   BWL

  Emigration file   1853-1882
  Immigration file   1891-1906
  <u>288</u>: Kommunalarchiv Ichenheim.   ADS 10:120

IDENSEN (parish)   NL

  Emigrants   1850
  <u>89</u>: I.1.10.A102.   ADS 4:286

IDSTEIN (Oberamt)   HEL

  Confiscation of the property of emigrants leaving
    without permission   1698
  <u>152</u>: Abt.133: Gen.XIVa: Nr. 21.   ADS 7:127

IDSTEIN (Amt)   HEL

  Nr. 1269: Arrest of listed persons who aided
          Napoleon [some emigrated thereafter to
          America]   1814-1840
  Nr. 990: Emigration   1817-1879
  Nr. 1214: Emigration to South America 1853-1856
  Nr. 358: Emigration & immigration   1875-1885
  Nr. 2313: Emigration & immigration   1855-1866
  Nr. 445: Releases from citizenship & emigration
          to America   1856-1879
  Nr. 1209: List of emigrants to USA   1817-1867
  <u>152</u>: Abt.229: Amt Idstein: files as above.
                                  ADS 7:132

IHRINGEN (village)   BWL

  Emigration files   1877-1943
  <u>288</u>: Kommunalarchiv Ihringen.   ADS 10:121

ILBENSTADT (village)   HEL

  Notations on emigrants to America   n.d.
  <u>188</u>: Ortsbuergerverzeichnis.   ADS 7:243

ILBESHEIM (Kreis Landau, Kurpfalz)   RPL

  Names of emigrants   18th century

<u>201</u>: Archivabt. Ausfautheiakten.   ADS 8:62

ILIENWORTH. *See* Oster-Ilienworth, Wester-Ilienworth

ILLINOIS (German Consulate). *See* Cincinnati, Ohio

ILLMENSEE (village)   BWL

  Emigration file   1896-1936
  <u>288</u>: Kommunalarchiv Illmensee.   ADS 10:121

ILLWANGEN (village)   BWL

  Emigration file   1874-1941
  <u>288</u>: Kommunalarchiv Illwangen.   ADS 10:121

IMBSHAUSEN (Kirchenkreis Hohnstedt)   NL

  List of emigrants to America   1849
  <u>90</u>: I.A.1.10: Paket No.1: Nr. 104.   ADS 4:286

IMMENREICH (village)   BWL

  Emigration file   1931-1937
  <u>288</u>: Kommunalarchiv Immenreich.   ADS 10:121

IMMENSTAAD (village)   BWL

  Emigration file   1823-1932
  Emigration to the USA & South America 1847-1911
  <u>288</u>: Kommunalarchiv Immenstaad.   ADS 10:121

IMPFLINGEN (village)   RPL

  Names of emigrants   18th century
  <u>201</u>: Archivabt. Ausfautheiakten.   ADS 8:62

INDIANA (German Consulate). *See* Cincinnati, Ohio

INDLEKOFEN (village)   BWL

  Emigration file   1831-1908
  <u>288</u>: Kommunalarchiv Indlekofen.   ADS 10:121

INGENHEIM (village)   BYL

  Request for notarization of an American marriage
    license and a baptismal certificate   1883
  <u>227</u>: Rep. MA 1921: A.V.V: Nr. 15036.   ADS 9:89

INGOLSTADT (Landratsamt)   BYL

  Emigration   1860-1913
  <u>229</u>: Rep. Landratsaemter: Ingolstadt: 2.12; 3.13.
                                  ADS 9:118

INSTERBURG (Landratsamt)   East Prussia (Gumbinnen)

Applications for emigration permits & releases
   from citizenship   1874-1927

51: Rep.18: Abt.VIII: Nr. 1, Bd. 1.   ADS 4:48

INSTERBURG (Kreis)   East Prussia (Gumbinnen)

Emigration [file missing?]   1853-1876

51: Rep.12: Abt.I, Titel 3: Nr. 8, Bd. 6.
                                            ADS 4:32

INZLINGEN (village)   BWL

Emigration files   1782-1865
Emigration to the USA   ? -1871
General summary of emigration to Canada   1854-1948
Emigration file   1875-1939

288: Kommunalarchiv Inzlingen.   ADS 10:121

IOWA   USA

Emigration to Iowa   1868

50: Rep.21a: Nr. 9005.   ADS 4:9

IRMENACH (Duchy of Zweibruecken)   RPL

Young men who have journeyed to America for a two-
   year period   1741

201: Zweibruecken III: Akt Nr. 3358.   ADS 8:60

ISENBURG (Ducal estate)   HEL

Emigration lists   19th century

155   ADS 7:190

ISERLOHN (Kreis)   RWL

Nr. 89: Emigration   1817-1853
Nr. 90: Emigration & expulsions   1868-1879
Nr. 96-99: Immigration   1843-1879
Nr. 101: Certificates of original domicile
                                           1864-1891

116: Arnsberg Reg.: Iserlohn: files as above.
                                            ADS 6:60

ITTENDORF (village)   BWL

Emigration file   1841-1935

288: Kommunalarchiv Ittendorf.   ADS 10:121

JACOBIDREBBER (parish)   NL

Notes on persons who emigrated to North America
   from the parish   1831-1846

91: III.A.KB5 [for information, apply to Landkreis
   Archiv, Hannover]   ADS 4:287

JESTETTEN (village)   BWL

Emigration file   1867-1950
Immigration file   1907-1941

288: Kommunalarchiv Jestetten.   ADS 10:121

JOELLENBECK. See Schildesche-Joellenbeck (Kreis
   Bielefeld)

JORK   NL

Emigration from Borstel   1858-1881

92: A.II. Akte 2.   ADS 4:287

JUELICH-BERG   RWL

Negotiations regarding citizens who wish to emigrate to the "new French establishment in
   America"   1764

115: Rep.B.2: Juel.-Berg Landesarchiv II: Nr. 31.
                                           ADS 6:17

JUELICH (Landratsamt)   RWL

Nr. 113: Emigration   1817-1880
Nr. 114: Emigration & immigration   1880-1900
Nr. 117: Certificates of domicile; passports
                                           1848-1891
Nr. 118: Certificates of domicile; passports
                                           1892-1900
Nr. 119: Certificates of domicile; passports
                                           1900-1904
Nr. 120: Emigration   1848-1873
Nr. 121: Emigration   1873-1894
Nr. 122: Emigration   1896-1903
Nr.   1: Emigration of Prussian subjects 1816-1848

115: Rep. G 29/22: Juelich: files as above.
                                           ADS 6:42

KADOLZBURG (Landgericht; Bezirksamt Fuerth)   BYL

Nr. 100: Files on emigration & immigration; ex-
         portation of personal property 1808-1819
Nr. 101: Files on emigration & immigration; ex-
         portation of personal property 1810-1849
Nr. 105: Files on emigration & immigration; ex-
         portation of personal property [summaries
         & statistics?]   1819-1833
Nr. 333: Emigration matters   1836

231: Rep. 212/7: Fuerth: files as above.
                                        ADS 9:128-129

KALTBRUNN (village)   BWL

Emigration to the USA   1854-1877
Emigration files   1875-1891

288: Kommunalarchiv Kaltbrunn.   ADS 10:121

KALTBRUNN (village; Kreis Wolfach)   BWL

Emigration file   1837-1940

288: Kommunalarchiv Kaltbrunn (Kr. Wolfach)
                                         ADS 10:122

KAMEN (city)   RWL

    Releases from Prussian citizenship; emigration;
       certificates of original domicile   1850
    <u>129</u>   ADS 6:96

KAMPEN (Kreis)   RWL

    Emigration applications   1852-1860
    <u>115</u>: Rep.D8: I.Abt.3: Nr. 567.   ADS 6:25

KANDEL (village; Landkreis Kandel, Kurpfalz)   RPL

    Names of emigrants   18th century
    <u>201</u>: Archivabt. Ausfautheiakten.   ADS 8:63

KANDEL (Landkreis, Kurpfalz). *See also* Freckenfeld,
Minfeld, Minderslachen

KAPPELRODECK (village)   BWL

    Emigration files (3 fascicles)   1800-1939
    <u>288</u>: Kommunalarchiv Kappelrodeck.   ADS 10:122

KARLSHOLZ (Gemeinde Sinbronn; Oberamt Ansbach) BYL

    Emigration affairs   1797-1798
    <u>231</u>: Rep. 165a: Ansbacher Oberamtsakten: Nr. 3006.
                                                   ADS 9:127

KARLSRUHE (Amt)   BWL

    Emigration permits   [18th century?]
    <u>260</u>: Abt.236: Innenmin., Wegzug: Auswand.: Nr.
       1687.   ADS 10:3

KARLSRUHE   BWL

    Emigration & emigration agents   1870-1900
    <u>260</u>: Rep.Abt.330: Polizeipraesidium Karlsruhe.
                                                   ADS 10:15

KARSAU (village)   BWL

    Emigration file   1844-1906
    <u>288</u>: Kommunalarchiv Karsau.   ADS 10:122

KASSEBRUCH (former courthouse)   NL

    Releases from citizenship   n.d.
    <u>52</u>: Han.Des.74: Hagen: Regiminalia Fach 41: Nr. 3.
                                                   ADS 4:69

KASSEL (Provincial Regierung)   HEL

    Death certificates for Hessian citizens dying
       abroad:

    Nr. 2: Vol. I   1823-1841
    Nr. 2a: Vol. II   1841-1848
    Nr. 2b: Vol. III   1849-1861
    Nr. 2c: Vol. IV   1862-1867
    <u>151</u>: Bestand 17(a): Gef.39: files as above.
                                                   ADS 7:69

    Miscellaneous estate & inheritance matters:
    Nr. 41: Vol. I   1832-1845
    Nr. 42: Vol. II   1846-1848
    Nr. 42a: Vol. III   1849-1863
    Nr. 42b: Vol. IV   1864-1867
    <u>151</u>: Bestand 17(a): Gef.36: files as above.
                                                   ADS 7:68

KASSEL (Kreis)   HEL

    Nr. 1: Naturalizations, Vols I, IV   1822-1839
    Nr. 1e: Naturalizations, Vols. V-XI 1839-1863
    <u>151</u>: Bestand 17(b): Gef.91: files as above.
                                                   ADS 7:69-70

KASSEL (city)   HEL

    Releases from citizenship   1838
    <u>151</u>: Bestand 16.II: Klasse 15: Nr. 13. ADS 7:66

    Nr. 1252: Foreign residents   1868-1926
    Nr. 517: Foreigners in Germany during World War I
                 2 vols.   1914-1916
    Nr. 738: Foreigners in Germany during World War I
                 Vol. 4   1918-1922
    Nr. 739: Foreigners in Germany during war & post-
                 war period   1918-1929
    Nr. 1214: Lists of foreigners in various Kreise
                                          1922-1923
    <u>151</u>: Bestand 165: files as above.   ADS 7:109

KATLENBERG [Katlenburg?]-LINDAU. *See* Gieboldehausen   NL

KATZENELNBOGEN (Niedergrafschaft)   HEL

    C 1: Emigration   1697-1805
    C 2: Emigration to Pennsylvania, Hungary, & Lithu-
         ania   1709-1787
    <u>152</u>: Abt. 300: Katzenelnbogen: Gen.XIV: files as
       above.   ADS 7:137

KAUB (Landratsamt Ruedesheim)   HEL

    List of miners who emigrated to North America 1873
    <u>152</u>: Abt.415: Ruedesheim: Nr. 45.   ADS 7:156

KEHDINGBUSCH (parish)   NL

    Various applications for releases from citizenship
                                          1865-1906
    Lists of emigrants (names, ages, professions, des-
       tinations)   n.d.
    <u>100</u>: Gem. Kehdingbusch: V. Nr. 6.   ADS 4:311

KEHLHEIM (Bezirksamt)   BYL

Emigration   [19th century]
225: Rep.168: Verz. 1.   ADS 9:15

KEHLHEIM (Landgericht)   BYL

Applications for emigration permits   1811-1846
225: Rep.168: Verz. 1.   ADS 9:16

KEILBERG (village)   BYL

Emigration & immigration   n.d.
253: Fach 24, fasz. 18.   ADS 9:173

KEMPEN (Landratsamt)   RWL

Nr. 422: Emigration   1837-1865
Nr. 582: Emigration   1866-1884
Nr. 35: Emigration   1885-1894
Nr. 595: Emigration   1894-1899
115: Rep. G29/23: Kempen: files as above. ADS 6:43

KEMPEN (Kreis)   RWL

Emigration   1873-1895
115: Rep.D8: Neuere Akt.: Abt.1,C3: Nr. 12072-
     12074.   ADS 6:27

KEMPFENBRUNN   HEL

Applications for emigration to Pennsylvania   1748
151: Bestand 80: Lit.A: Nr. 6 [Verschiedene]
                                          ADS 7:78

Emigration matters   1764-1767
151: Bestand 81: Rep.E: 55.Amt Bieber VII: Nr. 1.
                                          ADS 7:83

KEMPTEN (Bezirksamt)   BYL

Individual files (122 in number) regarding emigration to North America   1847-1868
230: Rep. Akten der Bezirksaemter: Abgabe 1950:
     B11a/1 ff.   ADS 9:123

The Kemptener Zeitung [newspaper] contains emigration notices, particularly for the period 1848-
   1852
244   ADS 9:155

KESSELSTADT (Amt)   HEL

For exemptions from emigration tax, see Hesse-
   Kassel [Kurhesse]

KIECHLINGSBERGEN (village)   BWL

Emigration file   1853-1927
288: Kommunalarchiv Kiechlingsbergen.   ADS 10:122

KILIANSSTAEDTEN (Amt)   HEL

For exemptions from emigration tax, see Hesse-
   Kassel [Kurhesse]

KINZIGTAL (village)   BWL

Emigration file   1829-1934
General file   1851-1937
Immigration file   1919-1931
288: Kommunalarchiv Kinzigtal.   ADS 10:122

KIPFENBERG   BYL

Emigration & immigration   1841
231: Rep.270/II: Reg. Mittelfranken: Kammer des
        Innern: Abgabe 1932: Nr. 465.   ADS 9:134

KIPPENHAUSEN (village)   BWL

Emigration to the USA   1820-1889
288: Kommunalarchiv Kippenhausen.   ADS 10:122

KIPPENHEIM (village)   BWL

General emigration file   1854-1912
Emigration file   1871-1932
Immigration & emigration file   1895-1950
288: Kommunalarchiv Kippenheim.   ADS 10:122

KIPPENHEIMWEILER (village)   BWL

Emigration file   1846-1938
Surveillance of enemy & neutral foreigners
                                       1914-1930
Germans abroad   1914-1941
288: Kommunalarchiv Kippenheimweiler.   ADS 10:122

KIRCHEN. See Efringen-Kirchen   BWL

KIRCHHAIN (town)   HEL

Naturalizations   1821-1865
151: Bestand 20: Verz.6, Acc.1909/9.19: Nr. 94.
                                          ADS 7:75

KIRCHHEIM (Landratsamt)   HEL

Releases from citizenship   1897
151: Bestand 180: Kirchheim: Nr. 63.   ADS 7:119

KIRCHHEIM (Oberamt)   BWL

Nr. 54-79: Lists of emigrants, by village, alpha-
              betically arranged   [19th century]
Nr. 28: Emigration Kirchheim (Kreisregierung Ulm);
           disposal of property by emigrants; loss &
           retention of Wuerttemberg citizenship
                                       1855-1891
261: F 176: Kirchheim: files as above.   ADS 10:20

KIRCHHEIMBOLANDEN (Bezirksamt)   RPL

Emigration files   1869-1910

<u>201</u>: Archivabt. Bezirksaemter: Kirchheimbolanden:
      Akt Nr. V,2.   ADS 8:64

KIRCHHEIMBOLANDEN (Landkreis). *See* Albisheim, Bischheim, Breunigweiler, Dandenfels, Dreisen, Marnheim,
Ruessingen, Sippersfel, Weierhof   RPL

KIRCHHOFEN (village)   BWL

Emigration file   1835-1867
Emigration to the USA   1847
Emigration & Germans abroad   1867-1948
General file   1882-1898

<u>288</u>: Kommunalarchiv Kirchhofen.   ADS 10:122

KIRNBACH (village)   BWL

General emigration file   1832-1853
Emigration file   1854-1912

<u>288</u>: Kommunalarchiv Kirnbach.   ADS 10:123

KIRSCHHAUSEN   HEL

Emigration   1834-1922

<u>149</u>: Rep.40,2: Ablief.2: Bd. 302.   ADS 7:23

KLEINKEMS   BWL

Emigration files   1530-1833

<u>260</u>: Rep.Abt.229: Spezialakten: Fasz. 53950-54023.
                                         ADS 10:11

Emigration file   1875-1931

<u>288</u>: Kommunalarchiv Kleinkems.   ADS 10:123

KLEVE (Kreis)   RWL

Emigration applications   1852-1861

<u>115</u>: Rep.D8: I.Abt.c: Nr. 570-573.   ADS 6:25

KLINGEN (village)   RPL

Names of emigrants   18th century

<u>201</u>: Archivabt. Ausfautheiakten.   ADS 8:62

KLUFTERN (village)   BWL

Emigration file   1904-1941

<u>288</u>: Kommunalarchiv Kluftern.   ADS 10:123

KNITTELSBACH (village; Oberamt Ansbach)   BYL

Emigration matters   1800-1805

<u>231</u>: Rep.165a: Ansbacher Oberamtsakten: Nr. 3007.
                                         ADS 9:127

KOBLENZ (Regierungsbezirk)   RPL

Nr. 5088: Summary of emigration & immigration
                                          1845-1866
Nr. 5089: Summary of emigration & immigration
                                          1845-1863
Nr. 5094: Emigration to America   1848-1852

<u>200</u>: Abt.441: Regierung Koblenz: files as above.
                                         ADS 8:32

Protestant church records [Evangelische Kirchen-
buecher] contain some material of interest to
German-American researchers

<u>200</u>: Abt.555: Evang. Kirchenbuecher.   ADS 8:45

The Lutheran Church records for many parishes contain marginal notations regarding emigrants to
America (only Links-Rheinischen Gebieten, i.e.,
areas west of the Rhine River)   before 1798

<u>206</u>   ADS 8:72

KOBLENZ (city)   RPL

Reports on foreigners   1830

<u>200</u>: Abt.403: Oberpraesidium Koblenz (Rheinpro-
      vinz): Nr. 2058.   ADS 8:19

Miscellaneous inquiries regarding emigrants, Germans abroad, etc., 5 vols.   1864-1922

<u>200</u>: Abt.623: Stadt Koblenz: Nr. 4489-4493.
                                         ADS 8:46

KOBLENZ-LAND (Landkreis)   RPL

Naturalizations   1850-1866

<u>200</u>: Abt.441: Regierung Koblenz: Nr. 15594; 15362;
      21987-25003; 24997.   ADS 8:28

KOCHEM (Kreis)   RPL

Naturalizations   1846-1914

<u>200</u>: Abt.441: Regierung Koblenz: Nr. 24993; 21979-
      21981.   ADS 8:28

KOELN   RWL

Applications for emigration (each volume contains
      name list)   1834-1909

<u>115</u>: Rep.D9III: Abt.E: Nr. 96-114.   ADS 6:34

KOELN (arrondissement)   RWL

Population movements   1801-1803; 1811-1812

<u>115</u>: Rep.D2.2: II.Division, 1.Bureau, 2.Bevoelker-
      ung: Nr. 112.   ADS 6:19

KOELN (Landkreis)   RWL

Emigration   1857-1888

<u>115</u>: Rep.D9: Abt.I.e: Nr. 253-257.   ADS 6:31

KOELN (city)   RWL

   Emigration of Germans & Swiss to America 1817-1819

   <u>130</u>: Rep.400: V.11.A: Nr. 4.   ADS 6:96

   Emigration   1887-1890

   <u>115</u>: Rep.D9: Abt.I.e: Nr. 250-252.   ADS 6:31

   List of foreigners   1822

   <u>200</u>: Abt.403: Oberpraesidium Koblenz (Rheinprovinz):
      Nr. 2057.   ADS 8:19

KOENIGSBERG   East Prussia

   Nr. 40: Issuance of emigration permits & certifi-
        cates of residency   1818-1869
   Nr. 42: Issuance of emigration permits & certifi-
        cates of residency   1817-1835
        [Mainly emigration to Poland, Russia, &
        other German states, but some to America]
   Nr. 36, Bd. 7: Various complaints by emigrants
        against certain emigration agents   1893-

   <u>51</u>: Rep.2,I^2: Titel 30: files as above.   ADS 4:26

KOETZTING (Bezirksamt)   BYL

   Emigration   [19th century]

   <u>225</u>: Rep.168: Verz. 1.   ADS 9:15

KOLLMARSREUTE (village)   BWL

   Emigration file   1857-1937
   General file   1859-1937

   <u>288</u>: Kommunalarchiv Kollmarsreute.   ADS 10:123

KOENIGSFELD (village)   BWL

   Emigration file   1865-1939
   Emigration to overseas countries   1921-1930

   <u>288</u>: Kommunalarchiv Koenigsfeld.   ADS 10:123

KOOGE. *See* Hedwigenkoog

KORBACH (town; Waldeck)   HEL

   A II 63: Passports   1769-
   A IV 1: Military files   1666-1830
   A IV 46: Recruitment for Waldeck Battalion for
        Holland or England   1776; 1785
   A IV 66: Recruitment for Waldeck Battalion for
        Holland or England   1776; 1785
   F 2:   Certificates of original domicile
                                      1851-1879
   F 20:   Collection of materials for genealogical
        research (family names, passports, re-
        ports)   n.d.
   <u>189</u>: files as above.   ADS 7:248

KORBACH. *See also* Waldeck   HEL

KORK (village)   BWL

   General emigration file   1871-1934
   Emigration file   1871-1937

   <u>288</u>: Kommunalarchiv Kork.   ADS 10:123

KORK (Amtsgericht). *See also* Odelshofen   BWL

KREFELD (Kreis)   RWL

   Emigration   1852-1858

   <u>115</u>: Rep.D8: Neuere Akt: Abt.1 C3: Nr. 11894.
                                            ADS 6:25

KREFELD (city)   RWL

   Nr. 3: Emigration   1849-1894
   Nr. 6: Special files of the Mayor regarding quer-
       ies from descendants of the Germantown
       (Pastorius) colony in Pennsylvania   n.d.

   <u>131</u>: Abt.I, Fach 8: files as above.   ADS 6:110

KRENKINGEN (village)   BWL

   Emigration file   1885-1953

   <u>288</u>: Kommunalarchiv Krenkingen.   ADS 10:123

KREUZNACH (Kreis)   RPL

   Nr. 25002: Emigration to America   1867
   Nr. 22002-22026: Naturalizations   1854-1914

   <u>200</u>: Abt.441: Regierung Koblenz: files as above.
                                         ADS 8:27-29

KROPP BEI SCHLESWIG   SHL

   The *Predigerseminar* [seminary for preachers] edu-
      cated many Lutheran ministers for service in
      America; files   1882-1938

   <u>20</u>   ADS 1:55

KROZINGEN. *See* Bad Krozingen

KRUMBACH (village)   BWL

   Emigration file   1853-1929

   <u>288</u>: Kommunalarchiv Krumbach.   ADS 10:123

KRUMBACH (Bezirksamt)   BYL

   Individual files (218 in number) regarding emigra-
      tion to America & North America   1839-1910

   <u>230</u>: Rep. Akten der Bezirksaemter: Krumbach: Nr.
      3186; 3229-.   ADS 9:123

KRUMBACH (village)   HEL

   Emigration   1834-1922

   <u>149</u>: Rep.40,2: Ablief.2: Bd. 302.   ADS 7:23

KUEBELBERG (village; Landkreis Waldmohr, Kurpfalz)
RPL

Names of emigrants   18th century

201: Archivabt. Aufautheiakten.   ADS 8:63

KUENZELSAU (Oberamt)   BWL

Nr. 691-691a: Emigration files, by village
                                    19th century
Nr. 692:   Lists [of emigrants]   1817-1823
Nr. 693-713: Emigration files, chronologically
                      arranged   1844-1869

261: F 177: Kuenzelsau: files as above. ADS 10:21

KUERNBACH   HEL

Emigration   1863-1892

149: Rep.40,2: Ablief.2: Bd. 303.   ADS 7:23

KULMBACH (Bezirksamt)   BYL

Nr. 1: Emigration to North America   1832-1861
Nr. 3: Emigration to North America   1854-1860

223: Rep.K.13/Verz.VII: files as above. ADS 9:7

Individual cases of emigration (destinations not
    given)   1854-1914

223: Rep.K.13/Verz.IX: Nr. 18-818.   ADS 9:7

KULMBACH   BYL

Emigration files   middle of 19th century
245   ADS 9:156

KUERZELL (village)   BWL

Emigration file   1899-1930

288: Kommunalarchiv Kuerzell.   ADS 10:123

KUESSNACH (village)   BWL

Emigration file   1860-1941

288: Kommunalarchiv Kuessnach.   ADS 10:123

KUHBACH (village)   BWL

Emigration file   1880-1938
Surveillance of foreigners   1921-1934

288: Kommunalarchiv Kuhbach.   ADS 10:124

KURPFALZ (Electoral Palatinate)   RPL

Emigration, particularly to Pennsylvania, Prussia,
    Cayenne, Spain, & Bavaria   1658-1909

260: Rep.Abt.77: Pfalz Gen.: Nr. 29.   ADS 10:11

File on emigrants to Pennsylvania from Kurpfalz,
    Heidelberg   1709

260: Abt.236: Innenmin., Wegzug: Auswand. 6735.

ADS 10:4

Some special emigration cases to America 1746-1793

201: Archivabt. Geheimrats-Akten I: Akt Nr. 224-
    1/3.   ADS 8:56

File on emigrants to Cayenne Island [French Gui-
    ana] & to Prussian territories; herein all for-
    eign countries are included   1763-1764

260: Abt.236: Innenmin., Wegzug: Auswand.: Nr.
    6743.   ADS 10:5

Emigration of Palatine subjects to other countries,
    particularly to the Cayenne Island [French Gui-
    ana]   1764-1804

260: Abt.236: Innenmin., Wegzug: Auswand.: Nr.
    6744.   ADS 10:5

Prohibited emigration of female persons & informa-
    tion on the confiscation of property in special
    cases   1792-1796

260: Abt.236: Innenmin., Wegzug: Auswand.: Nr.
    6557.   ADS 10:4

The Heimatstelle Pfalz has a card catalog contain-
    ing many thousands of names, mainly with birth-
    places & destinations, of Palatine emigrants to
    America (and other countries). The information
    therein has been collected from all possible
    sources, published & unpublished. Most nine-
    teenth century information has been gleaned from
    personal letters. The Heimatstelle publishes
    Schriften zur Wanderungsgeschichte der Pfaelzer
    [Articles on the Migration History of the Pala-
    tines] based partially upon material in this
    card catalog & partly upon parish registers, etc.

220   ADS 8:91

There are numerous Familienakten [family files],
    inventories, lists of personal property, guard-
    ianships, etc., which contain information on
    emigrants from the following communities: Dann-
    stadt, Edenkoben, Freinsheim, Hessheim, Lachen,
    Nannweiler, Obersuelzen, Ottersheim (Kreis
    Kirchheimbolanden), Stetten, Weisenheim a.S.,
    Westheim, Wolfsten (the latter containing let-
    ters from emigrants)   n.d.

201: Above-mentioned communities.   ADS 8:55

Notarized certifications often include exhibits &
    other documentation containing powers of attor-
    ney from emigrants in America; such powers of
    attorney are important in that they give the
    American residence of the emigrant
                                mainly 19th century

201: Notariate.   ADS 8:66

The Ausfautheiakten (lists of properties, inven-
    tories, guardianships), nearly all from the 18th
    century, contain many applications for emigra-
    tion permits & powers of attorney for relatives
    remaining in Germany.

201: Archivabt. Ausfautheiakten.   ADS 8:62

KUSEL (Bezirksamt)   RPL

Files on emigration permits, tax, & military obli-
    gations   1818-1911

<u>207</u>: Nr. 132-07.   ADS 8:74

Nr. 706: Emigration; passport applications
                                              1869-1870
Nr. 650: Passport applications   1925-1926
Nr. 713: Passport applications   1914
Nr. 682: Passport applications   1927-1928
Nr. 849: Passport applications   1920-1921
Nr. 853: Emigration   1886
Nr. 893: Applications for certificates of original
         domicile & for Bavarian [?] citizenship
         (letter S only)   1925-1930
Nr. 103: Emigration, 2 fascicles   1883; 1885
Nr. 206: Emigration of conscientious objectors [to
         war]   1833-1865
Nr. 451: Emigration; releases from Bavarian citi-
         zenship   1881
Nr. 457: Emigration; releases from Bavarian citi-
         zenship   1892
Nr. 465: Emigration; releases from Bavarian citi-
         zenship   1884
Nr. 466: Emigration; releases from Bavarian citi-
         zenship   1875-1879
Nr. 631: Emigration [internal & external]
                                             1892-1893
Nr. 922: Passport applications   1922
Nr. 937: Passport applications, 2 fascicles   1930
Nr. 1081: Passport applications   1924
Nr. 1037: Applications for certificates of orig-
          inal domicile & citizenship   1905-1906
Nr. 974: Confiscation of the property of deserters
         & conscientious objectors [to war]
                                             1840-1896
Nr. 1351: Jewish emigration to the USA   1936-1940

<u>201</u>: Archivabt. Bezirksaemter: Kusel: files as
         above.   ADS 8:65

LACHEN (village). See Kurpfalz, Neustadt (Oberamt)

LADENBACH (Amt)   HEL

For exemptions from emigration tax, see Hesse-
   Kassel [Kurhesse]

LAGE (Amt; by Detmold)   RWL

Certificates of domicile, Vol. 3   1883-1905
<u>114</u>: Detmold: Fach 4, Nr. 5.   ADS 6:15

LAGE   RWL

Nr. 2: Certificates of citizenship   1880
Nr. 3: Certificates of original domicile   1857-1867
<u>132</u>: XIV B: files as above.   ADS 6:111

Nr. 2: Lists of emigrants & immigrants   1883
Nr. 3: Lists of emigrants & immigrants   1857-1880
Nr. 4: Lists of emigrants & immigrants   1881-1885
Nr. 5: Lists of emigrants & immigrants   1886
<u>132</u>: XIV C.2: files as above.   ADS 6:111

LAHR (town)   BWL

Emigration of criminals? [or illicit emigration?]
   to America   n.d.

<u>260</u>: Rep.Abt.211: Stadt Lahr.   ADS 10:11

LAMBSBORN (village; Landkreis Homburg, Kurpfalz) RPL

Names of emigrants   18th century
<u>201</u>: Archivabt. Ausfautheiakten.   ADS 8:63

LAMBSHEIM (village). See Neustadt (Oberamt)   RPL

LAMSTEDT/NIEDERELBE   NL

List of emigrants from the parish of Lamstedt,
   including Warstede   1840-1857

<u>92</u>: HS.8: Chronik von Pastor Zeidler.   ADS 4:287

As of 1957, Professor Wilhelm Klenk, Mittelstenahe
bei Lamstedt, had an abstracted emigrant list
for the district of Lamstedt

                                                 ADS 4:291

LANDAU/ISAR (Bezirksamt)   BYL

Emigration   [19th century]
<u>225</u>: Rep.168, Verz. 1.   ADS 9:15

Emigration   1896-1933
<u>225</u>: Rep.164, Verz. 9: Nr. 6.   ADS 9:14

LANDAU/ISAR (Landgericht)   BYL

Applications for emigration permits   1813-1846
<u>225</u>: Rep. 168, Verz. 1.   ADS 9:16

LANDAU/PFALZ (town)   RPL

Nr. 73: Emigration files   1817-1934
Nr. 100: Emigration permits   n.d.
<u>208</u>: N. A.: files as above.   ADS 8:75

LANDSHUT   BYL

Emigration & immigration   1799-1805
<u>225</u>: Rep.47.a.Saal XIII: Verz.13: Fasz.4: Nr. 1-
         109.   ADS 9:13

LANDSHUT (Bezirksamt)   BYL

Emigration & exportation of property   1806-1888
<u>225</u>: Rep.164, Verz.10: Nr. 19-.   ADS 9:14

Emigration   [19th century]
<u>225</u>: Rep.168, Verz. 1.   ADS 9:15

LANDSHUT (Magistrat)   BYL

Applications for emigration permits   1817-1846
<u>225</u>: Rep.168, Verz. 1.   ADS 9:16

LANDSHUT (Landgericht)   BYL

    Applications for emigration permits   1820-1846
    <u>225</u>: Rep.168, Verz. 1.   ADS 9:16

LANDSHUT (city)   BYL

    Emigration   [19th century]
    <u>225</u>: Rep.168, Verz. 1.   ADS 9:15

    Emigration files   latter half of 19th century
    <u>246</u>   ADS 9:157

LANDSTUHL (mayor's office)   RPL

    Applications for emigration permits   1846-1884
    Memorandum regarding secret [illicit] emigration
      & the disposition of the property of such emi-
      grants   1846
    List of emigrants to America who returned to
      Landstuhl   1855
    <u>209</u>   ADS 8:76

LANDSTUHL (Landkreis Sickingen). *See* Ellerstadt RPL

LANGENAU (village)   BWL

    Emigration   1858-1912
    <u>288</u>: Kommunalarchiv Langenau.   ADS 10:124

LANGENBRAND (village)   BWL

    Emigration of 7 families to the USA   1857 (1858)
    Emigration to the USA   1860
    <u>288</u>: Kommunalarchiv Langenbrand.   ADS 10:124

LANGENDIEBACH (Amt)   HEL

    For exemptions from emigration tax, *see* Hesse-
    Kassel [Kurhesse]

LANGENHAGEN. *See* Hannover-Langenhagen   NL

LANGENSALZA. *See* Bad Langensalza

LANGENSCHWALBACH (Amt)   HEL

    Nr. 317: Emigration & immigration   1816-1845
    Nr. 1031: Emigration & immigration   1824-1859
    Nr. 2914: Emigration to America   1834; 1852
    Nr. 1314: Emigration applications   1849-1858
    Nr. 727: Emigration & immigration   1852-1868
    Nr. 2793: Emigration & immigration   1870
    Nr. 1477: Naturalizations & releases from citizen-
      ship   1832-1872
    Nr. 1295: Naturalizations & releases from citizen-
      ship   1817-1820
    Nr. 1284: Passports   1833-1859
    <u>152</u>: Abt.231: Langenschwalbach: files as above.
                        ADS 7:133

Nr. 1031: Marriage applications [emigrants?]
                        1824-1859
Nr. 1295: Marriage applications   1817-1820
Nr. 1294: Marriage applications   1852-1858
    <u>152</u>: Abt.231: Langeschwalbach: files as above.
                        ADS 7:133

LANGENSCHWALBACH (Landratsamt)   HEL

    Nr. 1347: Releases from citizenship   1870-1913
    Nr. 1458: Emigration agents   1874-1886
    Nr. 1215: Emigration agents   1881-1920
    Nr. 1388: Emigration agents   1920-1938
    Nr. 1472: Immigration   1891-1921
    Nr. 1292: Immigration   1922-1938
    Nr. 1217: Releases from citizenship   1885-1920
    Nr. 1253: Naturalizations   1912-1920
    Nr. 1353: Certificates of citizenship, 2 vols.
                        1898-1920
    Nr. 1412: Extradition to foreign countries
                        1891-1937
    Nr. 1279: Emigration & naturalization (Jews)
                        1938-1943
    <u>152</u>: Abt.418: Langenschwalbach: files as above.
                        ADS 7:158

LANGENSCHWALBACH (baths)   HEL

    Treatment & guest lists   1867-1896
    <u>152</u>: Abt.405: Wiesbaden: Rep.405/I,4: Nr. 1217.
                        ADS 7:148

LANGENSTEIN (Herrschaft)   BWL

    Emigration of subjects   [18th? & 19th centuries]
    <u>273</u>   ADS 10:76

LANGENTHAL   HEL

    Emigration   1863-1891
    <u>149</u>: Rep.40,2: Ablief.2: Bd. 304.   ADS 7:23

LANGENWINKEL (village)   BWL

    Emigration file   1853-1936
    <u>288</u>: Kommunalarchiv Langenwinkel.   ADS 10:124

LAUENAU (Amt)   NL

    Nr. 2: Releases from citizenship   1772?-1839
    Nr. 3: Releases from citizenship   1840-1859
    <u>52</u>: Han.Des.74: Springe: III.F.: files as above.
                        ADS 4:78

LAUENBURG (Ratzeburg Regierung)   SHL

    Nr. 3431: Files on the settlement in America of
          useless [Lauenburger] subjects   1852
    Nr. 3369: Files on the settlement in America of
          useless [Lauenburger] subjects 1853-1857
    <u>2</u>: Abt.210: files as above.
    DLC has?   ADS 1:21

LAUENBURG. *See also* Holstein & Lauenburg   SHL

LAUENSTEIN (Amt)   NL

  Emigration permits   1829

  <u>52</u>: Han.Des.: Lauenstein: II. Fach 16, Nr. 1.
                               ADS 4:73

  Emigration   1868

  <u>52</u>: Han.Des.74: Lauenstein: I. Fach 3, Nr. 114.
                                ADS 4:73

LAUF (village)   BWL

  Emigration file   1855-1949

  <u>288</u>: Kommunalarchiv Lauf.   ADS 10:124

LAUFEN (village)   BWL

  Emigration file   1792-1915

  <u>288</u>: Kommunalarchiv Laufen.   ADS 10:124

LAUFENSELDEN (Amt)   HEL

  Nr. 1314: Emigration applications   1849-1858
  Nr. 1295: Marriage & emigration applications; nat-
        uralizations & releases from citizenship
                            1817-1820
  Nr. 1294: Marriage & emigration applications; nat-
        uralizations & releases from citizenship
                            1852-1858

  <u>152</u>: Abt.231: Langenschwalbach: files as above.
                                ADS 7:133

LAUPHEIM (Oberamt)   BWL

  Nr. 419: Lists [of emigrants]   1815-1871
  Nr. 420-452: Emigration files, chronologically ar-
          ranged   1810-1875

  <u>261</u>: F 178: Laupheim: files as above.   ADS 10:21

  XI 5 b: Emigration   1844-1910
  XII 1: Relations with foreign countries 1812-1870

  <u>262</u>: Wue 65/18: Laupheim: files as above.
                                ADS 10:38

LAUSHEIM (village)   BWL

  Emigration file   1837-1935

  <u>288</u>: Kommunalarchiv Lausheim.   ADS 10:124

LAUTENBACH (village; Kreis Offenburg)   BWL

  Emigration file   1832-1933

  <u>288</u>: Kommunalarchiv Lautenbach, Kr. Offenburg.
                                ADS 10:124

LAUTENBACH (village; Kreis Rastatt)   BWL

  Emigration file   1913-1939

<u>288</u>: Kommunalarchiv Lautenbach, Kr. Rastatt.
                                ADS 10:124

LEER (town)   NL

  25:709: Applications for emigration permits 1844-
  25:720: Emigration matters   1844-
  25:921: Passports issued   1823-1845
  25:993: List of emigrants   1860-1894
  25:936: Emigration of persons subject to military
         service   1852-1909
  25:796: Emigration of persons subject to military
         service   1891-1893
   9:1072: Seamen who have died, or were killed,
         abroad   1838-1913
   3:930: Secret escape of persons subject to mili-
         tary service   1868
   3:797: Emigration of persons subject to military
         service, Vol. 1   1871-1881
                 Vol. 2   1883-1891

  <u>93</u>: files as above   ADS 4:287-288

LEER   NL

  Deaths abroad of Leer citizens   1871-1932

  <u>50</u>: Rep.31a:XXVIII: Nr. 790.   ADS 4:13

  Nr. 14: Information regarding the acceptance of
        other citizenships by former Leer citizens
                            1869-1908
  Nr. 15: Granting & loss of citizenship   1929-1935
  Nr. 16: Issuance of certificates of citizenship
        & residency, Vol. I   1929-1935
                 Vol. II   1935-1939

  <u>50</u>: Rep.31a: V: files as above.   ADS 4:13

LEGELSHURST (village)   BWL

  Emigration file   1861-1949

  <u>288</u>: Kommunalarchiv Legelshurst.   ADS 10:124

LEHEN (village)   BWL

  Emigration file   1840-1875

  <u>288</u>: Kommunalarchiv Lehen.   ADS 10:124

LEHENGERICHT (village)   BWL

  Emigration file   1864-1938

  <u>288</u>: Kommunalarchiv Lehengericht.   ADS 10:125

LEIBERSTUNG (village)   BWL

  Emigration file   1822-1937

  <u>288</u>: Kommunalarchiv Leiberstung.   ADS 10:125

LEIBERTINGEN (village)   BWL

  Emigration to the USA   1850; 1854
  Emigration to the USA   1903

  <u>288</u>: Kommunalarchiv Leibertingen.   ADS 10:125

LEIMERSHEIM (parish)   RPL

The Catholic parish family book contains marginal
notations regarding emigrants to America
                                              1688-1735

201: Archivabt. Kirchenbuecher.   ADS 8:63

LEITISHOFEN. *See* Menningen   BWL

LEMFOERDE (Amt)   NL

Nr. 3: Emigration permits   1825-1840
Nr. 5: Emigration permits   1841-1858
52: Han.Des.74: Diepholz: Abt.III,F: files as
above.   ADS 4:66

LENNE (parish)   NL

List of emigrants   1905
94: II.Le.A105 [apply via Landkreis-Archiv Hann-
over]   ADS 4:288

LENNEP (Kreis)   RWL

Emigration applications   1852-1861
115: Rep.D8: I.Abt.e: Nr. 576-579.   ADS 6:25

LENNEP (Landratsamt)   RWL

Nr. 255: Emigration & immigration   1850-1853
Nr. 256: Emigration & immigration   1856-1859
Nr. 230: Emigration & immigration   1859-1864
Nr. 240: Emigration & immigration; deserters
                                    1883-1891
Nr. 241: Expulsions   1890-1897
Nr. 242: Expulsions   1897-1903
115: Rep. G29/25: Lennep: files as above. ADS 6:43

LENZKIRCH (with Stollenhof)   BWL

Emigration files   1312-1863
260: Rep.Abt.229: Spezialakten: Lenzkirch: Fasz.
59922-59999.   ADS 10:11

LEONBERG (Oberamt)   BWL

Nr. 486: Emigrant lists; emigration in general
                                    1817-1872
Nr. 489-515: Emigrants, alphabetical list, by village   19th century
261: F 179: Leonberg: file as above.   ADS 10:21

LEONBERG (town)   BWL

A number of letters from emigrants in America
                              [19th century?]
274   ADS 10:77

LESSENICH (canton)   RWL

Population movements   1801-1802; 1811-1812
115: Rep.D2.2: II.Division, 1.Bureau, 2.Bevoelker-
ung: Nr. 114.   ADS 6:19

LEUSTETTEN (village)   BWL

Emigration to the USA   1851-1919
288: Kommunalarchiv Leustetten.   ADS 10:125

LEUTKIRCH (Oberamt)   BWL

Emigration files, chronologically arranged
                                    1805-1889
261: F 180: Leutkirch: 285-315.   ADS 10:21

Releases from [Wuerttemberg] citizenship 1860-1908
262: Wue 65/19: Leutkirch: ZR. 167.   ADS 10:38

LICHTENBERG. *See* Hanau-Lichtenberg   HEL

LIEBENBURG (Amt)   NL

Nr. 220: General material on emigration 1831-1864
Nr. 221: Reports on emigration   1868-1885
52: Han.Des.74: Liebenburg: IV.G.3: files as above.
                                             ADS 4:73

LIEBESBACHERHOF. *See* Marzell   BWL

LIEDERN (Amt)   RWL

Emigration   1819-1853
116: Muenster Reg.: Borken: Nr. 4.   ADS 6:66

LIEL (village)   BWL

Emigration file   1834-1944
288: Kommunalarchiv Liel.   ADS 10:125

LIENHEIM (village)   BWL

Emigration file   1853-1943
288: Kommunalarchiv Lienheim.   ADS 10:125

LIERBACH (village)   BWL

Emigration file   1817-1950
288: Kommunalarchiv Lierbach.   ADS 10:125

LIMBACH (village; Landkreis Homburg, Kurpfalz) RPL

Names of emigrants   18th century
201: Archivabt. Ausfautheiakten.   ADS 8:63

LIMBURG (Amt)   HEL

Nr. 105: Naturalizations; releases from citizen-
          ship; emigration (special files for
          Dehrn, Eschhofen, Heringen, Hintermeil-
          ingen, Limburg, Mensfelden, Muehlen)
                                          1817-1886
Nr. 1360: Register of naturalized persons & emi-
          grants   1825-1839
Nr. 1562: Naturalizations & releases from citizen-
          ship; emigration   1842-1867

152: Abt.232: Limburg: files as above.
                                    ADS 7:133-134

LINDAU (Bezirksamt)   BYL

Nr. 1499; 1506: Emigration & immigration; exporta-
          tion of personal property
                                          1807-1862
Nr. 750: Emigration to countries overseas
                                          1862-1933

230: Rep. Akten der Bezirksaemter: Lindau: Abgabe
     1954: files as above.   ADS 9:123

LINDENFELS (Kreis)   HEL

Nr. 1: List of emigrants   1855-1859
Nr. 4: List of emigrants   1870

149: Rep.40,2: Ablief.2: Bd. 284: files as above.
                                          ADS 7:22

LINGEN (Amt)   NL

Nr. 1: Emigration & immigration   1800-1813
Nr. 3: Emigration permits & releases from citizen-
       ship   1824-1840
Nr. 6: Emigration to America   1834-1848
Nr. 9: Emigration certificates   1856-1866
Nr. 11: Emigration certificates   1866-1867
Nr. 11a: List of emigrated persons   1871-1906
Nr. 15: Emigration & immigration   1915-1926

54: Rep.122,VI: Lingen: files as above.
    DLC has No. 6.   ADS 4:136-137

Emigration permits   1871-1879
54: Rep.116,I: Nr. 6076.   ADS 4:120

Releases from citizenship   1880-1886
54: Rep.116,I: II.4.A.1e: Nr. 79.   ADS 4:121

LINGEN (Kreis)   NL

Releases from citizenship   1887
54: Rep.116,I: Nr. 13421.   ADS 4:123

Documents regarding occupations of German nationals abroad   1885-1921
54: Rep.122,VI,F.82.1 [or I.F. Nr. 25] ADS 4:137

LINGEN (Landratsamt)   NL

Nr. 69: List of foreigners   n.d.
Nr. 108: Return of emigrated Germans from foreign
         countries   n.d.

54: Rep.122,VI.A: Landratsamt Lingen: files as
    above.   ADS 4:137

LINGEN (town)   NL

Nr. 144: Emigration files   1834-1860
Nr. 145a: Acceptance of foreign citizenship by
          [former] residents   1863-1864
Nr. 149: Emigration of persons subject to military
         service, 2 vols.   1869-1926
Nr. 150: Emigration files   1881-1883
Nr. 558: Emigration permits   1828-1834

54: Depositum 29: Stadtarchiv Lingen: files as
    above.   ADS 4:147

Nr. 153: Administration of the estates of foreign-
         ers dying here   1860-1862
Nr. 162: Disbursements to residents of Lingen from
         estates of persons dying abroad 1898-1907

54: Depositum 29: Stadtarchiv Lingen: files as
    above.   ADS 4:147

Emigration permits [for permits before 1871, see
   Osnabrueck]   1871-1879
54: Rep.116,I: Nr. 6082.   ADS 4:120

Releases from citizenship   1880
54: Rep.116,I: Nr. 13435.   ADS 4:121

LINGEN (Vogtei)   NL

Emigration permits & emigration to America
                                          1822-1851
54: Rep.122,VI.b: Lingen: A: Nr. 4.
    DLC has.   ADS 4:137

LINNICH (canton)   RWL

Population movements   1801-1807
115: Rep.D2.2: II.Division, 1.Bureau, 2.Bevoelker-
     ung: Nr. 100.   ADS 6:19

LINX (village)   BWL

Emigration file   1853-1937
288: Kommunalarchiv Linx.   ADS 10:125

LIPBURG (village)   BWL

Emigration file   1851; 1896-1944
288: Kommunalarchiv Lipburg.   ADS 10:125

LIPPE (principality)   RWL

Various exit-tax matters   1786-1827
114: Lipp.Reg.: Aelt.Regis.: Fach 207: Nr. 28.
                                          ADS 6:3

Nr. 7: Emigration & immigration, 7 vols. 1860-1895
Nr. 8: Emigration certificates, 4 vols. 1824-1879
Nr. 14: Special matters regarding domicile of per-
        sons, 6 vols.   1840-1900

Nr. 16: Emigration permits, 41 vols [especially
    vols. 28-41]   1822-1924

<u>114</u>: Lipp.Reg.: Aelt.Regist.: Fach 146: files as
    above.   ADS 6:1-2

Nr. 3: Applications for emigration permits
          1856-1866
Nr. 8: Applications for certificates of domicile
          1863-1912
Nr. 9: Determination of the citizenship of emi-
    grants from Lippe to America   1863-1900
Nr. 17: Applications for releases from citizenship
          1884-1913

<u>114</u>: Lipp.Reg.: Ministerium: IV.Abt.Reg.4: files
    as above.   ADS 6:9-10

Passports issued   1854

<u>124</u>: D.Akten: Fach 76: Nr. 3; 4.   ADS 6:85

Nr. 3: Issuance of passports & certificates of
    domicile, 2 vols.   1872-1924
Nr. 5: Certificates of domicile, 10 vols.
          1901-1924

<u>114</u>: Lipp.Reg.: Regis.I: Abt.D, Fach 47: files as
    above.   ADS 6:4

Nr. 5: Issuance of passports & certificates of
    domicile, Vols. III, IV   1925-1943; 1944
Nr. 6: Denied petitions for passports, etc., of
    non-citizens in Lippe   1925-1934
Nr. 10: Requests for releases from citizenship, 2
    vols.   1878-1946
Nr. 13: Passports, arranged by number, Vols. VI-
    XXV   n.d.
Nr. 14: Certificates of domicile, arranged by num-
    ber   n.d.
Nr. 15: Releases from citizenship, arranged by
    number   n.d.

<u>114</u>: Lipp.Reg.: Abt.Innern: Gruppe XXV, Titel 2:
    files as above.   ADS 6:6

Marriages [of citizens] abroad   1855-1928

<u>114</u>: Lipp.Reg.: Ministerium: IV.Abt.Reg.9: Nr. 2.
          ADS 6:10

Visits of foreigners   1905

<u>124</u>: D.Akten: Fach 57: Nr. 1; 2.   ADS 6:85

Death certificates from & to foreign governments,
    3 vols.   1857-1909

<u>114</u>: Lipp.Reg.: Ministerium: II.Abt.   ADS 6:9

Applications of citizens for intercession in
    inheritance matters abroad, 2 vols.   1821-1924

<u>114</u>: Lipp.Reg.: Regis.I: Abt.F, Fach 111: Nr. 1.
          ADS 6:4

Intercession in inheritance matters abroad, Vols.
    III, IV   1925-1938; 1939

<u>114</u>: Lipp.Reg.: Abt. Innern: Gruppe II, Titel 1,
    Nr. 9.   ADS 6:5

Nr. 1: Estates of citizens dying abroad & the
    transmittal of inheritances, 4 vols.
          1867-1931
Nr. 3: Searches for missing citizens abroad
          1871-1928
Nr. 4: Miscellaneous requests from abroad, 2 vols.
          1874-1930

<u>114</u>: Lipp.Reg.: X.Abt.: Auswaert.Ang.: 11: files
    as above.   ADS 6:14

LITTLE ROCK (Arkansas)   USA

Warning against emigration to St. Joseph's Catholic
    Colony [from files of government in Koenigsberg,
    East Prussia]   1881

<u>51</u>: Rep.2,I^{2} Titel 30, Nr. 36, Bd. 4; *see also*
    Rep.12: Abt.I, Titel 3, Nr. 10, Bd. 5.
          ADS 4:25

Government warning against the Catholic St. Joseph
    Colony on the Arkansas River, near Little Rock.
    Colony attempts to recruit emigrants from Ger-
    many   1881

<u>115</u>: Rep. G29/15: Euskirchen: Nr. 156.   ADS 6:39

LOERRACH (with Stetten; Tuellingen; Turmringen) BWL

Alphabetical list of emigrants & immigrants
          1803-1937
Emigration file, Stetten   1888-1926
Emigration file, Tuellingen   1904-1930
Emigration file, Turmringen   1932

<u>288</u>: Kommunalarchiv Loerrach.   ADS 10:125

LOEWENBERG. *See* Vanselow, -- [surname index]
          Silesia

LOHR (Amt)   BYL

Emigration tax imposed   1580-1788

<u>232</u>: Rep.77: Aschaffenburger Archivreste: G.3209.
          ADS 9:137

LOHRHAUPTEN (Amt)   HEL

Emigration [reverse pages contain names]   1751

<u>151</u>: Bestand 81: Rep.Ba: Gef.9: Nr. 8.   ADS 7:80

Prohibition against emigration of persons worth
    more than 100 gulden; List of families who have
    emigrated between 1765 & 13 Feb 1767   1767?

<u>151</u>: Bestand 80: Lit.A: Nr. 14.   ADS 7:79

Emigration matters   1764-1767

<u>151</u>: Bestand 81: Rep.E: 55.Amt Bieber VII: Nr. 1.
          ADS 7:83

LOONENBURG (congregation in America)

Mentioned in correspondence   1754

<u>39</u>: Geistl.Minis.: II,7: Protokoll lit.B, 1746-
    1758: p.335(III); p.340(III).   ADS 2:473

LOONENBURG. *See also* Lunenburg, Nova Scotia   CDN

LOUISIANA (state)   USA

Transit & sojourn of emigrants recruited for
    Louisiana & Florida by businessman Kasimir Kurz

of Diebach ter Bacharach on behalf of Wils & Company, Amsterdam   1805

<u>158</u>: Frankfurter Ratsakten: Ugb.A.9.ad: Nr. 9.
    DLC has (original in <u>158</u> destroyed) ADS 7:198

LUDWIGSBURG   BWL

Monthly lists of foreigners in Ludwigsburg & Stuttgart [*see also* Wuerttemberg]   1817

<u>263</u>: Rep.E2: Polizeiminist.: 21,6.   ADS 10:52

LUDWIGSBURG (Oberamt)   BWL

Nr. 203: Emigration lists   1815-1900
Nr. 204-241: Emigration files, chronologically ar-
      ranged   1824-1890

<u>261</u>: F 181: Ludwigsburg: files as above. ADS 10:21

LUDWIGSBURG (city)   BWL

Emigration files & lists of citizens   1806-1890
<u>275</u>   ADS 10:78

LUDWIGSTEIN (Amt)   HEL

Emigration to South Carolina   1753

<u>151</u>: Bestand 17(b): Gef.104a: Nr. 43.   ADS 7:73

LUEBBECKE (Kreis)   RWL

Nr. 54: Residency of foreigners   1852-1896
Nr. 55: Emigration, immigration, & transit migra-
        tions   1896-1931

<u>116</u>: Arnsberg Reg.: Luebbecke: files as above.
                                                 ADS 6:63

Nr. 142-159: Emigration   1879-1924
Nr. 42-43: Issuance of certificates of original
            domicile   1892-1924
Nr. 83: Certificates of citizenship   1899-1909

<u>116</u>: Minden: Praes.-regist.: I.A.: files as above.
                                              ADS 6:50; 52

LUEBECK   SHL

Ministry in Berlin: Passports & certificates of
   residency [Heimatsscheine]   1861-1867

<u>1</u>: Rep.19: Aeltere Reg.S.1 [Pass, Heimat]
                                                 ADS 1:5

Nr. 2a: Emigrant paupers   n.d.
Nr. 3: Russian emigrants, reports on individual
        transports, 2 vols.   1906-1908

<u>1</u>: Rep.49$\frac{1}{}$.b: 1: files as above   ADS 1:7

LUEBECK. *See also* Hamburg (Hanseatic), Ministry of
   the Four Cities to the Bundestay in Frankfurt/Main

LUECHOW (Amt)   NL

Expediting ship passengers to overseas ports 1852-
<u>52</u>: Han.Des.74: Luechow: H.? Fach 400: Nr. 6.
                                                 ADS 4:73

LUEDINGHAUSEN (Landratsamt)   RWL

Nr. 12: Emigration (general)   1817-1866
Nr. 13: Emigration   1819-1843
Nr. 14-17: Emigration   1842-1871
Nr. 679: Foreigners   1895-1908
Nr. 741: Foreigners   1921-1922
Nr. 260: Foreigners (except Poles)   1896-1920
Nr. 681: Foreigners   1921-1926
Nr. 745: Expulsions of foreigners   1921-1925
Nr. 625: Expulsions of foreigners   1921-1925

<u>116</u>: Muenster Reg.: Luedinghausen: files as above.
    DLC has 12 & 13.   ADS 6:66

LUEDINGHAUSEN (Kreis)   RWL

Emigration   1888-1928
<u>116</u>: Muenster-Reg.: Nr. 1692-1693.   ADS 6:57

LUENEBURG   NL

Nr. 956: Emigration to America, Vol. I 1854-1863
Nr. 957: Emigration to America   1849-1853

<u>52</u>: Han.Des.80: Lueneburg I: Regiminalreg. FF:
    files as above.   ADS 4:92

Information on emigration & loss of citizenship
                                        1873-1879

<u>52</u>: Han.Des.8): Lueneburg I: Regiminalreg.: U:
    Nr. 797.   ADS 4:92

"Norddeutsche in aller Welt: Auswanderungen aus
   dem Lueneburgischen" [North Germans in All the
   World: Emigration from the Lueneburg Area] in
   Norddeutsche Familienkunde (1958) Heft 1, pp.
   23-25.   ADS 4:294

LUETZENKIRCHEN (Amt?)   RWL

Reasons for emigration given for several emigrants
                                               1805

<u>115</u>: Berg. Apanagialreg.: A.: Nr. A 29. ADS 6:22

LUNDEN   SHL

Nr. 413: Citizenship   1875-1885
Nr. 414: Emigration & immigration   1868-1889
Nr. 415: Emigration & immigration   1889-1904
Nr. 419: Passports   1869-1877
Nr. 420: Passports   1878-1883
Nr. 421: Passports   1884-1889

<u>22</u>: files as above.   ADS 1:56

LUNENBURG (Nova Scotia)   CDN

German Lutheran congregation in   1777-1788

<u>500</u>: Halle: Missionsbibliothek des Waisenhauses:
    Unverzeichnet.   Not in ADS

LUNENBURG (Nova Scotia). *See also* Loonenburg   CDN

LUTTER a. B. (Amt)   NL

  Emigration to America:
    Vol. 2   1846-1850
    Vol. 3   1850-1852
    Vol. 4   1855-1857
    Vol. 5   1861-1870
    Vol. 6   1871-1876

  <u>56</u>: L Neu Abt.129A: Gr.11: Nr. 11.   ADS 4:186

MADEMUEHLEN. *See* Dillenburg   HEL

MAHLBERG (village)   BWL

  Emigration file   1847-1950
  <u>288</u>: Kommunalarchiv Mahlberg.   ADS 10:126

MAINBURG   BYL

  Emigration   [19th century]
  <u>225</u>: Rep.168, Verz. 1.   ADS 9:15

MAINBURG (Bezirksamt)   BYL

  Emigration   1852-1930
  <u>225</u>: Rep.164, Verz.11: Nr. 318-404.   ADS 9:14

MAINE (state)   USA

  Data on Waldo Colony [in Volume I of series]
                       1750-1766
  <u>152</u>: Abt.172: Nr. 2664.
     DLC has.   ADS 7:128

MAINWANGEN (village)   BWL

  Emigration file   1907-1940
  Germans abroad   1924-1946
  <u>288</u>: Kommunalarchiv Mainwangen.   ADS 10:126

MAINZ (city)   RPL

  Emigration register   1856-1860; 1865-1875
  <u>210</u>   ADS 8:77

MAISACH (village)   BWL

  Emigration to the USA   1881-1895
  Emigration file   1940-1950
  Correspondences with emigrants   1939
  <u>288</u>: Kommunalarchiv Maisach.   ADS 10:126

MALECK (village)   BWL

  Emigration file   1858-1925
  <u>288</u>: Kommunalarchiv Maleck.   ADS 10:126

MALLERSDORF (Bezirksamt)   BYL

  Emigration   [19th century]
  <u>225</u>: Rep.168, Verz. 1.   ADS 9:15

  Releases from Bavarian citizenship 1832-1918
  <u>225</u>: Rep.164, Verz.12: Nr. 645-724. ADS 9:14

MALLERSDORF (Landgericht)   BYL

  Applications for emigration permits 1817-1846
  <u>225</u>: Rep.168, Verz. 1.   ADS 9:16

MALSBURG (village)   BWL

  Emigration file   1876-1950
  <u>288</u>: Kommunalarchiv Malsburg.   ADS 10:126

MALTERDINGEN (village)   BWL

  Emigration file   1888-1937
  <u>288</u>: Kommunalarchiv Malterdingen. ADS 10:126

MANNWEILER (village). *See* Kurpfalz   RPL

MANSFELD (Grafschaft)   Prussia

  File on English & Russian colonists 1753-1766
  <u>500</u>: Magdeburg: Koenigliche Staatsarchiv: Graf-
     schaft Mansfeld: Rep.A.32a, Tit.XXI: Nr. 11.
                       Not in ADS

MAPPACH (village)   BWL

  Emigration file   1853-1943
  <u>288</u>: Kommunalarchiv Mappach.   ADS 10:126

MARBACH (Oberamt)   BWL

  Nr. 1-3: Emigration files, chronologically arranged
                       [before 1811?]
  Nr. 5-27: Emigration files   1811-1869
  Nr. 4: Emigration lists   1819-1869
  <u>261</u>: F 182: Marbach: files as above. ADS 10:21

MARBURG (Landratsamt)   HEL

  Nr. 2: Releases from citizenship, City of Marburg
                       1865-1872
  Nr. 11: Releases from citizenship   1873-1883
  Nr. 12: Releases from citizenship   1884
  Nr. 4: Passport register   1870
  Nr. 8: Passport register   1871
  Nr. 14: Passport register   1871-1873
  Nr. 15: Passport register   n.d.
  Nr. 16: Passport register   1868
  <u>151</u>: Bestand 180: Marburg: Acc.1927/15: files as
     above.   ADS 7:119

Releases from Kurhessian citizenship from various
   villages of the Landratsamt   n.d.

151: Bestand 180: Marburg: Acc.1927/15: Nr 62; 69;
   85; 93; 114; 120; 130; 144; 170; 183; 207;
   238; 275; 294; 304; 315; 327.   ADS 7:119

Nr. 40: Emigration & immigration   1877-1925
Nr. 49: Emigration & releases from citizenship;
   returns   1868-1924
Nr. 50: Releases from citizenship   1894-1908
Nr. 51: Releases from Prussian citizenship
                                           1909-1924

151: Bestand 180: Marburg: Acc.1937/47: files as
   above.   ADS 7:120

Nr. 55: Certificates of original domicile & citizenship   1836-1915
Nr. 56: Certificates of original domicile
                                           1904-1907

151: Bestand 180: Marburg: Acc.1937/47: files as
   above.   ADS 7:120

MARCKENBACH. *See* Dillenburg   HEL

MARKT ERLBACH (Bezirksamt Neustadt a.d.A.). *See*
Neustadt an der Aisch   BYL

MARKTOBERDORF (Bezirksamt)   BYL

Individual files (186 in number) regarding emigration to North America and to America
                                       ca. 1846-1882

230: Rep. Akten der Bezirksaemter: Marktoberdorf:
   Nr. 2085-.   ADS 9:124

MARIENBERG (Amt)   HEL

Nr. 331: Naturalizations & releases from citizenship   1842-1866
Nr. 471: Summary of emigrants for 1866   1867

152: Abt.233: Marienberg: files as above.
                                           ADS 7:134

MARKOEBEL (Amt)   HEL

For exemptions from emigration tax, *see* Hesse-
Kassel [Kurhesse]

MARNHEIM (village; Landkreis Kirchheimbolanden, Kurpfalz)   RPL

Names of emigrants   18th century

201: Archivabt. Ausfautheiakten.   ADS 8:63

MARZELL (with Liebesbacherhof)   BWL

Emigration files   [17th & 18th centuries]

260: Rep.Abt.229: Spezialakten: Marzell: Fasz.
   65304-65335.   ADS 10:11

Emigration file   1871-1944

288: Kommunalarchiv Marzell.   ADS 10:126

MASHOLTE (parish). *See* Rietberg   RWL

MAUCHEN (village)   BWL

Emigration file   1931-1950

288: Kommunalarchiv Mauchen.   ADS 10:126

MAULBRONN (Oberamt)   BWL

Nr. 258: Immigration files   1870-1872
Nr. 287: Lists [of emigrants?]   1833-1854
Nr. 288-309: Emigration files, chronologically ar-
   ranged   1832-1870

261: F 183: Maulbronn: files as above.   ADS 10:21

MAULBURG (village)   BWL

Emigration file   1794-1944

288: Kommunalarchiv Maulburg.   ADS 10:126

MAYEN (Kreis)   RPL

Naturalizations   1856-1914

200: Abt.441: Regierung Koblenz: Nr. 24998; 22027-
   22028.   ADS 8:29

MECHTERSHEIM (village; Bezirksamt Speyer)   RPL

Emigration   1835-1865

201: Archivabt. Bezirksaemter: Speyer (Abgabe
   1907): Akt Nr. 4.   ADS 8:65

MECKENHEIM (village)   RWL

Emigration permits   1886-

134: A 23.   ADS 6:114

MECKLENBURG

Negotiations regarding transshipment of ten pardoned prisoners from Mecklenburg to America 1846

39: Bestand Senat: Cl.VIII, No.X, Jg.1846 (Register).   ADS 2:149

Protest of the Mecklenburg government regarding
   freed prisoners to be transshipped to America
                                                1850

39: Bestand Senat: Cl.VIII, No.X, Jg.1850 (Register).   ADS 2:157

Protest of the United States Consul regarding
   transhipment in Hamburg of Mecklenburger prisoners   1850

39: Bestand Senat: Cl.VIII, No.X, Jg.1850 (Register) No. 9. [Auswanderer-Verhaeltnisse]
                                           ADS 2:156

MECKLENBURG (Province)

Nearly all the church records (christening, marriage, confirmation, death), with few exceptions,
for the entire Province of Mecklenburg from ca.
1600 to 1876 (inauguration of the civil registry offices) were brought westward near the end
of the second world war & are now to be found in
Ratzeburg

31

MEDENBACH. *See* Dillenburg   HEL

MEERHOLZ (Amt)   HEL

Emigration   1816
151: Bestand 86: Nr. 9962.   ADS 7:86

MEHLBEK (Kreis Steinburg)   SHL

Passports   1811-1856
23: Nr. 99.   ADS 1:23

MEIBORSEN [*also spelled* Meiborssen]   NL

Emigration of 3 residents of Meiborsen to South
Carolina   1753
52: Han.Des.74: Polle: I.XV: Nr. 2.   ADS 4:74

MEISENHEIM (Kreis)   RPL

Nr. 7-9: Emigration   1882-1888
Nr. 11-13: Naturalizations   1866-1915
Nr. 112-117: Certificates of original domicile,
citizenship, marriage   1867-1911
Nr. 118: List of citizens released   1867-1911
200: Abt.471: Meisenheim: files as above. ADS 8:44

Naturalizations   1886-1914
200: Abt.441: Regierung Koblenz: Nr. 22030-22031;
23493.   ADS 8:29

MEISENHEIM (village)   BWL

Emigration file   1774-
Emigration; genealogical research files; lists of
addresses of emigrants   1816-1934
288: Kommunalarchiv Meisenheim.   ADS 10:126

MELDORF   SHL

Nr. 14: Passports   1811-1833
Nr. 72: Fees for releases from citizenship
1851-1861
24: Acta XI: files as above.   ADS 1:57

MELDORF. *See also* Thiessen, L. [surname index]

MELLE (Amt)   NL

Nr. 10: Emigration to North America 1846-1888
Nr. 11: Emigration to North America 1837-1865
Nr. 12: Lists of emigrants   1868-1870
Nr. 32a: Lists of emigrants; emigration permits
issued to persons subject to military
service   1866-1868; 1871-1882
Nr. 36: Releases from citizenship   1871
Nr. 37: Releases from citizenship   1872
Nr. 38: Releases from citizenship   1873
Nr. 39: Releases from citizenship   1874
54: Rep.122,VII.B.1: Han.Melle: Fach 159: files as
above.   ADS 4:140

Releases from citizenship   1881-1882
54: Rep.116,I: II.4.A.1e: Nr. 89.   ADS 4:122

MELLE (Kreis)   NL

Nr. 107: Releases from citizenship   1887-1889
Nr. 120: Releases from citizenship   1890-1891
54: Rep.116,I: II.4.A.13: files as above.
ADS 4:123-124

Releases from citizenship   1892
54: Rep.116,I: Nr. 13422.   ADS 4:124

MELLE (town)   NL

Emigration permits   1871-1879
[for permits before 1871, *see* Osnabrueck]
54: Rep.116,I: Nr. 13433.   ADS 4:120

MELLE. *See also* Groenenberg-Melle   NL

MELSUNGEN (Kreis)   HEL

Naturalizations, 2 vols.   1822-1855
151: Bestand 17(b): Gef.93: Nr. 1. ADS 7:69

MELSUNGEN (Landratsamt)   HEL

Applications for readmission to citizenship by
persons previously released   1841-1860
151: Bestand 180: Melsungen: Acc.1890/19: Verz. A:
Nr. 675.   ADS 7:120

Foreign residents in the various villages of the
Kreis   19th century
151: Bestand 180: Melsungen: Acc.1878/2: Nr. 31-.
ADS 7:120

MEMEL (Kreis)   East Prussia

Questionnaires regarding removal of farm workers
to West Prussia & Germany [includes emigrants to
America with name, family members, date, destination, statistical data]   1890-1914
51: Rep.18: Abt.VIII: Nr. 244.   ADS 4:49

MEMEL (Landratsamt)   East Prussia

   Emigration applications   1905-1908
   <u>51</u>: Rep.81: Abt.VIII: Nr. 249.   ADS 4:50

MEMMINGEN (Bezirksamt)   BYL

   Emigration to overseas countries   1839-1879
   <u>230</u>: Rep.Akten der Bezirksaemter: Memmingen: Nr.
     3433-3443; 3448.   ADS 9:124

MENGEN (with Bechtoldskirch)   BWL

   Emigration of Mennonites   1530-1862
   <u>260</u>: Rep.Abt.229: Spezialakten: Mengen: Fasz.
     66555-66804.   ADS 10:11

MENGEN (village)   BWL

   Emigration file   1852-1941
   <u>288</u>: Kommunalarchiv Mengen.   ADS 10:127

MENGERINGHAUSEN   HEL

   Releases from citizenship   n.d.
   <u>151</u>: Bestand 180: Arolsen: Acc.1937/76,5: Nr. 566.
                           ADS 7:113

MENNINGEN (with Leitishofen)   BWL

   Emigration files   1551-1883
   <u>260</u>: Rep.Abt.229: Spezialakten: Menningen: Fasz.
     66805-66827.   ADS 10:11

MENSFELDEN. *See* Limburg   HEL

MEPPEN (Amt)   NL

   Emigration permits   1871-1879
   [for permits before 1871, *see* Osnabrueck]
   <u>54</u>: Rep.116,I: Nr. 6077.   ADS 4:120

   Nr. 71: Releases from citizenship   1880-1883
   Nr. 101: Releases from citizenship   1884-1888
   <u>54</u>: Rep.116,I: II.4.A.1e: files as above.
                     ADS 4:121-123
   Passport issuance, Vols. I; II   n.d.
   <u>54</u>: Rep.122.VIII: Reg. Meppen: Nr. 1941-1942.
                         ADS 4:141

MEPPEN (Kreis)   NL

   Releases from citizenship   1889-1892
   <u>54</u>: Rep.116,I: II.4.A.1e: Nr. 116.   ADS 4:123

   Releases from citizenship   1893
   <u>54</u>: Rep.116,I: Nr. 13413.   ADS 4:124

MEPPEN. *See also* Brickweidt, Wilhelm

MERDINGEN (village)   BWL

   Emigration files   1845-1919
   Emigration to the USA   1850-1864
   Emigration from Merdingen over the last 30 years
     [manuscript?]   1932

MERENBERG (Amt)   HEL

   Emigration tax?   1752-1800
   <u>152</u>: Abt.162: Merenberg: Nr. 22.   ADS 7:127

MERGENTHEIM (Oberamt)   BWL

   Nr. 321-321a: Emigration lists   1835-1896
   Nr. 322: Emigration files, with retention of
          [Wuerttemberg] citizenship   1849-1902
   Nr. 323: Emigration files   1870-1902
   Nr. 320: General emigration files   1863-1865
   <u>261</u>: F 184: Mergentheim: files as above. ADS 10:22

MERZHAUSEN (village)   BWL

   Emigration files   1842-1927
   <u>288</u>: Kommunalarchiv Merzhausen.   ADS 10:127

MERZIG (Landkreis)   RPL

   Emigration   1852-1888
   <u>200</u>: Abt.442: Regierung Trier: Nr. 8660-8662; 185-
     187; 6054-6059.   ADS 8:36

   Emigration   19th century
   <u>200</u>: Abt.701: Nachlaesse, etc.: Nr. 975. ADS 8:54

   Naturalizations   1874-1924
   <u>200</u>: Abt.442: Regierung Trier: Nr. 8657-8659.
                         ADS 8:40

METTENBERG (village)   BWL

   Emigration file   1881-1937
   <u>288</u>: Kommunalarchiv Mettenberg.   ADS 10:127

METTMAN (Landratsamt)   RWL

   Nr. 231: Emigration & immigration   1923-1926
   Nr. 176: Emigration & immigration   1926-1927
   Nr. 174: Emigration & immigration   1927-1928
   Nr. 244: Emigration & immigration   1928-1929
   <u>115</u>: Rep. G29/29: Mettman: files as above.
                         ADS 6:43

MICHIGAN (northern). *See* Allardt, B. H.   USA

MIETERSHEIM (village)   BWL

General emigration file   1829-1939
Emigration files   1884-1939

<u>288</u>: Kommunalarchiv Mietersheim.   ADS 10:127

MIMMENHAUSEN (village)   BWL

Emigration files   1891-1944
List of overseas Germans   1938-1939

<u>288</u>: Kommunalarchiv Mimmenhausen.   ADS 10:127

MINDEN (Regierung)   RWL

Nr. 101: Emigration, general, Vol. I   1816-1850
Nr. 102: Emigration, general, Vol. III   1884-1931
Nr. 103: Emigration, general, Vol. IV   1884-1931
Nr. 104: Emigration, general (including emigration
         permits for Prussian citizens, 1816-1827)
                                           1816-1865
Nr. 106-108: Emigration   1861-1930
Nr. 111: Investigation of persons subject to mili-
         tary service emigrating without permits
                                           1834-1866
Nr. 93; 96: Emigration [or exit] tax
                                  1816-1839 [1846]
Nr. 414; 416: Death certificates   1866-1900

Emigrants:

A	missing	O	no entries
B	to 1927	P	to 1931
C	to 1927	Q	no entries
D	to 1927	R	to 1931
E	to 1930	Sch	to 1929
F	to 1929	St	to 1930
G	no entries	T	to 1926
H	to 1926	U	no entries
I	missing	V	no entries
J	missing	W	to 1930
K	to 1931	X	no entries
L	to 1929	Y	no entries
M	no entries	Z	no entries
N	to 1929		

<u>116</u>: Minden: Praes.regist.: I.A.: Nr. 188-210.
                                           ADS 6:51

Claims of inland Germans against [persons] abroad
& vice versa   1826-1931

<u>116</u>: Minden: Praes.-regist.: I.P.: Nr. 317.
                                           ADS 6:54

Nr. 301-302: Foreign reports 1874-1931; 1872-1895
Nr. 318: Extradition to & from Germany   1906-1931
Nr. 319: Searches for missing persons   1907-1914

<u>116</u>: Minden: Praes.-regist.: I.P.: files as above.
                                           ADS 6:55

Searches for missing persons 1829-1834; 1842-1921

<u>116</u>: Minden: Praes.-regist.: I.L.: Nr. 251-253.
                                           ADS 6:54

Inheritances from abroad of persons in Germany
             1857-1858; 1875-1892; 1911-1926

<u>116</u>: Minden: Praes.-regist.: I.L.: Nr. 222-230.
                                           ADS 6:54

Nr. 125: Citizens residing abroad   1817-1885
Nr. 126: Emigration of Prussian citizens 1835-1892

<u>116</u>: Praes.-regist.: files as above.   ADS 6:49

Nr. 6: Certificates of original domicile, Vol. I
                                           1825-1846
Nr. 7: Certificates of original domicile, Vol. II
                                           1847-1895
Nr. 8: Certificates of original domicile, Vol. III
                                           1896-1913
Nr. 9: Certificates of original domicile & citizen-
       ship, Vol. II   1895-1911

<u>116</u>: Praes.-regist.: files as above.   ADS 6:50

MINDEN (Kreis)   RWL

Nr. 160-173: Emigration   1881-1924
Nr. 44-49: Certificates of original domicile
                                           1890-1914
Nr. 50: Certificates of original domicile
                                           1921-1922
Nr. 84: Certificates of original domicile & citi-
        zenship   1908-1910
Nr. 85-86: Certificates of citizenship   1913-1919

<u>116</u>: Minden: Praes.-regist.: I.A.: files as above.
                                        ADS 6:50; 52

Nr. 55: Emigration & immigration   1896-1931
Nr. B 33: Emigration & immigration   1817-1856
Nr. A 13: Annual summary of emigration & immigra-
          tion   1845-1858

<u>116</u>: Arnsberg Reg.: Minden: files as above.
                                           ADS 6:63

Passports   1862-1935

<u>116</u>: Minden: Praes.-regist.: I.P.: Nr. 314.
                                           ADS 6:54

MINDEN (Nevada)   USA

"Deutsche Auswanderer aus Halle in Westfalen
 gruendeten in Nevada eine Stadt Minden" [German
 Emigrants from Halle in Westphalia Found the
 Town of Minden, Nevada" *Ravensberger Blaetter*,
 38 Jg., p. 23-.   ADS 6:76

MINDERSLACHEN (village; Landkreis Kandel, Kurpfalz)
                                                        RPL

Names of emigrants   18th century

<u>201</u>: Archivabt. Ausfautheiakten.   ADS 8:63

MINFELD (village; Landkreis Kandel, Kurpfalz)   RPL

Names of emigrants   18th century

<u>201</u>: Archivabt. Ausfautheiakten.   ADS 8:63

MINSELN (village)   BWL

Emigration files   1828-1944
General files   1907-1938
Immigration files   1904-1938

<u>288</u>: Kommunalarchiv Minseln.   ADS 10:127

MISELOHE (Amt)   RWL

   Emigration to America   1805
   <u>115</u>: Berg. Apanagialreg.: A: Nr. A26.   ADS 6:22

MISSAU (village; Landkreis Waldmohr, Kurpfalz)   RPL

   Names of emigrants   18th century
   <u>201</u>: Archivabt. Ausfautheiakten.   ADS 8:63

MITLECHTERN   HEL

   Emigration   1863-1891
   <u>149</u>: Rep.40,2: Ablief.2: Bd. 304.   ADS 7:23

MITTELBUCHEN (Amt)   HEL

   For exemptions from emigration tax, *see* Hesse-
   Kassel [Kurhesse]

MITTELFRANKEN (region)   BYL

   Nr. 320-: Emigration   n.d.
   Nr. 473: Emigration   1815
   Nr. 5218: Emigration   1856-1857
   <u>231</u>: Rep.271/V: Reg. [Mittelfranken]: Kammer der
      Finanzen: Abgabe 1937: files as above.
                        ADS 9:135

   Emigration & immigration files from various towns
    & Landgerichte [also individual cases]
                        19th century
   <u>231</u>: Rep.270/I: Regierung Mittelfranken, Kammer
      des Innern: Abgabe 1900: Nr. 4978-5024.
                        ADS 9:131

   Card file on emigrants from Mittelfranken [mater-
    ial gleaned from the *Mittelfraenkischen Intelli-
    genzblaetter*]   1837-1874
   <u>250</u>   ADS 9:164

   Nr. 664: Bavarians residing in America; financial
         condition   1850
   Nr. 667: Returning emigrants from America   1856
   Nr. 687-702: Emigration & immigration from all
         Bezirksaemter, except Ansbach & Weis-
         senburg   1862
   Nr. 737-754: Naturalizations & losses of citizen-
         ship in the cities & Bezirksaemter
                       1871-1904
   Nr. 736: Emigration to Canada   1904
   <u>231</u>: Rep.270/II: Reg. Mittelfranken: Kammer des
      Innern: Abgabe 1932: files as above.
                     ADS 9:134-135

   There are finance & tax files, beginning in 1806,
    having to do with deposits required in emigra-
    tion cases, tax overpayments, tax indebtedness.
    In addition, there are emigration applications,
    etc.   [19th century]
   <u>231</u>: Rep.225: Finanz (Rent-) Aemter.   ADS 9:134

MITTELRHEINKREIS (region)   BWL

   Nr. 656: Applications of several 1850 conscripts
         for permission to emigrate to America
                          [1851?]
   Nr. 658: Applications from various communities for
         financial for emigrants to America, Part
         I   1851-1854
   Nr. 659: State aid for the emigration of various
         persons, Part II   1855-1859
   Nr. 660: Grants of state aid for the emigration of
         various persons   1860-1864
   <u>260</u>: Abt.236: Innenmin., Wegzug: Auswand.: files
      as above.   ADS 10:2

MITTELRHEINPROVINZ. *See* Niederrheinprovinz

MITTERSFELS (Landgericht)   BYL

   Applications for emigration permits   1808-1846
   <u>225</u>: Rep.168, Verz. 1.   ADS 9:16

MOBILE (Alabama)   USA

   Hamburg Consulate in Mobile: Births & deaths 1871
   <u>39</u>: Bestand Senat: Cl.VI, No.16p: Vol.4b: Fasc. 8d,
      Invol. 1.   ADS 2:55

MOEGGINGEN (village)   BWL

   Emigration files   1855-1936
   <u>288</u>: Kommunalarchiv Moeggingen.   ADS 10:127

MOENCHEN-GLADBACH (Landratsamt)   RWL

   Emigration   1820-1914
   <u>115</u>: Rep. G29/18: Moenchen-Gl.: Nr. 625; 196; 628;
      145; 115; 518.
      DLC may have Nr. 625.   ADS 6:40

MOENCHHOF. *See* Neuenheim   BWL

MOENCHWEILER (village)   BWL

   Emigration files   1819-1921
   Emigration to the USA   1850-1892
   <u>288</u>: Kommunalarchiv Moenchweiler.   ADS 10:127

MOERLENBACH   HEL

   Emigration   1850-1898
   <u>149</u>: Rep.40,2: Ablief.2: Bd. 305.   ADS 7:23

MOERS (Kreis)   RWL

   Emigration applications   1852-1860
   <u>115</u>: Rep. D8: I.Abt.e: Nr. 583-585.   ADS 6:25

Emigration   1861-1989

<u>115</u>: Rep.D8: Neuere Akt.: Abt.1, C3: Nr. 12086-
       12090.   ADS 6:27

MOERSBACH (village; Landkreis Homburg, Kurpfalz) RPL

Names of emigrants   18th century

<u>201</u>: Archivabt. Ausfautheiakten.   ADS 8:63

MONSCHAU (Landratsamt)   RWL

Nr. 353: Emigration & immigration   1891-1926
Nr. 339: Emigration   1817-1843
Nr. 340: Emigration   1844-1850

<u>115</u>: Rep. G29/31: Monschau: files as above.
       DLC may have Nr. 339, 340.   ADS 6:44

MONTABAUR (Landratsamt)   RPL

Nr. 115: Reports on immigration & emigration
                            1866-1891; 1905
Nr. 137: Immigration & emigration   1890-1910

<u>152</u>: Abt.419: Montabaur: files as above. ADS 7:159

MONTABAUR (town)   RPL

Register containing a number of emigration cases
       to America   1890-1916

<u>212</u>: Abt.8: Nr. 7.   ADS 8:80

MONTJOIE (canton)   RWL

Population movements   1801-1807

<u>115</u>: Reg.D2.2: II.Division, 1.Bureau, 2.Bevoelker-
       ung: Nr. 101.   ADS 6:19

MOOS (village; Kreis Buehl)   BWL

Emigration files   1826-1907
General files   1870-1922

<u>288</u>: Kommunalarchiv Moos, Kr. Buehl.   ADS 10:128

MOOS (village; Kreis Konstanz)   BWL

Emigration to the USA   1857-1899

<u>288</u>: Kommunalarchiv Moos, Kr. Konstanz. ADS 10:128

MOSELLAND (region)   RPL

"Die Auswanderung aus dem mosellaendischen Raum:
I. Teil: Darstellung," [The Emigration from the
Moselland Region: Part I: Outline] Manuscript
                                             n.d.

<u>200</u>: Abt.701: Nachlaesse, etc.: Nr. 958. ADS 8:53

MUCKENSCHOPF (village)   BWL

Emigration file   1832-1924

<u>288</u>: Kommunalarchiv Muckenschopf.   ADS 10:128

MUEHLEN. *See* Limburg   HEL

MUEHLENBACH (village)   BWL

Emigration files   1859-1939

<u>288</u>: Kommunalarchiv Muehlenbach.   ADS 10:128

MUEHLHAUS (village)   BWL

Emigration file (contains only certificates of
       original domicile)   1838-1855

<u>288</u>: Kommunalarchiv Muehlhaus.   ADS 10:128

MUELHEIM (Kreis)   RWL

Emigration   1857-1881

<u>115</u>: Rep.D9: Koeln: Abt.I.e: Nr. 258-259. ADS 6:32

Emigration   1874-1920

<u>115</u>: Rep.D8: Neuere Akt.: Abt.1, 3C: Nr. 11965-
       11968.   ADS 6:26

MUELHEIM/RHEIN (Landratsamt)   RWL

Nr. 346: Emigration & immigration   1913-1921
Nr. 298: Emigration & immigration   1921-1926
Nr. 135: Emigration & immigration   1926-1930
Nr. 423: Emigration & immigration   1930-1933

<u>115</u>: Rep. G29/32: Muelheim: files as above.
                                             ADS 6:44

MUELHOFEN (village)   RPL

Names of emigrants   18th century

<u>201</u>: Archivabt. Ausfautheiakten.   ADS 8:62

MUELHOFEN (village)   BWL

Emigration to the USA   1858-1936

<u>288</u>: Kommunalarchiv Muehlhofen.   ADS 10:128

MUENCHWEIER (village)   BWL

General emigration file   1832-1949
Emigration files   1847-1933

<u>288</u>: Kommunalarchiv Muenchweier.   ADS 10:128

MUENSINGEN (Oberamt)   BWL

Nr. 163: Emigration files of town of Muensingen
                                        1813-1818
Nr. 164-196: Emigration & immigration, alphabetic-
             ally arranged   19th century
Nr. 197-207: Emigration files of Oberamt Muensingen,
             chronologically arranged   1849-1880

<u>261</u>: F 185: Muensingen: files as above. ADS 10:22

Emigration files from various villages (15 fasc.)
                                        19th century

<u>262</u>: Wue 65/20: Muensingen: ZR.   ADS 10:38

MUENSTER (Oberpraesidium)   RWL

  Nr. 2666 I: Emigration & immigration of citizens &
            persons subject to military service,
            Vol. I   1817-1866
  Nr. 2666 II: Emigration & immigration of citizens
            & persons subject to military service,
            Vol. II   1852-1893
  Nr. 2667: Emigration & immigration, Vol. VII
                                          1905-1907
  Nr. 2878: Affairs of, & requests from, citizens
            abroad, Vol. I   1810-1867
  Nr. 2879: Affairs of, & requests from, citizens
            abroad, Vol. II   1867-1874

  116: Oberpraesidium in Muenster: files as above.
       DLC has Nr. 2666 I.   ADS 6:47

MUENSTER (Regierung)   RWL

  Nr. 181: Emigration to America, Vol. I   1828-1833
  Nr. 130: Emigration to America, Vols. II-XLI
                                          1833-1851
  Nr. 796: Name list of Prussian subjects who have
           secretly emigrated to North America
                                          1834-1839
  Nr. V-9-1 to V-9-4: Estates of Prussian citizens
                     dying abroad   1819-1922
  Nr. 4227: Estates of Prussian citizens dying
            abroad   1921-1928
  Nr. 1663: Foreign citizens dying in Muenster Re-
            gierung   1819-1922

  116: Muenster Reg.: files as above.
       DLC has No. 181.   ADS 6:56

MUENSTER (Kreis)   RWL

  Nr. 1694-1695: Emigration   1885-1929
  Nr. 3403: List of certificates of citizenship
            issued   1921-1923

  116: Muenster Reg.: files as above.   ADS 6:56

MUENSTER (city)   RWL

  Emigration   1855-1862; 1888-1929

  116: Muenster Reg.: Nr. 132, I-II; 1666-1669.
                                         ADS 6:57

MUENZENBERG. See Hanau-Muenzenberg   HEL

MUGGENSTURM (village)   BWL

  Emigration to the USA   1846-1893
  Emigration files   1847-1918

  288: Kommunalarchiv Muggensturm.   ADS 10:128

MUNDELFINGEN (village)   BWL

  Emigration file   1853-1943
  Emigration to the USA   1853-1867
  Germans abroad, addresses   1907-1946
  Emigrant list   1932

  288: Kommunalarchiv Mundelfingen.   ADS 10:128

MUNDINGEN (village)   BWL

  Emigration file   1832-1844
  Emigrant list   1848-1882

  288: Kommunalarchiv Mundingen.   ADS 10:129

MUNZINGEN (village)   BWL

  Emigration files   1825-1911

  288: Kommunalarchiv Munzingen.   ADS 10:129

MUTTERSTADT   RPL

  Contract record contains information on emigrants
     to Pennsylvania   1714-1747

  201: Archivabt. Briefprotokolle: Nr. 254. ADS 8:63

MUTTERSTADT (village). See Neustadt (Oberamt)   RPL

MYSLOWITZ. See Ratibor   Upper Silesia

NABBURG (Bezirksamt)   BYL

  Nr. 206: Emigration to North America, 3 parts 1844
  Nr. 207-374: 168 files on emigrants to North Amer-
               ica & other countries   1816-1849
  Nr. 375-437: Exportation of property   1828-1863
  Nr. 532-626: Emigration & immigration; emigration
               tax; clearness [of all obligations]
               for emigration   1793-1833

  222: Rep.129: Nabburg: files as above.   ADS 9:3

NAGOLD (Oberamt)   BWL

  Nr. 280: Emigration lists   1849-1884
  Nr. 282-289: Emigration files, chronologically
               arranged; financial aid to emigrants;
               emigration & immigration lists
                                         1843-1887
  Nr. 279: Emigration   1832-1895
  Nr. 290-313a: Certificates of original domicile,
               with supporting documentation
                                         1880-1904
  Nr. 315-343: Passports   1872-1904
  Nr. 314: List of passports issued   1829-1852

  261: F 186: Nagold: files as above.   ADS 10:22

  Nr. 1040.32: Releases from [Wuerttemberg] citizen-
               ship   1870-1900
  Nr. 6135: Emigration   1897-1937

  262: Wue 65/21: Nagold: AZ: files as above.
                                         ADS 10:38

NAILA (Bezirksamt)   BYL

  Individual emigration cases to North America
                                         1847-1869

  223: Rep.K.16/Verz.IV: Nr. 258; 264; 267. ADS 9:7

NAPERVILLE, Illinois, St. John's Lutheran Church.
   See van Oven family [surname index]   USA

NASSAU (Dreiherrisch [tripartite]) HEL

Nr. VII a 9: Recruits to America with the Hessian
troops 1771-1783
Nr. VIII e 11: Emigration to Pennsylvania & releases from serfdom [manumissions]
1709-1724
Nr. VIII e 14: Emigration tax 1724-1758
Nr. VIII e 28: Manumissions 1764-1808

152: Abt.350: Nassau-Dreiherrisch: Gen.: files as
above.
DLC has Nr. VIII e 11. ADS 7:139

NASSAU (Province). *See also* Hesse-Hanau HEL

NASSAU-SAARBRUECKEN (Principality) RPL

Nr. 3520: Levy of emigration tax 1729-1736
Nr. 4064: Emigration of subjects 1764-1784

200: Abt.22: Fuerstentum Nassau-Saarbruecken:
files as above. ADS 8:9

NASSAU-USINGEN HEL

Forced auctions of emigrants' property 1749-1764

152: Abt.131: Gen.XIV c: Nr. 17.
DLC has. ADS 7:127

NASTAETTEN (Amt) HEL

Investigation of the deaths of former citizens who
have emigrated to Texas 1848-1849

152: Abt.236: Nastaetten: Nr. 1027. ADS 7:134

NAUHEIM (village; Landkreis Gross-Gerau) HEL

Reports regarding the issuance of certificates of
release [from Hessian citizenship] to emigrants
to America 1849-1852
List of emigrants from the community (giving birth
year, names of children, & other comments; about
145 persons in all) 1852-1883
List of emigrants during the year 1854 with information on financial aid given them, age, etc.
1854

180: XI, Nr. 4. ADS 7:233

List of emigrants based upon community accounting
records; compiled by Georg Diel, a local researcher; contains 188 names of emigrants who
were given financial assistance by the community
in order to emigrate 1709-1883

180 ADS 7:233

NAUHEIM (Amt) HEL

For exemptions from emigration tax, *see* Hesse-
Kassel [Kurhesse]

NAUMBURG. *See* von Oldershausen, August F. J. W. O.

NECKARSTEINACH HEL

Emigration 1859-1912
149: Rep.40,2: Ablief.2: Bd. 306. ADS 7:23

NECKARSTEINACH. *See also* Heppenheim (Kreis) HEL

NECKARSULM (Oberamt) BWL

Nr. 42,2: List of emigrants 1826-1855
Nr. 43-57: Emigration files, chronologically arranged 1853-1870

261: F 187: Neckarsulm: files as above. ADS 10:22

NECKARSULM (town) BWL

Reports from emigrants & motives for emigration
[19th century]

279: Neuer Nachlasszuwachs 1958. ADS 10:82

NEHEIM-HUESTEN RWL

Nr. I.140: Lists of emigrants with & without emigration permits (contains family &
military-service information) 1831-1871
Nr. I.903: Passports for four emigrating [famil-
ies?] 1859-1886
Nr. II.153: Passports (one only for USA) 1938-1939
Nr. II.221: Three emigrants to USA 1897-1927

135: files as above. ADS 6:116

NERESHEIM (Oberamt) BWL

Emigration files, chronologically arranged
1857-1913

261: F 188: Neresheim: Nr. 1095-1100. ADS 10:23

NESSELBRUNN (village; Landratsamt Marburg) HEL

Releases from citizenship 1823-1865

151: Bestand 180: Marburg: Acc.1937/47: Nr. 52.
ADS 7:120

NESSELWANGEN (village) BWL

Emigration to overseas countries 1848-1882

288: Kommunalarchiv Nesselwangen. ADS 10:129

NETHERLANDS, THE

Emigration & immigration to The Netherlands; exportation of personal property 1833

229: Rep.Reg. Akten von Oberbayern: Kammer des
Innern: Nr. 1794.10. ADS 9:116

NEUBURG an der Donau (Bezirksamt) BYL

Nr. 3695b: Emigration to North America 1819-1854

Nr. 3382-: Individual files (about 200 in number) regarding emigration to America & North America   1846-1890

230: Rep.Akten der Bezirksaemter: Neuburg: files as above.   ADS 9:124

NEUENBUERG (Oberamt)   BWL

Nr. 130: List of emigrants   1817-1871
Nr. 131: Disposal of property; lists of Wuerttemberg citizens living in Switzerland   1834-1872
Nr. 132-151: Emigration files   1801-1879

261: F 189: Neuenbuerg: files as above.   ADS 10:23

AZ 1040.32: Releases from [Wuerttemberg] citizenship   1880-1909
AZ 6135: Emigration   1851-1938

262: Wue 65/22: Neuenbuerg: Acc.49/1957: files as above.   ADS 10:38

NEUENHAUS (Amt)   NL

Emigration permits   1871-1879
[for permits before 1871, see Osnabrueck]

54: Rep.116,I: Nr. 6078.   ADS 4:120

Nr. 70: Releases from citizenship   1880-1883
Nr. 99: Releases from citizenship   1884-1887
Nr. 114: Releases from citizenship   1888-1896

54: Rep.116,I: II.4.A.1e: files as above.   ADS 4:121-123

Releases from citizenship   1897
54: Rep.116,I: Nr. 13415.   ADS 4:124

NEUENHEIM (with Moenchhof)   BWL

Emigration files   1463-1858

260: Rep.Abt.229: Spezialakten: Neuenheim: Fasz. 73221-73374.   ADS 10:11

NEUENKIRCHEN (parish)   NL

Nr. 1: Various emigration applications & permits   1829-1853
Nr. 3: Lists of immigrants & foreigners (many German-Americans)   1868-1874
Nr. 4: Lists of immigrants & foreigners   1896-1926
Nr. 5: List of emigrants (names, ages, family members, future residence)   1868-1874

100: Kirchspielsgericht Neuenkirchen: Fach 8: files as above.   ADS 4:312-313

NEUERSHAUSEN (village)   BWL

Emigration file   1868-1931

288: Kommunalarchiv Neuershausen.   ADS 10:129

NEUFRA (village; Oberamt Gammertingen)   BWL

Emigration to North America   1852

262: Ho 193: Gammertingen: II 12880.   ADS 10:35

NEUFRACH (village)   BWL

Emigration to overseas countries; lists of emigrants to the USA   1853-1944

288: Kommunalarchiv Neufrach.   ADS 10:129

NEU-GUELZE (Mecklenburg). See Rehagen, J. J. H, alias J. Brockmueller [surname index]

NEUHAUS/OSTE (Amt)   NL

Fraudulent emigration; investigations & extraditions of illegal emigrants   1868

52: Han.Des.74: Neuhaus/Oste: V.B.1: Nr. 10.   ADS 4:73

In the Amtsgericht (district court) there are estate files containing much information on emigrants, powers of attorney, notarized statements, etc. The information is not easily accessible, unless the names of persons are given.   ADS 4:291

NEUHOF (Amt)   HEL

Applications for emigration permits   1805-1810

151: Bestand 98: Fulda X: Acc.1875/27: Nr. 68-75.   ADS 7:91

NEUKIRCHEN (Amt?)   RWL

Reasons for emigration given for several emigrants   1805

115: Berg. Apanagialreg.: A.: Nr. A29.   ADS 6:22

NEUSATZ (village)   BWL

Emigration file   1851-1938
Emigration from the Waldmatt community   1828-1888

288: Kommunalarchiv Neusatz.   ADS 10:129

NEUSS (Landratsamt)   RWL

Nr. 187: Emigration & immigration   1851-1857
Nr. 14: Emigration & immigration   1852-1864
Nr. 163: Emigration & immigration   1865-1875
Nr. 142: Emigration & immigration   1876-1888
Nr. 176: Emigration & immigration   1888-1895
Nr. 215: Emigration & immigration   1900-1907
Nr. 41: Emigration & immigration   1907-1911
Nr. 278: Population lists (includes emigrants & immigrants)   1854-1855
Nr. 279: Population lists   1857-1859
Nr. 280: Population lists   1864-1867
Nr. 281: Population lists   1867-1871

115: Rep. G29/34: Neuss: files as above.   ADS 6:44-45

NEUSS (Kreis)   RWL

Emigration applications   1852-1860
115: Rep.D8: I.Abt.e: Nr. 583-585.   ADS 6:25

NEUSTADT (Oberamt)   RPL

Records of the former Oberamt Neustadt contain
emigration permits to Pennsylvania for residents
of the villages of Edenkoben, Lachen, Lambsheim,
Mutterstadt, Weidenthal   175-1754; 1764
201: Neustadter Oberamtsprotokolle: Akt Nr. 1711.
                                           ADS 8:56

NEUSTADT AM RUEBENBERGE (Amt)   NL

Emigration of persons subject to military service
                                               1870
52: Han.Des.74: Neustadt a. Rbge.: VII.B.6b: Nr. 1.
                                           ADS 4:74

Nr. 8: Searches abroad for heirs to deceased resi-
        dents & their heirs   1871
Nr. 7: Searches for missing persons, Vols I-IV
                                          1871-1906
52: Han.Des.74: Neustadt a. Rbge.: II.10.a: files
as above.   ADS 4:74

NEUSTADT a.d. AISCH (Bezirksamt)   BYL

Nr. 449: Emigration & immigration, Neustadt &
          Markt Erlbach   1851-1860
Nr. 468: Lists of German emigrants to America 1844
Nr. 736-758: Immigration & emigration applications
                                      1853-1871
231: Rep.212/13: Neustadt a.d. A.: files as above.
                                           ADS 9:129

Applications for release from Bavarian citizenship
                                          1902-1914
231: Rep.212/13/IV: Neustadt a.d. A.: Abgabe 1951:
Nr. 41.   ADS 9:132

NEUSTADT a.d. WALDNAAB (Bezirksamt)   BYL

Nr. 1810-1835: Individual applications for permits
                to emigrate to various foreign
                countries, including America
                                      1804-1859
Nr. 1803: Emigration & immigration; exportation of
           property   1817-1822
Nr. 1804: Applications & negotiations for pass-
           ports for the purpose of emigration to
           America   1836-1845
Nr. 1806: Reports of all communities within the
           administrative area regarding emigration
           to other countries during the year 1853
                                      1853-1855
222: Rep.132: Neustadt a.d. W. N.: files as above.
                                           ADS 9:3

NEUWEIER (village)   BWL

Emigration file   1834-1937

288: Kommunalarchiv Neuweier.   ADS 10:129

NEUWIED (Kreis)   RPL

Naturalizations   1829-1914
200: Abt.441: Regierung Koblenz: Nr. 22032-22041;
      23572; 24047.   ADS 8:29

NEUWIED (Landratsamt)   RPL

Certificates of original domicile & citizenship
                                          1907-1938
200: Abt.475: Neuwied: Nr. 1084; 1489; 1066; 1269;
      1287; 1523.   ADS 8:45

NEUWIED. See also Wied (Principality)

NEUWIED (Texas)   USA

Emigrant letter   1850
154: CA4b1: Nr. 2, pp. 59-68.
      DLC has.   ADS 7:165

NEW DORTMUND   USA

"Der Traum von New Dortmund" [The Dream of New
  Dortmund] Beitraege zur Geschichte Dortmunds
  und der Grafschaft Mark, Vol. 54.   ADS 6:86

NEW MEHRING (colony)   USA?

"Gruendung der Kolonie Neu-Mehring und Auswanderung der Kolonisten nach Amerika" [Founding of
  the New Mehring Colony and the Emigration of
  Colonists to America] typescript   n.d.
200: Abt.701: Nachlaesse, etc.: Nr. 968. ADS 8:53

NEW ORLEANS (Louisiana)   USA

Hamburg Consulate in New Orleans: Death certificates; searches for missing Hamburg citizens in
  the United States; legal matters   1859-1868
39: Bestand Senat: Cl.VI, No.16p: Vol.4b: Fasc. 1f
      [Nachforschungen]   ADS 2:47

Death certificates   1866
39: Bestand Senat: Cl.VIII, No.X, Jg.1866 (Regis-
      ter) III.9.   ADS 2:191

German consular report from New Orleans regarding
  estate matter [ADS description gives no names]
                                              1870
39: Bestand Senat: Cl.VIII, No.X, Jg.1870 (Regis-
      ter) 8 [Nordamerika]   ADS 2:209

Hamburg Consulate in New Orleans: Death certificates; searches for missing Hamburg citizens in
  the United States; estates   1871-1893
39: Bestand Hamburg: Cl.VI, No.16p: Vol.4b: Fasc.
      1g: Invol. 2-13 [Totenscheine]   ADS 2:47

Death certificates sent to Hamburg   1872
Two death certificates   1872
Consular report on civil registry matter [mar-
    riage?]   1872

39: Bestand Senat: Cl.VIII, No.X, Jg.1872 (Regis-
    ter; Journal)   ADS 2:220-222

Consular report on births & deaths   1874

39: Bestand Senat: Cl.VIII, No.X., Jg.1874 (Regis-
    ter) 1 [Nordamerika]   ADS 2:227

Transmittal of death certificates [ADS description
    does not give names]   1880

39: Bestand Senat: Cl.VIII, No.X, Jg.1880 (Regis-
    ter) 42 [Nordamerika]   ADS 2:245

List of matters dealt with by the Royal [Wuerttem-
    berg] Consulate in New Orleans (no. 98)
                                    [19th century]

263: Minist. d. Ae.: 1910 eingekommene Akten: Verz.
    58, 42.   ADS 10:41

Lists of Hamburg ships arriving in the port of
    New Orleans   1854; 1855; 1856

42: Bstand H516: Hamburg Consulate in New Orleans
    (1)(Anlage).   ADS 2:530

NEWTON (congregation). *See* Rhynbeck (congregation;
New York)   USA

NEW YORK   USA

Hamburg Consulate in New York: Births, death cer-
    tificates; estates   1866

39: Bestand Senat: Cl.VIII, No.X, Jg.1866 (Regis-
    ter) III.1-2; 8; 11; 14-17; 19; 21; 28.
                                    ADS 2:191

Hamburg Consulate in New York: Estate matters 1867

39: Bestand Senat: Cl.VIII, No.X, Jg.1867 (Regis-
    ter) C.1; C.21 [Nordamerika]   ADS 2:193

Hamburg Consulate in New York: Death certificates;
    estates; extradition cases   [1868?]

39: Bestand Senat: Cl.VI, No.16p: Vol.4b: Fasc. 21,
    Invol. 2-6; 9-11; 13-23.   ADS 2:50

General Consulate in New York: Estate matters 1870

39: Bestand Senat: Cl.VIII, No.X, Jg.1870 (Regis-
    ter) 62 [Nordamerika]   ADS 2:213

Hamburg Consulate in New York: Searches for miss-
    ing persons; etates   1870-1893?
39: Bestand Senat: Cl.VI, No.16p: Vol.4b, Fasc. 2m:
    Invol. 4-14; 16-55.   ADS 2:51

German consular report from New York on estate
    matters   1871

39: Bestand Senat: Cl.VIII, No.X, Jg.1871 (Regis-
    ter) 41 [Nordamerika]   ADS 2:217

General Consulate: Transmittal of death certifi-
    cates   1881

39: Bestand Senat: Cl.VIII, No.X, Jg.1881 (Regis-
    ter) 8.   ADS 2:246

List of [emigrants & seamen] who died aboard Ham-
    burg vessels at sea or in New York harbor   1866

42: Bestand H516: Konsulats-Berichte 1867: Band 2,
    M-Z (New York)(1)(c)(Anlage).   ADS 2:561

Deaths aboard emigrant ships   1868

39: Bestand Senat: Cl.VIII, No.X, Jg.1868 (Regis-
    ter) C.13 [Nordamerika]   ADS 2:198

German Consulate in New York: Births & deaths, ab-
    stracts from ship registers   1869

39: Bestand Senat: Cl.VIII, No.X, Jg.1869 (Regis-
    ter) 7 [Nordamerika]   ADS 2:201

Consular report on births & deaths aboard emigrant
    ships in year 1874   1875

39: Bestand Senat: Cl.VIII, No.X, Jg.1875 (Regis-
    ter) 6 [Nordamerika]   ADS 2:229

Births & deaths [aboard ship] for 1875   1876

39: Bestand Senat: Cl.VIII, No.X, Jg. 1876 (Regis-
    ter) 25 [Nordamerika]   ADS 2:233

Hamburg guardianship matters in New York   1860

39: Bestand Senat: Cl.VIII, No.X, Jg.1860 (Regis-
    ter) 3 [Vormundschaft]   ADS 2:178

German consular report from New York on guardian-
    ship matter   1871

39: Bestand Senat: Cl.VIII, No.X, Jg.1871 (Regis-
    ter) 28 [Nordamerika]   ADS 2:216

List of Hamburg seamen who deserted [in America]
    totaling 384 men; detailed list missing [but
    might be found among Consulate's papers]   1864

42: Bestand H516: Konsulats-Berichte 1865: Band 2,
    M-Z (New York)(1)(C)(Anlage)   ADS 2:553

Report from New York on deserting & remaining crew
    members of *S.S. Schiller* which sank   1875

39: Bestand Senat: Cl.VIII, No.X, Jg.1875 (Regis-
    ter) 19 [Nordamerika]   ADS 2:230

Annual report & membership list of the German so-
    ciety in New York   1859

48: B.13b: Nr. 6.   ADS 3:19

Membership list of German Society, New York   1871

39: Bestand Senat: Cl.VIII, No.X, Jg.1871 (Regis-
    ter) 14 [Nordamerika]   ADS 2:215

Lists of Hamburg ships in New York port
    1847-1849; 1856-1857; 1859-1864; 1866-1868

42: Bestand H516: Konsulats-Berichte [various
    years](New York)(1-4)(A, B, C)(Anlagen)
        ADS 2:515; 517; 520; 534-536; 540; 546; 549;
                                    553; 561; 563-564

NEW YORK (state) *See also* Moravian Brethren   USA

NIBLER (village [possibly Nibl]; Oberpfalz)   RPL

Emigration   1817-1819

222: Regierung: Kammer der Finanzen: Rep.118, II.
    Teil: Nr. 9713.   ADS 9:2

NIEDERBRONN (Grafschaft Hanau-Lichtenberg)   RPL

Emigrants to the New World presently at the Ket-
terichhof, Lemberg   1884

<u>201</u>: Archivabt. Grafschaft Hanau-Lichtenberg: Akt
Nr. 1090.   ADS 8:67

NIEDERBUEHL (village)   BWL

Emigration file   1848-1940

<u>288</u>: Kommunalarchiv Niederbuehl.   ADS 10:129

NIEDEREGGENEN (village)   BWL

Emigration file   1858-1942

<u>288</u>: Kommunalarchiv Niedereggenen.   ADS 10:129

NIEDERFELL (village)   RPL

Naturalizations & releases from citizenship
                                   1829-1893

<u>200</u>: Abt.655.7: Brodenbach: Nr. 196.   ADS 8:50

NIEDERHAUSEN (village)   BWL

Emigration file   1785-1933

<u>288</u>: Kommunalarchiv Niederhausen.   ADS 10:129

NIEDERHESSEN (Province)   HEL

Nr. 9: Naturalizations in Province of Niederhesse
       with Schaumburg, 8 vols.   1831-1849
Nr. 13: Releases from citizenship, 3 vols.
                                   1827-1847
Nr. 22: Releases from citizenship   1822
Nr. 20: Naturalizations, 7 vols.   1852-1866?

<u>151</u>: Bestand 16.II: Klasse 15: files as above.
                                       ADS 7:66

NIEDER-LIEBERSBACH   HEL

Emigration   1860-1910

<u>149</u>: Rep.40,2: Ablief. 2: Bd. 307.   ADS 7:23

NIEDERRHEIN (Grandduchy, 1816-1822)   RPL

Nr. 867: Files on the emigration of Prussian &
         other subjects [emigration to Russian
         Poland only?]   1816-1817
Nr. 868: Files on the emigration of Prussian &
         other subjects [to other countries?],
         Vol. II   1817-1821

<u>200</u>: Abt.402: Oberpraesidium Koblenz (Provinz
Grossherzogtum Niederrhein 1816-1822): files
as above.   ADS 8:17

NIEDERRHEINPROVINZ [Lower Rhine Province]

Emigration to America   1816-1819

<u>500</u>: Berlin-Dahlem: Gen. Staatsarchiv: Auswaertige
Amt: Abt.III, Rep.I: Auswanderung-Generalia:
Nr. 2, Vol. II.   Not in ADS

NIEDERRIMSINGEN (village)   BWL

Emigration file   1817-1932

<u>288</u>: Kommunalarchiv Niederrimsingen. ADS 10:129

NIEDERSCHOFFHEIM (village)   BWL

Emigration file   1864-1949

<u>288</u>: Kommunalarchiv Niederschoffheim. ADS 10:130

NIEDERUNG (Kreis)   East Prussia (Gumbinnen)

Emigration [file missing?]   1832-1877

<u>51</u>: Rep.12: Abt.I, Titel 3: Nr. 8, Bd. 7. ADS 4:32

NIEDERWASSER (village)   BWL

Emigration file   1852-1918

<u>288</u>: Kommunalarchiv Niederwasser.   ADS 10:130

NIEDERWEILER (village)   BWL

Emigration file   1889-1937

<u>288</u>: Kommunalarchiv Niederweiler.   ADS 10:130

NIEDERWIHL (village)   BWL

Emigration file   1832-1932

<u>288</u>: Kommunalarchiv Niederwihl.   ADS 10:130

NIMBURG (village)   BWL

Emigration tables   1852-1952
Emigration files   1890-1920

<u>288</u>: Kommunalarchiv Nimburg.   ADS 10:130

NOEGGENSCHWIEL (village)   BWL

Emigration file   1852-1942

<u>288</u>: Kommunalarchiv Noeggenschwiel.   ADS 10:130

NOERDLINGEN (Bezirksamt)   BYL

Nr. 790/3: Emigration to North America 1831-1837
Nr. 938: Emigration to various countries 1871
Nr. 1301: Emigration & immigration (contains 23
          permits to emigrate to America)
                                   [19th century]
Nr. 1302: Emigration (contains 61 emigration per-
          mits to America)   1868-1870

<u>230</u>: Rep. Akten der Bezirksaemter: Noerdlingen:
files as above.   ADS 9:124

NOERDLINGEN (town)   BYL

   Emigration files   [19th century?]
   <u>249</u>   ADS 9:163

NOERTERSHAUSEN (village)   RPL

   Naturalizations & releases from citizenship
                                   1849-1891
   <u>200</u>: Abt.655.7: Brodenbach: Nr. 219.   ADS 8:50

NONNENWEILER (village)   BWL

   Emigration file   1850-1940
   <u>288</u>: Kommunalarchiv Nonnenweiler.   ADS 10:130

NORDEN (Landratsamt)   NL

   Nr. 2227: Issuance of citizenship releases for
            purpose of emigration   1881-1883
   Nr. 2172: Issuance of citizenship releases for
            purpose of emigration   1884-1888
   <u>50</u>: Rep.32a: 6: files as above.   ADS 4:13

NORDERDITHMARSCHEN. *See* Dithmarschen, Norder

NORDERGOESHARDE   SHL

   Nr. 149: Emigration to America   1868-1888
   Nr. 395: Emigration; immigration; citizenship
                              1868-1882
   Nr. 397: Passports   1869-1887
   <u>16</u>: files as above.   ADS 1:50-51

NORDERWOEHRDEN (Kreis Norderdithmarsch)   SHL

   Nr. 355: Passports   1822-1868
   Nr. 357: Certificates of citizenship   1874-1891
   Nr. 358: Releases from citizenship   1892
   Nr. 359: Citizenship   1871-1896
   Nr. 360: Emigration & immigration   1863-1887
   Nr. 362: Searches for missing persons   1881-1888
   Nr. 556: Actions against persons subject to mili-
          tary service who have emigrated 1863-1871
   Nr. 559: Persons subject to military service who
          have been released from citizenship
                              1887-1889
   Nr. 561: Searches for persons subject to military
          service   1870-1899
   <u>25</u>: files as above.   ADS 1:58

NORDHALDEN (village)   BWL

   Licit & illicit emigration   1853-1882
   <u>288</u>: Kommunalarchiv Nordhalden.   ADS 10:130

NORDRACH (village)   BWL

   Emigration file   1858-1912
   <u>288</u>: Kommunalarchiv Nordrach.   ADS 10:130

NORDSCHWABEN (village)   BWL

   Emigration file   1833-1912
   <u>288</u>: Kommunalarchiv Nordschwaben.   ADS 10:130

NORDSTRAND (rural area). *See* Helgoland   SHL

NORSINGEN (village)   BWL

   Emigration file   1841-1888
   <u>288</u>: Kommunalarchiv Norsingen.   ADS 10:130

NORTHEIM   NL

   Fach 9: Releases from citizenship   n.d.
         (Seite 771, Ablage 294)
   Fach 12: Emigration matters   n.d.
         (Seite 774, Ablage 297)
   <u>98</u>: Des.2, Vol.I: Abschn.III.A: files as above.
                                   ADS 4:293

NORTHEIM   NL

   Fach 9: Applications for releases from citizenship
         (Ablage 235)   n.d.
   <u>98</u>: Des.2, Vol.II: Abschn.III.A: file as above.
                                   ADS 4:293

NUERNBERG (Bezirksamt)   BYL

   Emigration & immigration; emigration tax
                        1822; 1848-1869
   <u>231</u>: Rep.212/14: Nuernberg: 13.Abgabe 1929: 1-4.
                                   ADS 9:132

   Releases from citizenship (including individual
         files & names)   1852-1889
   <u>231</u>: Rep.212/14: Nuernberg: 14.Abgabe 1929: Nr.
      223-231.   ADS 9:132

NUERNBERG (city)   BYL

   File on emigration   1807-1808
   <u>231</u>: Rep.137: A.A.-Akten: 1865.   ADS 9:127

   II/16/3: Older magistrates' files [emigration]
                             [19th century]
   II/2: Main files [personal data on emigrants]
                                19th century
   <u>251</u>: Rep.C.16: Verz. der Auswaertigen Akten: alpha-
      betisch.   ADS 9:168

   Naturalizations & losses of citizenship 1871-1889
   <u>231</u>: Rep.270/II: Reg. Mittelfranken: Kammer des
      Innern: Abgabe 1932: Nr. 729.   ADS 9:134

NUERTINGEN (Oberamt)   BWL

   List of emigrants   1806-1823
   <u>261</u>: F 190: Nuertingen.   ADS 10:23

NUERTINGEN (Kreis)   BWL

Emigration files   [19th century?]
276   ADS 10:79

Heimatbuch [citizens' registry book] contains
   numerous references to emigrants [19th century]
277   ADS 10:80

NUSSDORF (village)   BWL

Emigration file   1870-1929
288: Kommunalarchiv Nussdorf.   ADS 10:130

OBER-ABTSTEINACH (village)   HEL

Emigration   1860-1910
149: Rep.40,2: Ablief.2: Bd. 307.   ADS 7:23

OBERALPFEN (village)   BWL

Emigration file   1854-1950
288: Kommunalarchiv Oberalpfen.   ADS 10:131

OBERBALDINGEN (village)   BWL

Emigration file   1881-1943
288: Kommunalarchiv Oberbaldingen.   ADS 10:131

OBERBAYERN (Upper Bavaria)   BYL

Nr. 1159.1: Emigration & manumission taxes
                                       1599-1699
Nr. 1159.2: Emigration & manumission taxes
                                       1701-1777
Nr. 1160.3: Emigration & manumission taxes for
      Pfalzbayern [Bavarian Palatinate]
                                       1778-1789
Nr. 1160.4: Emigration & manumission taxes for
      Pfalzbayern   1790-1808
Nr. 1163.16: Special church tax? [legata pia] 1787
Nr. 422: Emigration   1804-1816
Nr. 423: Emigration   1817
Nr. 424: Emigration   1818
Nr. 425.46-48: Emigration   1819-1823
Nr. 428.53: Emigration   1809-1814
Nr. 429-430: Emigration, A-K, L-Z   n.d.
229: Rep. G.R.: General-Registratur: files as
      above.   ADS 9:110-111

Nr. 3215.17: List of emigrants   1806
Nr. 475.1: Emigration & immigration applications;
      exportation of personal property
                                       1817-1831
Nr. 1541: Emigration & immigration; export of per-
      sonal property; confiscations   n.d.
Nr. 1091.100: Passports   n.d.
229: Rep.G.L.: Gerichtsliteralien: files as above.
                                      ADS 9:112-114

Manumission & emigration taxes: The archive contains a large number of files from 1600-1818, at
   least. Although these files are listed in ADS,

it seems clear that further arrangement would be
   necessary before analysis by genealogical researchers can be undertaken
229: Rep.G.L.:Gerichtsliteralien.   ADS 9:112-113

Nr. 1167.22: Emigration of Bavarian citizens 1806
Nr. 1245.95-149: Emigration to North America
                                       1846-1853
Nr. 1246.150-309: Emigration to North America
                                       1846-1853
Nr. 1247.310-472: Emigration to North America
                                       1846-1853
Nr. 1248.473-681: Emigration to North America
                                       1846-1853
Nr. 1249.682-699: Emigration to North America
                                       1846-1853
Nr. 1794.6: Bavarian emigration   1853
Nr. 1794.6a: Bavarians in foreign service
                                       1812-1813
Nr. 1744.17: Deserters & unlawful persons   1812
Nr. 3801.16922: Reports to local police regarding
                court sentences given to Bavarian
                citizens abroad   1822-1914
Nr. 1805.149: Bavarians dying abroad   1852
Nr. 1805.154: Death certificates for Bavarian
                citizens dying abroad 1857-1878
Nr. 1794.37: Returning emigrants from North Amer-
                ica   1856
229: Rep. Reg. Akten von Oberbayern: Kammer des
      Innern: files as above.   ADS 9:115-117

Whereabouts of Bavarians abroad & foreigners in
   Bavaria; searches & investigations of such persons; issuance of birth & death certificates,
   court citations, & declarations of missing persons   [19th century?]
229: Rep. Reg. Akten von Oberbayern: Kammer des
      Innern: Nr. 1268.664-837; 1269.838-987.
                                        ADS 9:116

OBERBRECHEN (village; Kreis Limburg)   HEL

Applications for release from Nassau citizenship
                                            1853
152: Abt.232: Limburg: Nr. 1421.   ADS 7:134

OBERDURLACH (Amt)   BWL

Emigration of various citizens   1771
260: 216A.   ADS 10:1

OBEREGGENEN (village)   BWL

Emigration file   1880-1951
288: Kommunalarchiv Oberreggenen.   ADS 10:131

OBEREGGINGEN (village)   BWL

Emigration file   1890-
Immigration file   1919-
288: Kommunalarchiv Obereggingen.   ADS 10:131

OBEREUTERSBACH (village)   BWL

   Emigration   1853-1899

   <u>288</u>: Kommunalarchiv Obereutersbach.   ADS 10:131

OBERFELL (village)   RPL

   Naturalizations & releases from citizenship
                            1815-1888

   <u>200</u>: Abt.655.7: Brodenbach: Nr. 195.   ADS 8:50

OBERFRANKEN (Upper Franconia)   BYL

   Emigration to North America & other countries
                            1837-1924

   <u>223</u>: Rep.K.3.A.I: Nr. 2101-2632.   ADS 9:9

OBERFRANKEN. *See also* Bamberg   BYL

OBERGLADBACH (Amt)   HEL

   Naturalizations & releases from citizenship; emigration   1854-1863

   <u>152</u>: Abt.231: Langenschwalbach: Nr. 5.   ADS 7:133

OBERGLOTTERTAL (village)   BWL

   Emigration file   1874-1909
   Emigration file   1899-1933

   <u>288</u>: Kommunalarchiv Oberglottertal.   ADS 10:131

OBERHARMERSBACH (village)   BWL

   Emigration file   1845-1951

   <u>288</u>: Kommunalarchiv Oberharmersbach.   ADS 10:131

OBERHAUSEN (village; Kreis Emmendingen)   BWL

   Emigration files   1855-1940
   General emigration file; list of emigrants
                            1892-1942

   <u>288</u>: Kommunalarchiv Oberhausen.   ADS 10:131

OBERHAUSEN. *See* Essen-Oberhausen   RWL

OBERHESSEN (province)   HEL

   Nr. 1: Naturalizations, 4 vols.   1827-1859
   Nr. 1a: Naturalizations   1828
   Nr. 2: Releases from citizenship   1832-1853
   Nr. 4: Matters dealing with the property [estates?]
      of former Kurhessian citizens resident
      abroad   1821

   <u>151</u>: Bestand 16.II: Klasse 16: files as above.
                             ADS 7:66

OBERISSIGHEIM (Amt)   HEL

   For exemptions from emigration tax, *see* Hesse-
   Kassel [Kurhesse]

OBERLAUCHRINGEN (village)   BWL

   Emigration file   1864-1898

   <u>288</u>: Kommunalarchiv Oberlauchringen.   ADS 10:131

OBERLAUDENBACH   HEL

   Emigration   1859-1890

   <u>149</u>: Rep.40,2: Ablief.2: Bd. 308.   ADS 7:23

OBERLUSTADT (Herrschaft)   RPL

   Nr. 65: Inventories & property divisions 1752-1760
   Nr. 66: Inventories & property divisions 1761-1770
   Nr. 68: Inventories & property divisions 1783-1784
      [contains details on family members of
      persons emigrating to Pennsylvania]

   <u>201</u>: Archivabt. Johanniterorden: files as above.
                             ADS 8:61

OBERMUENSTERTAL (village)   BWL

   Emigration file   1835-1846
   Emigration file, alphabetically arranged 1854-1941
   General file   1854-1914
   Emigration to the USA   1861-1933

   <u>288</u>: Kommunalarchiv Obermuenstertal.   ADS 10:131

OBERNDORF (Oberamt)   BWL

   Nr. 94: List of emigrants   1817-1819
   Nr. 95-122: Emigration files, alphabetically ar-
          ranged by village   19th century

   <u>261</u>: F 191: files as above.   ADS 10:23

OBERNDORF (town)   BWL

   Newspaper clippings regarding persons from Oberndorf in America   [19th & 20th century?]

OBERNDORF (village; Kreis Rastatt)   BWL

   Emigration file   1851-1888

   <u>288</u>: Kommunalarchiv Oberndorf.   ADS 10:134

OBERPFALZ (Upper Palatinate)   BYL

   Nr. 11375: Emigration to America & Brazil; secret
          emigration   1824-1875
   Nr. 11526-11529: Emigration to North America
                         1832-1868

   <u>222</u>: Regierung: Kammer des Innern: Rep.116: files
      as above.   ADS 9:1

OBERPFALZ. *See also* Amberg   BYL

OBERRHEINKREIS   BWL

   Nr. 661-662: Financial aid for the emigration of
      various persons from villages in the
      Kreis, Fasz. I, II   1851-1857

Nr. 663: State aid for the transportation of various persons to America, Part III
1858-1864

260: Abt.236: Innenmin., Wegzug: Auswand.: files
as above.   ADS 10:2

OBERRODENBACH (village)   HEL

Petitions for emigration permits   1804-1820

151: Bestand 81: Rep.E: 16.Gemeinde Oberrodenbach:
Nr. 13 [Gesuche]   ADS 7:82

OBERROSPHE (village; Landratsamt Marburg)   HEL

Releases from citizenship   1824-1889

151: Bestand 180: Marburg: Acc.1937/47: Nr. 53.
ADS 7:120

OBERROTTWEIL (village)   BWL

Emigration file   1853-1874
Emigration to the USA   1870-1871; 1884

288: Kommunalarchiv Oberrottweil.   ADS 10:132

OBERSCHAFFHAUSEN. *See* Boetzingen   BWL

OBERSCHOPFHEIM (village)   BWL

Emigration file   1870-1949

288: Kommunalarchiv Oberschopfheim. ADS 10:132

OBERSCHOSCHEL (congregation)   RPL

Marginal entries on emigration to be found in
either the Lutheran or Reformed parish regis-
tries   18th century
217   ADS 8:86

OBERSIMONSWALD (village)   BWL

Emigration file   1818-1907
Emigration file   1851-1911

288: Kommunalarchiv Obersimonswald. ADS 10:132

OBERSTE[I]NWEILER (village)   BWL

Emigration file   1893-1938

288: Kommunalarchiv Obersteinweiler. ADS 10:132

OBERSUELZEN (village). *See* Kurpfalz   RPL

OBERUHLDINGEN (village)   BWL

Emigration file   1852-1934
Emigration & immigration   1910-1944

288: Kommunalarchiv Oberuhldingen.   ADS 10:132

OBERWEIER (village; Kreis Lahr)   BWL

Emigration file   1790-1936

288: Kommunalarchiv Oberweier.   ADS 10:132

OBERWEIER (village; Kreis Rastatt)   BWL

Emigration file   1834-1895

288: Kommunalarchiv Oberweier.   ADS 10:132

OBERWESEL (city)   RPL

Censuses (lists of residents with birthplaces &
year; also contains notations on emigration)
1798-1812

200: Abt.631: Stadt Oberwesel: Nr. 123.   ADS 8:49

OBERWIHL (village)   BWL

Emigration file   1839-1939
Emigration to South Carolina   1890-1939

288: Kommunalarchiv Oberwihl.   ADS 10:132

OBERWINDEN (village)   BWL

Emigration file   1892-1932

288: Kommunalarchiv Oberwinden.   ADS 10:132

OBERWOLFACH (village)   BWL

Emigration file   1845-1914

288: Kommunalarchiv Oberwolfach.   ADS 10:132

OCHSENFURT (Bezirksamt)   BYL

Emigration   1826-1837

232: Rep. Ochsenfurt: Nr. 145.
DLC has.   ADS 9:138

ODELSHOFEN (village)   BWL

Emigration file   1817-1884
Emigration in general & from Amtsgericht Kork
1828-1936
Germans abroad   1932-1939

288: Kommunalarchiv Odelshofen.   ADS 10:133

ODERNHEIM a. GL. (village)   RPL

List of persons who wish to emigrate to America
ca. 1740

214: VIII: Akt Nr. 9.   ADS 8:82

ODISHEIM (parish)   NL

Applications for releases from citizenship
1868-1905
Emigration list (with destinations) 1925-1936

Emigration lists   1882
List of persons emigrating without releases from
    citizenship   1890-1897
Persons emigrating to America   1900-1904
Lists of emigrants   1925-1930; 1932-1933; 1935

<u>100</u>: Kirchspielsgericht Odisheim: I.6: Nr. 3.
                                            ADS 4:314-315

OEDSBACH (village)   BWL

Emigration file   1882-1938

<u>288</u>: Kommunalarchiv Oedsbach.   ADS 10:133

OEFINGEN (village)   BWL

Emigration file   1888-1933

<u>288</u>: Kommunalarchiv Oefingen.   ADS 10:133

OEHNINGEN (village)   BWL

Emigration file   1844-1913

<u>288</u>: Kommunalarchiv Oehningen.   ADS 10:133

OENSBACH (village)   BWL

Emigration file   1861-1950

<u>288</u>: Kommunalarchiv Oensbach.   ADS 10:133

OEREL (parish; near Bremervoerde)   NL

Emigrants from parish (name, birthdate, parents,
    year of emigration, reason, where)   1824-1883
<u>63</u>   ADS 4:240

Emigration lists of Bremervoerde & the parish of
    Oerel   n.d.
    (in possession of Oberstudienrat i.R., Dr. Died-
    rich Mueller, Kehdinger-Muehren 30, Stade,
    Niedersachsen)   ADS 4:336

OEHRINGEN (Oberamt)   BWL

Nr. 426: List of emigrants   1849-1870
Nr. 427-440: Emigration files   1853-1900
Nr. 441: Naturalizations & losses of citizenship
                                          1856-1859

<u>261</u>: F 192: Oehringen: files as above. ADS 10:23

OESEDE (Vogtei)   NL

Emigration permits   1836-1852

<u>54</u>: Rep.122,V.4: Iburg: Nr. 13.
    DLC has.   ADS 4:136

OETIGHEIM (village)   BWL

Emigration to the USA   1901-1937

<u>288</u>: Kommunalarchiv Oetigheim.   ADS 10:134

OFFENBACH. *See* Dillenburg   HEL

OFFENBURG (village)   BWL

Material on emigration for Offenburg & Hofweier,
    particularly for the period 1847-1855. There is
    also a Buergerbuch [register book of citizens]
    beginning in 1837 which gives information on
    emigrants who were former residents

<u>288</u>: Kommunalarchiv Offenburg.   ADS 10:133

OHIO, German Consulate in. *See* Cincinnati   USA

OHLSBACH (village)   BWL

Emigration file   1851-1938

<u>288</u>: Kommunalarchiv Ohlsbach.   ADS 10:133

OHREN (village; Kreis Limburg)   HEL

Application by town council for a loan of 4000
    florins to aid the emigration of 14 poor famil-
    ies to America; list of prospective emigrants
    appended   1856

<u>152</u>: Abt.232: Limburg: Nr. 1560.   ADS 7:134

OHRENSBACH (village)   BWL

Emigration file   1890-1927

<u>288</u>: Kommunalarchiv Ohrensbach.   ADS 10:133

OLDENBURG   NL

Emigration   1826

<u>53</u>: Bestand 31: Oldenburg: 31-12-7-22.
    DLC has.   ADS 4:109

Releases from citizenship; emigration to America
                                          1830-1858

<u>53</u>: Bestand 31: Oldenburg: 31-13-63-11.
    DLC has.   ADS 4:110

Releases from citizenship; emigration; transporta-
    tion at public expense   1845-1856

<u>53</u>: Bestand 30: Luebeck: 30-14-2-2.   ADS 4:108

B.V.3: Passports   1869-
B.V.5: Emigration   n.d.
B.V.6: Emigration   n.d.
B.V.7: Citizenship; certificates of residency; re-
    leases   n.d.

<u>26</u>: files as above.   ADS 1:59

"Zur Geschichte der Auswanderung aus dem alten
    Damme (Oldb.), insbesondere nach Nordamerika
    in den Jahren 1830-1880" [History of the Emi-
    gration from Alten Damme (Oldb.) Particularly
    to North America, in the Years 1830-1880]
    *Oldenburger Jahrbuch des Landesvereins fuer
    Geschichte und Heimatkunde*, Vols. 46 & 47 (1942-
    1943), pp. 164-297 [includes 44 pages of emi-
    grants & the year of departure]   ADS 4:295

OLDENBURG. *See also* Hamburg (Hanseatic) Ministry in Washington, D.C.

OLETZKO (Kreis). *See* Treuburg (Kreis)
                                East Prussia (Gumbinnen)

OLFEN (Amt: Kreis Luedinghausen)   RWL

 Nr. 15: Emigration   1825-1862
 Nr. 16: Emigration   1832-1896

 116: Muenster Reg.: Luedinghausen: files as above.
                                        ADS 6:66

OFFINGEN (village)   BWL

 Emigration to the USA   1837
 Emigration files   1853-1942

 288: Kommunalarchiv Oensbach.   ADS 10:133

ORB (Landratsamt)   HEL

 Nr. 39: Emigration   1803
 Nr. 872: Trips to North America   1836-1838
 Nr. 1211: Certificates of original domicile, Vol.I
                                     1840-1845
 Nr. 1213: Certificates of original domicile,
           Vol. II   1847-1849
 Nr. 36: List of special files on emigration &
         immigration   1846
 Nr. 37: General file on emigration & immigration
                                     1851-1861
 Nr. 38: General file on emigration & immigration
                                          1814
 Nr. 41: Secret [illicit] emigration to America
                                         1851-
 Nr. 42: Emigration to America   1851
 Nr. 46: Prohibited emigration to North America
                                          1857
 Nr. 48: Notations regarding sequestered properties
         of persons emigrating secretly [illicitly]
                                          1860
 The Verzeichnis [list] of Acc.1886/9 contains, on
   pages 24-37, the names of 306 emigrants to North
   America from various villages in Kreis Orb n.d.

 151: Bestand 112: Acc.1886/9 (Orb): files as
      above.   ADS 7:96-97

 Files of Otto Trautmann, emigration agent   1887

 151: Bestand 180: Gelnhausen: Acc. 1958/36B: Nr.
      77.   ADS 7:118

 Passports   1843-1862

 151: Bestand 112: Acc.1903/9: LIV: Nr. 14.
                                     ADS 7:98

ORKE (Bezirk)   HEL

 Naturalization petitions   1867

 151: Bestand 82: Acc.1904/13.   ADS 7:84

ORSCHWEIER (village)   BWL

 Emigration file   1880-1942

288: Kommunalarchiv Orschweier.   ADS 10:133

ORSOY (town)   RWL

 Nr. 237: Emigration applications   1836-1837
 Nr. 415: Emigration applications   1846-1849
 Nr. 417: Passports   1814-1848
 Nr. 283: Passports   1821-1825
 Nr. 263: Passports   1825-1834
 Nr. 64: Emigration (for Budberg)   1855-1880

 115: Rep. G29/42: Waldbroel?: files as above.
                                     ADS 6:46

ORTENBERG (village)   BWL

 Emigration file   1852-1949
 Immigration & emigration   1945-1946

 288: Kommunalarchiv Ortenberg.   ADS 10:134

ORTENBERG (Amt)   HEL

 Emigration [reverse pages contain names]   1751

 151: Bestand 81: Rep. Ba: Gef. 9: Nr. 8. ADS 7:80

 Emigrants to Nova Scotia & Pennsylvania   1752

 151: Bestand 86: Nr. 51974.   ADS 7:88

 Emigrants   1741

 151: Bestand 86: Nr. 2085.   ADS 7:89

ORTENBERG (town)   HEL

 Several emigration cases are listed   1933-

 191: Melderegister.   ADS 7:252

OSNABRUECK (Regierung)   NL

 Nr. 763-810: Issuance of emigration permits, 47
              vols.   1823-1870
 Nr. 4242-4244: Emigration to Brazil & North Amer-
                ica   1826-1853
 Nr. 4247-4250: Information collected by the
                government on persons who have emi-
                grated to America, Vols. I, III, IV,
                V   1834-1847
                [*See* II.7.K.Nr.3 for Volume II to
                this series.]
 Nr. 13499 (Nr. 183): List of persons to whom re-
                leases from citizenship were
                issued   n.d.
 Nr. 13456 (Nr. 65) Requests for extradition n.d.
 Nr. 13500 (Nr. 185-186): Expulsion proceedings;
                        extraditions   n.d.
 Nr. 13457: Alphabetical index to expulsions of
            foreigners from German territory   n.d.

 54: Rep.116,I: files as above.
     DLC has Nr. 4247-4250.
                    ADS 4:118-120; 124; 129-130

 Information collected by the government regarding
   persons who have emigrated to America, Vol. II
   [This volume is part of a series to be found
   under Rep.116,I: Nr. 4247-4250 above.]

 54: Rep. 116,I: II.7.K. Nr. 3.   ADS 4:124

Nr. 173: Proceedings regarding desertions
[Lauf. Nr. 2493]   1769-1779

50: Depositum I Ostfries. Land: LVII: files as
above.   ADS 4:15

OSTRACH (Oberamt)   BWL

Emigration & immigration   1854-1862

262: Ho 198: Ostrach: I 9428.   ADS 10:36

OTTENHOFEN (village)   BWL

Emigration file   1834-1935
Illicit emigration   1850-1867
Germans abroad   1933-1934

288: Kommunalarchiv Ottenhofen. ADS 10:134

OTTERNDORF   NL

Emigration papers 1829-1853; 1831-1863; 1863-1872

100: Magistrat Otterndorf: Fach 207: Nr. 22; 23;
24.   ADS 4:303

Part of a list of emigrants (name, age, destination)   1871
Card entries for emigrants (name, age, destination
1880; 1881; 1883

100: Magistrat Otterndorf: AR Fach 112: Nr. 4.
ADS 4:300

Applications for releases from citizenship (about
60 in all); occasional letters from relatives in
America are included   1821-1862

100: Magistrat Otterndorf: Fach 140: Nr. 2, Vol. I.
ADS 4:302

Investigations of persons subject to military service who have emigrated to America   1853; 1855

100: Magistrat Otterndorf: Fach 140: Nr. 1.
ADS 4:301

Financial aid for the transportation of persons to
America   1863-1866

52: Han.Des.80: Stade: P, Titel 32: Nr. 7.
ADS 4:93

Passport register [contains names, ages, profes-
sions, birthplaces, & date, German residence,
reason for trip; not all passports for emigration to USA]   1826-1858

100: Magistrat Otterndorf: Fach 207: Nr. 21; 25.
ADS 4:302

OTTERNDORF (Amt)   NL

Applications for releases from citizenship (about
150 applications, often with letters from relatives already in America)   1873-1880
Lists of emigrants (name, birthplace & date, destination)   1863-1866; 1869-1907

100: Amt Otterndorf: Loc.163: Nr. 1; 2.
ADS 4:317-318

OTTERNDORF. See also Osterende-Otterndorf, Westerende-Otterndorf; Oster-Ilienworth, Wester-Ilienworth   NL

OTTERSDORF (village)   BWL

Emigration file   1847-1938

288: Kommunalarchiv Ottersdorf. ADS 10:134

OTTERSHEIM (village; Landkreis Germersheim, Kurpfalz)
RPL

Names of emigrants   18th century

201: Archivabt. Ausfautheiakten. ADS 8:63

OTTERSHEIM (Kreis Kirchheimbolanden). See Kurpfalz

OTTERSTADT (village: Bezirk Speyer)   RPL

Emigration   1816-1852

201: Archivabt. Bezirksaemter: Speyer (Abgabe 1907):
Akt Nr. 5.   ADS 8:65

OTTERSWEIER (village)   BWL

Emigration file   1829-1934
Emigration from Hatzenweier community 1828-1911

288: Kommunalarchiv Ottersweier. ADS 10:134

OTTILIENBERG. See Eppingen   BWL

OTTWEILER (Herrschaft; Principality of Nassau-Saarbruecken)   RPL

Report on citizens & serfs who have emigrated &
corresponding emigration tax levied   1733-1738

200: Abt.22: Fuerstentum Nassau-Saarbruecken: Nr.
3860.   ADS 8:9

OTTWEILER (Landkreis)   RPL

Nr. 203-204; 6126-6128: Emigration   1865-1922
Nr. 8588-8616: Naturalizations   1866-1924

200: Abt.442: Regierung Trier: files as above.
ADS 8:36; 40

OWINGEN (village)   BWL

Emigration file   1843-1919

288: Kommunalarchiv Owingen.   ADS 10:134

PADERBORN (Kreis)   RWL

Nr. 174-178: Emigration   1877-1904; 1914-1924
Nr. 51-56: Certificates of original domicile
1863-1924
Nr. 87-89: Certificates of citizenship   1899-1924

116: Minden: Praes.-regist.: I.A.: files as above.
ADS 6:50-52

PALATINATE. *See* Kurpfalz, Oberpfalz, Pfalz, Pfalz-bayern, Pfalz-Zweibruecken, Rheinpfalz

PANKER. *See* Hessenstein (Herrschaft)　　　　SHL

PAPENBURG (Amt)　　　　NL

Releases from citizenship　　　　1880
54: Rep.116,I: Nr. 13436.　　　　ADS 4:121

PAPENBURG (town)　　　　NL

Emigration permits　　　　1871-1879
　[for permits before 1871, *see* Osnabrueck (Regierung)]
54: Rep.116,I: Nr. 6084.　　　　ADS 4:120

PASSAU (Bezirksamt)　　　　BYL

Emigration & immigration; exportation of property
　　　　　　　　　　　　　　1804-1859
225: Rep.164: Verz.13: Nr. 20-640.　　ADS 9:14

Emigration　　　　[19th century]
225: Rep.168: Verz. 1.　　　　ADS 9:15

PASSAU (Landgericht I)　　　　BYL

Applications for emigration permits
　　　　　　　　1814-1837; 1842-1846
225: Rep. 168: Verz. 1.　　　　ADS 9:16

PASSAU (Landgericht II)　　　　BYL

Applications for emigration permits　1840-1845
225: Rep.168: Verz. 1.　　　　ADS 9:16

PASSAU (town)　　　　BYL

Emigration　　　　[19th century]
225: Rep. 168: Verz. 1.　　　　ADS 9:15

PASSAU (Stadtmagistrat)　　　　BYL

Applications for emigration permits　1817-1844
225: Rep.168: Verz. 1.　　　　ADS 9:16

PELLWORM (rural area). *See* Helgoland　　SHL

PENNSYLVANIA (state)　　　　USA

Data on a number of women, children, & movable
　property taken to the "Island of Pennsylvania"
　by emigrants　　　　16 Mar 1709
200: Abt.613: Stadt Bacharach: Nr. 52, Blatt 43 u.
　44.　　　　ADS 8:45

[Lutheran] Church accounting records
　　　　　　　　1743-1753; 1769-1771
500: Halle: Missionsbibliothek des Waisenhauses:
　Abteilung H IV (Nordamerika) Fach F, Nr. 17;
　Nr. 20.　　　　Not in ADS

Correspondence with [Lutheran] pastors in Pennsyl-
　vania　　　　1745-1765
500: Halle: Missionsbibliothek des Waisenhauses:
　Abteilung H IV (Nordamerika) Fach B: Nr. 1;
　3-7a; 8-10; Fach C: Nr. 11-12.　Not in ADS

Mission correspondence with [Lutheran] ministers
　in Philadelphia　　　　[years below]
500: Halle: Missionsbibliothek des Waisenhauses:
　Abteilung H IV (Nordamerika): Korrespondenz
　mit Pennsylvanien 1762-1768; 1769-1770; 1771-
　1772; 1773-1775; 1774; 1778-; 1783-1787;
　1792-1795.　　　　Not in ADS

Correspondence with [Lutheran pastors]
　　　　1763-1771; 1785; 1787-1790; 1798-1800
Accounts with [Lutheran pastors]　　1796
500: Halle: Missionsbibliothek des Waisenhauses:
　Abteilung H IV (Nordamerika) Unverzeichnet.
　　　　Not in ADS

PETERSTAL. *See* Bad Peterstal

PFAFFENHOFEN (Landratsamt)　　　　BYL

Lists of emigrants & immigrants; exportation of
　personal property　　　　1852-1853
229: Rep. Landratsaemter: 2.10.　　ADS 9:119

PFAFFENWEILER (village)　　　　BWL

Emigration file　　　　1871-1937
288: Kommunalarchiv Pfaffenweiler.　ADS 10:134

PFALZBAYERN (Bavarian Palatinate). *See* Oberbayern

PFALZ-ZWEIBRUECKEN (Duchy)　　　　RPL

Sales made by emigrants to America　　1777
200: Abt.24: Fuerstentum (Herzogtum) Pfalz-Zwei-
　bruecken: Nr. 1569.　　　　ADS 8:10

PFALZ (Palatinate). *See* Kurpfalz, Oberpfalz, Pfalz-
　Zweibruecken, Rheinpfalz

PFALZGRAFENWEILER. *See* Freudenstadt　　BWL

PFARRKIRCHEN (Landgericht)　　　　BYL

Applications for emigration permits　1817-1847
225: Rep.168: Verz. 1.　　　　ADS 9:16

PFARRKIRCHEN (Bezirksamt)   BYL

　Emigration & immigration   1869
　<u>225</u>: Rep.164: Verz.14: Nr. 934a (Nr. 2–87).
　　　　　　　　　　　　　　　　　　　ADS 9:14

　Emigration   [19th century]
　<u>225</u>: Rep.168: Verz. 1.   ADS 9:15

PFERDSBACH (village; Kreis Buedingen)   HEL

　Emigration of entire village   1847
　<u>149</u>: Rep.139/1: Konv.135: Fasz. 6.   ADS 7:10

PFOHREN (village)   BWL

　Emigration file   1877–1941
　Emigration file   1890–1933
　Germans abroad   1928–1931
　<u>288</u>: Kommunalarchiv Pfohren.   ADS 10:134

PFORZHEIM (Amt)   BWL

　Emigration files   1769–1787
　<u>260</u>: Nr. 505.   ADS 10:1

PFULLENDORF (village)   BWL

　Emigration to overseas countries   1818–1858
　Emigration files   1868–1929
　<u>288</u>: Kommunalarchiv Pfullendorf.   ADS 10:134

PHILADELPHIA (Pennsylvania)   USA

　Hamburg Consul in Philadelphia: Estate matters
　　　　　　　　　　　　　　　　　　　　　 1867
　<u>39</u>: Bestand Senat: Cl.VIII, No.X, Jg.1867 (Regis-
　　ter) C.7 [Nordamerika]   ADS 2:193

　Hamburg Consul in Philadelphia: Transmittals of
　　death certificates; legal decisions, etc.   n.d.
　<u>39</u>: Bestand Senat: Cl.VI: Nr. 16p: Vol. 4b: Fasz.
　　12h: Invol. 1–8; 10–12; 13–16.   ADS 2:58

　List of matters dealt with by the Royal [Wuerttem-
　　berg] Consulate in Baltimore (No. 97)
　　　　　　　　　　　　　　　　　　[19th century]
　<u>263</u>: Minist. d. Ae.: 1910 eingekommene Akten: Verz.
　　58, 41.   ADS 10:41

　List of matters dealt with by the Royal [Wuerttem-
　　berg] Consulate in Philadelphia (No. 98)
　　　　　　　　　　　　　　　　　　[19th century]
　<u>263</u>: Minist. d. Ae.: 1910 eingekommene Akten: Verz.
　　58, 42.   ADS 10:41

　List of Hamburg ships in the port of Philadelphia
　　　　　　　　　　　　　　　　　　　　　 1864
　<u>42</u>: Bestand H516: Konsulats-Berichte 1865: Band 2,
　　M–Z (Philadelphia)(1)(Anlagen)   ADS 2:555

PHILADELPHIA (St. Michaelis Evangelical Lutheran
　　Church)   USA

　Short history of the church & congregation   1758
　<u>500</u>: Halle: Missionsbibliothek des Waisenhauses:
　　Abteilung H IV (Nordamerika) Fach A, Nr. 11.
　　　　　　　　　　　　　　　　　　　 Not in ADS

PHILIPPSRUHE (castle?). *See* Hessen-Rumpenheim   HEL

PILKALLEN (Kreis)   East Prussia (Gumbinnen)

　Emigration [file missing?]   1843–1865; 1865–1881
　<u>51</u>: Rep.12: Abt.I, Titel 3: Nr. 8, Bd 8; Bd. 9.
　　　　　　　　　　　　　　　　　　　 ADS 4:32

PIRMASENS (Bezirksamt)   RPL

　Nr. 41: Emigration to America   1835–1915
　Nr. 42: Emigration to other countries (also to
　　California)   1813–1920
　<u>201</u>: Archivabt. Bezirksaemter: Pirmasens: Abgabe
　　1952: files as above.   ADS 8:65

PLANKSTATT   BWL

　Emigration files   1570–1862
　<u>260</u>: Rep.Abt. 229: Spezialakten: Plankstatt: Fasz.
　　83249–83333.   ADS 10:11

PLANTLUENNE (Vogtei)   NL

　Emigration to America   1834–1847
　<u>54</u>: Rep.122,VI.c: Plantluenne: V.A.: Nr. 6.
　　DLC has.   ADS 4:137

PLAUEN (Vogtland). *See* Schmidt, Karl

PLITTERSDORF (village)   BWL

　General emigration file   1840–1954
　Emigration files   1908–1937
　Emigration files   1873
　Emigration files   1903–1944
　<u>288</u>: Kommunalarchiv Plittersdorf.   ADS 10:135

PLOEN und AHRENSBOEK (Aemter)   SHL

　Passports, registrations, residency   1775–1825
　<u>2</u>: Abt.108: Nr. 67.   ADS 1:18

PLOEN (Landratsamt)   SHL

　Nr. 176: Emigration & immigration   1869–1928
　Nr. 272: Emigration & immigration   1877–1915
　Nr. 273: Emigration & immigration   1916–1928
　Nr. 180: Releases from citizenship   1867–1920
　Nr. 280: Releases from citizenship, Vol. I
　　　　　　　　　　　　　　　　　　　　 1882–1893

Nr. 281: Releases from citizenship, Vol. II
1893-1904
Nr. 282: Releases from citizenship, Vol. III
1904-1917
Nr. 283: Releases from citizenship, Vol. IV
1917-1923
Nr. 189: Expulsions; persons dying abroad, Vol. I
1869-1914
Nr. 190: Expulsions; persons dying abroad, Vol. II
1914-1927
Nr. 332: Files on persons dying abroad   1869-1915

2: Abt.320 (Ploen): files as above.   ADS 1:27

POLLE (Amt)   NL

Releases from citizenship   1846-1867

52: Han.Des.80: Polle: B.u.VI.4: Nr. 216. ADS 4:84

PRAUNHEIM (Amt)   HEL

For exemptions from emigration tax, see Hesse-
Kassel [Kurhesse]

PRECHTAL (village)   BWL

Emigration files, 6 vols.   1820-1946
288: Kommunalarchiv Prechtal.   ADS 10:135

PRUEM (Landkreis)   RPL

Emigration to America   1856; 1858-1910
200: Abt.442: Regierung Trier: Nr. 200; 1506-1507;
6087-6902; 8653-8654.   ADS 8:42

Manuscript on emigration from Kreis Pruem, written
by Joseph Mergen   1853
200: Abt.701: Nachlaesse, etc.: Nr. 964. ADS 8:53

PRUSSIA

File on Prussian citizens emigrating to North America via Hamburg   1802-1804
500: Magdeburg: Koenigliche Staatsarchiv: Rep.A.8:
Magdeburg: Kammer Nr. 921.   Not in ADS

Emigration to America   1819-1845
500: Berlin-Dahlem: Geh. Staatsarchiv: Auswaertige
Amt III: Rep.I, Auswanderung-Generalia: Nr. 2,
Vols. III-XI.   Not in ADS

Petitions of various individuals for releases from
citizenship, emigration, etc., Vol. 1 1834-1903
113: Rep.90: Nr. 152.   ADS 5:4

Correspondence between American & German courts in
estate matters   1823-1888
199: Bestand P 135: Preus. Justizministerium: Nr.
8640.   ADS 8:8

Citizenship matters are also to be found in Repositur 77, Nr. 861-1029. At the time of the ADS
inventory, these files were being recatalogued &
were not accessible.

113   ADS 5:4

PYRMONT. See Bad Pyrmont

QUAKENBRUECK (town)   NL

Nr. 1594: Emigration permits   1824-1835
Nr. 1595: Emigration to America   1832-1861
Nr. 1604: Emigration files   1864-1882
Nr. 1605: Applications for emigration permits n.d.

54: Depositum 50b: Stadtarchiv Quakenbrueck: IV:
files as above.   ADS 4:148

Nr. 6085: Emigration permits   1871-1879
Nr. 13434: Releases from citizenship   1880

54: Rep.116,I: files as above. ADS 4:120-121; 148

RABENSCHIEDT. See Dillenburg   HEL

RADERACH (village)   BWL

Emigration file   1841-1911
288: Kommunalarchiv Raderach.   ADS 10:135

RAGNIT (Kreis)   East Prussia (Gumbinnen)

Emigration [file missing?]   1856-1883
51: Rep.12: Abt.I, Titel 3, Nr. 8, Bd. 10.
ADS 4:32

RAIN (Landgericht; Bezirksamt Neuburg a.D.)   BYL

Summary of emigration & immigration   1828-1829
230: Rep. Akten der Bezirksaemter: Neuburg: Nr.
1086.   ADS 9:124

RAMHOLZ (Amt)   HEL

Emigration   1831-1835
151: Bestand 86: Nr. 5575.   ADS 7:86

RAMMERSWEIER (village)   BWL

Emigration file   1845-1938
288: Kommunalarchiv Rammersweier.   ADS 10:135

RAMSBACH (village)   BWL

Emigration file   1829-1942
Emigration to the USA   1829-1866
Emigration & returning emigrants   1911-1939
Genealogical research (emigrants)   1936-1942
288: Kommunalarchiv Ramsbach.   ADS 10:135

RARITAN (congregation; New York)   USA

Call for a pastor for the German congregation at
Raritan, New York   7 Mar 1732
39: Geistl.Minis: II,6: Protokoll lit.A, 1720-1745:
P.74f(4).   ADS 2:468

RARITAN (congregation). See also Wolff, Johann Aug.

RATIBOR   Upper Silesia

Entry refusal for emigrants with less than 400
  Marks each at [the border] at Ratibor & Myslowitz   1887

<u>39</u>: Bestand Senat: Cl.VIII, No.X, Jg.1887 (Regis-
  ter) 7 [Auswanderer-Deputation]   ADS 2:258

RATINGEN (town)   RWL

Passports   1802-1805

<u>115</u>: Rep.C.620: III. Akten 8. Polizei: Nr. 49.
                                          ADS 6:22

RATZEBURG (town)   SHL

Tit. I: Emigration   1803-1872
Tit. II: Emigration permits   1868-1873
Tit. III: Passports   1828-1867
Tit. III, Nr. 2: Collection of information regard-
          ing emigrants, Vol. I   1853

<u>29</u>: I. Teil I: files as above.   ADS 1:60-61

Nr. 5: Passports   1842-1862
Nr. 7-8: Passports deposited   1861-1862
Nr. 9: Passports; certificates of domicile 1870
Nr. 10: Passports; certificates of domicile 1850
Nr. 11: Aid to useless persons for the purpose of
        emigration to America   1853
Nr. 13: Deposited passports   1863-1864
Nr. 15: Deposited passports   1851-1866

<u>30</u>: IV.D.: files as above.   ADS 1:62

Nr. 28: Certificates of domicile; passports 1860-
Nr. 29: Certificates of domicile; passports 1861-
Nr. 30: Certificates of domicile; passports
                                        1860-1870
Nr. 31: Certificates of domicile; passports for
        emigration to America   1873
Nr. 32: Certificates of domicile; passports
                                        1869-1871

<u>30</u>: I. B.: files as above.   ADS 1:62

Applications [by private persons] for diplomatic
  assistance in matters of inheritances, missing
  persons, etc., Vol. I   1870

<u>29</u>: I. Teil II. Tit.Ib: Nr. 5.   ADS 1:61

Expulsions of persons from the German Reich & the
  Prussian state   1880-1903

<u>2</u>: Abt.320 (Ratzeburg): Nr. 426.   ADS 1:28
  DLC may have.

Releases from citizenship   1903-1924

<u>2</u>: Abt.320 (Ratzeburg): Nr. 450.
  DLC may have.   ADS 1:28

RAUENTAL (village)   BWL

Emigration file   1827-1942

<u>288</u>: Kommunalarchiv Rauental.   ADS 10:135

RAVENSBURG (Oberamt)   BWL

Nr. 135: List of emigrants   1833-1877
Nr. 149-156: Emigration files, A-Z   1807-1874
Nr. 157: Emigration, in general, & individual
         cases   1920

<u>261</u>: F 193: Ravensburg: files as above. ADS 10:23

RECHBERG (village)   BWL

General emigration file   1854-1890
Emigration files   1844-1939

<u>288</u>: Kommunalarchiv Rechberg.   ADS 10:135

RECKLINGHAUSEN (Kreis)   RWL

Emigration   1893-1929

<u>116</u>: Muenster-Reg.: Nr. 1696-1698.   ADS 6:57

RECKLINGHAUSEN (Landkreis)   RWL

List of certificates of citizenship issued
                                        1914-1923

<u>116</u>: Muenster-Reg.: Nr. 3404.   ADS 6:56

RECKLINGHAUSEN (city)   RWL

Nr. 3394: List of certificates of citizenship is-
          sued   1902-1909
Nr. 1665: Emigration   1902-1929

<u>116</u>: Muenster-Reg.: files as above.   ADS 6:56-57

REES (Kreis)   RWL

Emigration   1852-1859

<u>115</u>: Rep. D8: I.Abt.e: Nr. 587-589.   ADS 6:25

Nr. 12131: Emigration   1860-1861
Nr. 12133-12135: Emigration   1864-1870
Nr. 12137-12138: Emigration   1873-1879

<u>115</u>: Rep. D8: Neuer Akt.: Abt. 1, C3: files as
  above.   ADS 6:27

REGEN (Bezirksamt)   BYL

Emigration   [19th century]

<u>225</u>: Rep. 168: Verz. 1.   ADS 9:15

REGEN (Landgericht)   BYL

Emigraton & applications for emigration permits
                                        1805-1860

<u>225</u>: Rep. 164: Verz. 15: Fasz. 3, N4. 70. ADS 9:14

Applications for emigration permits   1814-1845

<u>225</u>: Rep. 168: Verz. 1.   ADS 9:16

REGENSBURG (city)   BYL

    Fasz. 2: Emigration   1818-1871
    Fasz. 6: Emigration & immigration   1818-1871
    Fasz. 8: Emigration to overseas countries   n.d.
    Fasz. 13: Renunciation of Bavarian citizenship n.d.

    253: Fach 24: files as above.   ADS 9:173

    Fasz. 5: Exportation of personal property by in-
        heritance & emigration   1811
    Fasz. 6: Miscellaneous emigration files   n.d.

    253: Fach 10: files as above.   ADS 9:173

    Fasz. 10: Marriages of foreigners   1870-1897
    Fasz. 18: Marriages of foreigners   n.d.

    253: Fach 240: files as above.   ADS 9:173

    Taxation of emigration & immigration, 3 vols.
                              1868-1884

    222: Regierung: Kammer des Innern: Rep. 116: Nr.
        11525.   ADS 9:1

REHAU (Bezirksamt)   BYL

    Individual cases of emigration   1811-1899
    223: Rep. K.18/Verz. VI: Nr. 25; 299; 301; 501.
                              ADS 9:8

REHBURG (Amt)   NL

    Releases from citizenship   1834
    52: Han.Des.80: Rehburg: B.v.VI.4: Nr. 157.
                              ADS 4:84

REICHENAU (village)   BWL

    Emigration & immigration   1848
    Emigration files   1770-1935

    288: Kommunalarchiv Reichenau.   ADS 10:135

REICHENBACH (village; Kreis Lahr)   BWL

    Emigration file   1833-1949

    288: Kommunalarchiv Reichenbach, Kr. Lahr.
                              ADS 10:136

REICHENBACH (village; Kreis Wolfach)   BWL

    Emigration file   1834-1941
    Emigration to the USA   1837-1887
    General emigration & returning emigrants 1874-1934

    288: Kommunalarchiv Reichenbach, Kr. Wolfach.
                              ADS 10:136

REICHENBACH. See also Heppenheim (Kreis)   HEL

REICHENTAL (village)   BWL

    Emigration to the USA   1840
    288: Kommunalarchiv Reichental.   ADS 10:136

REISEN   HEL

    Emigration   1859-1890
    149: Rep.40,2: Ablief.2: Bd. 308.   ADS 7:23

REMETSCHWIEL (village)   BWL

    Emigration file   1858-1935
    288: Kommunalarchiv Remtschwiel.   ADS 10:136

RENDSBURG   SHL

    Nr. 2b: Passports   1788-1789
    32: file as above.   ADS 1:64

RETTIGHEIM   BWL

    Emigration files   1540-1820
    260: Rep.Abt.229: Spezialakten: Fasz. 85899-85939.
                              ADS 10:11

REUSS (Aeltere Linie [Older Line]; Principality).
    See Thueringen

REUTE (village)   BWL

    Emigration to the USA   1852-1854
    Emigration files   1865-1938
    Germans abroad   1918-1944

    288: Kommunalarchiv Reute.   ADS 10:136

REUTLINGEN (Oberamt)   BWL

    Nr. 20: List of emigrants   1827-1880
    Nr. 23-31: Emigration files, A-Z   1853-1870
    Nr. 10-18: Citizenship matters, A-Z   19th century
    Nr. 19: [Emigration from?] Eningen   1802-1842
    Nr. 8; 10: List of emigrants   1815-1872
    Nr. 6: Naturalizations & losses of citizenship
                              1834-1891
    Nr. 28-34: Emigration files; list of stateless
        persons   1872-1927

    261: F 194: Reutlingen: files as above.
                             ADS 10:23-24

    Emigration files   n.d.
    262: Wue 65/27: Retulingen: A 83; 93-97 (reper-
        torium S 45/46).   ADS 10:38

REUTLINGEN   BWL

    Card catalog of emigrants from Reutlingen
                           [19th century]
    280   ADS 10:83

REUTLINGEN (Kreis). See also Rottweil   BWL

REUTLINGEN (Kreisregierung). See also Tuebingen.
                                   BWL

REZAT (Kreis General-Kommissariat) BYL

Nr. 1863: Emigration files (tables) 1812-1813
Nr. 1664: Emigration & immigration files; exportation of personal property 1816-1817

231: Rep.137: A.A.-Akten: files as above.
ADS 9:127

RHEINBACH (Kreis) RWL

Emigration 1857-1880

115: Rep.D9: Koeln: Abt.I.e: Nr. 260. ADS 6:32

RHEINBACH (Landratsamt) RWL

Nr. 433: Emigration & immigration 1888-1931
Nr. 12: Emigration & immigration 1845-1878
Nr. 13: Emigration 1856-1880
Nr. 431: Emigration 1880-1894
Nr. 430: Emigration 1895-1921
Nr. 673: Emigration 1930-1932

115: Rep.G29/37: Rheinbach: files as above.
ADS 6:45

RHEINBISCHOFSHEIM (village) BWL

Emigration file 1939; 1945

288: Kommunalarchiv Rheinbischofsheim. ADS 10:136

RHEINGAUKREIS (Landratsamt Ruedesheim) HEL

Foreigners residing in Kreis 1873-1904

152: Abt.415: Ruedesheim: Nr. 18. ADS 7:156

RHEINPFALZ RPL

Nr. 357: Emigration to North America & Brazil;
measures to be taken against emigrants,
particularly against those who encourage
such emigration, Conv. I 1816-1828
Nr. 358: Measures adopted by this, & other, governments against emigration to North & South
America, Conv. II 1829-1835
Nr. 359: Measures adopted by this, & other, governments against emigration to North & South
America, Conv. III 1836-1838

227: Rep. MA 1890: A.St.: files as bove. ADS 9:54

Nr. 368: Secret [illicit] emigration; herein proof
that special travel documents being pro-
vided to Bavarian citizens traveling via
French ports to America 1851-1861
Nr. 371: Secret [illicit] emigration to America
1852-1859
Nr. 373: Bavarian emigrants via port of Antwerpen
1853-1856
Nr. 375: Emigration of retired soldiers to America
without permission 1853-1855
Nr. 382: Emigration to the USA 1855-1856
Nr. 383: Emigration via Liverpool & other English
ports 1855-1860

227: Rep. MA 1890: A.St.: files as above.
ADS 9:54-55

Nr. 3301: Impoverished condition of South German
emigrants in LeHavre 1831
Nr. 3305: Embarkation of Bavarian emigrants without passports at the port of LeHavre
1836

227: Rep. Bayer. & Pfaelz. Gesandtschaft Paris:
files as above. ADS 9:49

Inquiries [requests for information on individuals]
1818-1862

227: Rep. MA 1898: A.St.: Nr. 3471-3763. ADS 9:58

Unbound fascicle of passport applications & visas
issued to Bavarian citizens 1808-1849

227: [Rep. Kasten Schwarz] MA 10000-10014.
ADS 9:41

For certain emigration & manumission taxes, *see*
Oberbayern

Extracts from emigration files, mainly from Abt.
403 (Oberpraesidium Koblenz), Abt. 463 (Landratsamt Koblenz), Abt. 500 (Landratsamt Zell), Abt.
560, Abt. 254 (Catholic parish of Rehbach),
Abt. 555,27 (Reformed parish Gemuenden). Names
of the emigrants have been entered in the Emigrant File. There is a plan to publish Rhenish
emigration statistics, probably by Joseph
Scheben, Rheinbach [18th & 19th centuries]

RHEINPFALZ. *See also* Bavaria, Kurpfalz, Oberbayern,
Pfalz, Pfalzbayern, Pfalz-Zweibruecken

RHEINPROVINZ RPL

Nr. 6772: Reception of would-be emigrants to America from Holland in the Koblenz prison
1817
Nr. 24647: Emigration & immigration (continuation
of Nr. 24656, not listed in ADS)
1888-1895
Nr. 28058-28067: Preparation of certificates of
release [from citizenship], alphabetical 1914-1923
[Nr. not given]: Issuance of certificates of original domicile [several hundred
files arranged by Kreis until
1914 & thereafter alphabetical to
1937; in addition, there are almost 100 files on certificates of
citizenship 1923-1944 arranged
alphabetically]
Nr. 5081-5082: Renaturalization of former emigrants
1831-1865
Nr. 28074-28093: Naturalizations, alphabetically
arranged 1914-1934
Nr. 28094-28133: Naturalizations, alphabetically
arranged 1914-1932
Nr. 35422: Naturalizations 1942
Nr. 28134-28227: Naturalizations, by year
1918-1938
Nr. 27996-28000; 28024: Expulsion of foreigners
[lists in Nr. 27997] 1924-1933
Nr. 5269; 5286; 5289: Governmental intercession by
foreign governments (some
estate matters in USA)
1850; 1853-1869

200: Abt.441: Regierung Koblenz: files as above.
ADS 8:27; 31-32

Nr. 7180-7183: Emigration, 4 vols.   1826-1913
Nr. 7184-7190: Administration of emigration by the government (therein: emigrants subject to military duty)   1867-1868
Nr. 7191-7198: Emigration (vols. 1 & 2 destroyed)   1846-1912

200: Abt.403: Oberpraesidium Koblenz (Rheinprovince): files as above.   ADS 8:17

Nr. 6787: Issuance of certificates of original domicile   1845-1916
Nr. 3813: Matters of expulsion   1825-1838
Nr. 3819: Petitions from deserters for renaturalization in Prussia   1833-1834
Nr. 13409-13412: Consular affairs (detailed)   1821-1826
Nr. 8774: Marriages between foreigners & Prussian subjects   1833-1892
Nr. 12155: Death certificates requested abroad   1839-1858
Nr. 3926: Issuance of death certificates 1826-1858
Nr. 2173: Investigation of citizens living abroad   1840

200: Abt.403: Oberpraesidium Koblenz (Rheinprovinz): files as above.   ADS 8:17-19

"Quellen zur rheinischen Auswandererforschung in den Staatsarchiven Koblenz und Duesseldorf - Ein Nachtrag" [Sources for Rhenish Emigration Research in the State Archives at Koblenz and Duesseldorf: An Addendum" [typescript, 21 pages]   n.d.

200: Abt.701: Nachlaesse, etc.: Nr. 969.   ADS 8:53

RHEINPROVINZ (Prussian)   RWL

Emigration to America   1843

115: B. Provinzialstaendische Verf.: e: Nr. 376.   ADS 6:35

RHEINWEILER (village)   BWL

Emigration file   1888-1934
Emigration file   1890-1913

288: Kommunalarchiv Rheinweiler.   ADS 10:136

RHEYD.   See Gladbach-Rheyd   RWL

RHYNBECK (congregation, New York state)   USA

German congregations at Newton, Rhynbeck, Auf den Camp, etc., mentioned in letter   6 Oct 1730

39: Geistl.Minis.: II,6: Protokoll lit. A, 1720-1745, p.54(3) & p.55.   ADS 2:468

RICKENBACH (village)   BWL

Emigration file   1862-1950
Foreigners   1945-1949

288: Kommunalarchiv Rickenbach.   ADS 10:136

RICKLINGEN (Amt)   NL

Releases from citizenship   1854-1858

52: Han.Des.80: Ricklingen: B.w.VI.4: Nr. 130.   ADS 4:85

RIEDBOEHRINGEN (village)   BWL

Emigration file   1851-1950
Emigration to the USA   1851-1861

288: Kommunalarchiv Riedboehringen.   ADS 10:136

RIEDERN a. S. (village)   BWL

Emigration file   1882-1939

288: Kommunalarchiv Riedern a. S.   ADS 10:136

RIEDHEIM (village)   BWL

Emigration file   1861-1943

288: Kommunalarchiv Riedheim.   ADS 10:137

RIEDLINGEN (village)   BWL

Emigration file   1840-1932

288: Kommunalarchiv Riedlingen.   ADS 10:137

RIEGEL (village)   BWL

Emigration file, alphabetically arranged 1870-1910
List of emigrants   1880-1914

288: Kommunalarchiv Riegel.   ADS 10:137

RIELASINGEN and ARLEN (village)   BWL

Emigration to the USA   1854-1935
Emigration files   1890-1925

288: Kommunalarchiv Rielasingen und Arlen.   ADS 10:137

RIETBERG (parish)   RWL

Village chronicle for the parishes of Rietberg & Masholte contains list of emigrants   1841
137   ADS 6:122

RIMBACH   HEL

Emigration   1859-1913

149: Rep.40,2: Ablief.2: Bd. 309.   ADS 7:23

RINTELN (Regierungsbezirk)   HEL

Naturalizations, 4 vols.   1849-1863

151: Bestand 16.II: Klasse 15: Nr. 17.   ADS 7:66

RIPPOLDSAU. *See* Bad Rippoldsau

RITZEBUETTEL   HH

File on the payment of the tithe [Zehnten-Pfennig]
on inheritances & when leaving the country 1584-

39: Bestand Amt Ritzebuettel I: I, Fach 5, vol. A.
ADS 2:460

Vol. D: Files on citizens who requested releases
from citizenship [as a prerequisite to
emigration](Nos. 1-304)   1747-1890
Vol. E: Claims, orders, & negotiations regarding
persons releases from citizenship   n.d.
Vol. F: Claims of the Ritzebuettel office regarding citizens who have left the area without proper releases   n.d.
Vol. G: Emigration of native-borne citizens without obtaining appropriate releases, & matters of family members left behind   n.d.
Vol. H: Special & doubtful cases, certificates of
domicile of persons born or domiciled here
n.d.

39: Ritzebuettel I: II: Fach 13: files as above.
ADS 2:461-462

Passports   1732-1855

39: Ritzebuettel I: IV, Fach 2, vol. N. ADS 2:463

Transports of criminal prisoners to America   n.d.

39: Ritzebuettel I: XII, Fach 8, vol. E. ADS 2:464

Expulsion of persons from state of Hamburg
1907-1912

39: Ritzebuettel II: Abt.VIII, A, Nr. 93.
ADS 2:464

The Stadtarchiv Cuxhaven is preparing an emigrant
file [Auswanderer-Kartei] as part of a catalog
of Ritzebuettel residents [information as of
October 1957]   ADS 4:249

RODHEIM (Amt)   HEL

Emigration [reverse pages contain names]   1751

151: Bestand 81: Rep.Ba: Gef. 9: Nr. 8.   ADS 7:80

ROGGENBEUREN (village)   BWL

Emigration to the USA; Germans abroad   1835-1937
Emigration files   1841-1942

288: Kommunalarchiv Roggenbeuren.   ADS 10:137

ROHRBACH BEI LANDAU (village)   RPL

Names of emigrants   18th century

201: Archivabt. Ausfautheiakten.   ADS 8:62

ROSSBACH (Amt)   HEL

For exemptions from emigration tax, *see* Hesse-
Kasse [Kurhesse]

ROTENBURG (Kreis)   HEL

Naturalizations, 4 vols.   1822-1854
151: Bestand 17(b): Gef. 93: Nr. 2.   ADS 7:69

ROTENBURG (Landratsamt)   HEL

Releases from citizenship   1847-1865

151: Bestand 180: Rotenburg: Acc.1887/16: Nr. 50.
ADS 7:121

Nr. 95: Emigration & immigration   1852-1882
Nr. 96: Emigration contracts approved   1857
Nr. 100: Various emigration matters   1922-1923
Nr. 846: Persons under police observation
1829-1907
Nr. 102: Passport register   1849-1896
Nr. 104: Passport register   1897-1921
Nr. 105: Passport register   1897
Nr. 107: Passport register   1920-1923
Nr. 46-61; 63-69: Naturalizations & releases from
citizenship   1856-1920
Nr. 62: Renaturalizations of former German citi-
zens   1884

151: Bestand 180: Rotenburg: Acc.1935/52: files as
above.   ADS 7:121

ROTENBURG (Amt)   NL

Nr. 1: Files on inheritances   1731-1759
Nr. 2: Files on inheritances   1760-1811
Nr. 3-23: A series of files on emigration applications, etc., are listed but are shown as
missing in the Staatsarchiv Hannover
[possibly still in Rotenburg?] 1828-1876

52: Han.Des.74: Rotenburg: I.A.1.i: Fach 21: files
as above.   ADS 4:75-76

Nr. 2: File on citizens dying abroad   1826-
Nr. 3: Files on citizens & foreigners & their
estates   1827-
Nr. 4: Searches for missing persons for purpose of
declaring them dead [estate matters]   1896

52: Han.Des.74: Rotenburg: IV.L.3: files as above.
ADS 4:78

Files on the resettlement of recidivists & dangerous persons in America   1835-1855

52: Han.Des.74: Rotenburg: IV.H.6: Fach 553: Nr. 2.
ADS 4:78

Nr. 26: Emigration lists   1865-
Nr. 29: Emigration certificates   1872-1875
Nr. 30: Releases from citizenship for persons born
in 1876-1880   1891-
Nr. 31: Releases from citizenship for persons born
in 1881-1885   1892-
Nr. 33: Releases from citizenship for persons born
in 1886-1890   1890-
Nr. 34: Releases from citizenship for persons born
in 1891-1895   1907-

52: Han.Des.74: Rotenburg: I.A.1.i: Fach 22: files
as above.   ADS 4:77

Files on emigration & military papers for persons
living abroad   1884

52: Han.Des.74: Rotenburg: IV.H.4.a: Fach 542: Nr.
13.   ADS 4:77

ROTH (village)   BYL

    List of emigrants   1900-1920
    Names of six emigrating Jews   1933-1934
    <u>254</u>   ADS 9:176

ROTHENBURG o.d. TAUBER (Bezirksamt)   BYL

    Nr. 992-993: Emigration & immigration from the
        newly-organized communities 1810-1868
    Nr. 996-1035: Individual cases of emigration to
        the USA   1854-1855
    <u>231</u>: Rep.212/15: Rothenburg o.d. T.: files as
        above.   ADS 9:129

    Emigration & immigration, Vol. V   1846
    <u>231</u>: Rep.270/II: Reg. Mittelfranken: Kammer des
        Innern: Abgabe 1932: Nr. 30.   ADS 9:134

    Applications for emigration permits   19th century
    <u>255</u>   ADS 9:177

ROTTENBURG (Landgericht)   BYL

    Applications for emigration permits   1842-1846
    <u>225</u>: Rep.168: Verz. 1.   ADS 9:16

ROTTENBURG (Bezirksamt)   BYL

    Nr. 36-54: Emigration to America   1846-1849
    Nr. 401-: Emigration to America   1849-1908
    <u>225</u>: Rep. 164: Verz. 16: files as above. ADS 9:14

    Emigration   [19th century]
    <u>225</u>: Rep.168: Verz. 1.   ADS 9:15

ROTTENBURG (Oberamt)   BWL

    Nr. 199: List of emigrants   1818-1849
    Nr. 202-220: Emigration files, chronologically ar-
        ranged   1812-1874
    Nr. 229: Emigration with retention of [Wuerttem-
        berg] citizenship   1819-1847
    Nr. 230: Emigration with retention of [Wuerttem-
        berg] citizenship   1848-1859
    Nr. 200: Certificates of original domicile for
        persons residing in Switzerland 1835-1840
    Nr. 231: Immigration lists   1860-1874
    Nr. 232-234: Naturalizations   1841-1874
    Nr. 235: Stateless persons   1819-1858
    Nr. 221-228: Emigration files   1871-1900
    <u>261</u>: F 196: Rottenburg: files as above. ADS 10:24

ROTTHALMUENSTER (Landgericht)   BYL

    Applications for emigration permits   1840-1846
    <u>225</u>: Rep.168: Verz. 1.   ADS 9:16

ROTTWEIL (Oberamt)   BWL

    Nr. 41: Naturalizations   1850-1858
    Nr. 44-49: Emigration files   1803-1885

    Nr. 50-52: Emigration files for Kreis Reutlingen
                                          1847-1890
    Nr. 53-54: Lists of emigrants   1821-1852
    Nr. 48: Naturalizations   1851-1858
    Nr. 55: Passports & emigration visas   1860-1874
    <u>261</u>: F 197: Rottweil: files as above. ADS 10:24-25

    Nr. 1040: Loss of [Wuerttemberg] citizenship
                                          1805-1870
    Nr. 6135: Emigration   1805-1823
    <u>262</u>: Wue 65/30: Rottweil: Acc.11/1950: AZ: files
        as above.   ADS 10:39

ROTTWEIL (town)   BWL

    List of emigrants   1802-1914
    <u>281</u>   ADS 10:85

ROWAN (North Carolina)   USA

    Gerh[ard] Rintelmann [makes gift?] to the church
        at Rowan, North Carolina   1772
    <u>39</u>: Bestand Senat: Cl.VIII, No.X, Jg.1772 (Regis-
        ter)[Gratificationes]   ADS 2:98

RUDOLSTADT (Principality). *See* Thueringen. Schwarzburg-Rudolstadt

RUEBENBERGE (region). *See* Neustadt am Ruebenberge
                                            NL

RUEDESHEIM (Amt)   HEL

    Emigration to America, Australia, Hungary, Brazil,
        & Russia   1845-1867
    <u>152</u>: Abt.238: Ruedesheim: Nr. 97.   ADS 7:135

    Releases from citizenship & emigration:
        Fasz. I: 1877-1882
        Fasz. II: 1883-1885
    <u>152</u>: Abt.238: Ruedesheim: Nr. 523. ADS 7:135

RUEDESHEIM (Landratsamt)   HEL

    Nr. 33: Emigration   1867-1879
    Nr. 106: Investigation of emigrants & immigrants
                                          1868-1921
    Nr. 31: Emigration of men of service age 1869-1877
    Nr. 167: Emigration   1876-1893
    Nr. 71: Illicit emigration, etc.   1878-1898
    <u>152</u>: Abt.415: Ruedesheim: files as above.
                                          ADS 7:156

RUEDESHEIM. *See also* Rheingaukreis   HEL

RUEMMINGEN (village)   BWL

    Emigration files (2 fascicles)   1871-1936
    <u>288</u>: Kommunalarchiv Ruemmingen.   ADS 10:137

RUESSELSHEIM (Landkreis Gross-Gerau)   HEL

   Applications for emigration permits   1832-1895
   List of emigrants (18 persons)   1854-1861
   Register of departures for 1867 contains four
     cases of emigration to America   1867
   Register of departures for 1873 contains one case
     of emigration to America   1873

   <u>182</u>: XI,4.   ADS 7:235

RUESSINGEN (village; Landkreis Kirchheimbolanden,
Kurpfalz)   RPL

   Names of emigrants   18th century

   <u>201</u>: Archivabt. Ausfautheiakten.   ADS 8:63

RUESSWIHL (village)   BWL

   Emigration files (4 fascicles)   1837-1934
   Germans abroad   1890-1947

   <u>288</u>: Kommunalarchiv Ruesswihl.   ADS 10:137

RUHRORT (Kreis)   RWL

   Emigration   1887-1897

   <u>115</u>: Rep.D8: Neuere Akt.: Abt.1,C3: Nr. 12148.
                            ADS 6:27

RUHRORT (city)   RWL

   Emigration   1816-1907

   <u>115</u>: Rep.G24/4: Nr. 99-102.   ADS 6:31

RUMBACH (village; Kreis Dahn, Kurpfalz)   RPL

   Names of emigrants   18th century

   <u>201</u>: Archivabt. Ausfautheiakten.   ADS 8:63

RUMPENHEIM (Amt)   HEL

   For exemptions from emigration tax, *see* Hesse-
   Kassel [Kurhesse]

RUMPENHEIM.   *See* Hessen-Rumpenheim   HEL

RUNKEL.   *See* Wied-Runkel (Grafschaft)

RUSCHWEILER (village)   BWL

   Emigration file; Germans abroad   1849-1874
   Emigration files   1887-1943

   <u>288</u>: Kommunalarchiv Ruschweiler.   ADS 10:137

RUST (village)   BWL

   Emigration files (3 fascicles)   1852-1949
   Germans abroad   1925-1938

   <u>288</u>: Kommunalarchiv Rust.   ADS 10:137

RUTHE.   *See* Fraenckel, Simon [surname index]   NL

SAAR (Departement)

   Planned emigration of numerous inhabitants of the
   Saar Departement, their detention and return to
   villages of origin, Year XIII of the [French?]
   Republic. Contained therein are letters regarding 199 emigrants to America from Saarbruecken
   region who were detained at Amsterdam, returned
   via Koeln [Cologne] to Saarbruecken   n.d.

   <u>200</u>: Abt.256: Rhein- und Moseldepartement, Prae-
     fektur Koblenz: Nr. 68.   ADS 8:13

SAARBRUECKEN (Landkreis)

   Emigration [to all parts of world]   1855-1924

   <u>200</u>: Abt.442: Regierung Trier: Nr. 208-209; 6098-
     6099; 1602; 6125; 6129.   ADS 8:36-37

   Emigration to America   1863-1910

   <u>200</u>: Abt.442: Regierung Trier: Nr. 6099-6100; 1478;
     6121; 4442.   ADS 8:37

   Naturalizations   1857-1868; 1891-1923

   <u>200</u>: Abt.442: Regierung Trier: Nr. 1501-1504;
     8618-8644.   ADS 8:40

SAARBRUECKEN (city)

   Nr. 202: Emigration   1910-1924
   Nr. 8551-8556: Naturalizations   1911-1924

   <u>200</u>: Abt.442: Regierung Trier: files as above.
                             ADS 8:40-42

SAARBRUECKEN.   *See also* Nassau-Saarbruecken (principality)

SAARBURG (Amt)

   Emigration   1654-1726

   <u>200</u>: Abt.1C: Kurtrier: Nr. 4878.   ADS 8:9

SAARBURG (Landkreis)

   Emigration to America   1856-1892

   <u>200</u>: Abt.442: Regierung Trier: Nr. 179-184; 8652;
     6131-6133.   ADS 8:42

SAARBURG (Kreis)

   Manuscript on emigration from Kreis Saarburg, 2
    parts, written by Joseph Mergen   1952

   <u>200</u>: Abt.701: Nachlaesse, etc.: Nr. 964. ADS 8:53

SAARLAUTERN (Landkreis)

   Naturalizations   1859-1923

   <u>200</u>: Abt.442: Regierung Trier: Nr. 8575-8582.
                             ADS 8:40

SAARLOUIS (Landkreis)

Emigration [to all parts of the world]   1860-1924

<u>200</u>: Abt.442: Regierung Trier: Nr. 206-207; 8583-
      8587.   ADS 8:37

Emigration to America   1888-1908

<u>200</u>: Regierung Trier: Nr. 6093-6094; 6095-6097;
      6130; 6134.   ADS 8:37

ST. LOUIS (Missouri)   USA

Brunswick Consulate: registers for legalization of
    documents   1855-1863; 1863-1868

<u>56</u>: L Neu Abt.19F: St. Louis: Nr. 1; 4. ADS 4:183

List of matters dealt with by the Royal [Wuerttem-
    berg] Consulate in St. Louis   1858-

<u>263</u>: Minist. d. Ae.: 1910 eingekommene Akten: Verz.
      58,33.   ADS 10:41

German Consulate in St. Louis: Court decisions,
    estates, etc.   1875-1892

<u>39</u>: Bestand Senat: Cl.VI, No.16p: Vol. 4b, Fasc.
      23, Invol. 1-12.   ADS 2:60

SALEM (town)   BWL

Emigration to the USA   1840-1945
Immigration files   1900-1944
Engagement of minor girls for American cloisters
                                           1892-1942

<u>288</u>: Kommunalarchiv Salem.   ADS 10:138

SALZUFLEN. *See* Bad Salzuflen

SAND (village)   BWL

General emigration file   1897-1939

<u>288</u>: Kommunalarchiv Sand.   ADS 10:138

SANDWEIER (village)   BWL

Emigration files   1901-1932
General emigration matters   1871-1941

<u>288</u>: Kommunalarchiv Sandweier.   ADS 10:138

SAN FRANCISCO (California)   USA

Guardianship [persons unnamed in ADS description]
                                                1854

<u>39</u>: Bestand Senat: Cl.VIII, No.X, Jg.1854 (Regis-
      ter)   ADS 2:166

Hamburg Consulate in San Francisco: Estates, etc.
                                           1854-1857

<u>39</u>: Bestand Senat: Cl.VI, No.16p: Vol.4b: Fasc.
      10c.   ADS 2:55

Transmittal of death certificates, etc. 1858-1861

<u>39</u>: Bestand Senat: Cl.VI, No.16p: Vol.4b: Fasc.
      10e [Uebersendung]   ADS 2:55

German consular report from San Francisco regard-
    ing estate matter [names not given in ADS de-
    scription]   1870

<u>39</u>: Bestand Senat: Cl.VIII, No.X, Jg.1870 (Regis-
      ter) 27 [Nordamerika]   ADS 2:210

German consular report from San Francisco regard-
    ing estate matter [names not given in ADS de-
    scription]   1871

<u>39</u>: Bestand Senat: Cl.VIII, No.X, Jg.1871 (Regis-
      ter) 15 [Nordamerika]   ADS 2:215

Hamburg Consulate in San Francisco: Searches for
    missing persons; legal matters   n.d.

<u>39</u>: Bestand Senat: Cl.VI, No.16p: Vol.4b: Fasc.
      10k, Invol. 2-6; 10; 12-25.   ADS 2:56

List of north German vessels to San Francisco [29
    entries]   1869

<u>42</u>: Bestand H516: Konsulats-Berichte 1870: (San
      Francisco)(1).   ADS 2:570

SANKT GEORGEN (village)   BWL

Emigration file   1820-1945
Germans abroad; list of emigrants   1920-1939

<u>288</u>: Kommunalarchiv St. Georgen.   ADS 10:141

SANKT GOAR (Kreis)   RPL

Naturalizations   1847-1910

<u>200</u>: Abt.441: Regierung Koblenz: Nr. 22042-22045.
                                             ADS 8:30

SANKT GOARSHAUSEN (Amt)   HEL

Naturalizations & releases from citizenship   1867

<u>152</u>: Abt.240: St. Goarshausen: Nr. 697. ADS 7:135

SANKT ULRICH (village)   BWL

Emigration file   1837-1921

<u>288</u>: Kommunalarchiv St. Ulrich.   ADS 10:141

SANKT WENDEL (Landkreis)   RPL

Nr. 9371; 8648-8649: Emigration to America
                                           1816-1889
Nr. 188-190; 6061-6070: Emigration to America
                                           1856-1906
Nr. 210; 6060; 9161-9162; 9306: Emigration [to all
        parts of the world]   1858-1923
Nr. 9307-9308; 9163-9164: Naturalizations
                                           1896-1924

<u>200</u>: Abt.442: Regierung Trier: files as above.
                                      ADS 8:37-38; 40; 42

SANKT WILHELM (village)   BWL

  Emigration file   1841-1877

  <u>288</u>: Kommunalarchiv St. Wilhelm.   ADS 10:142

SASBACH (village)   BWL

  Emigration file (3 fascicles)   1827-1944
  Illicit emigration   1851-1863

  <u>288</u>: Kommunalarchiv Sasbach.   ADS 10:138

SASBACH (village; Kreis Emmendingen)   BWL

  Emigration files   1859-1935

  <u>288</u>: Kommunlarchiv Sasbach, Kr. Emmendingen.
                                               ADS 10:138

SASBACHWALDEN (village)   BWL

  Emigration file (2 fascicles)   1817-1931
  Illicit emigration   1850-1858
  Emigration to the USA   1861-1897
  General files   1903-1944

  <u>288</u>: Kommunalarchiv Sasbachwalden.   ADS 10:138

SAULDORF (village)   BWL

  Emigration to the USA   1868-1902

  <u>288</u>: Kommunalarchiv Sauldorf.   ADS 10:138

SAULGAU (Oberamt)   BWL

  Nr. 97: List of emigrants; emigration files
                                               1816-1871
  Nr. 117-132: Emigration files, alphabetically ar-
       ranged by village   1806-1839
  Nr. 98: [Emigration] with retention of [Wuerttem-
       berg] citizenship   1828-1857
  Nr. 104-112: Emigration & immigration   1850-1859
  Nr. 96: Lists of naturalizations; emigration files
       alphabetically arranged (therein a volume
       entitled "Amerikaauswanderungen [Emigra-
       tion to America] 1872-1913" 1840-1915

  <u>261</u>: F 198: Saulgau: files as above.   ADS 10:25

SAVANNAH (Georgia)   USA

  Hamburg Consulate in Savannah: Court orders, etc.
                                               1871-1893

  <u>39</u>: Bestand Senat: Cl.VI, No.16p: Vol. 4b: Fasc.
    13e: Invol. 1-5.   ADS 2:58

SAXONY

  Nr. 1186a-d: Emigration of Saxon citizens
                                               1832-1844
  Nr. 1204: Emigration & colonization plans
                                               1841-1854
  Nr. 288: File on emigration of Zweig-Verein Kamenz
                                               1849

  <u>500</u>: Dresden: Saechsisches Hauptstaatsarchiv: Min.
    d. Innern: Files as above [films] Not in ADS

  Emigration affairs   1788-1831

  <u>500</u>: Dresden: Saechsisches Hauptstaatsarchiv: Loc.
    31519: Vols. XIII, XIV, XV; Loc. 31520: 16-24
    [films]   Not in ADS

  Emigration of Saxon citizens   1829-1835

  <u>500</u>: Dresden: Saechsisches Hauptstaatsarchiv: Loc.
    31613; 31614 [film]   Not in ADS

  File on overseas emigration   1848-

  <u>500</u>: Dresden: Saechsisches Hauptstaatsarchiv: Loc.
    31715 [film]   Not in ADS

  Miscellaneous Saxon emigration matters   n.d.

  <u>500</u>: Dresden: Saechsisches Hauptstaatsarchiv:
    Hausarchiv Johann: Nr. 20 A-B [film]
                                       Not in ADS

  Estate [& inheritance] matters of Juedefeinde
    [Jewish enemies]   1807

  <u>500</u>: Dresden: Saechsisches Hauptstaatsarchiv:
    Amtsgericht Freital-Doehlen: Nr. 1141 [film]
                                       Not in ADS

SAXONY-COBURG-GOTHA (Principality)

  Nr. 134/1: Public assistance for emigration to
      America   1834-1852
  Nr. 134/2: Applications for financial aid to emi-
      grate to America   1852-1856
  Nr. 134/3: Approvals [of applications for] public
      aid to emigrate to America   1841-1873

  <u>500</u>: Coburg: Bayer. Staatsarchivalien: Abt.: Loc.
    D, Tit.IV, 2: files as above. *Also includes*
    Rep.C., Verz.III.4f: Nr. 39 [films]
                                       Not in ADS

SAXONY-WEIMAR-EISENACH   Thueringen

  Public aid to a number of emigrants to America
                                               1851-1853

  <u>500</u>: Weimar: Thueringisches Staatsarchiv: XIX.C.c.
    59, 60. [film]   Not in ADS

SAYN-ALTENKIRCHEN (und HACHENBURG)(Reichsgrafschaft)
                                          RPL

  Files on emigration tax   1742-1802

  <u>200</u>: Abt.30: Reichsgrafschaft Sayn-Altenkirchen
    (u. Hachenburg): Nr. 4-7.   ADS 8:10

SCHABENHAUSEN (village)   BWL

  Emigration files   1843-1941

  <u>288</u>: Kommunalarchiv Schabenhausen.   ADS 10:138

SCHADECK. *See* Westerburg und Schadeck (Herrschaft)

SCHALLBACH (village)   BWL

  Emigration file   1873-1941

  <u>288</u>: Kommunalarchiv Schallbach.   ADS 10:138

SCHAPBACH (village)   BWL

   Emigration files (4 fascicles)   1853-1946

   288: Kommunalarchiv Schapbach.   ADS 10:139

SCHAUMBURG   HEL

   Nr. 9: Naturalizations in Province of Niederhessen
      with Schaumburg, 8 vols.   1831-1849
   Nr. 13: Releases from citizenship, 3 vols.
                                       1827-1847
   Nr. 22: Releases from citizenship   1822

   151: Bestand 16,II: Klasse 15: files as above.
                                       ADS 7:66

SCHAUMBURG (Kreis)   HEL

   Nr. 21: Temporary residencies for foreigners [Vol.
      I]   1834-1838
   Nr. 31: Temporary residencies for foreigners [Vol.
      II]   1839-1843
   Nr. 40: Temporary residencies for foreigners [Vol.
      III]   1844-1847
   [also Nr. 40] Temporary residencies for foreigners
      Vol. IV   1848-

   151: Bestand 17(g): Gef. 41: files as above.
                                       ADS 7:73

SCHAUMBURG   NL

   Nr. 3: Emigration to America   1833-1847
   Nr. 4: Emigration to America   n.d.
   Nr. 5: Applications for releases from citizenship
                                       1846-1849
   Nr. 7: Applications for releases from citizenship,
      2 vols.   1849-1862
   Nr. 8: Releases from citizenship   1849-1862
   Nr. 1: Immigration from Kreis Schaumburg, Vols. 17,
      18, 19   1840-1849

   52: Schaumburg Des.H.2: Rinteln: Neue Reg.: II,
      Klasse 1: files as above.   ADS 4:98

SCHAUMBURG (Grafschaft)   NL

   Emigrants to America   1853-1866

   103   ADS 4:330

SCHEINFELD (Emigration collection point [Lager]) BYL

   Emigration & immigration   1853-1861

   231: Rep.270/II: Reg. Mittelfranken: Kammer des
      Innern: Abgabe 1932: Nr. 646.   ADS 9:134

SCHENKENZELL (village)   BWL

   Emigration file   1872-1921
   Emigration in the last fifty years, prepared dur-
      ing period   1929-1944

   288: Kommunalarchiv Schenkenzell.   ADS 10:139

SCHERZHEIM (village)   BWL

   Emigration file   1850-1940

   288: Kommunalarchiv Scherzheim.   ADS 10:139

SCHERZINGEN (village)   BWL

   Emigration file   1857-1905

   288: Kommunalarchiv Scherzingen.   ADS 10:139

SCHIFFERSTADT (village; Bezirk Speyer)   RPL

   Nr. 8: Emigration   1833-1850
   Nr. 9: Emigration   1851-1854

   201: Archivabt. Bezirksaemter: Speyer (Abgabe
      1907): files as above.   ADS 8:65

SCHILDESCHE-JOELLENBECK (Kreis Bielefeld)   RWL

   Nr. 7: Emigration   1825-1893
   Nr. 25: Passports   1817-1886

   116: Arnsberg Reg.: Bielefeld: files as above.
                                       ADS 6:63

SCHILLINGSFUERST (Oberamt)   BYL

   Emigration & immigration, Vol. II   1825

   231: Rep.270/II: Reg. Mittelfranken: Kammer des
      Innern: Abgabe 1932: Nr. 26.   ADS 9:134

SCHLANGENBAD (baths)   HEL

   Treatment & guest lists   1867-1896

   152: Abt.405: Wiesbaden: Rep.405/I,4: Nr. 1217.
                                       ADS 7:148

SCHLATT (village)   BWL

   Emigration file   1875-1935
   Immigration & returning emigrants   1908-1935

   288: Kommunalarchiv Schlatt.   ADS 10:139

SCHLATT UNTER KRAEHEN (village)   BWL

   Emigration file   1852-1932

   288: Kommunalarchiv Schlatt unter Kraehen.
                                       ADS 10:139

SCHLESWIG and HOLSTEIN (Duchies)   SHL

   Files on the emigration of persons liable for
      military service   1868

   2: Abt.59^{3}: Nr. 921 [Militaer-Pflichtiger]
                                       ADS 1:12

SCHLESWIG-HOLSTEIN   SHL

  Emigration assistance; death declarations for
    missing persons   1852-1871; 1903; 1909
  18: Nr. 458.   ADS 1:52

  Releases from Prussian citizenship, Vols. 1, 3-6
                                             1894-1926
  2: Abt.301: Nr. 2960 [Entlassungen]   ADS 1:22

SCHLESWIG (Prussian Government of)   SHL

  Nr. 16928: Gendarmerie files on illicit emigration
             to America of persons subject to mili-
             tary service   1867
  Nr. 16892-16896: Applications for emigration 1868
  Nr. 16890: Files on the residences of American
             citizens in Schleswig-Holstein who emi-
             grated from the Province [many files]
                                             1879-1902
  2: Abt.309: files as above.   ADS 1:23

SCHLESWIG (Landratsamt)   SHL

  Nr. 64: Files on emigration [main files] 1872-1928
  Nr. 65: Files on emigration   1920-1927
  Nr. 66: Files on emigration   1927-1929
  Nr. 67: Files on emigration   1930-1938
  Nr. 70: Releases from Prussian citizenship
                                          1901-1920
  2: Abt.320 (Schleswig): files as above.   ADS 1:29

SCHLESWIG (Cathedral rectory)   SHL

  Emigration files   1766
  2: Abt.169: Nr. 181.   ADS 1:20

SCHLIENGEN (village)   BWL

  Emigration files (3 fascicles)   1865-1944
  288: Kommunalarchiv Schliengen.   ADS 10:139

SCHLUECHTERN (Kreis)   HEL

  Petitions for releases from Kurhessian citizen-
    ship   1837-1866
  151: Bestand 82: Reg.Hanau: Nr. 96.   ADS 7:84

  Petitions for naturalization   1851-1866
  151: Bestand 82: Acc.1904/13: Nr. 116.   ADS 7:84

SCHLUECHTERN (Landratsamt)   HEL

  Nr. 236: Marriages of foreigners   1826-1882
  Nr. 650: Emigration tax   1827-1843
  151: Bestand 180: Schluechtern: Acc.1905/36: files
       as above.   ADS 7:121

  Nr. 1: Certificates of citizenship; releases from
         citizenship; naturalizations   1905-1907
  Nr. 2: Certificates of citizenship; releases from
         citizenship; naturalizations   1908

  Nr. 3: Passports   1904-1915
  151: Bestand 180: Schluechtern: Acc.1937/101:
       files as above.   ADS 7:122

SCHLUECHTERN (Amt)   HEL

  For exemptions from emigration tax, see Hesse-
    Kassel [Kurhesse]

SCHMALKALDEN (Kreis)   HEL

  Applications for emigration permits   1822-1837
  151: Bestand 100: Nr. 6925.   ADS 7:94

SCHMALKALDEN (Regierungsbezirk)   HEL

  Naturalizations   1849
  151: Bestand 16,II: Klasse 17: Nr. 7.   ADS 7:67

SCHMIEDEBERG. See Kopisch, Adam [surname index]

SCHMIEHEIM (village)   BWL

  General file; Germans abroad   1849-1914
  Illicit emigration   1852-1859
  Jewish emigration   1922-1948
  Emigration files, alphabetically arranged
                                           1870-1919
  288: Kommunalarchiv Schmieheim.   ADS 10:139

SCHOENBERG (village)   BWL

  Emigration file   1854-1937
  288: Kommunalarchiv Schoenberg.   ADS 10:139

SCHOENINGEN   NL

  Emigration files   1845-
  105: Ratsakten Nr. 25: Nr 2; 3; 5.   ADS 4:332

  Information on 88 emigrants in Heimatbuch der Salz-
    stadt Schoeningen, Vol. IV (1947), pages 33-38
  105   ADS 4:332

SCHOETMAR (Amt)   RWL

  Fach 3, Nr. 6: Emigration, 11 vols.   1824-1873
  Fach 3, Nr. 17: Annual list of emigrants & immi-
                  grants   1847-1875
  Fach 22, Nr. 38: Passports, Vol. 2   1843-1887
  114: Schoetmar: files as above.   ADS 6:15

SCHORNDORF (Oberamt)   BWL

  Nr. 72-76: Emigration   1833-1856
  Nr. 77-81: [Emigration] files   1866-1880
  Nr. 87: Returning emigrants   1818-1866
  Nr. 96: Lists [of emigrants]   1819-1894
  261: F 199: Schorndorf: files as above. ADS 10:25

SCHUTTERN (village)   BWL

  Emigration file   1866-1941

  288: Kommunalarchiv Schuttern.   ADS 10:139

SCHUTTERTAL (village)   BWL

  Emigrant file; emigrant list   1891-1939

  288: Kommunlarchiv Schuttertal.   ADS 10:140

SCHUTTERWALD (village)   BWL

  Emigration to the USA   1817
  Emigration file   1830-1859
  Emigration & immigration   1860-1936
  List of emigrants   1908-1951

  288: Kommunalarchiv Schutterwald.   ADS 10:140

SCHUTTERZELL (village)   BWL

  Emigration file   1832-1929

  288: Kommunalarchiv Schutterzell.   ADS 10:140

SCHWABACH   BYL

  Emigration & immigration files (individual, ar-
    ranged by name)   middle 19th century

  256: III, 20.   ADS 9:179

SCHWABACH (Bezirksamt)   BYL

  Nr. 4265-4347: Emigration; applications for emi-
      gration permits; exportation of
      personal property (individual
      cases)   1806-1857
  Nr. 5554-5586: Emigration of individual persons to
      various countries   1836-1876
  Nr. 7968: Determination of the original domicile
      [Heimat] of Bavarians dying abroad
                                1846-1857
  Nr. 8164: Information on Americans living in Bav-
      aria   1907

  231: Rep.212/17: II: Schwabach: files as above.
                                ADS 9:130

SCHWABACH (Landratsamt)   BYL

  Nr. 8766: Emigration & immigration   1865-1912
  Nr. 8803: Emigration (general files, including
      USA)   1866-1885
  Nr. 8804: Applications for emigration permits
                                1857-1886
  Nr. 8805: Naturalizations (names, including from
      USA)   1902-1924
  Nr. 8807: Naturalizations & releases from citizen-
      ship   1886-1907
  Nr. 8808: Naturalizations & releases from citizen-
      ship   1907-1921
  Nr. 8809: Naturalizations   1919-1924

  231: Rep.212/17/III: Schwabach: Abgabe 1956: files
    as above.   ADS 9:132

SCHWABELWEIS (village)   BYL

  Emigration & immigration   n.d.

  253: Fach 24, Fasz. 16.   ADS 9:173

SCHWABEN (Province)   [BWL] BYL

  Annual reports of the General Commissariat regard-
    ing births, deaths, marriages, emigration &
    immigration from the various Kreise [it is not
    certain that the material herein is in enough
    detail to be of use to genealogical researchers]
                                   1807-1833

  230: Rep. Regierungsakten: Nr. 4865 I-II; Nr. 4866
    I-XIX; Nr. 4867 I-II; Nr. 4868 I-IX; Nr. 4869;
    4870.   ADS 9:122

SCHWALBACH. *See* Langenschwalbach

SCHWALENBERG (town)   RWL

  III. Nr. 1: Loss & gain of citizenship; certifi-
        cates of citizenship   1860-1928
  Nr. 522: Certificates of original domicile & citi-
      zenship   1866-1877
  Nr. 523: Emigration & immigration   1894-1924

  139: files as above.   ADS 6:124

SCHWANHEIM   HEL

  Emigration of several families with public assist-
    ance   1851-1854

  149: Rep.39/1: Konv.135, Fasz. 10.   ADS 7:10

SCHWANINGEN (village)   BWL

  Emigration file   1873-1938

  288: Kommunalarchiv Schwaningen.   ADS 10:140

SCHWARZACH (village)   BWL

  Emigration file   1812-1883

  288: Kommunalarchiv Schwarzach.   ADS 10:140

SCHWARZBURG-RUDOLSTADT (Principality). *See* Thueringen

SCHWARZENAU UEBER BERLEBURG (village)   RWL

  The Schwarzenau community has long had a close
    relationship with the Church of the Brethren,
    particularly in Pennsylvania. In celebration
    of 250 years since the original emigration from
    Schwarzenau to Pennsylvania a *Festschrift* [com-
    memorative publication] by Professor W. Hartnack
    was published in 1958. Archival materials were
    used in its preparation [title not otherwise
    given]

  140   ADS 6:125

SCHWARZENBERG (Principality)   BYL

Emigration & immigration   1808

231: Rep.270/II: Reg.Mittelfranken: Kammer des
    Innern: Abgabe 1932: Nr. 28.   ADS 9:134

SCHWARZENBORN (village; Landratsamt Marburg)   HEL

Releases from citizenship   1882-1899

151: Bestand 180: Marburg: Acc.1937/47: Nr. 54.
                        ADS 7:120

SCHWEIGHOF (village)   BWL

Emigration files (3 fascicles)   1868-1949

288: Kommunalarchiv Schweighof.   ADS 10:140

SCHWELM (Kreis)   RWL

Passports   1886-1909

116: Arnsberg: Schwelm: Vd: Nr. 1. ADS 6:60

SCHWENNINGEN (town)   BWL

Card file on persons from Schwenningen living
    abroad (mainly USA)
"Monatsblaettle" [Monthly News Sheet] for Schwen-
    ningen natives resident abroad
Newspaper clippings on persons from Schwenningen
    who have emigrated to the USA; correspondence
283   ADS 10:86-87

SCHWERIN   Mecklenburg

Transportation to America of various prisoners
    from the workhouse at Guestrow   1824-1856

500: Schwerin: Grossherzoglisches Geh. u. Haupt-
    archiv: Fruehere Landesarchiv in Rostock: XX,
    338, 203 [film]   Not in ADS

SCHWERTE   RWL

List of emigrants   n.d.
141   ADS 6:126

SCHWERZEN (village)   BWL

Emigration file   1847-1938

288: Kommunalarchiv Schwerzen.   ADS 10:140

SCOHARIE (congregation; New York)   USA

Letter from Trinity Church, London, thanking for
    sending Candidate Pastor Reimbold to replace
    Pastor Gerdes, deceased, & asking for a pastor
    for the Scoharie congregation in New York [state]
                 7 May 1742

39: Geistl.Minis.: II.6: Protokoll lit.A, 1720-
    1745: p.150(2).   ADS 2:469

SEEBACH (village)   BWL

Emigration files (2 fascicles)   1824-1933

288: Kommunalarchiv Seebach.   ADS 10:140

SEEFELDEN (village)   BWL

Emigration files (2 fascicles)   1831-1945

288: Kommunalarchiv Seefelden.   ADS 10:140

SEEKREIS   BWL

Nr. 664: Applications of citizens of various com-
        munities for financial assistance for
        emigration to America, Vol. I   1851-1853
Nr. 665: Financial assistance to emigrants to
        America, Vol. II   1854-1864

260: Abt.236: Innenmin., Wegzug: Auswand.: files
    as above.   ADS 10:2

SEELBACH (village)   BWL

Emigration file   1863-1948
General file; list of emigrants   1930-1936

288: Kommunalarchiv Seelbach.   ADS 10:140

SEESEN   NL

Emigration certificates   1828-1855
Files of the City Magistrate on emigration & immi-
    gration   1882-1889
Files of the City Magistrate on expulsion of per-
    sons in Seesen   1879-1886

56: Akten der Stadt Seesen [uncatalogued]
                 ADS 4:193

SEESEN (Amt)   NL

Emigration   1848-1849
Emigration, Vol. 3   1852
Emigration, Vol. 4   1855-1859
Emigration, Vol. 5   1860-1870
Emigration, Vol. 6   1871-1877

56: L Neu Abt.129A: Gr. 11: Nr. 10; 11.
              ADS 4:185-186

SEESEN (town)   NL

Emigration to America   1848-1859

56: L Neu Abt.129A: Gr. 11: Nr. 16. ADS 4:186

SEGEBERG (Amt)   SHL

Release from citizenship & military service
                 1848-1862

2: Abt.110: XXV, B VI, conv.XVI [Austritt]
               ADS 1:18

SEGEBERG (Landratsamt)   SHL

Emigration to North America   1872-1874

2: Abt.320 Segeberg: L 183.   ADS 1:29

SELBACH (village)   BWL

Germans abroad; immigration   1920-1945
Emigration files   1848-1944
Germans abroad   1933-1939

288: Kommunalarchiv Selbach.   ADS 10:140

SELIGENSTADT (town)   HEL

Miscellaneous material on emigrants to America
                                                1846

193   ADS 7:256

SELTERS (Amt)   HEL

Annual tables on naturalization & release from
    citizenship   1839-1872

152: Abt.241: Selters: Nr. 390.   ADS 7:136

SENTENHART (village)   BWL

Emigration files (2 fascicles)   1856-1940

288: Kommunalarchiv Sentenhart.   ADS 10:141

SESPENROTH (village; Amt Wallmerod)   HEL

Nr. 247: Dissolution of Sespenroth community due
         to emigration to America   1852-1855
Nr. 268: Petitions of residents not emigrating
         regarding disposition of their portion of
         community property upon dissolution
                                           1852-1854
Nr. 270: Domicile of persons not emigrating
                                           1853-1872
Nr. 277: Files of the discontinued community 1852
Nr. 276: Files of the discontinued community 1853

152: Abt.243: Wallmerod: files as above. ADS 7:136

SEXAU (village)   BWL

Emigration files (3 fascicles)   1891-1940

288: Kommunalarchiv Sexau.   ADS 10:141

SICKINGEN (Landkreis). See Landstuhl (Sickingen)
                                                      RPL

SIEDELSBRUNN   HEL

Emigration   1859-1899

149: Rep.40,2: Ablief.2: Bd. 310.   ADS 7:23

SIEGBURG   RWL

The Einwohnermeldeamt [Office of Civil Registra-
    tion] has registers of citizens moving in and
    out of Siegburg since 1843. Their card index

dates only from 1908, however

C III 2, 1 Gen.: Emigration & immigration of for-
    eigners   1739-1891

142   ADS 6:127

SIEGELAU (village)   BWL

Emigration files (2 fascicles)   1853-1899

288: Kommunalarchiv Siegelau.   ADS 10:141

SIEGEN (Kreis)   RWL

A Nr. 87: Releases from citizenship   1828-1838
A Nr. 88: Acceptances [& releases?] of Prussian
          citizenship, Vol. 1   1846-1852
                      Vol. 2   1863-1909
                      Vol. 3   1923-1932
A Nr. 89: Releases from citizenship   1920-1932
A Nr. 90: Certificates of original domicile &
          citizenship   1915-1931
A Nr. 92: Emigration   1910-1929
A Nr. 93: Emigration   1921-1932
A Nr. 94: Investigation of emigrants; death not-
          ices   1925-1932
B Nr. 15: Emigration & immigration   1933-1939
B Nr. 16: Extradition & expulsion   1933-1938
B Nr. 128: Prisoners of war (excluding Russians)
                                           1939-1944

116: Arnsberg Reg.: Siegen: files as above.
                                          ADS 6:60-61

A contract between mining supervisor J. J. Albrecht,
an employee of the English, and three Reformed
ministers [unnamed in ADS description]. Albrecht
was recruiting miners in Siegerland for the British colonies in North America & promised the ministers annual stipends "as soon as the mines are
able to meet their responsibilities to God" [to
support the ministers & their churches in America?]   18th century

143: Sub L 39.   ADS 6:128

SIEGKREIS   RWL

Emigration   1857-1889

115: Rep.D9: Koeln: Abt.I.e: Nr. 261-272. ADS 6:32

SIENSBACH (village)   BWL

Emigration to the USA (2 fascicles)   1848-1875
Emigration files   1912
Germans abroad   1933-1939

288: Kommunalarchiv Siensbach.   ADS 10:141

SIGMARINGEN (Oberamt)   BWL

Emigration from various villages   1852-1936

262: Ho 199: Sigmaringen: II 6, 24-79; see also
             Nachtrag lfd. Nr. 1365.   ADS 10:36

I 9268-9269: Births & deaths of Wurttemberg citi-
             zens abroad   1895-1905

I 9344-9345: Marriages abroad   1895-1925
I 9429-9432: Emigration & immigration   1853-1925
I 9412:   Germans abroad   1920-1925

<u>262</u>: Ho 199: Sigmaringen: files as above.
                   ADS 10:36

SIGMARINGEN.   *See also* Hohenzollern-Sigmaringen

SILESIA

File on emigration of Silesian citizens   1821-1880

<u>500</u>: Breslau: Preus. Staatsarchiv: Rep.200: O.P.:
       Acc.54/16: Nr. 2634, Vol. 2.   Not in ADS

SIMBACH (Landgericht)   BYL

Applications for emigration permits   1818-1846

<u>225</u>: Rep.168: Verz. 1.   ADS 9:16

SIMMERN (Subprefecture)   RPL

Nr. 346: Movement of persons   1804-1805
Nr. 347: Movement of persons   1805-1806
Nr. 352: Movement of persons   1808
Nr. 353: Movement of persons   1809
Nr. 120: Movement of persons   1812-1813
Nr. 542: Report as to reason for emigration
                           ca. 1804

<u>200</u>: Abt.261: Unterpraefektur Simmern: files as
       above.   ADS 8:14

SIMMERN (Kreis)   RPL

Naturalizations   1846-1913

<u>200</u>: Abt.441: Regierung Koblenz: Nr. 22046-22048.
                           ADS 8:30

[Publication?] by Diener & Siegel, *Auswanderer aus
dem Kreise Simmern*, 1955.   Contains the names of
all emigrants from Kreis Simmern from 1750 to
1955

<u>215</u>   ADS 8:84

SIMMERN (town)   RPL

Emigration & immigration   1908-1913

<u>200</u>: Abt.640: Simmern: Nr. 95.   ADS 8:49

SINGEN (town)   BWL

Emigration files (from about 60 villages)
                       [19th century]
<u>284</u>   ADS 10:89

SINGHOFEN (village?)   HEL

Manumissions   1785-1814

<u>152</u>: Abt.351: Nassau-Vierherrisch: Gen.VIIIe46.
                         ADS 7:139

SIPPERSFELD (village; Landkreis Kirchheimbolanden,
  Kurpfalz)   RPL

Names of emigrants   18th century

<u>201</u>: Archivabt. Ausfautheiakten.   ADS 8:63

SIPPLINGEN (village)   BWL

Emigration to the USA   1849-1857
Emigration files   1897-1926

<u>288</u>: Kommunalarchiv Sipplingen.   ADS 10:141

SITTARD (canton)   RWL

Population movements   1801-1807

<u>115</u>: Rep.D2.2: II.Division, 1.Bureau, 2.Bevoelker-
      ung: Nr. 102.   ADS 6:19

SITZENHOFEN (village)   BWL

Emigration file   1870-1949
Foreigners   1947-1950

<u>288</u>: Kommunalarchiv Sitzenhofen.   ADS 10:141

SOBERNHEIM (town)   RPL

Passports; emigration   1867-1910

<u>200</u>: Abt.642: Stadt Sobernheim: Nr. 721.   ADS 8:49

SOELDEN (village)   BWL

Emigration files; Germans abroad   1835-1937

<u>288</u>: Kommunalarchiv Soelden.   ADS 10:141

SOEGEL (Landratsamt)   NL

Nr. 13: Certificates & matters regarding issuance
         of passports to emigrants to America
                          1869-1870
Nr. 15: Emigration files   1889-1914

<u>54</u>: Landratsamt Soegel: VIII, H.7, Fach 165: files
     as above.   ADS 4:145

SOELLINGEN (village)   BWL

Emigration to the USA   1847-1920

<u>288</u>: Kommunalarchiv Soellingen.   ADS 10:141

SOEST (Kreis)   RWL

Nr. 100: Emigration & immigration permits & certi-
           ficates of original domicile   1856-1860
Nr. 32:   Emigration permits & certificates of
           original domicile   1817-1846
Nr. 129: Emigration permits   1879
Nr. 98; 117; 127-128; 130: Passports
                  1890-1906; 1920-1927

<u>116</u>: Arnsberg Reg.: Soest: files as above.
                         ADS 6:61

SOHREN (village; Grafschaft Sponheim)   RPL

Nr. 175: Petition of the inhabitants of Sohren who
emigrated to Pennsylvania in 1709 but who
thereafter returned, to be given back
their auctioned possessions   n.d.
Nr. 178: Confiscation of the property of serfs
from Sohren & Buechenbeuren who emigrated
to America without permission; also returns of portions of this property
1771-1778

200: Abt.33: Vordere Grafschaft Sponheim u. beide
Grafschaften gemeinschaftlich: files
as above.   ADS 8:11

SOLINGEN (Kreis)   RWL

Emigration applications   1852-1860

115: Rep.D8: I.Abt.e: Nr. 591-593.   ADS 6:25

SOLINGEN (Landratsamt)   RWL

Nr. 328: Emigration   1863-1873
Nr. 329: Emigration   1879-1892
Nr. 330: Emigration   1892-1896
Nr. 331: Census of October 1816 & changes to Janu-
ary 1817
Nr. 336: Releases from Prussian citizenship
1879-1884
Nr. 337: Releases from Prussian citizenship
1885-1888
Nr. 338: Certificates of domicile   1885-1892

115: Rep. G 29/41: Solingen: files as above.
ADS 6:45

SOLINGEN (city)   RWL

Emigration files   1831-

144: A-5-5.   ADS 6:129

SONTHOFEN (Bezirksamt)   BYL

Nr. 2795: Applications for emigration permits
(individual files)   1840-1890
Nr. 2989-: Emigration to overseas countries (contains 12 individual files on emigrants
to North America)   1880-1887

230: Rep. Akten der Bezirksaemter: Sonthofen:
files as above.   ADS 9:124

SORGE (village)   HEL

Certificates of original domicile   1849

151: Bestand 100: m: Nr. 233.   ADS 7:96

SOUTH CAROLINA   USA

Emigration of Hessian subjects from Ludwigstein &
Gregenstein (Aemter)   1753

151: Bestand 17(b): Gef. 104a: Nr. 43.   ADS 7:73

Emigration   1773

158: Frankfurter Ratsakten: Ugb.A.9: Nr.
DLC has (the original document in 158 has
been destroyed)   ADS 7:198

Emigration to South Carolina [from Oberwihl, Baden-
Wuerttemberg]   1890-1939

288: Kommunalarchiv Oberwihl.   ADS 10:132

SOUTH CAROLINA. *See also* Moravian Brethren

SPAICHINGEN (Oberamt)   BWL

Nr. 39: Lists of emigrants   1815-1892
Nr. 40-42: Emigration files, alphabetically ar-
ranged by village   19th century
Nr. 43-57: Emigration files   1842-1870

261: F 200: Spaichingen: files as above. ADS 10:25

Nr. 5-6: Emigration & emigration agents   1883-1938
Nr. 7-9: Germans abroad   1926-1937

262: Wue 65/32: Spaichingen: Acc.4/1956: A 54:
files as above.   ADS 10:39

SPALT. *See* Eichstaett (Kastenamt Spalt)   BYL

SPEYER (city)   RPL

Applications for emigration to North America; emigration with permission (1831-1852)   1832-1870

216: Bestand 3: Nr. 7.   ADS 8:85

Nr. 32: Emigration   1850-1853
Nr. 33: Emigration   1854-1880
Nr. 106: Emigration   1854-1880

201: Archivabt. Bezirksaemter: Speyer (Abgabe 1907):
files as above.   ADS 8:65

SPONHEIM (Grafschaft)   RPL

Manumissions of a number of emigrants to America
before 1773

201: Zweibruecken III: Akt Nr. 3358.   ADS 8:59

Files of Baden officials regarding expropriations
of properties [of emigrants]   1777-1794

200: Abt.33: Vordere Grafschaft Sponheim u. beide
Grafschaften gemeinschaftlich: Nr. 3766-3767.
ADS 8:11

SPONHEIM. *See also* Sohren   RPL

SPRINGE (Amt)   NL

Nr. 4: Files of the former Springe Magistrate re-
garding releases from citizenship 1838-1849
Nr. 5: Releases from citizenship, Vol. I 1821-1886
Nr. 6: Releases from citizenship, Vol. II   1864

52: Han.Des.74: Springe: III.F.: files as above.
ADS 4:78-79

Nr. 1: Files on emigration to America, Canada 1765
Nr. 2: Emigration certificates   1844

<u>52</u>: Han.Des.74: Springe: VIII.H.7: files as above.
ADS 4:79

Nr. 2: Files on resettlement of useless & dangerous persons to America   1835
Nr. 7: Files on resettlement of useless & dangerous persons to America, Vol. I   1836-1850
Nr. 8: Files on resettlement of useless & dangerous persons to America, Vol. II 1851-1862
Nr. 9: Files on resettlement of useless & dangerous persons to America, Vol. III   1863

<u>52</u>: Han.Des.74: Springe: VIII.G.5: files as above.
ADS 4:79

Releases from citizenship   1860-1870

<u>52</u>: Han.Des.80: Springe: B.y.VI.4: Nr. 423.
ADS 4:85

Files on citizenship grants & renunciations 1875

<u>52</u>: Han.Des.74: Springe: II.G: Nr. 2.   ADS 4:78

STADE   NL

Nr. 1: (RR 1001, P.246) Emigration   1727-1761
Nr. 2: (RR 1001, P.247) Emigration   1756
Nr. 3: (RR 1001, P.248) Emigration   1756-1768
Nr. 4: (RR 1001, P.249) Emigration   1784
Nr. 5: (RR 1002, P.250) Emigration   1786-1792
Nr. 7: (RR 1002, P.252) Emigration   1825-1836
Nr. 8: (RR 1002, P.253) Emigration   1824
Nr. 10: (RR 1002, P.255) Emigration   1832-1889
Nr. 14: (RR 1002, P.258) Emigration   n.d.
        (RR 1258, P.259) Emigration   n.d.
Nr. 15: (RR 1002, P.260) Emigration   1851
Nr. 16: (RR 1002, P.261) Emigration   1851-1868
Nr. 17: (RR 100s, P.262) Emigration   1852
Nr. 17a: (RR 1003, P.266) Emigration   1867-1871

<u>52</u>: Han.Des.80: Stade: P, Titel 85: Nr. 1.
ADS 4:93

Captain Friedrich Spengemann, "Wer wanderte vor
rund 100-120 Jahren aus der derzeitigen Land-
rostei Stade, bezw. dem Lande Hannover nach
Nordamerika aus?" [Who emigrated from . . .
Stade & Hannover to North America 100-120 years
ago?] (manuscript based upon the logs of Captain
Juergen Meyer of the emigrant ships *Isabella,
Pauline, Meta,* & *Uhland*)   1832-1849

<u>108</u>   ADS 4:335

Hans Mahrenholtz, "Auswanderung deutscher Kolonisten ueber Stade nach Amerika vor 200 Jahren,"
[Emigration of German Colonists via Stade Two
Hundred Years Ago] *Stader Jahrbuch* (1958), pp.
108-122

<u>108</u>   ADS 4:335

STADELHOFEN (village)   BWL

Emigration files   1893-1932

<u>288</u>: Kommunalarchiv Stadelhofen.   ADS 10:142

STADTSTEINACH (Oberamt)   BYL

Individual emigration cases   1865-1869

<u>223</u>: Rep.K.6.1: Verz.II: Nr. 118-119; 121-124;

126-128; 130-134; 137-138; 140; 143-
143; 147-148; 150-159; 163; 167;
169-178; 180-186; 188-191; 193-195;
197-198; 201-203.   ADS 9:5

Individual emigration cases with destinations
1865-1869

<u>223</u>: Rep.K.19: Verz.II: Nr. 118-213.   ADS 9:8

Emigration & immigration   1880-1908

<u>223</u>: Rep.K.19: Verz.XII: Nr. 25-26.   ADS 9:8

STAHRRINGEN (village)   BWL

Immigration   1872-1940

<u>288</u>: Kommunalarchiv Stahrringen.   ADS 10:142

STALLUPOENEN (Kreis)   East Prussia (Gumbinnen)

Emigration   1842-1874

<u>51</u>: Rep.12: Abt.I, Titel 3: Nr. 8, Bd. 11.
ADS 4:32

STATT WIEDORF. *See* Dillenburg   HEL

STAUFENBERG (village)   BWL

Emigration file   1837-1924
Germans abroad   1934

<u>288</u>: Kommunalarchiv Staufenberg.   ADS 10:142

STEGEN (village)   BWL

Emigration file   1857-1945

<u>288</u>: Kommunalarchiv Stegen.   ADS 10:142

STEIN (village)   BWL

Emigration to the USA   1850
Emigration files; Germans abroad   1907-1935

<u>288</u>: Kommunalarchiv Stein.   ADS 10:142

STEINACH (village)   BWL

Emigration files (2 fascicles)   1837-1933
Emigration over fifty years   (1932-1947)

<u>288</u>: Kommunalarchiv Steinach.   ADS 10:142

STEINAU (Amt)   HEL

Emigrants to Nova Scotia & Pennsylvania   1752

<u>151</u>: Bestand 86: Nr. 51974.   ADS 7:88

STEINBACH (village)   BWL

General emigration file   1828-1855
Emigration files (3 fascicles)   1818-1834

<u>288</u>: Kommunalarchiv Steinbach.   ADS 10:142

STEINBACH (Amt)   HEL

  Emigration   1770-1780

  <u>151</u>: Bestand 17(b): Gef.97: Nr. 3.   ADS 7:70

  For exemptions from emigration tax, *see* Hesse-
  Kassel [Kurhesse]

STEINBURG (Landratsamt)   SHL

  Nr. 1319: Main files on emigration   1868-1940
  Nr. 71: Emigration permits   1873-1880
  Nr. 72: Emigration permits   1880-1884
  Nr. 73: Emigration permits   1884-1888
  Nr. 74: Emigration permits   1888-1896
  Nr. 70: Searches for persons emigrating without
          permits   1890-1899
  Nr. 506: Detailed evidence regarding searches for
          persons subject to military service emi-
          grating without permits   1878-1882

  <u>2</u>: Abt.320 (Steinburg): files as above.
                                          ADS 1:30; 32

STEINBURG. *See also* Wilster   SHL

STEINENSTADT (village)   BWL

  Emigration files   1887-1947

  <u>288</u>: Kommunalarchiv Steinenstadt.   ADS 10:142

STEINFURT (Kreis)   RWL

  Nr. 405; 667; 1177; 1368; 1597; 630: Applications
          for emigration permits   1848-1925
  Nr. 1343; 360; 1291; 783; 1389: Issuance of emi-
          gration permits   1837-1934
  Nr. 946: Master list of releases from citizenship
                                            1856-1891
  Nr. 945: Passports   1868-1869
  Nr. 330: Passports   1881-1886
  Nr. 948: Passport journal   1851-1859
  Nr. 1843: Passport fees   1860-1879
  Nr. 382; 1292: Issuance of certificates of orig-
          inal domicile   1868-1925
  Nr. 1685-1686: Emigration   1895-1924

  <u>116</u>: Muenster-Reg.: files as above.   ADS 6:57; 67

STEINFURT (Landratsamt)   RWL

  Issuance of certificates of original domicile &
    citizenship   1881-1934; 1930-1934

  <u>116</u>: Muenster Reg.: Steinfurt: Nr. 934; 994; 1822;
      452; 1086; 1089; 309; 1652; 441; 894.
                                              ADS 6:67

STEINHORST   SHL

  Domicile; releases from citizenship   1847-1871
  <u>29</u>: I.Teil IV: Tit.IIa; IIb.   ADS 1:61

STEINMAUERN. *See* Astor, John Jacob [surname index]

STEINWEG (village)   BYL

  Emigration & immigration   n.d.
  <u>253</u>: Fach 24, Fasz. 17.   ADS 9:173

STEINWEILER (town)   RPL

  Names of emigrants   18th century
  <u>201</u>: Archivabt. Ausfautheiakten.   ADS 8:62

STEISSLINGEN (village)   BWL

  Emigration files (2 fascicles)   1901-1932
  Emigration to the USA   1911
  List of emigrants   1932

  <u>288</u>: Kommunalarchiv Steisslingen.   ADS 10:142

STERNBERG-BARNTRUP (Amt)   RWL

  Emigration   1881-1889
  <u>114</u>: Brake: Fach 7, Nr. 6.   ADS 6:15

STETTEN with GUENZGAU (village)   BWL

  Emigration file   1882-1948
  <u>288</u>: Kommunalarchiv Stetten.   ADS 10:143

STETTEN BEI LOERRACH (village). *See* Loerrach   BWL

STETTEN (village; Kreis Ueberlingen)   BWL

  Emigration file   1883-1941
  General file   1907-1950

  <u>288</u>: Kommunalarchiv Stetten, Kr. Ueberlingen.
                                              ADS 10:143

STETTEN (village). *See* Kurpfalz   RPL

STETTEN BEI ENGEN (village; Kreis Konstanz)   BWL

  Emigration files (3 fascicles)   1893-1939
  <u>288</u>: Kommunalarchiv Stetten bei Engen, Kr. Konstanz.
                                              ADS 10:143

STETTEN AM KALTEN MARKT (village)   BWL

  Emigration files (3 fascicles)   1848-1947
  <u>288</u>: Kommunalarchiv Stetten am Kalten Markt.
                                              ADS 10:143

STETTIN (city)   Silesia

  Files on emigration to America   1853-1854
  <u>500</u>: Stettin: Koeniglich Preussisches Staatsarchiv:
        Abt.1, Tit.2a, Sec.3: Nr. 10, Vol. 9.
                                            Not in ADS

STOLLENHOF. *See* Lenzkirch   BWL

STOLLHOFEN (village)   BWL

Emigration files (2 fascicles)   1851-1937

288: Kommunalarchiv Stollhofen.   ADS 10:143

STOLZENAU (Amt)   NL

Applications for releases from citizenship, Vol.
II [Vol. I may still exist]   1858-1859

52: Han.Des.80: Stolzenau: B.2.VI.4: Nr. 574.
                                      ADS 4:85

STRASSBERG (Oberamt)   BWL

Emigration from various villages   1827-1854

262: Nr. I 11972; 11974; 11116; 11314; 11747;
     12337; 12692; 13868; 14940.   ADS 10:37

Emigration to America   1839-1854

262: Ho 200: Strassberg: Nr. I 10674; 11978; 12409;
     12335; 13866; II 8703.   ADS 10:37

Emigration (individual cases; begun in the Sigmaringen series, thereafter in Gammertingen series)
                                           1848

262: Ho 200: Strassberg: Sigmaringen OA II 6, 200.
                                      ADS 10:37

STRAUBING (Bezirksamt)   BYL

Emigration   [19th century]

225: Rep.168: Verz. 1.   ADS 9:15

Emigration   1887-1925

225: Rep.164: Verz. 17: Nr. 25-.   ADS 9:14

STRAUBING (Landgericht)   BYL

Applications for emigration permits   1811-1846

225: Rep. 168: Verz. 1.   ADS 9:16

STRAUBING (Stadtmagistrat)   BYL

Applications for emigration permits   1808-1846

225: Rep. 168: Verz. 1.   ADS 9:16

STRAUBING (city)   BYL

Emigration   [19th century]

225: Rep. 168: Verz. 1.   ADS 9:15

STUEHLINGEN (village)   BWL

Emigration files (4 fascicles)   1907-1945
Emigration files, including Jewish   1907-1939

288: Kommunalarchiv Stuehlingen.   ADS 10:143

STUTTGART   BWL

Monthly lists of foreigners in Ludwigsburg &
Stuttgart [*see also* Wuerttemberg]   1817

263: Rep.E2: Polizeiminist.: 21,6: Nr. 23,4; 24,4.
                                      ADS 10:52

STUTTGART (Oberamt)   BWL

Nr. 98: Lists of emigrants   1817-1849
Nr. 99-117: Emigration files, chronologically ar-
            ranged   1806-1894
Nr. 118: Emigration from Gaisburg   1817-1899

261: F 202: Stuttgart: files as above. ADS 10:26

STUTTGART (Polizeipraesidium)   BWL

Nr. 170: Lists of emigrants   1849-1872
Nr. 171-199: Emigration files; Gaisburg; emigra-
             tion files II; passport files
                                      1808-1920
Nr. 200-214: Emigration files   1891-1915

261: F 201: Polizeipraesidium Stuttgart: files as
            above.   ADS 10:26

STUTTGART (city)   BWL

A number of letters from Stuttgart & Wuerttemberg
emigrants   [19th century?]

286   ADS 10:97

SUEDERENDE AUF FOEHR (parish)   SHL

Lorenz Braren, *Vollstaendiger Nachweis saemtlicher
Personen, die im Kirchspiel Suederende auf Foehr
gelebt haben, soweit Nachrichten ueber sie zu
erhalten waren: Ein grosser Prozentsatz der
Bevoelkerung ist nach Amerika ausgewandert und
pflegt noch jetzt sehr enge Beziehungen mit der
alten Heimat* (Foehr: Geschlechter-Reihen St.
Laurentii, 1949) [Complete directory of all
persons who lived in the parish of Suederende
auf Foehr, as far as known: A large portion of
the inhabitants emigrated to America & still
maintain contact with their places of origin.]

19: printed work.   ADS 1:53

SUEDERHASTED (Kreis Suederdithmarschen)   SHL

Nr. 189: Emigration & immigration   1875-1914
Nr. 191: Passports   1889-1915
Nr. 192: Searches for missing persons 1889-1916
Nr. 488: Emigration & immigration   1776-1868

33: Abt.II: files as above.   ADS 1:64-65

SUGGENTAL (village)   BWL

Emigration file   1931-1946

288: Kommunalarchiv Suggental.   ADS 10:143

SULINGEN   NL

Hans Mahrenholtz, "Auswanderer-Schicksale von Ein-
wohnern unserer engern und weiterer Heimat im
19. Jahrhundert" [The fate of emigrants . . .
in the nineteenth century] *"Unter der Baeren-
klaue" Heimatblaetter fuer das Sulinger Land:
Heimatbeilage der Sulinger Kreiszeitung*, Nr. 8-
10 (25 October 1957) & Nr. 12 (22 February 1958)

<u>52</u>: Han.Des.80: Hannover IA II B: Regiminalverwal-
     tung I 3 c: Nr. 421, Bl. 71-76.   ADS 4:336

SULZ (Oberamt)   BWL

Emigration   1814-1883

<u>262</u>: Wue 65/34: Sulz: Acc.6/1954: Nr. 186-188.
                                                  ADS 10:39

Nr. 81: Emigration list & index (for 1853-1865
        only)   1848-1897
Nr. 84-104: Emigration files, alphabetically ar-
        ranged by village   19th century
Nr. 105: Immigrant list   1851-1872

<u>261</u>: F 203: Sulz: files as above.   ADS 10:26

SULZBACH (Bezirksamt)   BYL

Nr. 863-903: Emigration   1870-1874
Nr. 946-948: Issuance of certificates of original
             domicile   1905-1916

<u>222</u>: Rep.139: Sulzbach:files as above.   ADS 9:4

SULZBACH (village)   BWL

General emigration file   1871-1940
Emigration files   1846-1944
Emigration to the USA
            1848; 1858; 1861; 1871-1872; 1874

<u>288</u>: Kommunalarchiv Sulzbach.   ADS 10:143

SULZBURG (village)   BWL

Emigration file   1851-1946

<u>288</u>: Kommunalarchiv Sulzburg.   ADS 10:143

SYKE (Amt)   NL

Nr. 464: Releases from citizenship, Vol. II
                                          1851-1859
Nr. 465: Releases from citizenship, Vol. III
                                          1860-1864
Nr. 466: Releases from citizenship, Vol. IV   1865
Nr. 467: Releases from citizenship, Vol. V   1866
Nr. 468: Releases from citizenship, Vol. VI
                                          1867-1870

<u>52</u>: Han.Des.80: Syke: B.aa.VI.4: files as above.
                                          ADS 4:85

Emigration & immigration [part of annual report of
     affairs]   1859-1867

<u>52</u>: Han.Des.74: Syke: A.VII: Nr. 43.   ADS 4:79

TAISERSDORF (village)   BWL

Emigration file   1902-1941

<u>288</u>: Kommunalarchiv Taisersdorf.   ADS 10:143

TANN (Amt)   HEL

Passports   1843-1848

<u>151</u>: Bestand 112: Acc.1903/9: LIV: Nr. 15.
                                          ADS 7:98

TANNHEIM (village)   BWL

Emigration file   1873-1945
Emigration file   1874-1937
General file   1911-1931

<u>288</u>: Kommunalarchiv Tannheim.   ADS 10:144

TANNENKIRCH (village)   BWL

Emigration files   1852-1944

<u>288</u>: Kommunalarchiv Tannenkirch.   ADS 10:144

TARNOWITZ. *See* Meyer, Riccardo [surname index]

TECKLENBURG   RWL

Letters from Tecklenburg emigrants, mainly from
     Cincinnati, Ohio, to their relatives   1831-1866

<u>133</u>: Abt.IV: Nr. 123.   ADS 6:113

TECKLENBURG (Kreis)   RWL

Emigration   1892-1927

<u>116</u>: Muenster Reg.: Nr. 1687-1689.   ADS 6:57

TECKLENBURG (Landratsamt)   RWL

Nr. 177; 271-272; 1540; 1912; 1904: Emigration
                         1837-1870; 1883-1939
Nr. 703: Reservists & militiamen emigrating with-
         out permits   1878-1881
Nr. 876: Deserters, general   1845-1914
Nr. 693: List of deserters   1813-1826
Nr. 1494: Certificates of original domicile &
          citizenship   1839-1937
Nr. 1475: Certificates of original domicile &
          citizenship   1931-1938
Nr. 1493: Citizenship certificates   1936-1939
Nr. 1476: Certificates of original domicile
                                          1938-1939
Nr. 1103: Passports   1928-1939
Nr. 1387: Passports   1916-1919
Nr. 1478: Passports   1923-1925
Nr. 273-274; 1531; 1533; 1562; 1650; 1785; 1895;
          1385: Immigration   1868-1939
Nr. 1516: Residency permits   1932-1938
Nr. 1712: Residency permits   1933-1940
Nr. 1947: Residency permits   1937-1939
Nr. 266: Foreigners   1899-1911
Nr. 1787: Foreigners   1911-1916

Nr. 1970: Foreigners   1916-1933
Nr. 1418: Foreigners   1922-1930
Nr. 1556: Foreigners   1930-1933
Nr. 951: Foreigners   1902
Nr. 267: Foreigners   1902-1914

116: Muenster Reg.: Tecklenburg: files as above.
                                         ADS 6:68

TETTNANG (Oberamt)   BWL

Nr. 58: Lists of emigrants   1835-1867
Nr. 59-77: Emigration files   1836-1870

261: F 204: Tettnang: files as above.   ADS 10:26

TEXAS   USA

File on the Swiss settlement in Texas proposed by
persons from Neufchatel & their emigration
thereto   1819-1868

500: Berlin-Dahlem: Geh.Staatsarchiv: Auswaertige
Amt III: Rep.I: Auswanderung ausser Europa
No. 2.   Not in ADS

THALEISCHWEILER (congregation)   RPL

Entries on emigration to be found in either the
Lutheran or Reformed parish registers
                                      18th century

217   ADS 8:86

THEDINGHUSEN. See Westen-Thedinghusen   NL

THUERINGEN (Principality of Reuss, Aeltere Linie)

Nr. 6; 7; 7a: Settlement of various individuals in
North America   1837-1840
Nr. 8: File on the transport of a number of con-
victs from here to North America   1842
Nr. 9: The accomplished settlement of various
individuals in North America   1847

500: Greiz: Thueringisches Staatsarchiv [Obere
Schloss]: a.Rep.A.Cap.XXVI.b: files as above.
                                    Not in ADS

Proposals for the regulation of releasing convicts
for the purpose of settlement in America   1856

500: Greiz: Thueringisches Staatsarchiv [Obere
Schloss] n.Rep.A.Cap.XXVII.b.Nr. 19.
                                    Not in ADS

THUERINGEN (Principality of Schwarzburg-Rudolstadt)

Emigration matters, Vol. I   1846-1897

500: Rudolstadt: Staatsarchiv: Kr.R.XI.C.: Nr. 1.
                                    Not in ADS

TIENGEN (village; Kreis Freiburg)   BWL

Emigration file   1873-1913

288: Kommunalarchiv Tiengen.   ADS 10:144

TIENGEN/HOCHRHEIN (village)   BWL

Emigration files (2 fascicles)   1853-1896
Jewish emigration   1933

288: Kommunalarchiv Tiengen/Hochrhein. ADS 10:144

TILSIT (Kreis)   East Prussia (Gumbinnen)

Emigration [file missing?]   1855-1876

51: Rep.12: Abt.I, Titel 3: Nr. 8, Bd. 12.
                                         ADS 4:32

TOENNING (Kreis Eiderstadt)   SHL

Files on citizenship, emigration, passports
                                        1880-1881

2: Abt.320 (Toenning): Nr. 68 (Paket 52-53).
                                         ADS 1:32

TOENNING   SHL

Nr. 305-309: Passports   1807-1868
Nr. 310-315: Visaed passports   1812-1868
Nr. 316-322: Various passport documents (marriage
certificates; certificates of domi-
cile, etc.)   1805-1868

34: Fb 1, Abt.I: files as above.   ADS 1:65

Nr. 31: Applications for releases from Prussian
(& Danish) citizenship   1857-1890
Nr. 174: Penalties for having failed to report [to
authorities] before emigrating   1867-1877

34: Fb 1, Abt.II: files as above.   ADS 1:66

Nr. 204-210: Papers on persons moving away
                                        1887-1904
Nr. 211-212a: Civil registry lists   1871-1899
Nr. 214: Lists of persons moving away   1843
Nr. 116-127: Mustering of sailors [engaged in the
export of cattle to England & other
countries]   1872-1924

34: Fb.2: Abt.II: files as above.   ADS 1:66-67

TONDERN (Amt)   SHL

Emigration matters   1753-1767

2: Abt.161: Nr. 455.   ADS 1:19

TRANZFELDE. See Neitzel, C. A. [surname index]

TRAUTSKIRCHEN (Patrimonialgericht)   BYL

Emigration files from Trautskirchen, estate of the
Barons [Freiherren] von Seckendorff   1817

231: Rep.212/13: Neustadt a.d.A.: Nr. 469.
                                         ADS 9:129

TREBUR (village; Landkreis Gross-Gerau)   HEL

Applications for emigration permits   1837-1854

Application for emigration to America with public
financial assistance   1846-1862
List of all emigrants to North America   1846
List of persons who have emigrated with public
financial assistance   ca. 1850
List of emigrants to America at public cost
1852-1853
List of persons who have emigrated with or without
permits   1862-1872
Applications for releases from citizenship
1867-1886

183: Abt.XI,4.   ADS 7:237

TREUBURG (Oletzko)(Kreis)   East Prussia (Gumbinnen)

Emigration   1852-1875

51: Rep.12: Abt.I, Titel 3: Nr. 8, Bd. 13.
ADS 4:33

TRIER (Regierung)   RPL

Nr. 4443: Summary of persons who have emigrated to
America   1843-1875
Nr. 10209-10214; 9743-9747: Passports   1854-1918
Nr. 1498: Death certificates of Germans dying
abroad   1870-1877
Nr. 1499: Death certificates of Germans dying
abroad   1877-1881
Nr. 1505: Lists of emigrants & immigrants
1871-1876
Nr. 3838: Persons who have emigrated without per-
mits   1837-1868
Nr. 6808: Letters from emigrants   1852-1863
Nr. 8013-8038; 11195: Inheritance matters (proof
of heirs; sworn statements, etc.)
1875-1933
Nr. 11779: Issuance of citizenship certificates
1931-1945
Nr. 11194: Estates in foreign countries (many
American)   1937-1942
Nr. 14262: Jewish affairs (not including property;
emigration; movements; expropriation of
gold & other precious metals) 1938-1941

200: Abt.442: Regierung Trier: files as above.
ADS 8:35-36; 39; 40-41

The Lutheran Church records for many parishes contain marginal notations regarding emigrants to
America [check also the parish records of the
Reformed Church in areas where they existed]
before 1798
206   ADS 8:72

TRIER (Regierungsbezirk)   RPL

Protestant church records [Evangelische Kirchenbuecher; apparently contain material of interest
to German-American researchers]

200: Abt.555: Evang. Kirchenbuecher.   ADS 8:45

TRIER (Praefektur; Saardepartement)   RPL

Correspondence among various offices regarding 35
emigrants to America, Year XI [of French Republic?]

200: Abt.276: Saardepartement, Praef. Trier: Nr.
1723.   ADS 8:14

TRIER (Landkreis)   RPL

Emigration to America   1855-1911

200: Abt.442: Regierung Trier: Nr. 172-178; 6107-
6116; 8651.   ADS 8:42-43

Josef Mergen, "Die Amerika-Auswanderung aus dem
Regierungsbezirk Trier" [Emigration to America
from Trier](typescript)   1952

200: Abt.701: Nachlaesse, etc.: Nr. 964. ADS 8:53

TRIER (city)   RPL

Nr. 171; 6103-6106; 8650: Emigration to America
1854-1907
Nr. 14267: Naturalization of foreigners & state-
less persons   1901-1908

200: Abt.442: Regierung Trier: files as above.
ADS 8:40-42

TROCHTELFINGEN (Oberamt)   BWL

Emigration in contravention of the regulations for
military recruits   1800

262: Ho 197: Trochtelfingen: Nr. I 13765.
ADS 10:36

TUEBINGEN (Oberamt)   BWL

Nr. 152-158: Emigration   1871-1900
Nr. 158: Financial aid to 1870 refugees from
France; residency abroad with retention
of Wuerttemberg citizenship; uncompleted
emigration; naturalizations of the former
Kreisregierung Reutlingen; naturalizations
(1867-1869); naturalizations (1818-1843;
1885-1900)
Nr. 122: List of immigrants   1853-1868
Nr. 126-129: Naturalizations   1853-1900

261: F 205: Tuebingen: files as above. ADS 10:27

TUEFINGEN (village)   BWL

Emigration, including USA   1866-1943

288: Kommunalarchiv Tuefingen.   ADS 10:144

TUELLINGEN BEI LOERRACH (village). See Loerrach BWL

TUNSEL (village)   BWL

Emigration file   1851-1922

288: Kommunalarchiv Tunsel.   ADS 10:144

TURMRINGEN BEI LOERRACH. See Loerrach   BWL

TUTSCHFELDEN (village)   BWL

Emigration file; list of emigrants   1855
Emigration files   1855-1938

288: Kommunalarchiv Tutschfelden.   ADS 10:144

TUTTLINGEN (Oberamt)   BWL

Nr. 101: Lists of emigrants   1812-1882
Nr. 191-222: Emigration files   19th century
Nr. 157-184: Witnesses [attestations of good char-
             acter]   1892-1904
Nr. 223: Emigration files   n.d.
Nr. 185: Attestations to citizenship certificates
                                                 1901
Nr. 150-156: Loss of certificates of original dom-
             icile   1884-1891

261: F 206: Tuttlingen: files as above. ADS 10:27

Emigration   1871-1915

262: Wue 65/37: Tuttlingen: Acc.5/1956: Nr. 467.
                                          ADS 10:39

UCHTE (Amt)   NL

Releases from citizenship   1851-1870

52: Han.Des.80: Uchte: B.66.VI.4: Nr. 269.
                                          ADS 4:85

UEBERLINGEN AM RIED (village)   BWL

Emigration, including USA   1872-1945

288: Kommunalarchiv Ueberlingen am Ried.
                                         ADS 10:144

UEFFELN (parish)   NL

List of emigrants to America   1834-1847

109: II.K.B.1: Kirchenbuecher 1670-1822. ADS 4:337

UFFENHEIM (Bezirksamt)   BYL

Emigration & immigration applications   1818-1866

231: Rep.212/18: Uffenheim: 16.Abgabe, 1927: Nr.
     215.   ADS 9:133

Nr. 484: Emigration to North America   1818-1860
Nr. 485: Emigration & immigration   1814-1861
Nr. 486: Emigration to North America & other far
         lands   1851-1865
Nr. 488: Bavarian emigration [statistical?]
                                        1853-1865

231: Rep.212/18: Uffenheim: files as above.
                                          ADS 9:130

Nr. 92: List of emigrants to overseas countries
                                     [19th century]
Nr. 93: List of persons who have emigrated without
        permits   [19th century]

231: Rep.212/18: Uffenheim: 18.Abgabe, 1930: files
     as above.   ADS 9:133

ULM (Oberamt)   BWL

Nr. 16-32: Emigration files, A-Z   19th century
[same Nr.] Emigration list   1849-1872

261: F 207: Ulm: files as above.   ADS 10:27

ULM (Kreisregierung). *See also* Kirchheim   BWL

ULM (village; Kreis Buehl)   BWL

Emigration file   1832-1926

288: Kommunalarchiv Ulm, Kr. Buehl.   ADS 10:144

ULM BEI OBERKIRCH (village)   BWL

Emigration file   1806-1937
Emigration file   1913-1932

288: Kommunalarchiv Ulm bei Oberkirch. ADS 10:144

ULRICHSTEIN (town)   HEL

List of emigrants   n.d.

194   ADS 7:257

UNITED STATES OF AMERICA

Hamburg General Consulate in USA: Numerous death
   certificates for Hamburg citizens who died in
   U.S. civil war   [ca. 1865]

39: Bestand Senat: Cl.VI, No.16p, vol.4a: Fasc. 7b,
    im ganzen 39 Nummern [total of 39 files]
    [Totenscheine]   ADS 2:46

Hamburg General Consulate in U.S.: Death certificates; searches for Hamburg citizens in U.S.,
   etc.   [ca. 1870?]

39: Bestand Senat: Cl.VI, No.16p: vol.4a: Fasc. 8e
    [Nachforschungen]   ADS 2:46

Hamburg General Consulate in USA: Death certificates for Hamburg citizens dying in America;
   legal affairs   n.d.

39: Bestand Senat: Cl.VI, No.16p: vol.4a: Fasc. 9,
    Invol. 2-11; 13-14 [Totenscheine]   ADS 2:46

UNNA (Kreis Hamm)   RWL

Nr. 39: Certificates of original domicile
                                        1833-1860
Nr. 58: Certificates of original domicile, corres-
        pondence regarding   1861-1886
Nr. 67: Certificates of original domicile; legiti-
        mation papers   1856-1875
Nr. 1177: Certificates of original domicile &
          citizenship   1919-1921
Nr. 540: Releases from citizenship   1874-1896
Nr. 538: Releases from citizenship   1896-1909
Nr. 1267: Hindering of emigration to America
                                        1826-1912
Nr. 1155: Emigration   1920-1929

116: Arnsberg Reg.: Hamm: 1: files as above.
                                       ADS 6:59-60

UNTERALPFEN (village)   BWL

Emigration file   1858-1934

288: Kommunalarchiv Unteralpfen.   ADS 10:145

UNTERBALDINGEN (village)   BWL

  Emigration files (4 fascicles)   1883-1946
  Germans abroad   1933-1944

  288: Kommunalarchiv Unterbaldingen.   ADS 10:145

UNTERDONAU (Kreis)   BYL

  Emigration & immigration; exportation of property
                          1831-1834

  225: Rep.168: Verz. 1: Fasz. 1773: Nr. 160.
                                ADS 9:15

UNTERENTERSBACH (village)   BWL

  Emigration file   n.d.

  288: Kommunalarchiv Unterentersbach.   ADS 10:145

UNTERFRANKEN   BYL

  Individual emigration files (to North America)
                          [19th century?]

  232: Rep. Regierung von Unterfranken: 1943/45
      Sachen: Nr. 9399-9407.   ADS 9:138

UNTERGLOTTERTAL (village)   BWL

  General emigration files   1880-1948

  288: Kommunalarchiv Unterglottertal.   ADS 10:145

UNTERHARMERSBACH (village)   BWL

  Emigration file   1890-1940

  288: Kommunalarchiv Unterharmersbach. ADS 10:145

UNTERIBENTAL (village)   BWL

  Emigration files   1853-1869

  288: Kommunalarchiv Unteribental.   ADS 10:145

UNTERLAUCHRINGEN (village)   BWL

  Emigration file   1840-1912

  288: Kommunalarchiv Unterlauchringen. ADS 10:145

UNTERMETTINGEN (village)   BWL

  Emigration file   1858-1943

  288: Kommunalarchiv Untermettingen.   ADS 10:145

UNTERMUENSTERTAL (village)   BWL

  Emigration to the USA   1832-1860

  288: Kommunalarchiv Untermuenstertal. ADS 10:145

UNTER-SCHARBACH   HEL

  Emigration   1859-1899

  149: Rep.40,2: Ablief.2: Bd. 310.   ADS 7:23

UNTER-SCHOENMATTENWAG   HEL

  Emigration   1859-1916

  149: Rep.40,2: Ablief.2: Bd. 311.   ADS 7:23

UNTERSIMONSWALD (village)   BWL

  Emigration file   1852-1941
  Emigration to the USA (Wisconsin) with supporting
    documents   1853-1854

  288: Kommunalarchiv Untersimonswald.   ADS 10:145

UNTERUHLDINGEN (village)   BWL

  Emigration file   1845-1942

  288: Kommunalarchiv Unteruhldingen.   ADS 10:145

UNZHURST (village)   BWL

  Emigration file   1880-1921
  Emigration file for Zell [village]   1832-1856

  288: Kommunalarchiv Unzhurst.   ADS 10:146

URACH (Oberamt)   BWL

  Emigration files, alphabetically arranged by vil-
    lage   [19th century]

  261: F 208: Urach: Nr. 27-33.   ADS 10:27

URBERG (village)   BWL

  Emigration file   1858-1936

  288: Kommunalarchiv Urberg.   ADS 10:146

URNAU (village)   BWL

  Emigration to the USA; Germans abroad 1825-1939

  288: Kommunalarchiv Urnau.   ADS 10:146

URSENSOLLEN (Oberpfalz)   BYL

  Emigration   1817-1819

  222: Regierung: Kammer der Finanzen: Rep.118: II.
      Teil: Nr. 9713.   ADS 9:2

USINGEN (Amt)   HEL

  Emigration from villages in Amt Usingen 1871-1877

  152: Abt.242: Usingen: Nr. 1068.   ADS 7:136

USINGEN (Landratsamt)   HEL

  Releases from Prussian citizenship   1893-1925
  <u>152</u>: Abt.420: Usingen: Nr. 16.   ADS 7:159

USINGEN. *See also* Nassau-Usingen   HEL

VAIHINGEN (Oberamt)   BWL

  Nr. 298-299: List of emigrants   1822-1894
  Nr. 300-351: Emigration & immigration,   by village
                                            1820-1915
  <u>261</u>: F 209: Vaihingen: files as above.   ADS 10:28

VALLENDAR (village)   RPL

  Emigration cases   1654-1726
  <u>200</u>: Abt.1C: Kurtrier: Nr. 4878.   ADS 8:9

VARENHOLZ (Amt)   RWL

  Emigration   1880-1894
  <u>114</u>: Brake: Fach 7, Nr. 8.   ADS 6:15

VEGESACK. *See* Bremen   HB

VERINGENSTADT (town)   BWL

  Emigration   1822
  <u>262</u>: Ho 199: Sigmaringen: Nr. I 11903.   ADS 10:36

VERSMOLD (Amt; Kreis Bielefeld)   RWL

  Emigration & immigration; trips to Holland
                                            1816-1873
  <u>116</u>: Arnsberg Reg.: Bielefeld: IV.   ADS 6:63

VIECHTACH (Bezirksamt)   BYL

  Emigration to North American states   1839-1869
  <u>225</u>: Rep.164: Verz.18: Fasz.90: Nr. 195.   ADS 9:14

  Emigration   [19th century]
  <u>225</u>: Rep.168: Verz. 1.   ADS 9:15

VIECHTACH (Landgericht)   BYL

  Applications for emigration permits   1817-1846
  <u>225</u>: Rep.168: Verz. 1.   ADS 9:16

VIERNHEIM   HEL

  Bd. 312: Emigration, A-J   1859-1899
  Bd. 313: Emigration, K-Z   1859-1914
  <u>149</u>: Rep.40,2: Ablief.2: volumes as above.
                                            ADS 7:23

According to information from the Magistrate of
the town of Viernheim, there are files for 452
emigration cases, all of whom apparently emigrated in 1852 to the USA. The list has been
published in the *Viernheimer Ortschronik* [date
of publication not given]

<u>195</u>   ADS 7:258

VILSBIBURG (Bezirksamt)   BYL

  Emigration   [19th century]
  <u>225</u>: Rep.168: Verz. 1.   ADS 9:15

VILSBIBURG (Landgericht)   BYL

  Applications for emigration permits   1824-1846
  <u>225</u>: Rep.168: Verz. 1.   ADS 9:16

VILSECK (Landgericht)   BYL

  Annual list of emigrants   1838-1860
  <u>222</u>: Rep.123: Amberg: Nr. 10.   ADS 9:2

VILSHOFEN (Bezirksamt)   BYL

  Emigration   [19th century]
  <u>225</u>: Rep.168: Verz. 1.   ADS 9:15

VILSHOFEN (Landgericht)   BYL

  Applications for emigration permits   1819-1846
  <u>225</u>: Rep.168: Verz. 1.   ADS 9:16

VIMBUCH (village)   BWL

  Emigration file   1838-1935
  <u>288</u>: Kommunalarchiv Vimbuch.   ADS 10:146

VOEGISHEIM (village)   BWL

  Emigration file   1826-1943
  <u>288</u>: Kommunalarchiv Voegisheim.   ADS 10:146

VOERDEN (Amt)   NL

  Nr. 6080: Emigration permits   1871-1879
            [for permits before 1871, *see* Osna-
            brueck]
  Nr. 9273: Releases from citizenship   1880
  <u>54</u>: Rep.116,I: files as above.   ADS 4:120-121

VOERDEN (Flecken)   NL

  Emigration matters   1835-1839
  <u>54</u>: Depositum 34: Archiv. Fl.Voerden: VII: Nr. 22.
                                            ADS 4:147

VOERSTETTEN (village)   BWL

   Emigration file   1864-1920

   <u>288</u>: Kommunalarchiv Voerstetten.   ADS 10:146

VOHENSTRAUSS (Bezirksamt)   BYL

   Applications for emigration   1840-1872

   <u>222</u>: Rep.141: Vohenstrauss: Nr. 2180-3252. ADS 9:4

VOIGTDING (parish)   NL

   List of emigrants   1874; 1880-1888

   <u>100</u>: Gem. Voigtding: V.: Nr. 5.   ADS 4:312

VOLKERTSHAUSEN (village)   BWL

   Emigration file   1860-1932

   <u>288</u>: Kommunalarchiv Volkertshausen.   ADS 10:146

VOLMARSTEIN (Amt; Kreis Hagen)   RWL

   Settlement & emigration   1842-1872

   <u>116</u>: Arnsberg Reg.: Hagen: A Nr. 74.   ADS 6:59

WACHAU (North? Carolina)   USA

   Herrnhuter [Moravian Brethren] reports on settlement in Wachau   1753-1772

   <u>500</u>: Herrnhut: Archiv: Rep.14.Ba: Nr. 2a; 2b; 2c;
   2d.   Not in ADS

WACHENHEIM (town)   RPL

   List of emigrants to America   1836

   <u>218</u>: loses Blatt [loose leaf].   ADS 8:88

WAGENSTADT (village)   BWL

   General emigration file   1845-1939
   Emigration files   1860-1935

   <u>288</u>: Kommunalarchiv Wagenstadt.   ADS 10:146

WAGENSTEIG (village)   BWL

   Emigration file   1871
   Emigration file   1948-1949

   <u>288</u>: Kommunalarchiv Wagensteig.   ADS 10:146

WAGSHURST (village)   BWL

   Germans abroad   1933-1934
   Emigration files (3 fascicles)   1860-1929
   Emigration & applications of girls from Baden to
   be placed in American convents   1914-1915

   <u>288</u>: Kommunalarchiv Wagshurst.   ADS 10:146

WAHLEN (village)   HEL

   Emigration   1857-1906

   <u>149</u>: Rep. 40,2: Ablief.2: Bd. 314.   ADS 7:23

WAIBLINGEN (Oberamt)   BWL

   Nr. 404: List of emigrants   1818-1900
   Nr. 405: Emigration files, general   1815-1891
   Nr. 406-430: Emigration files, chronologically ar-
         ranged (with emigration & immigration
         list)   1818-1895

   <u>261</u>: F 210: Waiblingen: files as above. ADS 10:28

WALBERTSWEILER (village; Oberamt Sigmaringen)   BWL

   Emigration   1861-1931

   <u>262</u>: Ho 199: Sigmaringen: Acc.7/1955: II Nr. 6; 80.
                        ADS 10:37

WALD-AMORBACH   HEL

   Petition of several residents for financial assistance in emigrating to America   1846

   <u>149</u>: Rep.39/1: Konv.135: Fasz. 4.   ADS 7:10

WALDBROEL (Kreis)   RWL

   Emigration   1857-1880

   <u>115</u>: Rep.D9: Koeln: Abt.I.e: Nr. 273. ADS 6:32

WALDBROEL (Landratsamt)   RWL

   Nr. 122: Immigration   1840-1843
   Nr. 124: List of foreigners   1896
   Nr. 124a: List of foreigners   1907-1911
   Nr. 124b: List of foreigners   1912-1925
   Nr. 125: List of foreigners   1925-1932
   Nr. 126: List of certificates of original domicile
        & citizenship   1899-1918
   Nr. 127: Passports   1901-1918
   Nr. 128: Passports   1918-1931

   <u>115</u>: Rep.G 29/42: Waldbroel: files as above.
                        ADS 6:46

WALDECK (principality)   HEL

   Nr. 3388: Passports   1729; 1812
   Nr. 3390: Passports   1776-1843

   <u>151</u>: Bestand 118: files as above.   ADS 7:106

   Nr. 2899: Census ["List of Souls"]   1725-1793
   Nr. 2900: Census ["List of Souls"]   1794-1828
   Nr. 2901: Census ["List of Souls"]   1808-1812
   Nr. 2774: Emigration tax [cabinet file on]
                     1747-1813

   <u>151</u>: Bestand 118: a: files as above.   ADS 7:98

According to information from the Evangelischen
Landeskirche von Kurhessen-Waldeck, there are
several volumes entitled "Entlassungen aus dem
Untertanenverband" [Releases from citizenship]
19th century

<u>190</u>   ADS 7:249

WALDKIRCH (village)   BWL

Emigration file   1865-1919

<u>288</u>: Kommunalarchiv Waldkirch.   ADS 10:147

WALDMATT (village). *See* Neusatz   BWL

WALDMICHELBACH   HEL

Emigration   1857-1906

<u>149</u>: Rep.40,2: Ablief.2: Bd. 314.   ADS 7:23

WALDMOHR (Landkreis Waldmohr, Kurpfalz)   RPL

Names of emigrants   18th century

<u>201</u>: Archivabt. Ausfautheiakten.   ADS 8:63

See also Altenkirchen, Elschbach, Kuebelberg,
Misau (villages)

WALDMUENCHEN (Bezirksamt)   BYL

Applications for emigration permits   1854

<u>222</u>: Rep.142: Waldmuenchen: Nr. 1090.   ADS 9:4

WALDSEE (Oberamt)   BWL

Nr. 134-136: Emigration lists & files (*also* emigration deposits)   1831-1893

<u>261</u>: F 211: Waldsee: files as above. ADS 10:28

WALDULM (village)   BWL

Emigration file   1847-1937
General emigration file; Germans abroad 1925-1939

<u>288</u>: Kommunalarchiv Waldulm.   ADS 10:147

WALLAU (village; Kreis Biedenkopf)   HEL

According to the local history society of Kreis
Biedenkopf, the Chronicle of Wallau Gemeinde
contains a list of emigrants during the nineteenth century

<u>196</u>   ADS 7:259

WALLBURG (village)   BWL

Emigration, alphabetically arranged   1832-1900

<u>288</u>: Kommunalarchiv Wallburg.   ADS 10:147

WALLDORF (village; Landkreis Gross-Gerau)   HEL

List of emigrants & their places of residence (contains the names of eight emigrants to USA)
1882-1923

<u>184</u>: Abt.XI,4.   ADS 7:238

WALLERSTAEDTEN (village; Landkreis Gross-Gerau) HEL

Applications for emigration permits to America
1838-1852
Annual report of emigrants & their property   1853
List of emigrants to North America   1853-1854
List of emigrants & the property taken with them
1853-1877
List of emigrants released [from citizenship] &
amount of property taken with them   1877

<u>185</u>: Abt.XI,4.   ADS 7:239

Entries for three emigrants to America   1877-1897

<u>185</u>: Abmelderegister 1875-1910.   ADS 7:239

Entries for 17 emigrants to America
19th & 20th centuries

<u>185</u>: Ortsbuergerregister 1782-1919.   ADS 7:239

WALSCHBRUNN (Pfalz?)   France or RPL

Emigration encouraged by French officials, in particular, the mayor of Walschbrunn, of Bavarian
subjects through France to America & their
embarkation at LeHavre [This Bavarian file evid-
ently made because, at the time, such emigration
would have been illicit.]   1828

<u>227</u>: Rep. Bayer. & Pfaelz. Gesandtschaft Paris:
Nr. 3298.

WALSHAUSEN (village; Landkreis Hornbach, Kurpfalz)
RPL

Names of emigrants   18th century

<u>201</u>: Archivabt. Ausfautheiakten.   ADS 8:63

WALSHEIM (village; Landkreis Hornbach, Kurpfalz) RPL

Names of emigrants   18th century

<u>201</u>: Archivabt. Ausfautheiakten.   ADS 8:63

WALSRODE   NL

Nr. b.43: Emigration certificates issued   1857
Nr. c.33: Transportation of asocial persons 1853-

[in Registratur des Magistrats. Inquire of the
Staatsarchiv Hannover F.A.91 for the present location of these files.]   ADS 4:338

WANGEN/ALLGAEU (Oberamt)   BWL

Emigration affairs   1919-1935

<u>262</u>: Wue 65/42: Wangen/Allgaeu: ZR 110. ADS 10:39

WANGEN (Oberamt)   BWL

   Nr. 43: Lists of emigrants   1827-1871
   Nr. 48-56: Emigration files, A-Z   1865-1932
   Nr. 43?: Financial aid to emigrants   1849-1875
   Nr. 57-58: Residency abroad with retention of
         [Wuerttemberg] citizenship   1842-1870

   <u>261</u>: F 212: Wangen: files as above.   ADS 10:28

WANGEN (village; Kreis Ueberlingen)   BWL

   Emigration file   1922-1923

   <u>288</u>: Kommunalarchiv Wangen, Kr. Ueberlingen.
                             ADS 10:147

WANGEN (village; Kreis Konstanz)   BWL

   Emigration files (3 fascicles)   1853-1942

   <u>288</u>: Kommunalarchiv Wangen, Kr. Konstanz.
                             ADS 10:147

WANNA (parish)   NL

   List of emigrants (name, age, destination)   1864
   List of emigrants (name, age, profession, destina-
      tion)   1867-1886

   <u>100</u>: Kirchspielsgericht Wanna: Fach 73, Nr. 2.
                             ADS 4:317

WARBURG (Kreis)   RWL

   Passports   1852-1895

   <u>116</u>: Minden: Praes.-regist.: I.P.: Nr. 316.
                             ADS 6:54

   Nr. 179-181: Emigration   1882-1924
   Nr. 57-59: Certificates of original domicile
                          1895-1924
   Nr. 90: Certificates of citizenship   1899-1923

   <u>116</u>: Minden: Praes.-regist.: I.A.: files as above.
                             ADS 6:50-52

WARENDORF (Landratsamt)   RWL

   Nr. 26: Emigration; immigration; naturalization
                          1836-1837
   Nr. 27: Emigration; immigration; naturalization
                          1850-1854
   Nr. 28-29: Certificates of original domicile
                          1820-1899
   Nr. 30: Releases from citizenship   1855-1900
   Nr. 32-33: Emigration; immigration; ex post facto
         naturalization   1872-1900
   Nr. 268: Citizens dying abroad & foreigners dying
        here   1825-1892
   Nr. 601: Passports   1841-1843

   <u>116</u>: Muenster Reg.: Warendorf: files as above.
                             ADS 6:69

WARENDORF (Kreis)   RWL

   Emigration   1885-1927

   <u>116</u>: Muenster Reg.: Nr. 1690-1691; 1673. ADS 6:57

WARSTEDE/NIEDERELBE. *See* Lamstedt/Niederelbe   NL

WASENWEILER (village)   BWL

   Emigration files (4 fascicles)   1839-1942

   <u>288</u>: Kommunalarchiv Wasenweiler.   ADS 10:147

WASSER (village)   BWL

   Emigration file   1870

   <u>288</u>: Kommunalarchiv Wasser.   ADS 10:147

WASSERTRUEDINGEN   BYL

   Emigration & immigration, Vol. IV   1852

   <u>231</u>: Rep.270/II: Reg. Mittelfranken: Kammer des
       Innern: Abgabe 1932: Nr. 25.   ADS 9:134

WEDDINGSTEDT (Kreis Norderdithmarschen)   SHL

   Applications of persons subject to military ser-
     vice for release from citizenship for the pur-
     pose of emigration   1851-1869

   <u>35</u>: Nr. 61 [Gesuche]   ADS 1:67

WEENER (Landratsamt)   NL

   Nr. 1737: Citizenship matters, letter C 1869-1903
   Nr. 2606: Citizenship matters, letter D 1842-1905
   Nr. 1318: Citizenship matters, letter F 1866-1906
   Nr. 1805: Citizenship matters, letter J 1866-1894
   Nr. 1878: Citizenship matters, letter K 1841-1903
   Nr. 1847: Citizenship matters, letter L 1817-1908
   Nr. 2386: Citizenship matters, letter M 1846-1905
   Nr. 2184: Citizenship matters, letter N 1843-1894
   Nr. 2183: Citizenship matters, letter O 1844-1898
   Nr. 677: Citizenship matters, letter P 1840-1900
   Nr. 1736: Citizenship matters, letter R 1862-1908
   Nr. 2659: Citizenship matters, letter S 1841-1909
   Nr. 1938: Citizenship matters, letter T 1865-1904
   Nr. 1726: Citizenship matters, letter U 1868-1901
   Nr. 1727: Citizenship matters, letter V 1865-1905
   Nr. 2187: Citizenship matters, letter W 1835-1907
   Nr. 1317: Citizenship matters, letter Z 1880-1909

   <u>50</u>: Rep.33: 4.b: files as above.   ADS 4:14

   Nr. 2304: Deaths abroad of Weener citizens
                           1829-1896
   Nr. 1253: Transportation of pardoned convicts to
        other parts of the world   1843-1846

   <u>50</u>: Rep.33: 4.a: files as above.   ADS 4:13

WEGSCHEID (Bezirksamt)   BYL

   Emigration   [19th century]

   <u>225</u>: Rep.168: Verz. 1.   ADS 9:15

   Releases from Bavarian citizenship   1822-1900

   <u>225</u>: Rep.164: Verz. 21: Nr. 59-132.   ADS 9:14

WEGSCHEID (Landgericht)   BYL

Applications for emigration permits
1815-1845; 1821-1838

<u>225</u>: Rep.168: Verz. 1.   ADS 9:16

WEHEN (Amt)   HEL

Emigration   1877-1882

<u>152</u>: Abt.244: Wehen: Nr. 1409.   ADS 7:136

WEHLAU (Landratsamt)   East Prussia

Bd. 1: Applications for releases from citizenship
[contains name, address, profession, when &
where born, details of military duty, emigration permits, arrangements for travel,
statistical data]   1879-1888
Bd. 1: Information on emigrants who left without
receiving release certificates [contains
some names, but mainly statistical]
1883-1887
Bd. 2: Certificates of release from citizenship &
permits to emigrate   1888-1924

<u>51</u>: Rep.18: Abt. VIII: Nr. 2: file as above.
ADS 4:53

WEHLAU (Landratsamt). *See also* Passenger list
(*Pannonia*, 23 May 1883)

WEHR (village)   BWL

Emigration file   1857-1858
Emigration to the USA   1861-1900
Germans abroad   1919-1933

<u>288</u>: Kommunalarchiv Wehr.   ADS 10:147

WEIDEN (canton)   RWL

Population movements   1801-1802; 1811-1812

<u>115</u>: Rep.D2.2: II.Division, 1.Bureau, 2.Bevoelker-
ung: Nr. 115.   ADS 6:19

WEIDENTHAL (village). *See* Neustadt (Oberamt) RPL

WEIER (village)   BWL

Emigration file   1873-1938

<u>288</u>: Kommunalarchiv Weier.   ADS 10:147

WEIERBACH. *See* Zell-Weierbach   BWL

WEIERHOF (village; Landkreis Kirchheimbolanden, Kurpfalz)   RPL

Names of emigrants   18th century

<u>201</u>: Archivabt. Ausfautheiakten.   ADS 8:63

WEIHER   HEL

Emigration   1856-1893

<u>149</u>: Rep.40,2: Ablief.2: Bd. 318.   ADS 7:23

WEILBACH (baths)   HEL

Treatment & guest lists   1867-1896

<u>152</u>: Abt.405: Wiesbaden: Rep.405/I,4: Nr. 1217.
ADS 7:148

WEILBURG (Herrschaft)   HEL

Nr. 4479: Regulations regarding serfdom & emigration   1607-1806
Nr. 4493: Emigration of subjects   1699-1792

<u>152</u>: Abt.150: Herrschaft Weilburg: files as above.
DLC has   ADS 7:127

WEILBURG (Amt)   HEL

Nr. 1371: Emigration   1816-1867
Nr. 1571: Summaries of naturalizations & releases
from citizenship   1839; 1846-1848
Nr. 1572: Summaries of naturalizations & releases
from citizenship   1854-1859
Nr. 1573: Summaries of naturalizations & releases
from citizenship   1862-1865

<u>152</u>: Abt.245: Weilburg: files as above. ADS 7:137

WEILERSBACH (village)   BWL

Emigration file   1902-1914

<u>288</u>: Kommunalarchiv Weilersbach.   ADS 10:147

WEILHEIM I (with Dietlingen)   BWL

Emigration files   1827-1902

<u>288</u>: Kommunalarchiv Weilheim I.   ADS 10:147

WEIMAR. *See* Saxony-Weimar-Eisenach

WEINGARTEN (village; Landkreis Germersheim, Kurpfalz)   RPL

Names of emigrants   18th century

<u>201</u>: Archivabt. Ausfautheiakten.   ADS 8:63

WEINSBERG (Oberamt)   BWL

Nr. 49-50: Immigration   1812-1859
Nr. 51-57: Emigraion files   1820-1858

<u>261</u>: F 213: Weinberg: files as above. ADS 10:28

WEINSBERG (town)   BWL

Reports from emigrants   [19th century?]

<u>279</u>: Neuer Nachlasszuwachs 1958.   ADS 10:82

WEISENBACH (village)   BWL

General files   1911-1937

<u>288</u>: Kommunalarchiv Weisenbach.   ADS 10:148

WEISENHEIM a. S. (village). *See* Kurpfalz   RPL

WEISSENBURG (Bezirksamt)   BYL

   List of emigration files (292 personal files)
                  mainly mid-19th century

   <u>231</u>: Rep.212/19: Weissenburg: Abgabe 1937, 1940.
                               ADS 9:133

WEISSENBURG (town)   BYL

   Emigration & immigration   1865

   <u>231</u>: Rep.270/II: Reg. Mittelfranken: Kammer des
      Innern: Abgabe 1932: Nr. 713.   ADS 9:134

   Material on emigrants to America has been arranged
     in a special card file   [19th century]
   <u>257</u>   ADS 9:180

WEISSENFELS BEI NAUMBURG/SAALE. *See* Schmidt, Karl
[surname index]

WEISWEIL (village)   BWL

   Emigration file   1852-1939
   Foreigners   1946-1948

   <u>288</u>: Kommunalarchiv Weisweil.   ADS 10:148

WEITENUNG (village)   BWL

   Emigration files   1850-1934
   List of emigrants   1853-1877

   <u>288</u>: Kommunalarchiv Weitenung.   ADS 10:148

WEIZEN (village)   BWL

   Emigration file   1826-1940

   <u>288</u>: Kommunalarchiv Weizen.   ADS 10:148

WELSCHENSTEINACH (village)   BWL

   Emigration files   1839-1909

   <u>288</u>: Kommunalarchiv Welschensteinach.   ADS 10:148

WELSCHINGEN (village)   BWL

   General emigration file   1894-1933

   <u>288</u>: Kommunalarchiv Welschingen.   ADS 10:148

WELZHEIM (Oberamt)   BWL

   Lists of emigrants   1818-1891

   <u>261</u>: F 214: Welzheim: Nr. 91.   ADS 10:28

WENNIGSEN (Amt)   NL

   Releases from citizenship   1850-1868

   <u>52</u>: Han.Des.80: Wennigsen: B.cc.VI.4: Nr. 624.
                             ADS 4:85

WERDENFELS (Grafschaft)   BYL

   Manumission & emigration taxes   1763-1765

   <u>229</u>: Rep. G.L.: Gerichtsliteralien: Nr. 4488.44.
                             ADS 9:115

WERTHEIM

   Emigration to America of Wertheim subjects hindered
                                   1773

   <u>158</u>: Frankfurter Ratsakten: Ugb.A.9: Nr. 4.
      DLC has [original in <u>158</u> destroyed] ADS 7:198

WERTINGEN (Bezirksamt)   BYL

   About 270 individual emigration files [destination
     not given therein]   ca. 1830-1900

   <u>230</u>: Rep. Akten der Bezirksaemter: Wertingen: Ab-
      gabe 1929: Nr. 7-.   ADS 9:124

WESEL (city)   RWL

   Emigration   1842-1862

   <u>146</u>: Caps. 209: Nr. 13; 19.   ADS 6:133

WESSELBURG (Kreis Norderdithmarschen)   SHL

   Nr. 91: Loss of citizenship   1857-1892
   Nr. 92: Emigration   1868-1878
   Nr. 94: Passports   1865-1891
   Nr. 95: Registration lists   1891-1906

   <u>36</u>: Acta VIII: files as above.   ADS 1:68

WESTEN-THEDINGHUSEN   NL

   Emigration files   1828-

   <u>52</u>: Han.Des.74: Westen-Thedinghusen: I.96: Nr. 3.
                              ADS 4:80

   Applications for releases from citizenship:
     Nr. 1:   1827-1833
     Nr. 2:   1834-1835
     Nr. 3:   1836
     Nr. 5:   1839
     Nr. 6:   Emigration permits   1846-
     Nr. 7: Lists of emigrants & immigrants   1858

   <u>52</u>: Han.Des.74: Westen-Thedinghusen: III.1y: files
     as above.   ADS 4:80-81

   Nr. 12: Passports issued   1826-
   Nr. 21: Visa register   1826-1834
   Nr. 26: Passports & certificates of residency,
         Vols. I & II   1834-
   Nr. 38: Visa register   1835-1838
   Nr. 40: Visa register   1839-1841
   Nr. 43: Visa register   1842-1844
   Nr. 47: Visa register   1844-1845

   <u>52</u>: Han.Des.74: Westen-Thedinghusen: I.56: files
     as above.   ADS 4:80

Death certificates received from abroad   1836

   52: Han.Des.74: Westen-Thedinghusen: I.96: Nr. 12.
                                                  ADS 4:80

WESTERBURG UND SCHADECK (Herrschaft)   HEL

Nr. 3: Serfdom; emigration tax; manumission pay-
    ments   1796-1799
Nr. 4: Emigration & immigration fees   1798

   152: Abt.339: Westerburg u. Schadeck: VIII.e:
      files as above.   ADS 7:138

WESTERBURG (Landratsamt)   HEL

List of persons released from Prussian citizenship
                                  1884-1895

   152: Abt.421: Westerburg: Nr. 6.   ADS 7:159

WESTERENDE-OTTERNDORF (parish)   NL

Applications for releases from citizenship
                                  1835-1862

   100: Kirchspielsgericht Westerende-Ott.: Loc.9:
      Nr. 21.   ADS 4:307

WESTERHEVER (church; Eiderstedt rural area)   SHL

Census of emigrants   1860

   2: zu Abt.163: II: Nr. 26 [Zaehlung]   ADS 1:20

WESTER-ILIENWORTH (parish)   NL

Lists of emigrants (names of males; number of per-
    sons, destinations)   1864-1875
Lists of immigrants (same data)   1864-1875
Lists of foreigners [mainly Americans] residing in
    parish (names, professions, religion, birthdates,
    places of residence, where from)   1896-1934

   100: Kirchspielsgericht Wester-Ilienworth: V.C.:
      Nr. 2.   ADS 4:309-310

WESTER-ILIENWORTH. *See also* Oster-Ilienworth   NL

WESTFALEN (Westphalia; duchy)   RWL

Emigration from the Duchy   1764-1766

   116: Herzogtum Westfalen: Landesarchiv Akten: III:
      Nr. 11a.   ADS 6:69

WESTHEIM (village). *See* Kurpfalz   RPL

WETTELBRUNN (village)   BWL

Emigration files (4 fascicles)   1847-1952

   288: Kommunalarchiv Wettelbrunn.   ADS 10:148

WETZLAR (Kreis)   HEL or RPL

Naturalizations   1859-1900

   200: Abt.441: Regierung Koblenz: Nr. 23322; 24100;
      23063; 22049; 24037; 22050-22056.   ADS 8:30

WETZLAR (city)   HEL

Much information on emigrants contained in news-
    paper collection   1767-

   198   ADS 7:261

WEYHERS (Amt)   HEL

Passports   1843-1862

   151: Bestand 112: Acc.1909/9: LIV: Nr. 16.
                                   ADS 7:98

WIECHS (village)   BWL

Emigration file   1868-1940

   288: Kommunalarchiv Wiechs.   ADS 10:148

WIED (Principality)

Reception of Mennonites   1680

   213: Schrank 26: Gefach 10: Fasc. 10.   ADS 8:81

Fasc. 1: Moravian Brethren congregation in Neuwied
                                    n.d.
Fasc. 2: Mennonite congregation in Neuwied   n.d.
Fasc. 3: Congregation of the Inspired in Neuwied
                                    n.d.
Fasc. 4: The Pietists from Teschen   n.d.
Fasc. 5: The Pietists from Sweden   n.d.
Fasc. 6: Society for the Unification of Religions
    in the Grafschaft Wied-Neuwied   n.d.
Fasc. 7: Religious sects in Grafschaft Wied-Runkel
                                    n.d.
Fasc. 8: The Johann Christ Edelmann sect in Neu-
    wied (anti-Christian)   n.d.
Fasc. 16: Correspondence with Zinsendorf   n.d.
      [All the above fascicles are from the
      18th century]

   213: Schrank 65, Gefach 11: files as above.
                                    ADS 8:81

Fasc. 4: Colonists & foreigners   1721
Fasc. 6-7: Foreigners   n.d.
Fasc. 8: French colonists   1732; 1733
Fasc. 9: Colonists   1734

   213: Schrank 26: Gefach 10: files as above.
                                    ADS 8:81

Naturalization fees, emigration taxes, etc. [par-
    ticularly valuable, because many religious groups
    stopped in Wied while en route to America]
                                [18th century]

   213: Schrank 70: Gefach 5 [*see also* Schrank 103:
      Gefach 97-98; *and* Schraenke 76-79]   ADS 8:81

Emigration   1828-1845

<u>200</u>: Abt.443: Fuerstlich Wiedsche Regierung Neu-
       wied: Nr. 81.   ADS 8:43

Naturalizations & releases from citizenship
                                   1827-1848

<u>200</u>: Abt.443: Fuerstlich Wiedsche Regierung Neu-
       wied: Nr. 82-83; 88-89; 121; 128.   ADS 8:43

WIED. *See also* Neuwied; Wied-Runkel

WIEDEN (village)   BWL

Emigration file   1836-1937

<u>288</u>: Kommunalarchiv Wieden.   ADS 10:148

WIEDENBRUECK (Kreis)   RWL

Nr. 182-187: Emigration   1880-1924
Nr. 60-62: Certificates of original domicile
                                 1863-1914
Nr. 91-92: Certificates of citizenship   1899-1924

<u>116</u>: Minden: Praes.-regist.: I.A.: files as above.
                               ADS 6:50-52

Nr. 661: Passports   1885-1902
Nr. 252: Lists of foreigners in Kreis   1818-1853
Nr. 362: Expulsions   1860-1907
Nr. 283: Expulsions   1907-1935
Nr. 709: Certificates of original domicile
                                 1828-1869
Nr. 209: Certificates of original domicile & citi-
       zenship   1920-1935
Nr. 306: Releases from Prussian citizenship
                                 1884-1905
Nr. 831: Emigration & immigration   1872-1910
Nr. 272: Emigration & immigration   1907-1938

<u>116</u>: Arnsberg Reg.: Wiedenbrueck: files as above.
                               ADS 6:64

WIEDORF. *See* Dillenburg   HEL

WIED-RUNKEL (Grafschaft)   HEL

Emigration tax   1787

<u>152</u>: Abt.335: Wied-Runkel: XIV.b.2.   ADS 7:138

WIED-RUNKEL. *See also* Neuwied; Wied

WIESBADEN (city)   HEL

Expulsion of young men from Frankfurt/Main who
   have taken on foreign citizenship in order to
   avoid military service   1869

<u>152</u>: Abt.405: Wiesbaden: Rep.405/I,5: Nr. 117.
                               ADS 7:148

WIESBADEN (Landratsamt)   HEL

Immigration & emigration of persons subject to
   military service; avoidance of service, etc.

<u>152</u>: Abt.422: Wiesbaden: Nr. 36.   ADS 7:159

WIESBADEN (Landesregierung)   HEL

Emigration   1845-1876

<u>152</u>: Abt.211: Wiesbaden LR: Nr. 11010.
   DLC has.   ADS 7:130

The compilers of the ADS did not list the large
collection in the Staatsarchiv Wiesbaden, entitled "Heirats-, Rezeptions- und Entlassungs-
sachen" [Marriages, naturalizations, & releases
from citizenship]. A listing of these holdings
remains to be undertaken.

<u>152</u>   ADS 7:130

WIESBADEN (Prussian Regierung)   HEL

Nr. 6927: Gains & losses of citizenship (with
          lists of emigrants)   1882-1887
Nr. 2928: List of foreigners living in Regierungs-
          bezirk (except Frankfurt & Wiesbaden)
                                  1934
Nr. 2926: List of foreigners [in Wiesbaden?] 1934
Nr. 3384: Estate matters (many cases of persons
          dying in the USA)   n.d.
Nr. 2710: Estate matters   n.d.

<u>152</u>: Abt.405: Wiesbaden: Rep.405/I,2: files as
   above.   ADS 7:140; 142

WIESBADEN (Regierung)   HEL

Correspondence with foreign governments regarding
   matters of individuals & passports, Vol. 4
                                 1928-1933

<u>152</u>: Abt.405: Wiesbaden: Rep.405/I,2: Nr. 8474.
                               ADS 7:147

Nr. 8659: Trials of reservists subject to military
          duty who have emigrated illegally or who
          refuse to serve   1868-1907
Nr. 2746: English & American citizens during the
          [first world] war   1914-1918
Nr. 4235: Surveillance of enemy (& alleged) spies
                                  1905-1921

<u>152</u>: Abt.405: Wiesbaden: Rep.405/I,5: files as
   above.   ADS 7:148

WIESBADEN (baths)   HEL

Treatment & guest lists   1867-1896

<u>152</u>: Abt.405: Wiesbaden: Rep.405/I,4: Nr. 1217.
                               ADS 7:148

WIESENFELD BEI COBURG. *See* Guntzel, F. W. [surname
index]   Saxony-Coburg-Gotha

WIESENTAL. *See* Zell-Wiesental   BWL

WILDGUTACH (village)   BWL

Emigration file   1861-1881

<u>288</u>: Kommunalarchiv Wildgutach.   ADS 10:148

WILDTAL (village)   BWL

   Emigration files   1847-1932

   <u>288</u>: Kommunalarchiv Wildtal.   ADS 10:149

WILLSTAETT (village)   BWL

   Emigration files   1817-1941

   <u>288</u>: Kommunalarchiv Willstaett.   ADS 10:149

WILSTER (Kreis Steinburg)   SHL

   Passport journal   1885-1933

   <u>37</u>: Gruppe III: Nr. 255.   ADS 1:69

WIMPFEN [*also called* Bad Wimpfen]   HEL

   Bd. 315: Emigration, A-G   1859-1891
   Bd. 316: Emigration, H-M   1859-1907
   Bd. 317: Emigration, N-Z   1858-1895

   <u>149</u>: Rep.40,2: Ablief.2: files as above. ADS 7:23

WIMPFEN   BWL?

   Much material on emigrants to USA   19th century

   <u>287</u>: Altregistratur.   ADS 10:98

WINDECKEN (Amt)   HEL

   Nr. 2085: Emigrants   1741
   Nr. 51974: Emigrants to Nova Scotia & Pennsylvania
                                         1752

   <u>151</u>: Bestand 86: files as above.   ADS 7:88-89

WINDECKEN (Amt)   HEL

   For exemptions to emigration tax, *see* Hesse-
   Kassel [Kurhesse]

WINDSBACH (Kammeralamt; Oberamt Ansbach)   BYL

   Emigration file   1805-1806

   <u>231</u>: Rep.165a: Ansbacher Oberamtsakten: Nr. 1720.
                                         ADS 9:127

WINDSHEIM (town)   BYL

   Emigration & immigration   [19th century]

   <u>258</u>: Tit.I, Fach 9, Akt Nr. 1 & 2. ADS 9:181

WINNWEILER (Oberamt). *See* Falkenstein (Grafschaft)

WINSEN/LUHE   NL

   Nr. 2: Releases from citizenship   1826-1879
   Nr. 3-5: Emigration to America   1837-1889

   <u>110</u>: Rep.II, Fach 52: files as above. ADS 4:339

WINTERSULGEN (village)   BWL

   Emigration file   1849-1938

   <u>288</u>: Kommunalarchiv Wintersulgen. ADS 10:149

WINTERSWEILER (village)   BWL

   Emigration   1835-1931

   <u>288</u>: Kommunalarchiv Wintersweiler. ADS 10:149

WINZINGEN (village)   RPL

   List of emigrants to America   n.d.

   <u>221</u>   ADS 8:81

WIPPERFUERTH (Landratsamt)   RWL

   Nr. 277: Emigration   1817-1875
   Nr. 193: Emigration   1884-1930
   Nr. 367: Passports   1916-1917

   <u>115</u>: Rep. G 29/42: Waldbroel?: files as above.
                                         ADS 6:46

WISCONSIN   USA

   Emigration to Wisconsin from Untersimonswald
                                         1853-1854

   <u>288</u>: Kommunalarchiv Untersimonswald. ADS 10:145

WISCONSIN   USA

   German consulate for Wisconsin. *See* Cincinnati,
   Ohio

WITTENHOFEN (village)   BWL

   Emigration to the USA   1852-1855
   Emigration files   1897-1936
   General file [emigrant] lists   1892-1942

   <u>288</u>: Kommunalarchiv Wittenhofen. ADS 10:149

WITTENTAL (village)   BWL

   Emigration files   n.d.

   <u>288</u>: Kommunalarchiv Wittental.   ADS 10:149

WITTENWEILER (village)   BWL

   Emigration file   1843-1949

   <u>288</u>: Kommunalarchiv Wittenweiler. ADS 10:149

WITTGENSTEIN (Kreis)   RWL

   Nr. 15: Emigration, 4 vols.   1844-1857
   Nr. 16: Emigration & immigration, 4 vols. 1832-1844

   <u>116</u>: Arnsberg Reg.: Wittgenstein: files as above.
                                         ADS 6:62

WITTGENSTEIN (Kreis). *See also* Arfeld   RWL

WITTLAGE (Amt)   NL

   Nr. 1210a: Register of passports, etc.   1865-1909
   Nr. 1211: Emigration certificates issued   n.d.
   Nr. 241: Emigration permits for persons who have
        not completed military service   1852
   Nr. 243: Emigration permits for persons who have
        not completed military service 1864-1869
   Nr. 242: Emigration   1853-1870
   Nr. 242a: Emigration   1859-1864
   Nr. 244: Emigration permits   1871-1874

   <u>54</u>: Rep.122,X: Wittlage: files as above. ADS 4:143

   Nr. 6081: Emigration permits [for permits before
        1871, *see* Osnabrueck]   1871-1879
   Nr. 13411: Releases from citizenship   1892

   <u>54</u>: Rep.116,I: files as above.   ADS 4:120; 124

   Nr. 74: Releases from citizenship   1880-1881
   Nr. 91: Releases from citizenship   1881-1883
   Nr. 93: Releases from citizenship   1882-1883
   Nr. 97: Releases from citizenship   1883-1885
   Nr. 105: Releases from citizenship   1885-1887
   Nr. 113: Releases from citizenship (Kreis)
                              1888-1891

   <u>54</u>: Rep.116,I: II.4.A.1e: files as above.
                         ADS 4:121-122

WITTLAGE-HUNTEBURG (Amt)   NL

   Nr. 12: Emigration & immigration   1824-1837
   Nr. 13: Emigration & immigration   1838-1852

   <u>54</u>: Rep.122,X.B: Wittlage-Hunteburg: I. Fach 17:
      files as above.   ADS 4:143

   Emigration files   1834-1848

   <u>54</u>: Rep.122,X.B: Wittlage-Hunteburg: I. Fach 92:
      Nr. 21.
      DLC has   ADS 4:144

   Transportation of various individuals to America
                              1845-1852

   <u>54</u>: Rep.122,X.B: Wittlage-Hunteburg: I. Fach 81:
      Nr. 26.
      DLC has   ADS 4:144

   Nr. 1: Collection of old passports   1814-1826
   Nr. 2: Collection of old passports   1822-1840
   Nr. 3: Passport & visa register   1814-1845
   Nr. 6: Collection of old passports   1840-1844
   Nr. 7: Collection of old passports   1844-1845
   Nr. 8: Collection of old passports   1846-1847

   <u>54</u>: Rep.122,X.B: Wittlage-Hunteburg: I. Fach 47:
      files as above.   ADS 4:144

   Register of passports issued   1865-1909

   <u>54</u>: Rep.:Landratsamt Wittlage-Hunteburg: I: Nr. 16
      (*see* Rep.122,X: Nr. 1210a)   ADS 4:145

WITTLICH (Landkreis)   RPL

   Emigration to America   1857-1910

   <u>200</u>: Abt.442: Regierung Trier: Nr. 194-195; 6077-
      6078; 6080-6086.   ADS 8:43

Emigration   19th century

   <u>200</u>: Abt.701: Nachlaesse, etc.: Nr. 976. ADS 8:54

   Manuscript on emigration from Kreis Wittlich, by
     Josef Mergen   1955

   <u>200</u>: Abt.701: Nachlaesse, etc.: Nr. 964. ADS 8:53

WITTLINGEN (village)   BWL

   Emigration files   1863-1941
   Foreigners   1926-1951

   <u>288</u>: Kommunalarchiv Wittlingen.   ADS 10:149

WITTMUND (Landratsamt)   NL

   Grants & renunciations of citizenship   1870-
     [There is much information on emigrants to
     America in this collection. The files are in-
     complete, having been lost in Wittmund. Avail-
     able are files for applicants whose surnames
     begin with B, E, H, J, N, O, P, Q, R, T, U, V]

   <u>50</u>: 34a? [24a?] 1: Nr. 3038.   ADS 4:14-15

WITTNAU (village)   BWL

   Emigration files   1837-1939

   <u>288</u>: Kommunalarchiv Wittnau.   ADS 10:149

WITZENHAUSEN (Kreis)   HEL

   Naturalizations, 6 vols.   1821-1867

   <u>151</u>: Bestand 17(b): Gef. 96: Nr. 1.   ADS 7:69

WITZENHAUSEN (Landratsamt)   HEL

   Nr. 30: Releases from citizenship   1854-1856
   Nr. 56: Emigration via Hamburg & Bremen, 2 vols.
                               1851-1879

   <u>151</u>: Bestand 180: Witzenhausen: Acc.1894/28: files
      as above.   ADS 7:123

   Emigration via Hamburg & Bremen   1882-1911

   <u>151</u>: Bestand 180: Witzenhausen: Acc.1938/65: Nr.
      11.   ADS 7:122

   Nr. 8: Persons emigrating without releases from
        citizenship   1885-1918
   Nr. 10: Naturalizations & releases from citizen-
        ship   1920-1928
   Nr. 59: Register of releases from Prussian citizen-
        ship   1860
   Nr. 61: Releases from citizenship & naturalization
                            1855-1923
   Nr. 100: Expulsion of foreigners   1912-1928
   Nr. 7: Certificates of original domicile 1900-1925
   Nr. 9: Certificates of original domicile 1900-1928
   Nr. 60: Certificates of original domicile
                             1836-1900
   Nr. 13: Reports of birth & death certificates to
        foreign governments   1840-1898

151: Bestand 180: Witzenhausen: Acc.1938/65: files
as above.   ADS 7:122

Resident foreigners   n.d.

151: Bestand 180: Witzenhausen: Acc.1885/18: Nr.
224.   ADS 7:122

WOELPE (Amt)   NL

Releases from citizenship   1851-1858

52: Han.Des.80: Woelpe: B.ee.VI.4: Nr. 234.
                                             ADS 4:85

WOELTINGERODE (Amt)   NL

Nr. 83: Lists of emigrants   1858-1887
Nr. 84: Lists of emigrants   1872-1879

52: Han.Des.74: Woeltingerode: II.6: files as
above.   ADS 4:81

WOERNERSBERG. *See* Freudenstadt   BWL

WOLFACH (town)   BWL

Emigration file   1854-1902

288: Kommunalarchiv Wolfach.   ADS 10:149

WOLFENBUETTEL (Duchy of Brunswick)   NL

Conv. 188: Releases from citizenship   n.d.
Conv. 231: Inquiries regarding Brunswick soldiers
           who were prisoners of war, killed, or
           missing in action in America [during
           American Revolution] 2 lists [may be
           missing]   n.d.
Conv. 294: Applications for emigration permits
           [Part a, to 1850; Part b, after 1850]

56: L Neu Abt.26, Gr.1, Fb.1: files as above.
                                             ADS 4:184

Nr. 3759: Releases of prisoners from house of cor-
          rection in Bevern for purpose of emigra-
          tion to America & funds approved for
          this purpose   1847-1872
Nr. 5360: Pension payments to veterans serving in
          America   1849-1879
Nr. 3270: Emigration: miscellaneous 1832-1877
Nr. 3271: Emigration   1846-1899
Nr. 3272: Emigration permits   1830-1867
Nr. 3273: Issuance of emigration permits to women
                                     1818-1869
Nr. 3275: Emigrants [without permits?] 1862-1864
Nr. 3276: Emigration of men subject to military
          service   1868-1871
Nr. 3277: Financial aid for emigration to America,
          3 vols.   1846-1901
Nr. 3280: Releases from citizenship 1872-1905
Nr. 3282: Expulsions from Duchy of Brunswick
                                     1850-1904
Nr. 3283: Expulsions from City of Braunschweig
          [Brunswick]   1851-1885
Nr. 3284: Payment of transportation costs for per-
          sons expelled from Duchy of Brunswick
                                     n.d.

56: L Neu Abt. 12A: Fb5: files as above. ADS 4:183

WOLFENBUETTEL (city)   NL

Nr. 50: Releases from citizenship   n.d.
Nr. 52: Emigration to America   1846-1864

56: L Neu Abt.127: Fb1: files as above. ADS 4:185

WOLFHAGEN (Kreis)   HEL

Naturalizations, 6 vols.   1822-1867

151: Bestand 17(b): Gef.96: Nr. 2.   ADS 7:69

WOLFHAGEN (Landratsamt)   HEL

Nr. 35: Emigration agents, special., 1 vol.
                                     1853-1887
Nr. 36: Emigration tax   1836-1846
Nr. 41: Marriages of foreigners [in Kreis] & of
        Germans abroad   1854-1869

151: Bestand 180: Wolfhagen: Acc.1897/31: Nr. 41.
                                             ADS 7:123

Nr. 34: Petitions for Prussian renaturalization
                                     1850-1904
Nr. 35: Emigration & citizenship matters 1885-1902
Nr. 27: Releases from citizenship   1862-1866
Nr. 28: Releases from citizenship   1858-1867
Nr. 61: Releases from Kurhessian citizenship
                                     1859-1861
Nr. 62: Releases from Kurhessian citizenship
                                     1832-1851
Nr. 64: Releases from Kurhessian citizenship
                                     1859-1883
Nr. 1: Certificates of citizenship   1891-1906
Nr. 30: Certificates of original domicile, support-
        ing papers   1878-1901
Nr. 32: Passports, supporting papers 1913-1925

151: Bestand 180: Wolfhagen: Acc.1939/28: files as
     above.   ADS 7:123

Releases from citizenship   1859-

151: Bestand 180: Wolfhagen: Acc.1957/41: Nr. 22.
                                             ADS 7:123

Passport register   1871-

151: Bestand 180: Wolfhagen: Acc.1957/41: Nr. 504.
                                             ADS 7:123

WOLFSTEIN (Bezirksamt)   BYL

Emigration to America   1852-1908

225: Rep.163: Verz. 22: Nr. 24-410.   ADS 9:13

Emigration   [19th century]

225: Rep.168: Verz. 1.   ADS 9:15

WOLFSTEIN (Landgericht)   BYL

Applications for emigration permits   1815-1846

225: Rep.168: Verz. 1.   ADS 9:16

WOLFSTEIN (village). *See* Kurpfalz   RPL

WOLLMESHEIM   RPL

Names of emigrants   18th century

<u>201</u>: Archivabt. Ausfautheiakten.   ADS 8:62

WOLPADINGEN (village)   BWL

Emigration files   1892-1933

<u>288</u>: Kommunalarchiv Wolpadingen.   ADS 10:150

WUERTTEMBERG   BWL

Transit via Frankfurt of Wuerttemberg emigrants to
America   1805

<u>158</u>: Frankfurter Ratsakten: Ugb.A.9: ad Nr. 10.
DLC has [original document in <u>158</u> destroyed]
ADS 7:198

Property dispositions; applications for emigration
permits; estates   1811-1813

<u>263</u>: Rep.E 1-13: Kabinettsakten III: Minist. des
Innern: 15, 1-4.   ADS 10:51

Nr. 24,1: Applications for emigration permits by
women & for the purpose of marriage
abroad   [1814?]
Nr. 24,2: Emigration of male citizens   [1814?]
Nr. 25,1-2: Emigration of male citizens   1815
Nr. 27: Applications for emigration   1816
Nr. 28,1: Applications for emigration   1817-1818
Nr. 33,5: Emigration   1819
Nr. 34,5: Emigration   1820-1821
Nr. 35,2: Emigration   1823-1832
Nr. 96: Monthly lists of foreigners in the twelve
Landvogteien [district offices] 1817-1819
Nr. 101: Landespolizei lists of foreigners 1823

<u>263</u>: Rep.E 10: Minist. des Innern: files as above.
ADS 10:52-53

Nr. 79-80: Emigration; measures taken to prevent
the emigration of persons subject to
military duty (also in other German
states)   1811-1820
Nr. 81-83: Emigration; measures taken to prevent
the emigration of persons subject to
military duty; certificates of original domicile (with list thereof)
1809-1853
Nr. 84: Emigration of [Wuerttemberg] subjects;
exemptions from military conscription
(with list)   1811-1855
Nr. 86-87: Emigration to various countries (with
list)   1817-1863
Nr. 89: Emigration; applications for releases from
citizenship (with list)   1827-1848

<u>263</u>: Rep.E 41-44: Minist. d. AA/II: Verzeichnis
63: files as above.   ADS 10:41

Bd. 10,1: Monthly lists of female citizens marry-
ing abroad   1811-1815
Bd. 10,2: Application for emigration permits, with
name lists   1816

<u>261</u>: Rep.E 141-142: Minist. des Innern: I: files
as above.   ADS 10:29

Emigration of certain persons   1848-1884

<u>261</u>: Rep.E 150-153: Minist. des Innern IV: Nr. 61.
ADS 10:32

Nr. 284: Emigration & dispensation applications,
special files, arranged by Oberaemter
[district offices]   to 1877
Nr. 285: Accidents to ships, etc., & support of
persons in such accidents   1848-1875
Nr. 287: Emigration to certain countries 1839-1868
Nr. 293: Releases of citizenship   1853-1858
Nr. 295: Special emigration files, arranged by
Oberaemter   1855-1880
Nr. 296: Payments of financial aid to Wuerttemberg
emigrants   n.d.

<u>261</u>: Rep.E 146-149: Minist. des Innern III: files
as above.   ADS 10:31

Legal action [in Rheinprovinz] against emigrants
from southern Germany to America   1827-1851

<u>200</u>: Abt.441: Regierung Koblenz: Nr. 5106-5109.
ADS 8:32

Emigration of convicts to America   [about 1860?]

<u>263</u>: Rep.E 46-48: Minist. des AA. III: Polizei:
Verz. 2, 886/11.   ADS 10:45

Confiscation of the property of deserters & of
persons who have emigrated with[out?] permission   [18th century?]

<u>263</u>: Rep.202: Geheimer Rat I (44. Kriegsangelegen-
heiten): Nr. 14.   ADS 10:51

Nr. 390: Applications by foreigners for temporary
residence, alphabetically by name
1806-1828
Nr. 391: Foreigners resident in Wuerttemberg, arranged by Oberamt [district office] 1830
Nr. 393-416: Professional applications by foreigners in Wuerttemberg   19th century
Nr. 417-483: Emigration matters, arranged by Oberamt   19th century

<u>261</u>: Rep.E 143-145: Minist. des Innern II: C.943-
43/77 to C.943.45/79.   ADS 10:29

Emigration with public assistances, Aalen to
Welzheim [alphabetical by Oberaemter]
19th century

<u>261</u>: Rep.E 143-145: Minist. des Innern II: Nr. 496-
559.   ADS 10:30

Residence of [Wuerttemberg] citizens abroad & the
residence of foreigners in Wuerttemberg   1810

<u>263</u>: Rep.E 41-44: Minist d. AA II: Verzeichnis 63:
Nr. 93.   ADS 10:42

Nr. 2,2: Foreigners resident in Wuerttemberg 1813
Nr. 7: Foreigners resident in the 12 Landvogteien
[district offices] of Wuerttemberg, Stutt-
gart, & Ludwigsburg   1813-1814
Nr. 20,3: Residence of individual foreigners in
Wuerttemberg   1815
Nr. 20,4: Residence of individual foreigners in
Wuerttemberg   1816
Nr. 20,8: Lists of foreigners in the 12 Landvogt-
eien [district offices]   1816

<u>263</u>: Rep.E 2: Polizeiminist.: files as above.
ADS 10:52

Applications from, & data on, foreigners 1850–1891

263: Rep.E 14–16: Kabinettsakten IV: Inlaender u.
Auslaender: Nr. 1465–1471.   ADS 10:53

Nr. 5: Foreign citizenship, individual cases
[20th century]
Nr. 15: Repatriation of [Wuerttemberg] citizens in
need of assistance from foreign countries
[20th century?]

263: Rep.E 131: Wuert. Staatsminist.: Abg. 1946
(in Ludwigsburg): files as above.   ADS 10:59

Mr. 324–387: Information regarding citizens abroad;
estates; issuance of baptismal, mar-
riage, & death certificates; certificates of original domicile arranged
by Oberaemter (Aalen to Ludwigsburg)
19th century

261: Rep.E 143–145: Minist. des Innern II: C.943.
43/77 to C.943.45/79.   ADS 10:29

Death certificates for citizens dying in America
1851–1876

261: Rep.E 146–149: Minist. des Innern III: Nr.
262.   ADS 10:30

The transmittal of Wuerttemberg inheritances to
heirs abroad   1910–1913

263: Rep.E 49–51: Minist. d. AA IV: Verz. 19: Nr.
13/33.   ADS 10:48

Estate matters   1911–1918

263: Rep.E 131: Wuert. Staatsminist. (im "Stal"):
F 19, 858.   ADS 10:49

Estates & wages due to Wuerttemberg soldiers who
died in America military service   1863–1895

263: Rep.E 49–51: Minist. AA IV: Verz. 19: 1/3;
1/3a–b.   ADS 10:48

Applications for assistance in obtaining American
[military?] pensions   1872–1910

263: Rep.E 49–51: Minist. AA IV: Verz. 19: 1/10.
ADS 10:48

Military service obligations of former Germans
1876–1903; 1903–1917

263: Rep.E 46–48: Minist. d. AA III: Polizei:
Verz. 2, 1094/1095.   ADS 10:47

Nr. 114: Extraditions to & from America 1874–1912
Nr. 125: Requests for extradition to & from America
1868–1899
Nr. 126: Requests for extradition to & from America
1900–1912

263: Rep.E 46–48: Minist. d. AA III: Verz. 1:
files as above.   ADS 10:43

WUERZBURG (Hochstift)   BYL

Emigration of various subjects to America   1764

232: Rep. Gruensfeld resp. Hochstift Wuerzburg:
Gebr.A.VI.G.128.
DLC has.   ADS 9:136

WUNSTORF   NL

Emigration & immigration   1852–

111: XIV, 5, Nr. 4; 5; 25.   ADS 4:346

WUTOESCHINGEN (village)   BWL

Emigration files (2 fascicles)   1830–1932

288: Kommunalarchiv Wutoeschingen.   ADS 10:150

WYHL (village)   BWL

Emigration files   1853–1949
Emigration to the USA   1854–1856
Emigration file   1908–1937

288: Kommunalarchiv Wyhl.   ADS 10:150

WYHLEN (village)   BWL

Emigration file   1817–1945
Emigration to the USA   1845–1871
Lists [of emigrants?]   1882–1932

288: Kommunalarchiv Wyhlen.   ADS 10:150

WYK AUF FOEHR   SHL

Nr. 1006: Applications for release from citizen-
ship (Danish)   1858–1862
Nr. 354: Passports   1778–1837
Nr. 355: List of passports issued   1803–1813

38: Verz. Nr. 1: files as above.   ADS 1:70

Nr. 1503: Release from [Prussian] citizenship
1867–1871
Nr. 1589–1594: Passports; sea passports 1867–1875
Nr. 1595–1596: Emigration   1867–1875

38: Verz. Nr. 2.1 (Landvogtei): files as above.
ADS 1:70

Nr. 4796: Passports (& photos)   1915–1924
Nr. 4792: Passports; certificates of domicile
1874–1922
Nr. 4327: Lists of persons departing   1893–1902
Nr. 4121: List of emigrants from Wyk   1865–1889
Nr. 4122: Emigration & immigration   1871–1884
Nr. 4128: Releases from Prussian citizenship
1885–1887
Nr. 4034: List of persons (many living abroad)
subject to military duty   1881

38: Verz. Nr. 3: Flecken Wyk: files as above.
ADS 1:71

WYK AUF FOEHR (Amtsbezirk Osterlandfoehr)   SHL

Nr. 1842: Expulsions   1872–1909
Nr. 1843: Emigration; releases from citizenship
1888–1912

38: Verz. Nr. 2.3 (Osterlandfoehr): files as above.
ADS 1:70

ZARTEN (village)   BWL

Emigration file   1833-1914
288: Kommunalarchiv Zarten.   ADS 10:150

ZASTLER (village)   BWL

Emigration file   1853-1931
288: Kommunalarchiv Zastler.   ADS 10:150

ZEISKAM (village; Landkreis Germersheim, Kurpfalz)
                                                    RPL

Names of emigrants   18th century
201: Archivabt. Ausfautheiakten.   ADS 8:63

ZEISKAM (parish)   RPL

The Reformed Church record contains marginal notations on emigrants to America   1709-1785
201: Archivabt. Kirchenbuecher.   ADS 8:63

ZEITSKOFEN (Herrschaftsgericht)   BYL

Applications for emigration permits   1815-1833
225: Rep.168: Verz. 1.   ADS 9:16

ZELL (Kreis)   RPL

Naturalizations   1843-1912
200: Abt.441: Regierung Koblenz: Nr. 22054-22056.
                                       ADS 8:30

ZELL (village). *See* Unzhurst   BWL

ZELL a. A. (village; Kreis Ueberlingen)   BWL

Emigration files   1807-1927
288: Kommunalarchiv Zell a. A., Kr. Ueberlingen.
                                       ADS 10:150

ZELLERFELD (Amt)   NL

Nr. 131: Passport register   1826-1844
Nr. 132: Passport register   1826-1831
Nr. 133: Passport register   1832-1836
Nr. 134: Passport register   1837-1843
52: Han.Des.74: Zellerfeld: K.E.I.: files as above.
                                       ADS 4:81

ZELLERFELD. *See also* Clausthal-Zellerfeld   NL

ZELL-WEIERBACH (village)   BWL

Emigration files   1840-1937
288: Kommunalarchiv Zell-Weierbach.   ADS 10:150

ZELL-WIESENTAL (village)   BWL

Emigration file   1843-1941
Emigration file   1852-1907
288: Kommunalarchiv Zell-Wiesental.   ADS 10:150

ZIENKEN (village)   BWL

Emigration files   1865-1926
288: Kommunalarchiv Zienken.   ADS 10:151

ZIMMERSRODE (village; Kreis Fritzlar)   HEL

Emigration tax   1840-1855
151: Bestand 180: Fritzlar: Acc.1931/33: Nr. 89.
                                       ADS 7:114

ZOTZENBACH   HEL

Emigration   1856-1893
149: Rep.40,2: Ablief.2: Bd. 318.   ADS 7:23

ZUELPICH (canton)   RWL

Population movements   1813
115: Rep.D2.2: II.Division, 1.Bureau, 2.Bevoelker-
     ung: Nr. 116.   ADS 6:19

ZUSMARSHAUSEN (Bezirksamt)   BYL

Nr. 2122: Emigration files (101 cases) 1821-1883
Nr. 2123: Emigration files (124 cases of emigra-
          tion to North America)   1862-1867
Nr. 2142: Information regarding Americans living
          in Bavaria   1907
230: Rep. Akten der Bezirksaemter: Zusmarshausen:
     files as above.   ADS 9:125

ZWEIBRUECKEN   RPL

Young men who have journeyed to America for a two-
  year period   1741-
201: Zweibruecken III: Akt Nr. 3358.   ADS 8:60

ZWEIBRUECKEN (Duchy)   RPL

Manumission protocols [Vol. 1]   1724-1749
  [Note: Volume II of the Manumissionsprotokolle,
  1750-1771, will be found in the Protestant
  Kirchsschaffeneiarchiv Zweibruecken; both volumes contain the names of many emigrants to
  America.]
201: Zweibruecken III: Akt Nr. 2025.   ADS 8:59

Correspondence regarding manumission & the regulation of emigration contains a large number of
  emigration cases to Pennsylvania (detailed)
                                       1732-1785

Among the receipts [Rechnungen] of the Duchy of Zweibruecken are the so-called Landschreiberei-rechnungen [receipts made in the various Aemter & Oberaemter] entitled "Abkauf der Leibeigen-schaft" [manumission from serfdom] and "Abkauf des 10.Pfennigs" [emigration tax]. Many of the names therein are those of emigrants to Pennsyl-vania   18th century

201: Rechnungen des Herzogtums Zweibruecken.
ADS 8:60

ZWEIBRUECKEN (arrondisement)   RPL

Emigration   1798-1813

201: Archivabt. Departement Donnersberg I: Akt.
Nr. 86.   ADS 8:63

ZWEIBRUECKEN. *See also* Pfalz-Zweibruecken (Duchy)

ZWINGENBERG. *See* Heppenheim (Kreis)   HEL

Subject Index

AMERICAN REVOLUTION. The archival material pertaining to Germans serving in the American Revolution has been arranged hereinafter according to the six sovereign powers represented. They were as follows:

German Troops in America

	Sent to	Returned from	Remained in*
Anhalt-Zerbst	1,152	984	168
Ansbach-Bayreuth	2,353	1,183	1,170
Braunschweig	5,723	2,708	3,015
Hessen-Hanau	2,422	1,441	981
Hessen-Kassel	16,992	10,492	6,500
Waldeck	1,225	505	720
Total	29,867	17,313	12,554

*Includes soldiers dying or killed in action.

For the German-American genealogical researcher, identification of the 12,554 men remaining in America is of primary importance. It is suggested that some soldiers listed in German records as killed in action or dying of wounds or sickness may, in fact, have been deserters. Treaties between Great Britain and Brunswick, Hessen-Hanau, and Waldeck called for the payment of "blood money" to the German princes in the event soldiers were killed, but not if they deserted. Consequently, it was to the advantage of the Germans to claim soldiers were killed rather than deserted.

Anhalt-Zerbst. No muster rolls for these military units have been found in West German archives.

Ansbach-Bayreuth. The following archival material has been found:

Regiment Eyb, later Seybotten, & Voit von Salzburg, in British service in America (orders, reports, accounts, etc.)   1777-1783

<u>228</u>: HS V 62.   ADS 9:109

Files on the [Ansbach-Bayreuth] troops sent to America in British service   1777

<u>223</u>: Rep. C.18.1: Contractanea Spiess: Nr. 26.
ADS 9:5

Salary advances paid to various Ansbach officers interned [it is not clear what internment is meant at this late date]   1794-1795

<u>234</u>: Ms.hist.486.
DLC has.   ADS 9:140

Various acts of clemency granted to mercenaries sent to America from Ansbach   1779-1784

<u>234</u>: Ms.hist.487.
DLC has.   ADS 9:140

A list of deserters in America and settlers in Canada (after 1783) is being prepared by this writer and will be avalable from Heritage House, Thomson, Illinois.

Bayreuth. See *Ansbach Bayreuth*

Braunschweig [Brunswick]. The following archival material has been found:

Muster rolls of Brunswick troops   1776-1783

<u>56</u>: N Abt.783: II: Nr. 54.
DLC has.   ADS 4:217

Lists of soldiers added to, & subtracted from, Brunswick forces [muster rolls of prisoners of war, deaths, wounded]   1776-1783

<u>56</u>: N Abt.783: II: Nr. 55.
DLC may have this.   ADS 4:217

Payrolls, accounting reports, receipts, forage & uniform payments for Brunswick troops:
vol. I   1776-1779
vol. II   1780-1782
vol. III   1783-1784

<u>56</u>: N Abt.783: II: Nr. 52.
DLC may have part of this.   ADS 4:217

Muster rolls of Brunswick troops sent to America
1777-1779

<u>56</u>: L Alt Abt.38B: Aeltere Mil.: Nr. 247.
DLC has.   ADS 4:169

Letter from Major (later Lieutenant Colonel) Georg
v. Rauschenplat at St. Antoine [Canada?] to General v. Riedesel containing muster roll of his
company   n.d.

<u>56</u>: N Abt.783: II: Nr. 31.   ADS 4:210

Reports from Captain H. Urban Cleve, on General
v. Riedesel's staff (Brunswick troops) contain
many lists, accounting & statistical appendices
[some muster rolls may be of genealogical interest; the contents are not described in ADS]
                                                   n.d.

<u>56</u>: N Abt.783: II: Nr. 12.   ADS 4:198

List of Brunswick officers sent to America
                                  January-May 1776

<u>52</u>: Han.Des.41.V.1: Nr. 4.
DLC has.   ADS 4:57

Brunswick troop list [to America or Gibraltar]
                                                  1776?

<u>86</u>: Nr. 25.   ADS 4:285

Muster rolls of the Convention troops [mainly
Brunswick troops in American prisoner of war
camp in Virginia]   1777-1783

<u>56</u>: L Alt Abt.38B: Aeltere Mil.: Nr. 249.
DLC has.   ADS 4:169

Payments made to family members [in Braunschweig]
of soldiers sent to America   1780-1783

<u>56</u>: L Alt Abt.38B: Aeltere Mil.: Nr. 250.
DLC has.   ADS 4:169

List of wounded & maimed soldiers of the American
Revolution [Brunswick troopers] who are pensioned at Campen   1783-1784

<u>56</u>: L Alt Abt.8: Bd.1: Campen: Gr. 25: Nr. 9.
                                           ADS 4:166

List of [Brunswick] veterans in Seesen 1783-1808

<u>56</u>: L Alt Abt.8: Bd.3: Seesen: Gr. 21: Nr. 2.
                                           ADS 4:167

Two lists of all the Brunswick troopers who were
killed in action, died of wounds or illness,
deserted, or otherwise did not return from
America [settlers]   1783

<u>56</u>: L Alt Abt.38B: Aeltere Mil.: Nr. 260.
DLC has.   ADS 4:171

Confiscation of the properties of Brunswick subjects sent abroad in military service; reports
on soldiers returning & those missing
                                          1773-1805

<u>56</u>: N Abt.34: Gr. IV: Nr. 3.   ADS 4:194

A list of deserters in America & of settlers in
Canada has been prepared by this writer & is
available from Heritage House, Thomson, Illinois

Cassel. See *Hessen-Kassel*

Hanau. See *Hessen-Hanau*

Hessen-Hanau. The following archival material has
been found:

Regiment Erbprinz: Foundation list of supernumeraries   n.d.

<u>151</u>: Bestand 13: A.6: Nr. 235.   ADS 7:60

Regiment Erbprinz, First Battalion: Foundation
list of recruits   1782

<u>151</u>: Bestand 13: A.6: Nr. 273.   ADS 7:62

Artillery Company: Foundation list from departure
from Hanau in 1776 to return from America in
1783   1776-1783

<u>151</u>: Bestand 12b: Rote Nr. 963.   ADS 7:55

Grenadier Battalion: Foundation list   n.d.

<u>151</u>: Bestand 12b: Rote Nr. 962.   ADS 7:55

Regiment Erbprinz, First Battalion: Tom.[?] III
Foundation list of men present in Canada n.d.

<u>151</u>: Bestand 12b: Rote Nr. 961.   ADS 7:55

Infantry Regiment Erbprinz, First Battalion, &
Artillery Company: Foundation list   1782

<u>151</u>: Bestand 12b: Rote Nr. 960a.   ADS 7:54

Regiment Erbprinz, First Battalion: Vol. II from
departure from Hanau in 1776 to return from
America in 1783 in six companies   1776-1783

<u>151</u>: Bestand 12b: Rote Nr. 959.   ADS 7:54

Regiment Erbprinz: [List not further described
in ADS]   n.d.

<u>151</u>: Bestand 12b: Rote Nr. 958.   ADS 7:54

Free Corps: Rote Nr. 954: vol. 3   1781
             Rote Nr. 955: vol. 4   1782
             Rote Nr. 956: officers & men   n.d.

<u>151</u>: Bestand 12b: files as above.   ADS 7:54

Jaegerkorps: Rote Nr. 950: Vol. 2   1777
             Rote Nr. 951: Foundation roll, vol.
             3   1783
             Rote Nr. 952: Hssars & Jaegerkorps
                                         1783; 1784
             Rote Nr. 953: Jaegerkorps recruits
                                              1780

<u>151</u>: Bestand 12b: files as above.   ADS 7:54

Jaegerkorps: Nr. 205: Command list of officers
                                              1780
             Nr. 206: Recruit lists from 1 Janu-
             ary 1776 to 20 March 1782

<u>151</u>: Bestand 13: A.6: files as above.
DLC has.   ADS 7:58

Jaegerkorps: Middle & lower staff positions 1777

<u>151</u>: Bestand 11: Verz. 2: OWS 1370: Nr. 561.
                                           ADS 7:49

Muster lists of all Hesse-Hanau troops in America
(name lists without places of origin)
1776-1783

151: Bestand 4: Kriegssachen: 415, Nr. 4.
DLC has.   ADS 7:35

Miscellaneous material relating to Hesse-Hanau
troops in America   1776-1782

500: Hanau: Geschichtsverein: Mss. Nr. 505; 561;
561a; 1608; 1608a; 2189; 2189a. Not in ADS

Hessen-Kassel. The following archival material has
been found:

Monthly reports of units serving in America: Contains location; name of company [by commander];
strength of unit by rank & rating; additions
(Hessians or non-Hessians); losses (transferred,
sick present or absent, dead of illnesses, prisoners, killed in action, wounded, deserted,
arrested); needed to complete complement; number ready for duty; supernumeraries; names are
shown for changes [in explanatory section of
the reports]; signed by regimental officers:

Leib-Infantry Regiment (including rank listing)
April 1776-May 1784

151: Bestand 12b: Bl. Nr. 8807.
DLC may have.   ADS 7:52

Infantry Regiment v. Wutgenau (after March 1777
Landgraf) includes rank lists   1776-1783

151: Bestand 12b: Bl. 8810.
DLC has.   ADS 7:53

Fusilier-Regiment Erbprinz: monthly list
1775-1780

151: Bestand 12b: Bl. 8814.
DLC has.   ADS 7:53

Infantry Regiment Erbprinz (after 1784 called
Prinz Wilhelm): monthly lists   1781-1785

151: Bestand 12b: Bl. 8815.
DLC has.   ADS 7:53

Regiment v. Mirbach (1770-1781) thereafter Jung
v. Lossberg (1781-1784) monthly lists
1770-1784
rank lists   1770-1776

151: Bestand 12b: Bl. 8819.
DLC has.   ADS 7:53

Regiment v. Truembach (1770-1778) thereafter
v. Bose (1779-1783) monthly lists   1776-1783
rank lists   1763-1776

151: Bestand 12b: Bl. 8826.
DLC has.   ADS 7:53

Fusilier-Regiment Alt v. Lossberg: monthly
lists [file appears to be missing] 1770-1783

151: Bestand 12b: Bl. 8833.   ADS 7:53

Grenadier-Regiment Rall, after October 1777 v.
Woellworth, after 1778 v. Lossberg's second
battalion: monthly lists   1776-1778
rank list   September 1776

151: Bestand 12b: Bl. 8834.   ADS 7:53

Grenadier-Battalion Block (1776-1777), thereafter
v. Lengerke (1777-1778): monthly lists
1776-1784
rank lists   1776-1784

151: Bestand 12b: Bl. 8836.
DLC has.   ADS 7:53

Grenadier-Battalion v. Minnigerode (1776-1779),
thereafter Third Grenadier-Battalion:
monthly lists   1776-1779

151: Bestand 12b: Bl. 8838.   ADS 7:53

Grenadier-Battalion Ko[e]hler (1776-1779); Graff
(1779-1781); Platte (1781-1784): monthly lists
with rank lists   1776-1784

151: Bestand 12b: Bl. 8839.   ADS 7:53

Regiment Prinz Carl: monthly lists   1770-1784
seniority rank lists   1770-1776

151: Bestand 12b: Bl. 8824.
DLC has.   ADS 7:53

Grenadier-Battalion Koehler [monthly lists?]
1776

151: Bestand 11: Verz.2: OWS 1374: Nr. 858.
ADS 7:50

Infantry Regiment v. Donop: monthly lists with
rank lists   1776-1784

151: Bestand 12b: Bl. 8841.   ADS 7:53

Infantry Regiment v. Knyphausen: monthly lists
1776-1783

151: Bestand 12b: Bl. 8842.   ADS 7:53

Fusilier-Regiment v. Ditfurth: monthly lists
1776-1783
rank lists   1765-1776

151: Bestand 12b: Bl. 8848.
DLC has.   ADS 7:53

Light Infantry Battalion v. Buelow [monthly
lists?]   n.d.

151: Bestand 12b: Bl. 8850.
DLC has.   ADS 7:53

Feldjaegerkorps: monthly lists   1776-1784

151: Bestand 12b: Bl. 8853.
DLC has.   ADS 7:53

Garnisons-Regiment v. Wilcke: monthly lists &
tables   1776-1783

151: Bestand 12b: Bl. 8862.
DLC has.   ADS 7:53

Garnisons-Regiment v. Wissenbach: monthly lists
1776-1780

151: Bestand 12b: Bl. 8868.   ADS 7:54

Garnisons-Regiment v. Knoblauch: monthly lists
1780-1783

151: Bestand 12b: Bl. 8869.   ADS 7:54

Garnisons-Regiment v. Bu[e]nau: monthly lists
1777-1783

151: Bestand 12b: Bl. 8873.   ADS 7:54

Garnisons-Regiment v. Buenau: [monthly list]
1777

151: Bestand 11: Verz.2: OWS 1374: Nr. 851.
ADS 7:50

Garnisons-Regiment v. Buenau: [monthly list]
1778

151: Bestand 11: Verz.2: OWS 1374: Nr. 853.
ADS 7:50

Garnisons-Regiment v. Buenau: [monthly list]
1783

151: Bestand 11: Verz.2: OWS 1374: Nr. 853.
ADS 7:50

Garnisons-Regiment v. Huyn (1780-1782), v. Benning (1782-1783): Monthly lists with rank lists
1776-1783

151: Bestand 12b: Bl. 8878.   ADS 7:54

Garnisons-Regiment v. Huyn: [monthly lists?]
1776

151: Bestand 11: Verz.2: OWS 1374: Nr. 831.
ADS 7:50

Grenadier-Battalion v. Linsingen: [monthly lists]
1776-1784

151: Bestand 12b: Bl. 8837.
DLC has.   ADS 7:53

Korps-Kompagnie Captain v. Francken: [monthly
lists]   1777

151: Bestand 11: Verz.2: OWS 1370: Nr. 563.
ADS 7:49

Korps-Kompagnie Captain Graf v. Wittgenstein:
[monthly lists]   1777

    151: Bestand 11: Verz.2: OWS 1370: Nr. 564.
ADS 7:49

Fusilier-Regiment v. Ditfurth: [monthly lists]
1773 [?]

151: Bestand 11: Verz.2: OWS 1373: Nr. 778.
ADS 7:49

Garnisons-Regiment v. Stein: [monthly lists]
1777

151: Bestand 11: Verz.2: OWS 1374: Nr. 814.
ADS 7:49

Garnisons-Regiment v. Seitz, thereafter v. Por-
beck & Kietzel: monthly lists   1776-1783

151: Bestand 12b: Bl. 8865.
DLC has.   ADS 7:54

Infantrie-Regiment v. Seitz: [monthly lists]
1778

151: Bestand 11: Verz.2: OWS 1374: Nr. 815.
ADS 7:49

Infantrie-Regiment v. Seitz: [monthly lists]
1780

151: Bestand 11: Verz.2: OWS 1374: Nr. 816.
ADS 7:49

Infantrie-Regiment v. Seitz: [monthly lists]
1781

151: Bestand 11: Verz.2: OWS 1374: Nr. 817.
ADS 7:49

Infantrie-Regiment v. Seitz: [monthly lists]
1782

151: Bestand 11: Verz.2: OWS 1374: Nr. 818.
ADS 7:49

Infantrie-Regiment v. Seitz: [monthly lists]
1783

151: Bestand 11: Verz.2: OWS 1374: Nr. 819.
ADS 7:49

Infantrie-Regiment v. Benning: [monthly lists]
1783

151: Bestand 11: Verz.2: OWS 1374: Nr. 832.
ADS 7:50

Grenadier-Battalion v. Platte: [monthly lists]
1783

151: Bestand 11: Verz.2: OWS 1374: Nr. 859.
ADS 7:50

[Entire Hessen-Kassel contingent]: rank & seniority lists for officers [one or two lists per
year]   1776-1779; 1781-1784

151: Bestand 11: Verz.2: Blaetter 66-73.
ADS 7:50

Grenadier-Regiment v. Woelwarth: rank list 1776

151: Bestand 4: Kriegssachen: 328: Nr. 152.
ADS 7:36

First Grenadier-Battalion: rank lists 1781-1782

151: Verz.13: Unverzeichnet 16.-18. Jhdt: Akten-
paket Nr. 7.   ADS 7:57

Third Grenadier-Battalion: rank lists 1781-1782

151: Verz.13: Unverzeichnet 16.-18. Jhdt.: Akten-
paket Nr. 7.   ADS 7:57

Fourth Grenadier-Battalion: rank lists 1781-1782

151: Verz.13: Unverzeichnet 16.-18.Jhdt.: Akten-
paket Nr. 7.   ADS 7:57

The *Mass- und Rangierbuecher* contain lists, by
rank, of the higher & subordinate officers of
the regiments & gives a summary view of the various companies. These records are arranged as
follows: officers, non-commissioned officers,
drummers, soldiers. For the enlisted me (including noncommissioned officers) hieght, age,
country of origin are given. In addition, there
are lists of new troopers added & of casualties
suffered. Aside from the monthly lists above,
these are the most important sources of information for regimental organization:

Fusilier Regiment Alt v. Lossberg, Quebec
October 1782
151: Bestand 11: Verz.2: OWS 1361: Nr. 204.
ADS 7:48

Nr. 205: Fusilier-Regiment v. Lossberg   1783
Nr. 213: Grenadier-Regiment v. Rall   1776
Nr. 214: Grenadier-Regiment v. Woellwarth 1777
Nr. 215: Grenadier-Regiment v. Woellwarth 1778
Nr. 216: Grenadier-Regiment v. Truembach 1779

151: Bestand 11: Verz.2: OWS 1361: files as
     above.   ADS 7:48

Nr. 217: Grenadier-Regiment d'Angelelli   1780
Nr. 218: Grenadier-Regiment d'Angelelli   1781
Nr. 219: Grenadier-Regiment d'Angelelli   1782
Nr. 220: Grenadier-Regiment d'Angelelli   1783

151: Bestand 11: Verz.2: OWS 1362: files as
     above.   ADS 7:48-49

Nr. 297: Leib-Infanterie-Regiment   1783
Nr. 313: Regiment Landgraf   1783

151: Bestand 11: Verz.2: OWS 1363: files as
     above.   ADS 7:49

Regiment v. Donop   1783

151: Bestand 11: Verz.2: OWS 1365: Nr. 388.
                                   ADS 7:49

Nr. 471: Regiment Prinz Carl   1783
Nr. 513: Regiment Jung v. Lossberg   1783

151: Bestand 11: Verz.2: OWS 1368: files as
     above.   ADS 7:49

Nr. 554: Grenadier-Battalion v. Linsing[en]
                                          1776
Nr. 555: Grenadier-Battalion v. Linsing[en]
                                          1783
Nr. 556: Battalion v. Lengerke   1783

151: Bestand 11: Verz.2: OWS 1369: files as
     above.   ADS 7:49

Lists [perhaps of transported recruits; ADS description of contents is unclear]   1777-1782

151: Bestand 13: A.6: Nr. 223; 224; 225.
     DLC has.   ADS 7:59

List of casualties   1776-1781
Possessions of soldiers killed in action or dead
   of illness   1776-1783

151: Bestand 12b: Bl. 8657; 8659a.   ADS 7:52

Four muster rolls of Regiment Lossberg [prepared
in New York]   1776-1783
Muster roll of the Grenadier Company   n.d.

In the possession of Schulrat a.D. Karl Vogt,
   Dingelstedtwall 30, Rinteln, Niedersachsen.
                                   ADS 4:330

List of Hessian officers   1776

52: Han.Des.41.V.2: Nr. 20.
    DLC has.   ADS 4:59

List of officers sent to America   1776

56: Hs.Abt.VI: Gruppe 18: Nr. 17. ADS 4:165

The Murhard'sche Bibliothek, Kassel, Winbergstr.
6, has a large & important collection of diaries & journals written by Hessian officers &
soldiers who participated in the American Revolution. Some of these items have been photocopied by the Library of Congress, but there
remains a large amount which has not been
photocopied. The material appears to be of
general historical, rather than of genealogical,
interest, however.   ADS 7:244-246

Repatriation of foreign [mainly Hessian] troops
returning from American duty; list with 8,662
names [310 pages; does not include Brunswick or
Anhalt-Zerbst troops]   1783-1784

52: Han. Des.41.V.5: Nr. 33.
    DLC has.   ADS 4:61

List of quarters assigned [at Blumenthal to Hessian troops en route to America]   1776

55: Han.Des.74: Blumenthal: II: Fach 58: Nr.
    15.   ADS 4:152

Lists of Hessian & other troops [in Braunschweig
collection]   1776-1782

56: N Abt.783: II: Nr. 56.
    DLC may have.   ADS 4:217

The Staatsarchiv Marburg has begun a series of
publications listing Hessian mercenaries serving in America. The first thereof is entitled
*Hessische Truppen im amerikanischen Unabhaengig-
keitskrieg (Hetrina)* [Hessian Troops in the American Revolution (Hetrina Project)], Veroeffentlichungen der Archivschule Marburgs, Institut
fuer Archivwissenschaft, no. 10 (Marburg, 1972).
This is a publication of major value to researchers. It contains names, birth year (com-
puted), birthplace & modern zip code number for
easy identification, rank or rating, military
unit, advancement date, & what happened to each
soldier (returned to Europe, settled in Canada,
deserted, died of wounds or illness, killed in
action). The first of these publications covers
Grenadier battalions v. Linsingen, Block, v.
Lengerke, v. Minnigerode, v. Loewenstein, Koehler, Graf, Platte, & the Garnison Regiments v.
Huyn, v. Benning, & v. Normann. Addition publications are to follow annually until all
Hessian & Waldeck military units have been completed.

Kurmainz [Electoral Mainz]. Although not one of
the principalities sending troops to America,
some subjects of Mainz were shanghaied into Hessian military service:

List of Kurmainzer subjects recruited for service
by Hessen-Hanau   1777-1778

232: Aschaffenburger Archivreste: M.R.A.Hessen-
     K.: K.359/966.
     DLC has.   ADS 9:137

Nassau. Although not one of the principalities
sending troops to America, some subjects of Nassau were shanghaied into Hessian military service:

Recruits [from Nassau] to America with the Hessian troops   1771-1783

152: Abt.350: Nassau-Dreiherrisch: Gen.VII.a.9.
                                   ADS 7:139

Waldeck. The following archival material has been found:

Nr. 942*: Muster rolls of the Third English-Waldeck Regiment 1776-1783

Nr. 945: Payrecords for 1782 recruits to the Third English-Waldeck Regiment 1782

Nr. 946: Vouchers & receipts to the above-described pay records 1782

Nr. 951: Account book, individual officers & men 1776-

Nr. 960*: Deserters & others who have fled their service in the Third English-Waldeck Regiment 1776-1787

Nr. 966*: Annual reports of the Third English-Waldeck Regiment 1776-1782

Nr. 968*: Strength reports of the Third English-Waldeck Regiment 1776-1783

Nr. 969: Desertion of two soldiers [unnamed in ADS description] from the Third English-Waldeck Regiment 1782

Nr. 970*: Muster rolls of the Third English-Waldeck Regiment 1776-1777; 1783

Nr. 973: Transfers of soldiers from the First & Second Holland-Waldeck Regiments to the Third English-Waldeck Regiment 1776-1777

Nr. 977: Passports & recommendations for soldiers of the Third English-Waldeck Regiment 1775

Nr. 978*: List of dead, retired, & deserted officers & men [of the Third English-Waldeck Regiment?] 1776-1783

Nr. 983: Personnel files of officers & men of the Third English-Waldeck Regiment 1776-1785

Nr. 987*: Gifts of mercy granted to family members in Waldeck of soldiers of the Third English-Waldeck Regiment 1776-1783

Nr. 997: Payrolls & expense accounts for the transport of recruits [replacements] for the Third English-Waldeck Regiment 1781

Nr. 999: Vouchers for above-described payrolls & expense accounts 1781

Nr. 1002: Transport of recruits [replacements] to Third English-Waldeck Regiment 1782-1783

Nr. 1003: Transport of recruits [replacements] to Third English-Waldeck Regiment 1776-1778

Nr. 1009-1053: There are many account books & vouchers available which may contain lists of names of genealogical interest. The individual numbers have not been listed in this *Handbook*, but, it seems clear, search of these materials is warranted.

151: Bestand 118: files as above.
DLC has files marked with asterisk.
ADS 7:98-106

Among the Hanxleden-Huyn papers there is likely to be material of value regarding the Waldeck contingent, commanded by a General Huyn. The material has not been put in order or catalogued

189: Nachlass Hanxleden-Huyn. ADS 7:248

Zerbst. See *Anhalt-Zerbst*

Miscellaneous. The following archival material, not otherwise classified, has been found:

Corp Company, Lt.Col. v. Creutzberg [material not otherwise described in ADS] 1777

151: Bestand 11: Verz.2: OWS 1370: Nr. 562.
ADS 7:49

Muster & release of foreigners [non-Hessians] from troop units returning from America 1784

151: Bestand 4: Kriegssachen: 328: Nr. 159.
ADS 7:36

Payment of pensions abroad, 2 vols. [probably particularly valuable for Canadian settlers] 1786-1867

151: Bestand 12b: Rote Nr. 1023. ADS 7:55

Research on ten invalid soldiers who served in America (detailed investigation, life histories, medical reports) 1806

259: XXXV, 153. ADS 9:182

There is said to be an exact abstract of the German Lutheran Christ Church & Military Church [Garnisonskirche] in New York, organized for the German troops in British service. There are 354 christening entries, 27 marriage entries, & 108 death entries 1778-1782

The abstract is in the possession of Herrn Lauterbach, director, Heimatbund Niedersachsen e.V., Altes Rathaus, Hannover ADS 4:278

Letters from Captain A. J. Thomae, commanding officer of first recruit transport, contains company muster roll 1777-1779

56: N Abt.783: II: Nr. 34. ADS 4:213

Recruit transport, 1777: The Newberry Library, Chicago, has a photocopy of a lengthy list said to have come from the papers of Colonel Rainsford in the British Museum. A diligent search in the British Museum fails to disclose the original manuscript from which the photocopy was made. An alphabetized list is being prepared (1974) by this writer and will be published by Heritage House, Thomson, Illinois. It contains about 2,000 names.

Special protocol regarding investigation of Trenton affair. Interrogations of two Jaegerkorps non-commissioned officers Hassel & Bauer 1778 (1782)

151: Bestand 4: Kriegssachen: 328: Nr.141-142.
DLC has. ADS 7:36

See also Captain -- Alberti, Senior; Major -- v. Seebisch; Vicekorporal -- Reichert (all of the Third English-Waldeck Regiment)

BAPTISTS, GERMAN (Dunkards; Dunkers; Taeufer; Wiedertaeufer). The following archival material has been
found:

Nr. 4331-4332: Prohibition against Pietists & Separatists   [17th century?]
Nr. 4333: Baptists in the Kurpfalz [Palatine Elec-
torate]   1566-1744
Nr. 4334: Dissolution of real property sales made
to Baptists, vol. I   1662-1763
Nr. 4336: Information regarding Baptists & Quakers
in the Palatinate, particularly the concessions
made to the Baptists   1651-1679
Nr. 4336a: Continuation of Nr. 4336   1680-1743

260: Abt.236: Innenmin.Wegzug: Auswand.: files as
above.   ADS 10:4

Material on   n.d.

235: Sachregister, Abg. 1923.   ADS 9:142

In Bavaria   1535-1811

229: Rep.G.R.: General-Registratur: Fasz. 1260.19.
                                             ADS 9:111

File on investigation & punishment of Hamburg Baptists & intercession from Baptist congregations
in England & America   1837; 1840-1841; 1843

39: Bestand Senat: Cl.VII, Lit.H$^{\underline{f}}$, No. 4, Vol. 10.
                                             ADS 2:67

Regarding the Hamburg Baptists (prohibited group),
specifically their leader J. G. Oncken   1840

39: Bestand Senat: Cl.VII, Lit.J$^{\underline{b}}$, No. 20, Vol. 18
gg.6.
DLC has.   ADS 2:74

Much correspondence from American members
                                    [19th century]
46   ADS 2:604

BERLIN DOCUMENT CENTER. Near the end of the second
world war, American military forces captured vast
quantities of German official records. Much of
this material was brought to the United States,
but personnel files, for the most part, remained
in Germany and were eventually brought together at
the following address:

        Berlin Document Center,
        1 Berlin-Zehlendorf,
        Wasserkaefersteig 1,
        Bundesrepublik Deutschlands

With some restrictions, most of the Berlin Document Center material is open to public inquiry.
Most of it pertains to persons--both German and
foreign--living in the twentieth century. It will
have some usefulness to persons immigrating to the
United States after 1952. A description of the
major archival groups follows:

NSDAP Master File   10,707,974 cards

Membership cards of the National Socialist German Workers' Party (NSDAP, commonly Nazi Party),
although of several different types, generally
contain the following basic data: exact name, date
and place of birth, date of admission into the Party, Party number, occupation, and addresses. Many

cards also have information concerning cancellation
of membership, and some have photographs attached.
The collection is believed to be about 90% complete.
In addition to the membership cards, there are a few
"warning cards" describing persons to whom membership has been denied.

NSDAP Membership Applications   600,000 forms

This collection, which is thought to be from 10%
to 15% complete, contains signed applications for
Nazi Party membership. They also bear the Party
number stamps, which were added at the time the
Party headquarters made out the membership cards.
Rejected applications are also included.

PK - Partei Korrespondenz [Party   1,482,225 files
membership correspondence files]

Originally limited to correspondence pertaining
to Party membership, this collection has been ex-
panded to include a wide variety of biographic
information which does not fall within the compass
of other active collections. This collection is
one of the more valuable sources of biographic
information.

SS Officers (Service Records)   61,465 dossiers

These records of the Allgemeine & Waffen SS officers in the rank of *Untersturmfuehrer* (second
lieutenant) and above are similar in many cases to
the 201 files of U.S. officers, but in some cases
are more comprehensive and contain materials of
general interest. The collection is about 70% complete.

RuSHA Rasse- und Siedlungs-   238,600 files
hauptamt [Racial and Settlement Main Office]

This collection, which has proved to be of great
value to the criminal police, contains racial &
health investigations of members of the SS and the
police who were already married or who were seeking
permission to marry. The dossiers contain elabor-
ate genealogical charts of the man & his wife or
fiancée, their handwritten personal history statements, & photographs in three poses. The women's
names are not indexed. The collection is probably
nearly complete.

Other SS Records   329,220 files

Information in various forms on enlisted members
of the SS, applicants for SS membership, & police
members.

SA (Storm Troops)   260,583 files

This collection includes disciplinary records,
personnel questionnaires, & personnel files; a
warning-card file maintained by the SA includes
the names of non-SA persons; & a card file on the
SA Standarte Feldherrnhalle (paramilitary unit).
Indexed in this collection are also a number of
personnel files on the NSKK (Motor Corps) drivers
for Organization Todt, and a small number of files
pertaining to members of the National Socialist
Flyers' Corps (SSPK).

OPG - Oberste Parteigericht   90,000 files
[Supreme Party Court]

This collection includes records of disciplinary
actions taken by the Supreme Party Court against

violators of Party regulations; material on contested membership & denial of admission to the Party; & personal data & pertinent information on the judges of the Party's *Kreis- und Gaugerichte* (district & regional courts), & of the OPG itself.

RWZ - Rueckwandererzentrale   16,585 files
   (Office of Returning Emigrants)

These files contain correspondence & miscellaneous material concerning the repatriation of German nationals living abroad during the Nazi period.

EWZ - Einwandererzentrale   2,106,827 files
   (Office of Immigration [records of ethnic Germans])

This collection has cards & documents pertaining to the screening, resettlement, & naturalization of ethnic Germans from areas occupied or controlled by the Third Reich. Of particular importance are the naturalization certificates & the property records. Photographs & physical descriptions are also available. The collection has usefulness to American genealogist because many of these ethnic Germans immigrated to the United States during the 1950s.

RKK - Reichskulturkammer   185,000 files
   (Reich Chamber of Culture)

This collection contains personal data, including questionnaires & political screening reports, on members & applicants of the various organizations embracing all vocational groups in the field of public information (writers, artists, actors, librarians, & theatrical producers).

Volksgerichtshof (People's Court)   50,315 files

These files consist of bills of indictment & verdicts relating to the notorious special treason trials of the Nazi era. Judges & defendants are identified. These records constitute only a small portion (possibly 10% to 15%) of all the Volksgerichtshof files.

NSLB - NS-Lehrerverbund   492,000 files
   (Nazi Teachers' Association)

The basic file cards usually contain the following data: name, address, date & place of birth, nature of teaching position held, NSLB number, date of entry, offices held, & membership in other organizations. There are also a number of folders of correspondence pertaining to such matters as efficiency ratings & disciplinary actions.

Reichsaerzteverzeichnis   72,000 cards
   (Register of Physicians)

Principally a card file with data similar to the information on the NSLB cards described above. The collection also includes 2,000 dossiers on police physicians.

Berlin Party Census, 1938   220,000 sheets

In July 1938 the NSDAP conducted a Party census by means of a questionnaire. Only the questionnaires of the Berlin census, which applied to all Party members located in Berlin in July 1938, are included in this collection. Scattered questionnaires from other *Gaue* (regions) are included in the Partei-Kanzlei Collection. The questionnaires, which are usually signed by the members, contain information on their activities in the Party & its affiliates.

Speer Collection (Reich Ministry   15,264 cards
   for Armament & War Production)

These records consist of index cards (containing basic biographic data) & personnel dossiers on some of the Ministry's employees.

Technische Nothilfe   2,141 files
   (Technical Emergency Aid)

This collection consists of miscellaneous documents on members of the Technische Nothilfe, an organization formed to assist in special emergencies.

RAD - Reichsarbeitsdienst   6,520 files
   (Reich Labor Service)

These are chiefly miscellaneous records on officers serving in the RAD in Gau Oberschlesian (the Upper Silesian district) & data on applicants from various areas of western Germany.

University Professors   44,553 files

This collection has index cards to the personnel files on professors under the supervision of the Reich Ministry of Education, correspondence of the Ministry with Party offices, & a card file indexing correspondence of the Deutsche Forschungsgemeinschaft (German Research Society). It also includes dossiers on professors transferred to the Berlin Document Center by the National Archives, Washington, D. C.

Hauptamt fuer Technik   21,870 files
   (Main Office for Technology)

These are chiefly miscellaneous one-name documents pertaining to the professional activities of the members of the NSDDT (National Socialist League of German Technology). Also included are a number of lists, all of which are indexed.

NSDAP Awards   22,000 lists

This collection contains primarily lists of persons recommended for the *Kriegsverdienstkreuz* (War Merit Cross). In addition, there are some lists of the award winners with a short description of their activities.

HJ - Hitler Youth Expulsions   2,856 cards

This small collection consists of cards & lists of the *Reichsjugendfuehrung* (Reich Youth Leadership) on Hitler Youth members expelled on criminal or moral grounds.

NSDAP Officials   58,866 lists

These lists of leading officials of the Party contain basic data (date & place of birth, position held).

Gestapo & Police Records   95,912 files

The collection includes principally arrest orders & other data on persons held in protective custody; data on Gestapo arrestees at Wuerzburg (mainly foreign workers); Osnabrueck police card name register.

Miscellaneous   350,000 files

This collection consists of personnel files, various lists, & materials previously incorporated in other collections; arranged in folders by subject

& indexed by name in the Research Section of the Berlin Document Center. There are, for example, 2,600 personnel files of the *Ahnenerbe*, a program under the control of the SS having to do with genetic inheritance; files on the *Reichsnaehrstand* (Reich food supply) & the *Reichstatthalter in Bayern* (Reich Governor General in Bavaria); lists of Party & SA functionaries.

Police & Army Collection   103,293 files

These files contain information on police, SD (security police), & army personnel. Of great importance are the *Befehlsblaetter der Sicherheitspolizei* (Official Orders of the Security Police), some of which are available in other archives but are indexed by name only at the Berlin Document Center.

Reich Officials & Others   8,331 files

Includes original personnel files & some miscellaneous information on officials of the Reich ministries of Education, Interior, & Economics, as well as on a number of Wehrmacht (military) members & government employees.

SS Women   7,837 dossiers

Dossiers on women employed by the SS offices & concentration camps in various capacities.

Persons Whose German Citizenship   30,611 cards
Was Annulled by the Nazi Regime

Names of persons, principally Jews, whose German citizenship was annulled by the Nazi regime are recorded on printed cards containing the following information: full name, date & place of birth, date of nullification, & date of publication in the *Deutsche Reichsanzeiger* (official gazette). These cards have been microfilmed in National Archives Microcopy T-355, "Name Index of Jews Whose German Nationality Was Annulled by the Nazi Regime (Berlin Document Center)," on nine microfilm rolls. Of particular interest is the last place of residence, including persons living in Germany, Austria, Czechoslovakia, the U.S.S.R., & Switzerland.

Geschaedigte Juden (Jewish   9,123 files
Property Damage)

These files consist primarily of one-name documents pertaining to Jews from Bavaria. They concern confiscation, emigration, & special problems of Jews. There are also some lists of persons detained at Columbia-Haus in Berlin.

NS Frauenschaft (National   4,363,127 cards
Socialist Women's Organization)

The collection consists of applications for membership in the NS Frauenschaft & in the Deutsche Frauenwerk.

SS Lists   271,371 lists

These miscellaneous lists concern transfer, promotions, appointments, etc., of SS personnel, chiefly Waffen SS.

BRETHREN, CHURCH OF THE (Conservative Dunkers). Material regarding emigration from the estates of Prince Isenburg to Pennsylvania & the founding of

the Church of the Brethren   1715

155   ADS 7:190

See also Schwarzenau ueber Berleburg [Geographic Index]

CIVIL WAR, UNITED STATES

Death notices of Bavarian citizens killed at the Battle of Chickamauga   1863-1864

231: Rep.212/6: Feuchtwangen: Nr. 1116.   ADS 9:128

Numerous death certificates of Hamburg citizens who died in U.S. Civil War   [ca. 1865]

39: Bestand Senat: Cl.VI, No.16p, vol.4a: Fasc.7b; im ganzen 39 Nummern.   ADS 2:46

Death certificates; searches for Hamburg citizens in the United States   [ca. 1870?]

39: Bestand Senat: Cl.VI, No.16p: vol.4a: Fasc. 8e [Nachforschungen]   ADS 2:46

Bavarian citizens killed in action with the Ninth Regiment, Ohio   1863

227: Rep.MA 1921: A.St.I: Nr. 1240.   ADS 9:72

List of Bremen citizens who were killed during the American civil war & buried in America   1867

48: B.13.a: Nr. 69.   ADS 3:6

Occasional mention of Hessian soldiers in U.S. Civil War   n.d.

151: Bestand 4: Fuerstliche Personalia, Landgraf Friedrich II.   ADS 7:28

Search for heirs to persons born in Germany who deposited money in New York banks during U.S. Civil War & who did not reclaim these funds thereafter   1900

39: Reichs- u. auswaert.Ang.: E.IV.71 [Erbschaften] ADS 2:307

See also Diez [Geographic Index]

DESERTERS

Desertions of German crew members in New York harbor   1904

39: Reichs- u. auswaert.Ang.: S.Ik.15 [Schiffahrt] ADS 2:337

File on four seamen [unnamed in ADS description] who deserted from a Danish ship in New York 1857

55: Han.Des.80: Reg.Stade: N.Titel 1: Nr. 8, RR 748.   ADS 4:155

List of seamen deserting from German merchant vessels from 1876-1885 [gives port of desertion, name, age, rank or rating, home port or birthplace, etc.]   1876-1885

55: Han.Des.80: Reg.Stade: N.Titel 1: Nr. 33, RR 749.   ADS 4:156

See also American Revolution

DUNKARDS, DUNKERS. *See* Baptists; Brethren, Church of the

DUTCH (NIEDERDEUTSCHEN) CONGREGATION IN NEW YORK.

Letters from Congregation regarding division of church   1756

<u>39</u>: Geistl.Minis.: II,7: Protokoll lit.B, 1746-1758: pp. 395; 397 (IV); 416; 417 (VI); 418 (V).   ADS 2:476-477

ESTATE MATTERS

Where separately listed in ADS, names of persons in Germany or America having estates of interest to persons in the other country have been listed in this *Handbook* under those names. In addition, attention is called to the large collection of papers in the Auswaertige Amt [Foreign Office], Bonn, dealing with such matters   1922-1944

<u>148</u>: R XI.   ADS 6:262-263

ETHNIC GERMANS (VOLKSDEUTSCHE). *See* Berlin Document Center: Einwandererzentrale

FOREIGN WORKERS IN GERMANY. *See* Berlin Document Center: Gestapo & Police Records; Einwandererzentrale

GENESSEE COLONY.

Secret attempts to recruit for the Genesse[e] Colony in North America   1792

<u>2</u>: Abt.65: Nr. 4860, Fasc. 1 [Heim. Werbung] DLC has [*see under* Kiel].   ADS 1:12

GENETIC STUDY (*Ahnenerbe*). *See* Berlin Document Center: Miscellaneous

GERMAN MERCENARIES IN AMERICA BEFORE 1776.

List of officers wounded or killed in America 1756 [The attention of researchers is also called to *Zeitschrift des Historischen Vereins fuer Niedersachsen* (1898), p. 290.]

<u>108</u>: N: Nr. 34a, Fach 159.   ADS 4:334

GYPSIES, RUSSIAN.

File on the transport of a group of Gypsies from Russia [to America]   1900-1901

<u>39</u>: Bestand Senat: Cl.VII, Lit.K^e: Nr. 9.z: Vol. 52 [Befoerderung]   ADS 2:91

Permit for a transport of 147 Russian Gypsies to pass through Hamburg for the purpose of emigration   1901

<u>39</u>: Reichs- u. auswaert.Ang.: C.IIIc.8 [Auswanderung]   ADS 2:313

HOCHTEUTSCHEN COMPAGNIE.

Emigration file on   1706

<u>500</u>: Dresden: Saechsischen Hauptstaatsarchiv: Loc. 2249 [film]   Not in ADS

HUGUENOTS.

The French emigrants & their settlement in Franconia [Fraenkischen Kreis]   1792

<u>227</u>: Rep.Kasten Schwarz: Nr. 14556.   ADS 9:38

See also Wied (Duchy of) [Geographic Index]

HUNGARIAN EMIGRATION.

Hamburg file   n.d.

<u>39</u>: Reichs- u. auswaert.Ang.: C.IIIc,27 [Auswanderung]   ADS 2:322

INSPIRED, CONGREGATION OF THE (*Inspirierten*). *See* Wied (Duchy of) [Geographic Index]

INTERNEES, CIVIL.

The Auswaertige Amt [Foreign Office], Bonn, has a large collection of material on civil internees. These materials have not been described in ADS.

<u>148</u>: Rechtsabteilung R XII.   ADS 6:264

JEWS.

Various reports regarding the emigration of Russian Jews [Koenigsberg files]   1882-1893

<u>51</u>: Rep.2,I^2: Titel 30: Nr. 43.   ADS 4:28

Liquidation of the expropriated properties of enemies of the Reich (particularly Jews) [in Siegen Kreis]   1939-1949

<u>116</u>: Arnsberg Reg.: Siegen: B Nr. 20.   ADS 6:61

See also Berlin Document Center: Persons Whose German Citizenship Was Annulled by the Nazi Regime; Geschaedigte Juden. *Also* Emden; Gemuenden (Gemeinde, HEL); Heppenheim (Kreis); Saxony [Geographic Index]

LANGENSALZER KOMPAGNIE. *See* Bad Langensalza (Saxony) [Geographic Index]

LOW GERMAN CLUBS IN AMERICA (PLATTDEUTSCHES VEREINE). *See* Boersmann, Martin [Name Index]

LUTHERAN CONGREGATIONS IN AMERICA.

Bound files regarding internal disputes among congregations & pastors & the ordination in Hamburg of pastors for American churches [110 pages]   1741-1758

<u>39</u>: Geistl.Minis.: III.A2d: Gebundene Akten: Nr. CLXXXIV, pp. 376 ff.   ADS 2:476

Correspondence with various pastors
1798-1800; 1796-1802; 1803-1806

500: Halle: Missionsbibliothek des Waisenhauses:
Abteilung H IV (Nordamerika): Unverzeichnet.
Not in ADS

Files on the petitions of several theology students
requestiong preservation of their [doctoral?]
candidacy rights in the event they are assigned
ministries in North America   1845

2: Abt.65: Nr. 390 [Kandidatenrechte].
DLC has [see under Kiel]   ADS 1:12

Unarranged files of correspondence from Lutheran
emigrants in various parts of America [this collection may prove valuable when adequately cata-
logued]   [19th century]

45: Mappe Amerika.   ADS 2:601

See also Salzburger Emigrants [Subject Index];
Ebenezer, Georgia; Pennsylvania [Geographic In-
dex]; -- Boltz; -- Gronau; -- Thilo; -- Lemcken;
Schultzen; -- Muehlenberg; -- Driessler [Name
Index]

MENNONITES. The following archival material has been
found:

Emigration from Eppingen with Ottilienberg
1322-1864

260: Rep.Abt.229: Spezialakten: Eppingen: Fasz.
25773-26131.   ADS 10:11

Emigration from Mengen with Bechtoldskirch

260: Rep.Abt.229: Spezialakten: Mengen: Fasz.
66555-66804.   ADS 10:11

Mennonites in Bavaria   1535-1811

229: Rep.G.R.: General-Registratur: Fasz. 1260.19.
ADS 9:111

Undescribed data on Mennonites & Separatists
1801-1816

263: Rep.E 31-32: Geheimer Rat II: H. Kirchen- und
Schulwesen: Nr. 367.   ADS 10:54

Personal letters from & to persons in the USA
1807-1846

219   ADS 8:89

Nr. 8241: Mennonite cemeteries   [18th century?]
Nr. 8242: Regarding Mennonites   [18th century?]

260: Abt.236: Innenmin.: Wegzug: Auswand.: files
as above.   ADS 10:5

Files on [Mennonites]   [17th & 18th centuries?]

260: Abt.236: Innenmin.: Wegzug: Auswand.: Nr.
4328; 4355; 4359.   ADS 10:4

Nr. 4211-4213: Protection money [Schutzgeld]
[18th century]
Nr. 4237-4239: Protection money   [18th century]
Nr. 4230: List of Mennonites   n.d.
Nr. 4231: Use of water & meadows by Mennonites
n.d.

260: Abt.236: Innenmin.: Wegzug: Auswand.: files as
above.   ADS 10:3

There is an uncatalogued collection of material on
the Mennonites in the Mennonitische Forschungs-
stelle, Goettingen, Calsowstrasse 4. ADS 4:266

See also Wied (Duchy of) [Geographic Index]

MEXICO, UNITED STATES WAR WITH

Bavarian citizens killed in   1853-1871

231: Rep.212/13: Neustadt a.d. Aisch: Nr. 470.
ADS 9:129

Bavarian citizens killed in   1853-1871

231: Rep.212/6: Feuchtwangen: Nr. 789. ADS 9:128

Transmittal of claims to the United States Government from heirs of soldiers from Kurhessen
[Electoral Hesse] killed in   1848-1861

151: Bestand 5: 9.a: Nr. 26.   ADS 7:44

Estates of soldiers born in Kurhessen who died in
American service in war with Mexico, 3 vols.

151: Bestand 5: 9.a: Nr. 27.
DLC has volume 2 only.   ADS 7:44

MORAVIAN BRETHREN (also called Unitas Fratrum;
Herrnhuter; Maehrische Bruederkirche). The following archival material has been found:

Establishment in Georgia   1733-1739; 1746
Bethlehem, Pennsylvania   1768-1842
Nazareth, Pennsylvania   1743-1855
Hope, New Jersey   1799-1801
Muskingum, Michigan   1812; 1823; 1827

500: Herrnhut: Archiv: Rep.14.A: Nr. 1; 50a; 50b;
50c; 51.   Not in ADS

Reports & diaries regarding settlement in Georgia
1734-1744
Additional materials [Nachtraege]   1734-1740
Letters from American workers & Brethren, mainly
addressed to [Count] Zinzendorf
1740-1742; 1743-1757; 1742-1756

500: Herrnhut: Archiv: Rep.14A: Nr. 6; 8; 29; 30;
31.   Not in ADS

News regarding expedition to South Carolina re-
ported by Brother Petrus Boehler & Georg Schulius
1738-1740
American letters from Petrus Boehler   1740-1763
Principal reports from America   1748-1756
Travel diary   1736-1756
Visits to Pennsylvania, Virginia, New England,
Maryland, New York   1742-1756

500: Herrnhut: Archiv: Rep.14.A: Nr. 9; 33; 35; 36.
Not in ADS

Miscellaneous Americana [undescribed]   1716-1783

500: Halle: Missionsbibliothek des Waisenhauses:
Abteilung H IV (Nordamerika): Unverzeichnet.
Not in ADS

Bavarians who have fled to other countries as a result of political movements   1850-1852

229: Rep.Reg.Akten von Oberbayern: Kammer des
    Inneren: 1154.57.   ADS 9:115

Political refugees & suspicious personalities, letters B-W, 19 volumes   1851-1863

227: Rep.MA 1957: Laufende Registratur: Nr. 841-
    859.   ADS 9:108

Treasonable letters & proclamations from New York
from the organization for the murder of German
princes   1847-1852

226: Rep.MInn 7: 45608.   ADS 9:19

See also Schurz, Karl; Weitling, Wilhelm [Name Index]

QUAKERS. *See* Baptists

PRESBYTERIAN CHURCH.

A file on Father -- Chiniquy, Roman Catholic priest
of St. Anne's Church [Chicago, Illinois], who
left the Church with 5,000 parishioners after a
disagreement with the Catholic Bishop of Chicago.
The congregation then joined the Presbyterian
Church [Many parishioners were apparently German-
Americans]   1856

125: C II 18.   ADS 6:87

PRUSSIAN EMIGRANTS TO AMERICA.

Case of Scheffler *versus* Taakes, *et al*, in Hamburg
regarding shanghaied Prussian subjects sent to
North America   1802-1805

500: Berlin-Dahlem: Geh.Staatsarchiv: Rep.XI, 21a:
    Conv. 3,3.   Not in ADS

REFORMED CHURCH. *See* Siegen [Geographic Index]

REPATRIATION OF GERMANS. *See* Berlin Document Center:
RWZ - Rueckwandererzentrale [Subject Index]

REPATRIATION OF GERMANS FROM THE UNITED STATES.

Repatriation from North America   1938-1942
Repatriation from the United States   1937-1941

148: Kult E: Rueckwand.: Aw10 adhIII.   ADS 6:267

RUSSIAN EMIGRANTS. The following archival material
has been found:

Emigration via port of Bremen   1891-

49: A.4.(m) [Auswanderungswesen]   ADS 3:110

A list of emigrants at Hamburg (gives names, number of persons, where crossed border, possession
of funds & ship tickets, length of stay in Hamburg)[Koenigsberg files]
5 September to 22 November 1893; 1-15 March 1893

51: Rep.2,I^2: Titel 30: Nr. 36, Bd. 7.   ADS 4:26

Hamburg report on emigrants [167 Russian emigrants
with tickets; 293 without tickets][Koenigsberg
files]   January-June 1893

51: Rep.12: Abt.I, Titel 3: Nr. 10, Bd. 13.
                                              ADS 4:41

Lists of emigrants to America from the Russian
Government of Kowno, via Hamburg   1899

39: Reichs- u. auswaert.Ang.: C.IIIc,15 [Auswander-
    ung]   ADS 2:300

Transport of 156 Russian emigrants   1902

39: Bestand Senat: Cl.VII, Lit.K^e, No.9.2, vol. 5$^{\underline{3}}$.
                                              ADS 2:91

Transport of 156 Russian emigrants (contagion reported by escort von Prostken)   1902

39: Reichs- u. auswaert.Ang.: A.III.7 [Auswander-
    ung]   ADS 2:322

Communication from Russian Ministry of Foreign Affairs to Russian Mission in Washington, dated
12 December 1853 (No.1) & 25 January 1854 (No.26)
giving a list of persons who are forbidden to return to Russia, 15 pages [Archives of the Russian
Mission in Washington]

Listed in Frank A. Golder, *Guide to Materials for
American History in Russian Archives* (Washington, D.C.: Carnegie Institution, 1937) II, 45.

ST. MARY'S COLONY (Arkansas).

Letter from Father Bernhard Hafkenscheid to Archbishop Reisach in Munich, dated 1 October 1849,
giving an account of his activities during the
disbandment of St. Mary's Colony   1849

240: AZ: Nordamerika.   ADS 9:151

SALZBURGER EMIGRANTS.

Emigrants from Salzburg & Berchtesgaden & immigration Nova Scotia   1704-1797

52: Han.Des.74: Northeim: K.H.III.Fach 144: Nr. 1.
                                              ADS 4:74

File on Salzburger emigrant who became Prussian
colonists   1732-1736

231: Rep.165a: Ansbacher Oberamtsakten: 1738.
                                              ADS 9:127

Correspondence regarding   1732-1739

158: Frankfurter Ratsakten: Ugb.A.9: ad Nr. 11.
    DLC has [originals in 158 have been destroyed]
                                              ADS 7:198

Passage through Hamburg of twelve emigrants, said
to be from Salzburg   29 August 1749

39: Geistl.Minis.: II,7: Protokoll lit.B, 1746-
    1758: pp.138-139.   ADS 2:200

Movement of so-called Salzburger emigrants through
Bavaria, with interrogations   1731-1737

229: Rep.G.R.: General-Registratur: Fasz. 1261.23.
ADS 9:111

March through Bezirksamt Heilsbronn* [Bavaria]
*So spelled. 1732–1733

231: Rep.212/9: Heilsbronn: Nr. 217. ADS 9:130

News from [Ebenezer, Georgia] 1733–

500: Halle: Missionsbibliothek des Waisenhauses:
Abteilung H IV (Nordamerika), Fach G, Nr. 1;
2; 4; 6; 7; 8; 9; 11. Not in ADS

See also Ebenezer, Georgia [Geographic Index];
Amann, Adolf [Name Index]

SCHWENKFELDERS. The following archival material has
been found:

Schwenkfeld controversy & heresy, as well as the
colloquy with Zwingli 1529

231: Rep.111a: Ansbacher Religionsakten: Tom.XII,
Fasc. I. ADS 9:126

Affairs of Schwenkfelder congregation at Harpersdorf 1718

500: Breslau: Preus.Staatsarchiv: Rep.135 (Hand-
schriften) E 57. Not in ADS

Miscellaneous file 1732–1740

500: Breslau: Preus.Staatsarchiv: Rep.39F: Schweid-
nitz-Jauer: X.19n. Not in ADS

File on petitions of [Schwenkfelder] Separatists to
emigrate to other parts of the world 1836–1838

500: Breslau: Preus.Staatsarchiv: Rep.14: P.A.X:
Nr. 28g, vol. II. Not in ADS

SEPARATISTS. See Baptists; Schwenkfelders

TEXAS GERMAN COLONIES. The following archival material has been found:

Nr. 1: Inquiries & applications for emigration
1847–1858
Nr. 2: Inquiries & applications 1843–1844
Nr. 2b: Inquiries & applications 1844–1845
Nr. 2c: Inquiries & applications 1845

154: CA 4b4: files as above. ADS 7:169

Emigrants to Texas (151 cases) 1844–

248: Kgl. Bayer. Polizey Direktion Muenchen: C II
a 3 0 213. ADS 9:160

Cash book containing names of emigrants [1844?]

154: CA 4d5: Nr. 6. ADS 7:186

Lists of emigrants 1844–1845

154: CA 4c3: Nr. 1. ADS 7:179

Land contracts with emigrants; promissory notes for
advances made to emigrants 1845–1846

154: CA 4c1: Nr. 3. ADS 7:177

Original [emigration] contracts 1847–1863

154: CA 4c1: Nr. 30.
DLC has. ADS 7:176

Information on emigrants to Texas who have died
there 1847–1849

154: CA 4c2: Nr. 9. ADS 7:178

List of ships from Bremen & Antwerp carrying 4,237
emigrants 1845

154: CA 4c1: Nr. 25b. ADS 7:176

List of emigrants to Texas under contract with the
Texas Verein but who did not sail with Verein
ships 1846

154: CA 4c1: Nr. 25a. ADS 7:176

Correspondence with general [shipping] agent in
Bremen, contains many name lists 1846

154: CA 4c1: Nr. 28. ADS 7:176

Nr. 9: Emigration contracts 1844–1846
Nr. 10a: Antwerp cash book 1845
Nr. 10b: Emigrants' disbursements 1846
Nr. 10c: Emigrants' disbursements 1846
Nr. 10d: Name list of persons aboard ship
6 July 1846

154: CA 4c1: files as above. ADS 7:174

Lists of emigrants boarding at Bremen 1844–1846

154: CA 4c1: Nr. 8. ADS 7:174

Ship lists 1846–1876

154: CA 4c1: Nr. 7 [may be missing] ADS 7:174

Emigrant name lists 1845–1848

154: CA 4b6: Nr. 1. ADS 7:171

Nr. 15: Would-be emigrants who have decided to
back out of decision to go 1845–1846
Nr. 16: Estate matters of colonists 1846

154: CA 4b5: files as above. ADS 7:171

Nr. 1: Inquiries from would-be emigrants 1845
Nr. 2: Inquiries from would-be emigrants 1845–1846
Nr. 2a: Inquiries from would-be emigrants
1846–1847

154: CA 4b5: files as above. ADS 7:170

For researchers interested in the history of the
Texas Verein, the Fuerst zu Solm-Braunfelsche-
Archiv, Schloss, Braunfels, is of major import-
ance. ADS 7:165

WALDO COLONY (Westminster, Massachusetts)

Original contract for the settlement of 1,000 families n.d.

213: Schrank y: Gefach 5: Fasc. 2.
DLC has. ADS 8:81

See also Maine [Geographic Index]

WORLD WAR II.

Death certificates [Germans & Americans] 1942-1945
148: R KR/VR: 28 Nr. 21: Paket Nr. 164.   ADS 6:248

The Auswaertige Amt [Foreign Office], Bonn, has
files of potential interest to genealogical researchers, including the U.S. black lists of firms
& individuals in Latin America, seized property of
individuals & the claims resulting therefrom, prisoners of war, prisoner graves

148   ADS 6:167-270

ZWEIG-VEREIN KAMENZ.   See Saxony [Geographic Index]